Proceedings of the 4th Asia-Pacific

BIOINFORMATICS

Conference

SERIES ON ADVANCES IN BIOINFORMATICS AND COMPUTATIONAL BIOLOGY

Vol. 1: Proceedings of the 3rd Asia-Pacific Bioinformatics Conference
Eds: Yi-Ping Phoebe Chen and Limsoon Wong

Vol. 2: Information Processing and Living Systems
Eds: Vladimir B. Bajic and Tan Tin Wee

Vol. 3: Proceedings of the 4th Asia-Pacific Bioinformatics Conference
Eds: Tao Jiang, Ueng-Cheng Yang, Yi-Ping Phoebe Chen and Limsoon Wong

Series on Advances in Bioinformatics and Computational Biology – Volume 3

Proceedings of the 4th Asia-Pacific

BIOINFORMATICS

Conference

Taipei, Taiwan 13 – 16 February 2006

editors

Tao Jiang
University of California, Riverside, USA

Ueng-Cheng Yang
National Yang-Ming University, Taiwan

Yi-Ping Phoebe Chen
Deakin University, Australia

Limsoon Wong
National University of Singapore, Singapore

ICP

Imperial College Press

Published by

Imperial College Press
57 Shelton Street
Covent Garden
London WC2H 9HE

Distributed by

World Scientific Publishing Co. Pte. Ltd.
5 Toh Tuck Link, Singapore 596224
USA office: 27 Warren Street, Suite 401-402, Hackensack, NJ 07601
UK office: 57 Shelton Street, Covent Garden, London WC2H 9HE

British Library Cataloguing-in-Publication Data
A catalogue record for this book is available from the British Library.

PROCEEDINGS OF THE 4TH ASIA-PACIFIC BIOINFORMATICS CONFERENCE

ISBN 1-86094-623-2

Printed by FuIsland Offset Printing (S) Pte Ltd, Singapore

PREFACE

High-throughput sequencing and functional genomics technologies have given us a draft human genome sequence and have enabled large-scale genotyping and gene expression profiling of human populations. Databases containing large numbers of sequences, polymorphisms, and gene expression profiles of normal and diseased tissues in different clinical states are rapidly being generated for human and model organisms. Bioinformatics is thus rapidly growing in importance in the annotation of genomic sequences, in the understanding of the interplay between genes and proteins, in the analysis the genetic variability of species, etc.

The Asia-Pacific Bioinformatics Conference series is an annual forum for exploring research, development, and novel applications of Bioinformatics. It brings together researchers, professionals, and industrial practitioners for interaction and exchange of knowledge and ideas. The Fourth Asia-Pacific Bioinformatics Conference, APBC2006, was held in Taipei 13–16 February, 2006. Taking advantage of the presence of APBC 2006 in Taipei, several related activities were also organized immediately before or after APBC 2006, including the Third Association of Asian Societies for Bioinformatics Symposium.

A total of 118 papers were submitted to APBC 2006. These submissions came from China, Hong Kong, India, Japan, Korea, Singapore, Taiwan, Australia, Belgium, France, Germany, Italy, Norway, Russia, UK, Canada, and USA. We assigned each paper to at least 3 members of the programme committee. Although not all members of the programme committee managed to review all the papers assigned to them, a total of 340 reviews were received. As a result, there were almost 2.9 reviews per paper on average, and more than 98% of the papers received at least 3 reviews.

A total of 35 papers (*i.e.* 30%) were accepted for presentation and publication in the proceedings of APBC 2006. Each accepted papers had at least 2 positive recommendations and no negative recommendations from their reviewers. Based on the affiliation of the authors, 1.80 of the accepted papers were from China, 4.50 were from Hong Kong, 3.00 were from India, 3.50 were from Japan, 0.75 were from Korea, 3.00 were from Singapore, 3.00 were from Taiwan, 2.00 were from Australia, 3.20 were from Canada, 7.25 were from USA, 1.00 were from France, 1.00 were from Germany, and 1.00 were from Norway. In addition to the accepted papers, the scientific programme of APBC 2006 also included 3 keynote talks, as well as tutorial and poster sessions. There is no doubt that the presentations covered a broad range of topics in bioinformatics and computational biology, and were of very high quality. We had a great time in Taipei, enhancing the interactions between many researchers and practioners of bioinformatics, and advancing bioinformatics into a more mature scientific discipline.

Lastly, we wish to express our gratitude to: the authors of the submitted papers, the members of the programme commitee and their subreferees, the members of the organizing committee, the keynote speakers, our generous sponsors, and supporting organizations for making APBC 2006 a great success.

Tao Jiang
Ueng-Cheng Yang
Yi-Ping Phoebe Chen
Limsoon Wong

16 February 2006

APBC2006 ORGANIZATION

General Co-Chairs

Yi-Ping Phoebe Chen (Deakin University)
Wen-Hsiung Li (University of Chicago)
Limsoon Wong (National University of Singapore)

Organizing Committee

Jorng-Tzong Horng (National Central University, co-chair)
Cheng-Yan Kao (National Taiwan University, co-chair)
Chih-Jen Chang (Chang-Gang University)
Chuan-Hsiung Chang (National Yang Ming University)
Jung-Hsien Chiang (National Cheng Kung University)
Yi-Fang Chung (National Yang Ming University)
Hsien-Da Huang (National Chiao Tung University)
Hsueh-Fen Juan (National Taiwan University)
Ming-Tat Kao (Academia Sinica)
Chang-Huain Hsieh (National Center for High Performance Computing)
Feng-Sheng Wang (National Chung Cheng University)

Tutorial Chair

Wen-Chang Lin (Academia Sinica)

Poster Chair

Chuan Yi Tang (National Tsing Hua University)

Programme Committee

Tao Jiang (University of California, Riverside, USA, and Tsinghua University, China; co-chair)
Ueng-Cheng Yang (National Yang Ming University, Taiwan; co-chair)
Tatsuya Akutsu (Kyoto University, Japan)
Vineet Bafna (University of California, San Diego, USA)
Paola Bonnizoni (Universita' degli Studi di Milano - Bicocca, Italy)
David Bryant (McGill University, Canada, and University of Auckland, New Zealand)
Kun-Mao Chao (Natonal Taiwan University, Taiwan)
Francis Chin (University of Hong Kong, SAR, China)
Ross Coppel (Monash University, Australia)
Michael Cummings (University of Maryland, USA)
Bhaskar DasGupta (University of Illinois, Chicago, USA)
Nadia El-Mabrouk (University of Montreal, Canada)
Janice Glasgow (Queens University, Canada)
Sridhar Hannenhalli (University of Pennsylvania, USA)
Wen-Lian Hsu (Academia Sinica, Taiwan)
Haiyan Huang (University of California, Berkeley, USA)
Ming-Jing Hwang (Academia Sinica, Taiwan)
John Kececioglu (University of Arizona, USA)
Chris Langmead (Carnegie Mellon University, USA)
Sang-Yup Lee (Korea Advanced Institute of Science and Technology, Korea)
Jinyan Li (Institute for Infocomm Research, Singapore)
Jing Li (Case Western Reserve University, USA)
Guohui Lin (University of Alberta, Canada)
Stefano Lonardi (University of California, Riverside, USA)
Henry Horng-Shing Lu (National Chiao Tung Uniersity, Taiwan)
Bin Ma (University of Western Ontario, Canada)
Shinichi Morishita (University of Tokyo, Japan)
Laxmi Parida (IBM T.J. Watson Research Center, USA)
Kunsoo Park (Seoul National University, Korea)
Christian Pedersen (University of Aarhus, Denmark)
Alexander Schliep (Max Planck Inst. for Mol. Genetics, Germany)
Shoba Ranganathan (Macquarie University, Australia)
Christian Schoenbach (RIKEN, Japan)
Larry Ruzzo (University of Washington, USA)
Lusheng Wang (City University of Hong Kong, SAR, China)
Wei Wang (University of North Carolina, Chapel Hill, USA)
Eric Xing (Carnegie Mellon University, USA)
Michael Zhang (Cold Spring Harbour Labs, USA)
Yang Zhong (Fudan University, China)
Xianghong Zhou (University of Southern California, USA)

Additional Reviewers

A. Abu-Zeid
S. Besenbacher
H.L. Chan
L. Chen
I. G. Costa
G. Della Vedova
R. Dondi
D. Dutta
C. Ferretti
R. Fraser
J. Fredslund
B. Georgi
R.S.C. Ho
W.K. Hon
H. Hu
Y. Huang
S. Jensen
H.C.M. Leung
T. Mailund
C. Rangel
W. Rungsarityotin
J. Schug
S. Sedfawi
T.Y. Sung
A. Tam
S. Taylor
S. Teng
C.L. Wang
J. Wang
L. Wang
K.P. Wu
K. Zhang
L. Zhuge
E. Zuveria

CONTENTS

ON THE INFERENCE OF REGULATORY ELEMENTS, CIRCUITS AND MODULES

WEN-HSIUNG LI

Department of Ecology and Evolution, University of Chicago, USA and Genomics Research Center, Academia Sinica, Taiwan

Advances in genomics have led to the production of various functional genomic data as well as genomic sequence data. This is particularly true in yeasts. Such data have proved to be highly useful for inferring regulatory elements and modules. I shall present studies that I have done with my colleagues and collaborators on the following topics. (1) Detection of transcription factors (including their interactions) involved in a specific function such as the cell cycle, (2) inference of the cis elements (binding sites and sequences) of a transcription factor, (3) reconstruction of the regulatory circuits of genes, and (4) inference of regulatory modules. In all these topics, we have developed methods and have applied them to analyze data from yeasts.

AUTOMATING THE SEARCH FOR LATERAL GENE TRANSFER

MARK A. RAGAN

Institute for Molecular Bioscience, University of Queensland and Australian Research Council (ARC) Centre in Bioinformatics St Lucia, Q 4072, Australia

Most genes have attained their observed distribution among genomes by transmission from parent to offspring through time. In prokaryotes (bacteria and archaea), however, some genes are where they are as the result of transfer from an unrelated lineage. To elucidate the biological origins and functional consequences of lateral gene transfer (LGT), we have constructed an automated computational pipeline to recognise protein families among prokaryotic genomes, generate high-quality multiple sequence alignments of orthologs, infer statistically sound phylogenetic trees, and find topologically incongruent subtrees (prima facie instances of LGT). This pipeline requires that we automate workflows, design and optimize algorithms, mobilise high-performance computing resources, and efficiently manage federated data. I will summarise results from the automated comparison of 422971 proteins in 22437 families across 144 sequenced prokaryotic genomes, including the nature and extent of LGT among these lineages, major donors and recipients, the biochemical pathways and physiological functions most affected, and implications for the role of LGT in evolution of biochemical pathways.

WHOLE GENOME OPTICAL MAPPING

MICHAEL S. WATERMAN
University of Southern California
1050 Childs Way, MCB 403E, Los Angeles, CA 90089-2910, USA

An innovative new technology, optical mapping, is used to infer the genome map of the location of short sequence patterns called restriction sites. The technology, developed by David Schwartz, allows the visualization of the maps of randomly located single molecules around a million base pairs in length. The genome map is constructed from overlapping these shorter maps. The mathematical and computational challenges come from modeling the measurement errors and from the process of map assembly.

ACCURACY OF FOUR HEURISTICS FOR THE FULL SIBSHIP RECONSTRUCTION PROBLEM IN THE PRESENCE OF GENOTYPE ERRORS

DMITRY A. KONOVALOV

School of Information Technology, James Cook University, Townsville, QLD 4811, Australia

The full sibship reconstruction (FSR) problem is the problem of inferring all groups of full siblings from a given population sample using genetic marker data without parental information. The FSR problem remains a significant challenge for computational biology, since an exact solution for the problem has not been found. The new algorithm, named SIMPSON-assisted Descending Ratio (SDR), is devised combining a new Simpson index based $O(n^2)$ algorithm (MS2) and the existing Descending Ratio (DR) algorithm. The SDR algorithm outperforms the SIMPSON, MS2, and DR algorithms in accuracy and robustness when tested on a variety of sample family structures. The accuracy error is measured as the percentage of incorrectly assigned individuals. The robustness of the FSR algorithms is assessed by simulating a 2% mutation rate per locus (a 1% rate per allele).

1 Introduction

Let population sample N be a collection $(X_1, X_2, ..., X_n)$ of n diploid genotypes

$$X_i = \left((x_{i1}, x'_{i1}), (x_{i2}, x'_{i2}), ..., (x_{iL}, x'_{iL})\right), \tag{1}$$

where each locus l is described by an unordered pair of alleles (x_{il}, x'_{il}) and L is the total number of loci which are assumed to be unlinked. Each locus l is a set of λ_l codominant alleles $\{a_{l1}, ..., a_{l\lambda_l}\}$. The full sibship reconstruction (FSR) problem is the problem of finding the *best* partition B from the set of available partitions $\{P_j\}$, where each P_j represents the partitioning of N into groups of full siblings without the availability of parental information. In order to find partition B, the partitions are ranked by a scoring function which is algorithm specific. Currently there are a number of heuristic FSR algorithms[1-7] employing a variety of scoring functions and techniques for searching the partition space $\{P_j\}$.

Some FSR algorithms[1,5,3] utilize the Mendelian rules of inheritance in determining the full sibling groups. For example, Butler *et al.*[3] devised the so-called SIMPSON algorithm which used the Simpson index

$$S = \frac{1}{n(n-1)} \sum_{k=1}^{r} g_k (g_k - 1) = -\frac{1}{(n-1)} + \frac{1}{n(n-1)} \sum_{k=1}^{r} g_k^2 \tag{2}$$

as the scoring function, where N is partitioned into r sib groups with group k containing g_k individuals. The SIMPSON algorithm is a brute force heuristic which searches for the best partition B by starting from all given genotypes being placed in different groups of size one. The algorithm then searches the available partition space by randomly moving one individual into a different group if the newly enlarged group passes the Mendelian sibship test. The test is passed if all individuals in the group could be generated from the same pair of parental genotypes strictly obeying the Mendelian rules of inheritance. The number of random moves (iterations) is limited by the algorithm's parameter, $T_n = 100000$. The SIMPSON formulation of the FSR problem (FSR-S) has the partition search space at least exponential in n,[8] limiting the applicable range of the SIMPSON algorithm or any other "random-walk" based algorithms for that matter. For example, even a relatively small sample of 10 individuals restricted to being either full siblings or unrelated is estimated to yield 115975 partitions.[6] The estimation is provided by the Bell number and is an upper bound of the actual partition space size.[8]

Another class of algorithms, notably the GRAPH[2] and DR[4] algorithms, use the pairwise likelihoods of Goodnight and Queller[9] in construction and assessment of the sib groups. The important difference between the Mendelian sibship test and likelihood-based tests is the ability of likelihoods to accommodate the presence of genotype errors. Essentially the Mendelian sibship test is likely to fail for a previously valid sib group[3] if even one allele is mutated, while the likelihood-based sibship tests are expected to be more robust.[7] The interest in the errors is not purely academic. The discovery of microsatellite markers revolutionized[10] conservation biology and molecular ecology as well as medical, forensic and population genetics, to name a few. However, markers may suffer from a wide range of error types with drastic consequences: a relatively *"small 1% error rate in allele calling would lead to almost a quarter of 12-locus genotypes containing at least one error"*.[11] In the important case of noninvasive genotyping the situation is even more error-prone due to the small amount of target DNA further affecting the reliability of polymerase chain reaction (PCR) to correctly amplify all alleles.[12] In addition, microsatellite markers could be highly susceptible to mutation.[13]

In this study we compare the two existing algorithms; the SIMPSON[3] algorithm representing the class of algorithms based on the Mendelian sibship test and the Descending Ratio[4] (DR) algorithm which is purely likelihood based. We show that the SIMPSON algorithm could be replaced by a more efficient new $O(n^2)$ algorithm, named the Modified SIMPSON (MS2) algorithm. We also present a new algorithm, named the SIMPSON-assisted Descending Ratio (SDR) algorithm, which combines the advantages of the MS2 algorithm when there are no genotype errors with the robustness of DR to the errors.

2 Method

2.1 *Accuracy*

Normally[3] a sample with known sib groups (partition A) is generated by simulation (each such simulation is called a FSR trial). The sample is then presented to an FSR algorithm yielding the *best* (according to the algorithm) partition B. The known partition A and reconstructed partition B are compared and the accuracy measure for the given trial (and sample structure) is calculated. The accuracy measure is then averaged over a number of trials, as large as one hundred[2] or as small as six[3]. However, the measures of accuracy were defined differently in the published algorithms making them difficult (if not impossible) to compare. For example, the following measures currently exist: the minimum number of moves $\xi(A,B)$ required to convert B into A;[3,2] the percentage of trials where $A=B$;[2] $(S_{fs|fs}-S_{fs|ur})/T_{fs}$, where $S_{fs|fs}$ is the total number of correctly reconstructed full-sib pairs, $S_{fs|ur}$ is the total number of incorrectly reconstructed full-sib pairs, and T_{fs} is the total number of full-sib pairs in A;[6] the number of full-sib families being completely recovered relative to the actual numbers in a sample.[7]

For this study, the *accuracy-error* is adopted as the accuracy measure. The error equals the percentage of incorrectly assigned individuals[14] $\xi=\xi(A,B)/n$ and is equivalent to the partition-distance which has known theoretical properties[15] and could be efficiently calculated via the maximum[15] or minimum[16] assignment problem for bipartite graphs. In addition the accuracy-error is directly comparable[8] to the $\xi(A,B)$ results of GRAPH[2] and the four algorithms studied by Butler *et al.*,[3] i.e. the AF,[1] Full Joint Likelihood (FJL[a]),[5] SC[5] and SIMPSON[3] algorithms. The available measures of accuracy compare the known partition A to the reconstructed partition B, while the ultimate goal of the FSR algorithms is to provide B together with its confidence level[17] for a given population sample with an *unknown* structure. While, at present, the assessment of the confidence levels for the FSR remains unexplored, the accuracy-error could provide consistent initial comparisons between the FSR algorithms.

2.2 *Simulations*

There are a number of sample family structures that are used for testing of the FSR algorithms. For example, while testing their GRAPH algorithm, Beyer and May[2] used four family distributions for the population sample of $n=50$ individuals with the following family sizes: (5×10),[b] (20,10,10,5,5), (30,5,5,5,5) and (40,5,2,2,1). They also used $n=500$ where all family sizes from their $n=50$ testing set were multiplied by 10. Butler *et al.*[3] used the (50×1),[c] (5×10), (25,10,10,4,1) and (45,1,1,1,1) family sizes for $n=50$ and (20×10), (5×40), (100,40,40,16,4) and (196,1,1,1,1) for $n=200$. The JW[7] algorithm was tested on the simulated samples with family sizes following Poisson or negative binomial distributions. The reconstructions of empirical data sets were also carried out to assess or illustrate the accuracy of the algorithms under consideration.[3,5,7] However, any conclusions drawn from what are normally a very limited number of empirical trials are statistically questionable and hence such cases are not considered here.

The fixed family sizes[2,3,5] are not scalable between different values of n while the distribution based[7] sizes may be prone to misinterpretation. Eventually it would be

[a] denoted by *Likelihood* in [3].
[b] Five families containing 10 full siblings each.
[c] Fifty unrelated individuals.

desirable to reach a consensus on family structure benchmarks that are easy to reproduce, exactly defined, and scalable to a wide range of n. The benchmarks could be used in the reporting of an algorithm's accuracy, allowing for consistent comparison between different algorithms. Two such benchmarks are proposed below and used for the testing of the FSR algorithms in this study:

- The *uniform* distribution benchmark (inspired by the (5×10)[3,2] and (50×1)[3] distributions) is defined by a partition $U_n(r,g)$, where r is the number of families (sib groups) and g is the size of each family, giving the population size $n = rg$. This benchmark tests how well an FSR algorithm performs as the amount of genetic information is gradually reduced: the number of families r increases maintaining the constant population sample size n and reducing each group size $g = n/r$.
- The *skewed* distribution is defined by $S_n(r,q)$, where q is the skewing factor such that group k contains $g_k = g_1 + q(k-1)$ full siblings and the size of the first group is given by $g_1 = n/r - q(r-1)/2$. This benchmark is essential since the accuracy of some FSR algorithms deteriorates as the skewing increases, e.g. GRAPH,[2] SC,[3] and FJL.[3]

Any allelic mutation in an individual genotype (Eq. 1) may lead to misclassification of that individual and is referred to as the *genotype error*. The error could be due to a variety of factors, e.g. mutation, plain human error,[18] PCR missprinting[11,7] and allelic dropout (null allele).[12] Most of the existing sources of error manifest themselves on the per allele basis, making it natural to specify the errors as the error rate per allele or locus.[7,11] In this study the following error model is used capturing the majority of the biologically occurring errors in one parameter, the locus error rate ε. The error is applied by collecting all available loci from all the individuals from a given sample, obtaining nL loci. Next, εnL different loci are randomly selected and one allele at each of the loci is mutated into a randomly chosen *different* (change into itself is prohibited) allele from the same locus. Since a common misprinting error is relatively small (between 0.3% and 11% per allele)[12] the mutation of both alleles at the same locus is omitted from consideration.

3 Algorithms

3.1 *The Modified SIMPSON (MS2) Algorithm*

Let $d_l(X,Y)$ be the number of alleles in an individual X which are not present in an individual Y at locus l. The locus $D_l(X,Y)$ and genotype $D(X,Y)$ distances could be defined by $D_l(X,Y) = \max\left(d_l(X,Y), d_l(Y,X)\right)$, $D(X,Y) = \min D_l(X,Y)$, respectively.[8] The Modified SIMPSON (MS) algorithm significantly improved the SIMPSON[3] heuristic in speed while maintaining low accuracy-error using the genotype distances and achieving $O(n^3)$ running time.[8] The following $O(n^2)$ algorithm, named

MS2, is derived from the original MS algorithm utilizing the local-minimum property of the Simpson index. The MS steps (1-4) remain unchanged:[8] steps (1) and (2) - calculate and sort the list of genotype distances in ascending order; step (3) - create a pool of unassigned individuals; step (4) – repeat this and the following steps; select the next unassigned individual from the list of distances until all individuals are assigned. The new MS2 steps: step (5) - place the next individual into the first group that passes the sibship test;[d] step (6) - sort the available sib groups in the descending order of their sizes.

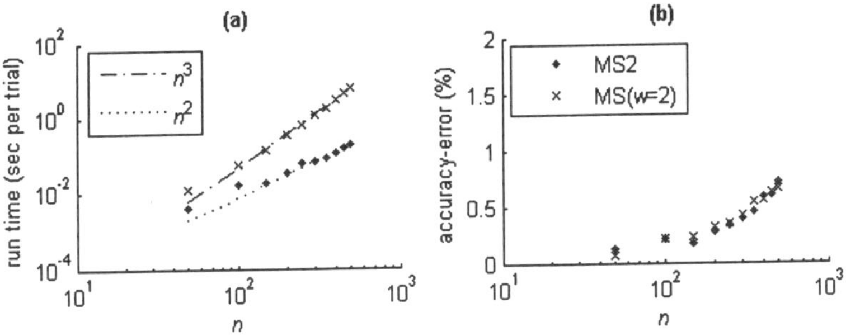

Figure 1. Runtime efficiency (in seconds per trial on a 3GHz PC) and the accuracy-error (%) of the MS (×) and MS2 (•) algorithms. Each FSR trial is performed on a freshly generated population sample genotyped for L=5 loci, each locus containing N_A=10 equifrequent alleles. The sample consists of r groups each containing 5 full siblings, giving the population size n=5r. The MS results are obtained with the window parameter w=2. The cubic and square powers of n are denoted by the dash-dot and dotted lines in the subfigure (a), respectively.

Figure 1(a) verifies that the complexity of the MS2 algorithm is $O(n^2)$, further improving the MS's $O(n^3)$. By the definition of the MS2 algorithm, the lower bound of its accuracy-error is the accuracy-error of the original MS algorithm when the MS's window parameter is $w=1$. Figure 1(b) indicates that any potential loss of accuracy could be insignificant. The efficiency improvement is due to the Simpson index (Eq. 2) which is maximized on the local scale by increasing the largest group. To illustrate that, let two available groups have sizes g and $g-1$. Assuming that the next individual could be added to both groups, the Simpson index is maximized by placing the individual into the g-group since $(g+1)^2+(g-1)^2>2g^2$. However the greedy method is still only a heuristic even on the local scale since two or more largest groups may have the same size. On the global scale this greedy approach has no guarantee in achieving the maximum value of the index, e.g. the partition with the group sizes (8,3,2) has a smaller index than the partition with (7,6) sizes.

Figure 1(b) verifies that the MS2 algorithm is as accurate as the MS algorithm However the MS2 algorithm is superior in run-time efficiency, e.g. Figure 1(a) shows that MS2 takes the same amount of computer time to reconstruct 500 individuals as for MS to

---

[d] The sibship test is performed on the newly created group containing the next individual and the existing group.

reconstruct 150 individuals. The absolute terms, MS2 requires only a fraction of a second to perform the full sibship reconstruction of 500 individuals on a 3GHz Pentium 4 PC.

3.2 *The SIMPSON-assisted Descending Ratio (SDR) Algorithm*

Figure 2 compares the DR,[4] SIMPSON[3] and MS2 algorithms. Figure 2(a) for 50 unrelated individuals stands out as a reminder that the Simpson index based formulation (FSR-S) is still only an approximation of the FSR problem. The MS2 correctly finds the partition with the largest Simpson index (as does SIMSPON) by placing the individuals in groups of size two or larger (any two individuals always pass the sibship test). While the Simpson index as the scoring function is biologically incorrect in this instance, the likelihood based DR algorithm makes sense biologically by becoming more accurate as the amount of genetic information increases (larger L). The DR results are obtained with the *null* and *primary* hypotheses[e] being the unrelated and diploid full-sibling relationships, respectively.

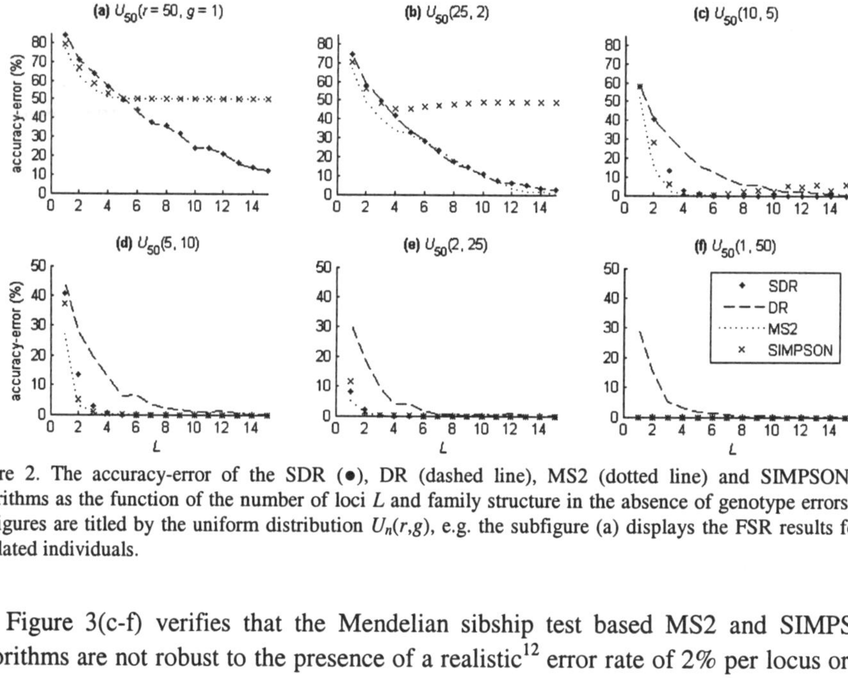

Figure 2. The accuracy-error of the SDR (•), DR (dashed line), MS2 (dotted line) and SIMPSON (×) algorithms as the function of the number of loci L and family structure in the absence of genotype errors. The subfigures are titled by the uniform distribution $U_n(r,g)$, e.g. the subfigure (a) displays the FSR results for 50 unrelated individuals.

Figure 3(c-f) verifies that the Mendelian sibship test based MS2 and SIMPSON algorithms are not robust to the presence of a realistic[12] error rate of 2% per locus or 1% per allele confirming the serious concern raised by Hoffman and Amos[11] who criticized the current common practice of reporting genotype inferred results without the error analysis. However in the absence of errors the MS2 and SIMPSON algorithms are more accurate than DR (Figure 2).

---

[e] *null* and *primary* are from the terminology of the KINSHIP [9] and KINGROUP [4] programs

The MS2 accuracy in the absence of genotype errors and the DR robustness to the errors prompts the following SIMPSON-assisted Descending Ratio (SDR) algorithm: step (1) - perform the reconstruction using MS2 algorithm; step (2) - retain one largest group with size 3 or larger; step (3) - assign the remaining unassigned individuals as per the DR[4] algorithm. Only one largest group is retained in step (2) since the MS2 (and hence MS and SIMPSON) algorithm tends to break up a true sib group into a number of smaller sib groups in the presence of mutated alleles.

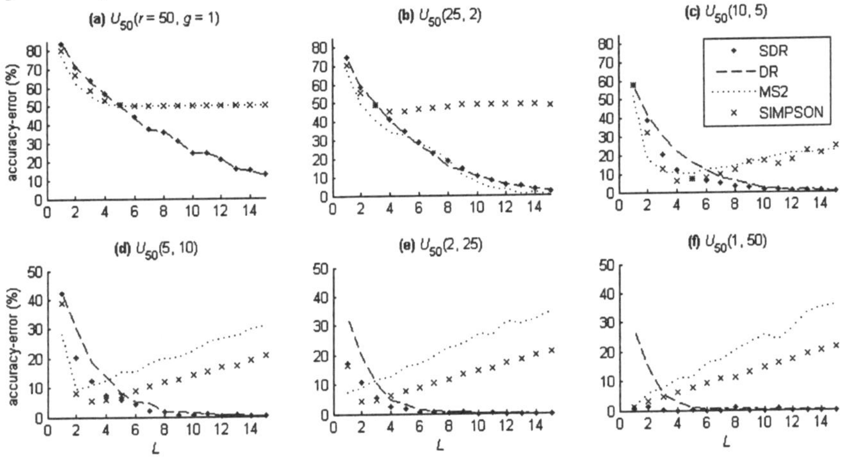

Figure 3. The same as in Figure 2 but with a 2% locus (1% allele) error rate applied to the generated population samples.

4 Results and Discussion

For this study the genotypes (Eq. 1) are considered with the same number of equifrequent alleles $\lambda_l = N_A = 10$ at each of the L loci. The number of loci L is chosen as a varying parameter since biologists would normally have a choice in the number of loci (e.g. microsatellite markers) but not their heterozygosity. Already having L as a parameter the variations in the number of alleles N_A are not considered since it is well understood that the increase in either N_A, L or both improves the accuracy of an FSR algorithm.[2,3] The SIMPSON results are calculated with $100\,000$ iterations. All presented results are averaged over 100 trials.

Figure 2 and Table 1 demonstrate that with 10 equifrequent alleles and in the absence of genotype errors: the SDR algorithm is as accurate as MS2 and SIMPSON from about $L = 5$ loci onwards; the MS2 and SIMPSON algorithms are essentially identical in accuracy. Figure 2(b) shows, however, that in the case of 25 families of two full siblings each, the MS2 algorithm is as accurate as DR while SIMPSON fails to distinguish correct sib groups.

In the presence of a 2% locus (1% allele) error rate (Figure 3): both MS2 and SIMPSON fail to deal with the errors, effectively arriving at proportionally worse partitions as the absolute number of errors increases with the increase of L; SDR is more accurate than the MS2, SIMPSON and DR algorithms, starting from about $L = 6$ loci; the SDR algorithm outperforms DR for all considered number of loci and family structures verifying the value of the MS2 preprocessing. The $O(n^2)$ cost of the MS2 preprocessing is negligible in comparison to the $O(n^3)$ cost[8] of the DR algorithm making SDR run in $O(n^3)$ and be feasible for practical applications.

Since SDR retains the largest sib group reconstructed by MS2, it may be expected that the effect of just one sib group should be proportionally small when a large number of groups is present, as in the case of 10 groups of 5 individuals each, see Figure 2(c). Surprisingly, Figure 2(c) demonstrates that the accuracy-error is reduced disproportionally, showing that the DR algorithm works significantly better if at least one "seed" sib group is supplied. This suggests a new approach which has a potential to resolve the current problem with the widely used KINSHIP[9] program. Using simulations, the program determines the pairwise likelihood ratios (the same ratios are used in the DR algorithm) for the given significance levels but then it is up to the user to manually assign individuals into sib groups based on their pairwise ratios. The problem arises when the same individual is significantly likely to be in the full sibling relationship with a number of individuals from different sib groups.[19] An algorithm similar to the SDR algorithm could accept all sib groups reconstructed by KINSHIP without conflict and then complete the reconstruction using the DR algorithm which, as shown here, becomes significantly more accurate once at least one seed group is supplied.

Figure 4 verifies that the SDR algorithm is robust to the mutation errors for skewed family structures. In particular, the accuracy-error SDR results in Figure 3(d) for 5 uniform groups are very similar to the results in Figure 4(b) for 5 skewed groups.

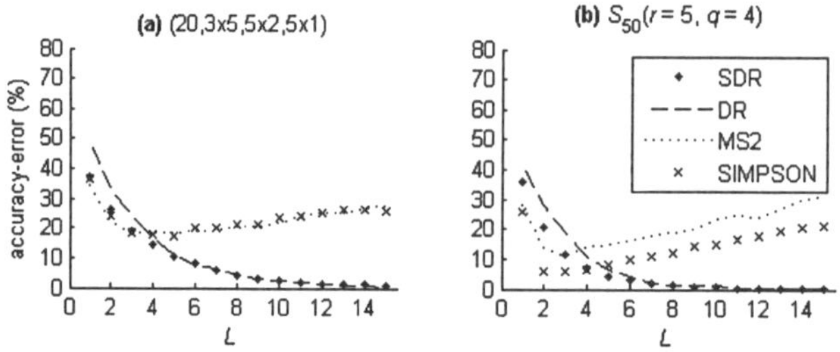

Figure 4. The same as in Figure 3 but for skewed family distributions: (a) 50 individuals distributed in 14 sib groups with (20, 5, 5, 5, 2, 2, 2, 2, 2, 1, 1, 1, 1, 1) sizes; (b) 50 individuals distributed in 5 sib groups with (2, 6, 10, 14, 18) sizes.

In conclusion, given a population sample without genotype errors and in the absence of unrelated individuals, the new MS2 $O(n^2)$ algorithm solves the FSR problem to the near-optimal level in speed and accuracy. On the other hand, the presented preliminary results suggest that the new SDR $O(n^3)$ algorithm could solve the FSR problem to a high level of accuracy even in the presence of unrelated individuals and genotype errors.

Table 1. The accuracy-error (percentage of incorrectly classified individuals) achieved by the DR, MS2, SDR and SIMPSON algorithms for 50 individuals uniformly distributed in r groups of g size each. The family distributions are denoted by (r, g). Each of the L loci is simulated with 10 equifrequent alleles.

Algorithm	$U_{50}(r,g)$	L=1	2	3	4	5	6	8	10	12	14
DR	(50,1)	84.6	71.2	63.8	57	50.7	43.3	35.6	23.9	20.9	13.6
DR	(25,2)	74.8	59.2	50.2	40.9	33.6	28.8	16.4	10.6	6	3.3
DR	(10,5)	58.5	40.6	31	23.7	16	12.5	5.4	2.9	1.9	1
DR	(5,10)	44.9	28.2	19.5	13.2	6.4	6.5	2.2	0.9	1.2	0.2
DR	(2,25)	31.1	19.2	9.7	3.9	4	1.6	0.6	0.4	1	0
DR	(1,50)	30	15.4	5.1	3.2	1.8	1.5	0.2	0.2	0.1	0
MS2	(50,1)	77	62.9	54.9	51.3	50.1	50	50	50	50	50
MS2	(25,2)	67	49.2	40.3	34.3	31.8	27.6	18.4	9.2	3.2	1.2
MS2	(10,5)	52.2	15	2.6	0.4	0.1	0	0	0	0	0
MS2	(5,10)	27.2	3.2	0.5	0	0	0	0	0	0	0
MS2	(2,25)	5	0.8	0.1	0	0	0	0	0	0	0
MS2	(1,50)	0	0	0	0	0	0	0	0	0	0
SDR	(50,1)	83.9	71.2	63.7	56.7	49.7	44.2	36.2	24.3	21.2	14
SDR	(25,2)	74.4	58.2	49.9	41.8	32.8	28	17.6	10.7	6.3	3.2
SDR	(10,5)	57.7	40.7	13.4	2.6	0.7	0	0	0	0	0
SDR	(5,10)	40.8	13.5	2.8	0.5	0	0	0	0	0	0
SDR	(2,25)	8.3	2.1	0.1	0	0	0	0	0	0	0
SDR	(1,50)	0	0	0	0	0	0	0	0	0	0
SIMPS	(50,1)	79.6	67.1	58.4	53.4	50.3	50	50	50	50	50
SIMPS	(25,2)	70.4	56.5	48.6	45.6	45.7	46.7	48	48.9	48.8	48.9
SIMPS	(10,5)	58	28.3	6.2	1.2	0.3	0	2.4	2.3	4.6	2.4
SIMPS	(5,10)	37.4	5	0.9	0.1	0	0	0	0	0	0
SIMPS	(2,25)	11.4	1.3	0	0	0	0	0	0	0	0
SIMPS	(1,50)	0	0	0	0	0	0	0	0	0	0

Acknowledgments

The author would like to thank Nigel Bajema, Marianne Brown, David Browning, Svetlana Frizen, Michael Henshaw, Dean Jerry and Bruce Litow for helpful discussions and assistance, as well as three anonymous reviewers for their constructive comments.

References

1. A. Almudevar and C. Field. Estimation of single-generation sibling relationships based on DNA markers. *Journal of Agricultural Biological and Environmental Statistics*, 4:136-165, 1999.
2. J. Beyer and B. May. A graph-theoretic approach to the partition of individuals into full-sib families. *Molecular Ecology*, 12:2243-2250, 2003.
3. K. Butler, C. Field, C. M. Herbinger and B. R. Smith. Accuracy, efficiency and robustness of four algorithms allowing full sibship reconstruction from DNA marker data. *Molecular Ecology*, 13:1589-1600, 2004.
4. D. A. Konovalov, C. Manning and M. T. Henshaw. KINGROUP: a program for pedigree relationship reconstruction and kin group assignments using genetic markers. *Molecular Ecology Notes*, 4:779-782, 2004.

5. B. R. Smith, C. M. Herbinger and H. R. Merry. Accurate partition of individuals into full-sib families from genetic data without parental information. *Genetics*, 158:1329-1338, 2001.
6. S. C. Thomas and W. G. Hill. Estimating quantitative genetic parameters using sibships reconstructed from marker data. *Genetics*, 155:1961-1972, 2000.
7. J. L. Wang. Sibship reconstruction from genetic data with typing errors. *Genetics*, 166:1963-1979, 2004.
8. D. A. Konovalov, N. Bajema and B. Litow. Modified SIMPSON $O(n^3)$ algorithm for the full sibship reconstruction problem. *Bioinformatics*:in press, 2005.
9. K. F. Goodnight and D. C. Queller. Computer software for performing likelihood tests of pedigree relationship using genetic markers. *Molecular Ecology*, 8:1231-1234, 1999.
10. G. Luikart and P. R. England. Statistical analysis of microsatellite DNA data. *Trends in Ecology & Evolution*, 14:253-256, 1999.
11. J. I. Hoffman and W. Amos. Microsatellite genotyping errors: detection approaches, common sources and consequences for paternal exclusion. *Molecular Ecology*, 14:599-612, 2005.
12. S. Creel, G. Spong, J. L. Sands, J. Rotella, J. Zeigle, L. Joe, K. M. Murphy and D. Smith. Population size estimation in Yellowstone wolves with error-prone noninvasive microsatellite genotypes. *Molecular Ecology*, 12:2003-2009, 2003.
13. H. Ellegren. Microsatellite mutations in the germline: implications for evolutionary inference. *Trends in Genetics*, 16:551-558, 2000.
14. T. Y. Berger-Wolf, B. DasGupta, W. Chaovalitwongse and M. Ashley. Combinatorial Reconstructions of Sibling Relationships. *6th International Symposium on Computational Biology and Genome Informatics (CBGI)*, Salt Lake City, Utah, 1252-1255, July 21-26, 2005.
15. D. Gusfield. Partition-distance: A problem and class of perfect graphs arising in clustering. *Information Processing Letters*, 82:159-164, 2002.
16. D. A. Konovalov, B. Litow and N. Bajema. Partition-distance via the assignment problem. *Bioinformatics*, 21:2463-2468, 2005.
17. A. Almudevar. A Bootstrap Assessment of Variability in Pedigree Reconstruction Based on Genetic Markers. *Biometrics*, 57:757-763, 2001.
18. P. T. O'Reilly, C. Herbinger and J. M. Wright. Analysis of parentage determination in Atlantic salmon (Salmo salar) using microsatellites. *Animal Genetics*, 29:363-370, 1998.
19. M. T. Henshaw, S. K. A. Robson and R. H. Crozier. Queen number, queen cycling and queen loss: the evolution of complex multiple queen societies in the social wasp genus Ropalidia. *Behavioral Ecology and Sociobiology*, 55:469-476, 2004.

INFERENCE OF GENE REGULATORY NETWORKS FROM MICROARRAY DATA: A FUZZY LOGIC APPROACH

PATRICK C.H. MA AND KEITH C.C. CHAN†

Department of Computing, The Hong Kong Polytechnic University, Hung Hom, Kowloon, Hong Kong SAR, China

Recent developments in large-scale monitoring of gene expression such as DNA microarrays have made the reconstruction of gene regulatory networks (GRNs) feasible. Before one can infer the structures of these networks, it is important to identify, for each gene in the network, which genes can affect its expression and how they affect it. Most of the existing approaches are useful exploratory tools in the sense that they allow the user to generate biological hypotheses about transcriptional regulations of genes that can then be tested in the laboratory. However, the patterns discovered by these approaches are not adequate for making accurate prediction on gene expression patterns in new or held-out experiments. Therefore, it is difficult to compare performance of different approaches or decide which approach is likely to generate plausible hypothesis. For this reason, we need an approach that not only can provide interpretable insight into the structures of GRNs but also can provide accurate prediction. In this paper, we present a novel fuzzy logic-based approach for this problem. The desired characteristics of the proposed algorithm are as follows: (i) it is able to directly mine the high-dimensional expression data without the need for additional feature selection procedures, (ii) it is able to distinguish between relevant and irrelevant expression data in predicting the expression patterns of predicted genes, (iii) based on the proposed objective interestingness measure, no user-specified thresholds are needed in advance, (iv) it can make explicit hidden patterns discovered for possible biological interpretation, (v) the discovered patterns can be used to predict gene expression patterns in other unseen tissue samples, and (vi) with fuzzy logic, it is robust to noise in the expression data as it hides the boundaries of the adjacent intervals of the quantitative attributes. Experimental results on real expression data show that it can be very effective and the discovered patterns reveal biologically meaningful regulatory relationships of genes that could help the user reconstructing the underlying structures of GRNs.

1 Introduction

Large-scale monitoring gene expression such as DNA microarrays [1,2] is considered to be one of the most promising techniques for reconstructing the gene regulatory networks (GRNs). A GRN is typically a complex biological system in which proteins and genes bind to each other and act as an input-output system for controlling various cellular processes. Since, living cells contain thousands of genes, each of which codes for one or more proteins. Many of these proteins in turn regulate the expression of some other genes through complex regulatory pathways to accommodate changes in different external environments or carry out the essential developmental programs. The key to understanding living processes is therefore to uncover the structures of these regulatory networks that underlie the regulations of cells.

† E-mail: {cschma,cskcchan}@comp.polyu.edu.hk

Previous attempts have been reported to inferring the underlying structures of GRNs such as the biochemically driven approaches [3,4], the Boolean network approaches [5], the Bayesian network approaches [6] and the data mining approaches [7-9]. However, these approaches have several limitations need to be overcome in order to effectively deal with the problem. For example, for the biochemically driven approaches, most of the biochemical reactions under participation of proteins do not follow linear reaction kinetics and also gene expression data seems not sufficient to globally understand regulatory networks at this level of detail [3,4]. For the Boolean network approaches, the validity of the pre-defined assumptions [5] and the values of the Boolean approach in general, have been questioned by a number of researchers, particularly in the biological community, where there is a perceived lack of connection between simulation results and empirically testable hypotheses [10]. For the Bayesian network approaches, the task of learning model parameters is NP-hard especially for high-dimensional data. Moreover, many parameters need to be estimated accurately and this requires a large amount of samples that may not always be readily available [6]. For the data mining approaches, clustering of gene expression data [7] only measures whether genes share a significant linear relationship with each other. The regulatory relationships such as which gene affects which other genes cannot be discovered. On the other hand, the crisp discretization procedures of the classification algorithms [8,9] such as C4.5 [11] do not take into account that values at the borderline between value categories may be very similar. This makes the classifiers less resilient to noise and some useful patterns exist at this borderline can be overlook.

Besides the above limitations, the patterns discovered by most of the existing approaches are not adequate for making accurate prediction on gene expression patterns in new or held-out experiments. Hence, it is difficult to compare performance of them or decide which approach is likely to generate plausible hypothesis. Therefore, we need an approach that not only can provide interpretable insight into the structures of GRNs but also can provide accurate prediction. For this reason, we propose a novel fuzzy logic-based approach in this paper. The rest of the paper is organized as follows. In Section 2, the proposed algorithm is described in details. The effectiveness of the proposed algorithm has been evaluated and compared through various experiments with real expression data. The experimental set-up, together with the results, is discussed in Section 3. Lastly, in Section 4, we give a summary of the paper.

2 The proposed algorithm

Fuzzy logic and fuzzy sets allow the modeling of language-related uncertainties by providing a symbolic framework for knowledge comprehensibility [12,13]. Fuzzy representation is becoming increasingly popular in dealing with problems of uncertainty, noise and inexact data. Recently, fuzzy logic has successfully been used for clustering gene expression data. For example, the fuzzy *k*-means algorithms [14,15] have been applied to discover the clusters of co-expressed genes so that genes have similar biological functions can be revealed. However, for the inference of GRNs, only limited studies have been proposed [16]. Due to the fact that there is a need to have an effective fuzzy logic-based algorithm, here, we propose such an algorithm and discuss the details in this section.

2.1. *Linguistic variables and linguistic terms representation*

Given a set of data D, each record r (experimental condition), is characterized by a set of attributes (genes), $A=\{A_1,\ldots,A_i,\ldots,A_n\}$. For any record, $r \in D$, $r[A_i]$ denotes the value in r for attribute A_i. Let $L=\{L_1,\ldots,L_i,\ldots,L_n\}$ be a set of linguistic variables such that $L_i \in L$ represents $A_i \in A$. For any quantitative attribute, A_i, let $dom(A_i)=[l_i,u_i]\subseteq \Re$ denote the domain of the attribute, where l_i and u_i represent the lower and upper bounds of A_i respectively. Moreover, A_i is represented by a linguistic variable, L_i, whose value is a linguistic term in $T(L_i)=\{l_{ij} \mid j=1,2,\ldots,s_i\}$ where l_{ij} is a linguistic term characterized by a fuzzy set, F_{ij}, that is defined on $dom(A_i)$ and whose membership function is $\mu_{F_{ij}}$. The degree of membership of the value in r with respect to F_{ij} is given by $\mu_{F_{ij}}(r[A_i])$. The degree to which r is characterized by l_{ij}, $\lambda_{l_{ij}}(r)$, is therefore defined as follows:

$$\lambda_{l_{ij}}(r)=\mu_{F_{ij}}(r[A_i]). \tag{1}$$

If $\lambda_{l_{ij}}(r)=1$, the attribute A_i of r is completely characterized by the linguistic term l_{ij}. If $\lambda_{l_{ij}}(r)=0$, the attribute A_i of r is not characterized by the linguistic term l_{ij}. If $0<\lambda_{l_{ij}}(r)<1$, the attribute A_i of r is partially characterized by the linguistic term l_{ij}. In the case where $r[A_i]$ is unknown, $\lambda_{l_{ij}}(r)=0.5$, which indicates that there is no information available concerning whether the attribute A_i of r is or is not characterized by the linguistic term l_{ij}.

2.2. *Discovering the interesting patterns*

Let $o(l_{ij})$ be the observed degree to which the records in the given dataset are characterized by the linguistic term l_{ij}. It is defined as follows:

$$o(l_{ij})=\sum_{r\in D}\lambda_{l_{ij}}(r)\cdot \tag{2}$$

Also, let $l_{ij} \Leftrightarrow l_{pq}$ be the association between the linguistic terms l_{ij} and l_{pq}. Then, the observed degree to which the records are characterized by this association, $o(l_{ij} \Leftrightarrow l_{pq})$, is defined as follows:

$$o(l_{ij} \Leftrightarrow l_{pq})=\sum_{r\in D}\min(\lambda_{l_{ij}}(r),\lambda_{l_{pq}}(r))\cdot \tag{3}$$

To decide whether an association, $l_{ij} \Leftrightarrow l_{pq}$, is interesting, it is objectively evaluated using the proposed objective interestingness measure, $d(l_{ij} \Leftrightarrow l_{pq})$. This measure reflects the differences in the observed $o(l_{ij} \Leftrightarrow l_{pq})$ and expected $e(l_{ij} \Leftrightarrow l_{pq})$ degrees to which the records are characterized by these linguistic terms. The objective interestingness measure, $d(l_{ij} \Leftrightarrow l_{pq})$, is defined as follows:

$$d(l_{ij} \Leftrightarrow l_{pq}) = \frac{z(l_{ij} \Leftrightarrow l_{pq})}{\sqrt{v(l_{ij} \Leftrightarrow l_{pq})}}, \tag{4}$$

where

$$z(l_{ij} \Leftrightarrow l_{pq}) = \frac{o(l_{ij} \Leftrightarrow l_{pq}) - e(l_{ij} \Leftrightarrow l_{pq})}{\sqrt{e(l_{ij} \Leftrightarrow l_{pq})}}, \tag{5}$$

$$e(l_{ij} \Leftrightarrow l_{pq}) = \frac{o(l_{ij}) \times o(l_{pq})}{\sum_{r \in D} \sum_{q=1}^{s_p} \sum_{j=1}^{s_i} \min(\lambda_{l_{ij}}(r), \lambda_{l_{pq}}(r))}, \tag{6}$$

$$v(l_{ij} \Leftrightarrow l_{pq}) = \left(1 - \frac{o(l_{ij})}{\sum_{r \in D} \sum_{j=1}^{s_i} \lambda_{l_{ij}}(r)}\right)\left(1 - \frac{o(l_{pq})}{\sum_{r \in D} \sum_{q=1}^{s_p} \lambda_{l_{pq}}(r)}\right). \tag{7}$$

If $d(l_{ij} \Leftrightarrow l_{pq}) > 1.96$ (i.e., the 95th percentile of the normal distribution) [17-19], we can conclude that the association $l_{ij} \Leftrightarrow l_{pq}$ is interesting. It means that it is more likely for a record to be characterized by both l_{ij} and l_{pq}.

2.3. *Prediction based on the discovered patterns*

Given that $l_{ij} \Leftrightarrow l_{pq}$ is interesting, the patterns can be constructed as follows:

$$w(l_{ij} \Leftrightarrow l_{pq}) = \log \frac{\Pr(l_{ij} \Leftrightarrow l_{pq})}{\Pr(l_{ij} \Leftrightarrow \neg l_{pq})}, \tag{8}$$

where

$$\Pr(l_{ij} \Leftrightarrow l_{pq}) = \frac{\sum_{r \in D} \min(\lambda_{l_{ij}}(r), \lambda_{l_{pq}}(r))}{\sum_{r \in D} \sum_{j=1}^{s_i} \min(\lambda_{l_{ij}}(r), \lambda_{l_{pq}}(r))},$$

$$\Pr(l_{ij} \Leftrightarrow \neg l_{pq}) = \frac{\sum_{r \in D} \sum_{u=1, u \neq q}^{s_p} \min(\lambda_{l_{ij}}(r), \lambda_{l_{pu}}(r))}{\sum_{r \in D} \sum_{u=1, u \neq q}^{s_p} \sum_{j=1}^{s_i} \min(\lambda_{l_{ij}}(r), \lambda_{l_{pu}}(r))}. \tag{9}$$

The term $\Pr(l_{ij} \Leftrightarrow l_{pq})$ can be considered as being the probability that a record is characterized by l_{ij} and l_{pq}, and the term $\Pr(l_{ij} \Leftrightarrow \neg l_{pq})$ can be considered as being the probability that a record is characterized by l_{ij} and l_{pu}, where $u \neq q$. Then, the term $w(l_{ij} \Leftrightarrow l_{pq})$ is a confidence measure that represents the uncertainty associated with $l_{ij} \Leftrightarrow l_{pq}$. It can be interpreted as being a measure of the difference in the information gain when a record that is characterized by l_{ij} is also characterized by l_{pq} as opposed to being characterized by other linguistic terms l_{pu}, where $u \neq q$.

Given a testing record r and it is characterized by n attribute values, $r[A_1],...,r[A_p],...,r[A_n]$, where $r[A_p]$ is the value that is to be predicted. Let l_p be a linguistic term with a domain of $T(L_p)$. The value of $r[A_p]$ is determined according to l_p. To predict $r[A_p]$, the discovered patterns are searched. If an attribute value, say $r[A_i]$, $A_i \neq A_p$, of r is characterized by the linguistic term in the antecedent of a pattern that implies l_{pq}, then it can be considered as providing some confidence that the value of l_p should be assigned to l_{pq}. By repeating this procedure, that is, by matching each attribute value of r against the discovered patterns, the value of l_p can be determined by computing the total confidence measure. Since each attribute of r may or may not provide a contribution to the total confidence measure and those that may support the assignment of different values. Therefore, the different contributions to the total confidence measure are measured quantitatively and then combined for comparison in order to find the most suitable value of l_p. For any attribute value, $r[A_i]$, $A_i \neq A_p$, of r, it is characterized by a linguistic term, l_{ij}, to a degree of compatibility, $\lambda_{l_{ij}}(r)$. Given the patterns those imply the assignment of l_{pq}, then, the confidence provided by $r[A_i]$ for such as assignment is as follows:

$$W_{l_{pq} r[A_i]} = w(l_{ij} \Leftrightarrow l_{pq}) \times \lambda_{l_{ij}}(r). \tag{10}$$

Suppose that among the $n-1$ attributes excluding A_p, only some combinations of them, $r[A_1],...,r[A_i],...,r[A_\beta]$, are found to match one or more patterns. Then, the total confidence measure of assigning the value of l_p to l_{pq} is given as follows:

$$TW_q = \sum_{i=1}^{\beta} W_{l_{pq}} r[A_i]. \tag{11}$$

Based on the above total confidence measure, in the case if $TW_q > TW_c$, where $q \neq c$. Then, l_p is assigned to l_{pq}.

3 Experimental results

3.1. *Experimental data*

For experimentation with real data, we used a set of gene expression data that contains a series of gene expression measurements of the transcript (mRNA) levels of S. cerevisiae genes [7,20]. In this dataset, the samples were synchronized by three different methods: α factor arrest, arrest of a cdc15, and cdc28 temperature-sensitive mutant. Using periodicity and correlation algorithms, a total of about 800 genes that meet an objective minimum criterion for cell cycle regulation were identified [7]. The expression data we used is available at [21]. Since gene expression can be described in a finite number of different states/patterns [22]. We therefore represented it in terms of three fuzzy sets: low (L), medium (M) and high (H). For any quantitative attribute A_i, the degree of membership of a record, $r[A_i]$, can be computed as follows [23] (in Fig. 1):

$$\mu_{low}(r[A_i]) = \begin{cases} 1, & if \;\; r[A_i] < Av_{i1} \\ \dfrac{P_{i2} - r[A_i]}{P_{i2} - Av_{i1}}, & if \;\; Av_{i1} \le r[A_i] < P_{i2} \\ 0, & otherwise \end{cases} \tag{12}$$

$$\mu_{medium}(r[A_i]) = \begin{cases} 0, & if \;\; r[A_i] < P_{i1} \\ \dfrac{r[A_i] - P_{i1}}{Av_{i2} - P_{i1}}, & if \;\; P_{i1} \le r[A_i] < Av_{i2} \\ \dfrac{P_{i2} - r[A_i]}{P_{i2} - Av_{i2}}, & if \;\; Av_{i2} \le r[A_i] < P_{i2} \\ 0, & otherwise \end{cases} \tag{13}$$

$$\mu_{high}(r[A_i]) = \begin{cases} 0, & if \;\; r[A_i] < P_{i1} \\ \dfrac{r[A_i] - P_{i1}}{Av_{i3} - P_{i1}}, & if \;\; P_{i1} \le r[A_i] < Av_{i3} \\ 1, & otherwise \end{cases} \tag{14}$$

where A_i is sorted in the ascending order of its values, P_{i1} is the value of A_i that exceeds one-third of the measurements and is less than the remaining two-thirds and P_{i2} is the value of A_i that exceeds two-thirds of the measurements and is less than the remaining one-third. And also, $A_{i_{max}}$ and $A_{i_{min}}$ denote the maximum and minimum values encountered along attribute A_i, and $Av_{i1} = \frac{A_{i\min} + P_{i1}}{2}$, $Av_{i2} = \frac{P_{i1} + P_{i2}}{2}$ and $Av_{i3} = \frac{P_{i2} + A_{i\max}}{2}$ [23].

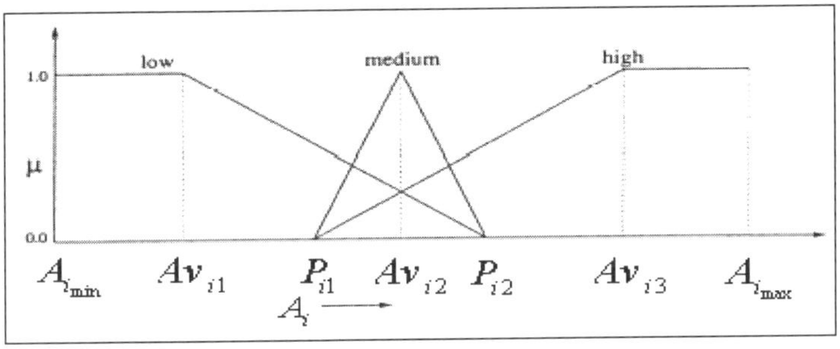

Figure 1. Membership function.

3.2. *Method of evaluating the results*

In this analysis, we chose the cdc15 experiment as the training set. Another two datasets: alpha and cdc28 experiments were used as the testing sets. For experimentation, we randomly selected 6 genes (CLN1, HTA1, HTB1, CLB1, CLN2, and CLB6) to evaluate the effectiveness of the proposed algorithm. Using the proposed algorithm, the patterns of these genes in the independent testing sets are predicted. Then, the predicted patterns are compared with the original patterns of these genes and the percentage of accurate prediction can therefore be determined.

3.3. *Results*

To evaluate the performance of the proposed method, we also compared it to the popular decision-tree based algorithm called C4.5 [11] as discussed in Section 1. Moreover, since one of the desirable features of the proposed algorithm is its feature selection capability, it is able to distinguish between relevant and irrelevant expression data. Therefore, for fair performance comparisons, we performed additional experiments to compare it to C4.5 with feature selection approach. There are many feature selection methods have been proposed for gene expression data such as filter and wrapper methods [24,25]. In this analysis, we adopted *t*-statistics measure [25]. Based on the *t*-statistics measure, the new subset of genes with largest *t*-values was obtained. The selection method of genes with largest *t*-values is as follows: (i) sorted the genes in descending order based on their *t*-values, (ii) initially, 5% (empirically set) of genes were selected from top of the rank list, (iii) the classification performance based on this subset of genes was measured by C4.5 (10-fold cross validation), (iv) added another 5% of genes from the rank list into this subset, (v) repeat steps (iii) and (iv) until the classification performance converged, (vi) the final subset of genes was selected.

In Tables 1 and 2, the comparisons of average prediction accuracy are showed. According to these tables, we found that the performance of C4.5 can be improved with the feature selection procedure. In addition, we also compared another well-known decision-tree based algorithm called FID [26] and trained the algorithm only using the significant features identified by C4.5 during the feature selection process as discussed above. FID is a fuzzy logic-based classifier that combines symbolic decision trees with approximate reasoning offered by fuzzy representation. It extends C4.5 by using splitting criteria based on fuzzy restrictions and using different inference procedures to exploit fuzzy sets. The experimental results of FID are also showed in Tables 1 and 2. According to these results, we found that the performance of the proposed algorithm is not only better than other popular algorithms and also the average prediction accuracy in each testing set is high. This indicates that the proposed algorithm is very effective in predicting gene expression patterns in the unseen samples.

Table 1. Result comparison (alpha dataset).

Gene	Proposed	C4.5	C4.5 + Feature selection	FID + Feature selection
CLN1	0.94	0.67	0.83	0.94
HTA1	0.89	0.61	0.78	0.83
HTB1	1	0.67	0.78	0.94
CLB1	0.94	0.67	0.83	0.94
CLN2	1	0.67	0.78	0.83
CLB6	0.89	0.72	0.83	0.83
Avg.	0.94	0.67	0.81	0.89

Table 2. Result comparison (cdc28 dataset).

Gene	Proposed	C4.5	C4.5 + Feature selection	FID + Feature selection
CLN1	0.88	0.65	0.76	0.88
HTA1	0.94	0.58	0.71	0.88
HTB1	0.94	0.53	0.65	0.88
CLB1	0.94	0.71	0.82	0.82
CLN2	0.94	0.71	0.82	0.94
CLB6	0.88	0.65	0.82	0.76
Avg.	0.92	0.64	0.76	0.86

3.4. *Biological interpretation*

In order to evaluate the biological significance of the discovered patterns, we tried to verify that any known regulatory relationships of genes could be revealed from them. In Fig. 2, it shows some of the discovered patterns (with high confidence measures, Section 2.3) represented in rules that reveal known regulatory relationships [27]. Based on the discovered relationships, we can then construct the gene interaction diagrams [28] as showed in Fig. 3 that might provide important clues in reconstructing the structures of the underlying GRNs. One of the appealing advantages of network reconstruction using the proposed algorithm is that the user can easily improve the classifier by adding new samples or experimental conditions and reproduce the architecture of a network consistent with the data. Since such iterative improvements can be part of an interactive process. Therefore, the proposed algorithm can be considered as a basis for an interactive expert system for gene network reconstruction.

R1: If CLN1=H then CLN2=H [*A*]
R2: If HTA1=L then HTB1=L [*A*]
R3: If FUS1=H then CLN1=H [*A*]
R4: If SPT21=H then HTA1=H [*A*]
R5: If FAR1=L then CLN2=H [*I*]
R6: If SPT16=H then CLN1=H [*A*]
R7: If RME1=H then CLN2=H [*A*]
R8: If CDC20=H then CLN1=L [*I*]

Figure 2. Patterns discovered (*A* - known activation relationships and *I* - known inhibition relationships).

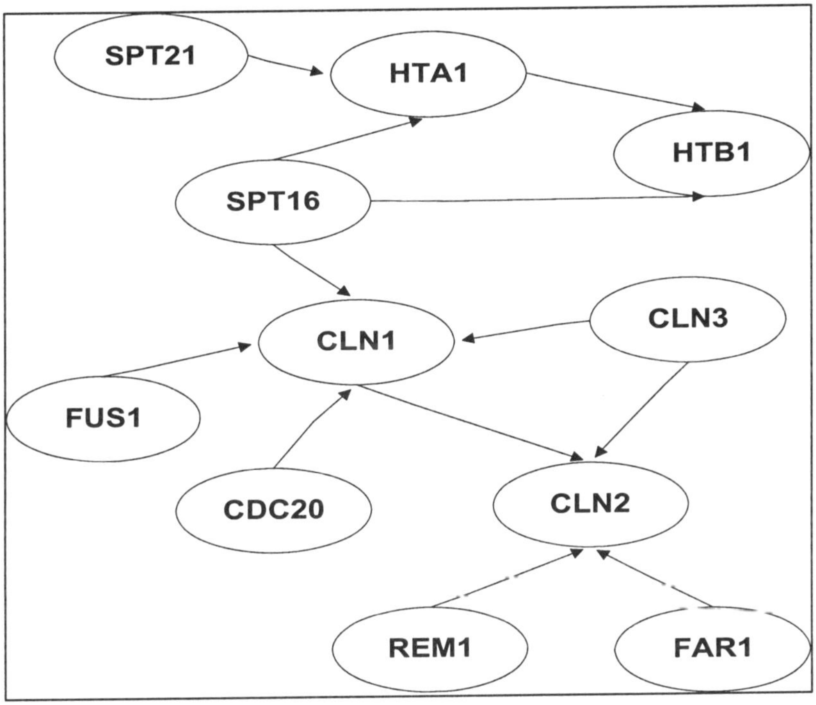

Figure 3. Gene interaction diagram discovered (12 known regulatory relationships involved). Solid lines correspond to activation relationships and broken lines correspond to inhibition relationships.

4 Conclusions

In this paper, we have presented a novel fuzzy logic-based approach for the inference of GRNs. The proposed algorithm is able to distinguish between relevant and irrelevant expression data in predicting the expression patterns of predicted genes without the need for additional feature selection procedures. And also, it is able to explicitly reveal the discovered patterns for possible biological interpretation. With the proposed objective interestingness measure, no user-specified thresholds are needed in advance. Experimental results on real expression data show that the proposed algorithm can be very effective and the discovered patterns reveal biologically meaningful regulatory relationships of genes that could help the user reconstructing the underlying GRNs.

References

1. M. Schena, D. Shalon, R.W. Davis and P.O. Brown. Quantitative monitoring of gene expression patterns with a complementary DNA microarray. *Science*, 270(5235):467—470, 1995.
2. D.J. Lockhart and E.A. Winzeler. Genomics, gene expression and DNA arrays. *Nature*, 405(6788):827—836, 2000.
3. J.C. Leloup and A. Goldbeter. Toward a detailed computational model for the mammalian circadian clock. *Proc. of the National Academy of Science, USA*, 100:7051—7056, 2003.
4. K.C. Chen, T.Y. Wang, H.H. Tseng, C.Y. Huang and C.Y. Kao. A stochastic differential equation model for quantifying transcriptional regulatory network in Saccharomyces cerevisiae. *Bioinformatics*, Advance Access published online on March, 2005.
5. T. Akutsu, S. Miyano and S. Kuhara. Identification of genetic networks from a small number of gene expression patterns under the boolean network model. *Pacific Sym. on Biocomputing*, 17—28, 1999.
6. B.E. Perrin, L. Ralaivola, A. Mazurie, S. Bottani, J. Mallet and F. Buc. Gene networks inference using dynamic bayesian networks. *Bioinformatics*, 19:138—148, 2003.
7. P.T. Spellman, G. Sherlock, M.Q. Zhang, V.R. Lyer, K. Anders, M.B. Eisen, P.O. Brown, D. Botstein and B. Futcher. Comprehensive identification of cell cycle-regulated genes of the yeast Saccharomyces cerevisiae by microarray hybridization. *Mol. Biol. Cell.*, 9(12):3273—3297, 1998.
8. L. Wong. *The Practical Bioinformatician.* World Scientific, 2004.
9. M. Middendorf, A. Kundaje, C. Wiggins, Y. Freund and C. Leslie. Predicting genetic regulatory response using classification. *Bioinformatics*, 20:232—240, 2004.
10. D. Endy and R. Brent. Modeling cellular behaviour. *Nature*, 409:391—395, 2001.
11. J.R. Quinlan. *C4.5: Programs for Machine Learning.* San Fran., CA: Morgan Kaufmann, 1993.
12. L.A. Zadeh. Fuzzy sets. *Inf. Contr.*, 8:338—353, 1965.
13. L.A. Zadeh. Fuzzy logic and approximate reasoning. *Synthese*, 30:407—428, 1975.
14. A.P. Gasch and M.B. Eisen. Exploring the conditional coregulation of yeast gene expression through fuzzy k-means clustering. *Genome Biol.*, 3(11): RESEARCH0059.1—0059.22, 2002.
15. C. Arima, T. Hanai and M. Okamoto. Gene expression analysis using fuzzy k-means clustering. *Genome Informatics*, 14:334—335, 2003.
16. P.J. Woolf and Y. Wang. A fuzzy logic approach to analyzing gene expression data. *Physiol Genomics*, 3:9—15, 2000.
17. K.C.C. Chan and A.K.C. Wong. A statistical technique for extracting classificatory knowledge from databases. *Knowledge Discovery in Databases*, G. Piatesky-Shapiro and W.J. Frawley, Eds. Menlo Park, CA:/Cambridge, MA: AAAI/MIT Press, 107—123, 1991.
18. P.C.H. Ma, K.C.C. Chan and D.K.Y. Chiu. Clustering and re-clustering for pattern discovery in gene expression data. *Journal of Bioinformatics and Computational Biology*, 3(2):281—301, 2005.
19. Y. Wang and A.K.C. Wong. From association to classification: Inference using weight of evidence. *IEEE Trans. Knowledge and Data Engineering*, 15(3):764—767, 2003.
20. R.J. Cho, M.J. Campbell, E.A. Winzeler, L. Steinmetz, A. Conway, L. Wodicka, T.G. Wolfsberg, A.E. Gabrielian, D. Landsman, D.J. Lockhart and R.W. Davis. A genome-wide transcriptional analysis of the mitotic cell cycle. *Mol. Cell.*, 2(1):65—73, 1998.
21. http://genome-www.stanford.edu/cellcycle
22. C. Creighton and S. Hansah. Mining gene expression databases for association rules. *Bioinformatics*, 19(1):79—86, 2003.
23. S. Mitra, K.M. Konwar and S.K. Pal. Fuzzy decision tree, linguistic rules and fuzzy knowledge-based network: generation and evaluation. *IEEE Trans. on Systems, Man and Cybernetics - Part C: Applications and Reviews*, 32:328—339, 2002.
24. M. Xiong, X. Fang and J. Zhao. Biomarker identification by feature wrappers. *Genome Res.*, 11:1878—1887, 2001.
25. Y. Su, T.M. Murali et. al. RankGene: Identification of diagnostic genes based on expression data. *Bioinformatics,* 19(12):1578—1579, 2003. Available: http://genomics10.bu.edu/yangsu/rankgene/.
26. C.Z. Janikow. Fuzzy decision trees: issues and methods. *IEEE Trans. on Systems, Man and Cybernetics - Part B: Cybernetics*, 28(1):1—14, 1998.
27. V. Filkov, S. Skiena and J. Zhi. Analysis techniques for microarray time-series data. *In Proceedings of RECOMB*, 124—131, 2001.
28. J.M. Bower and H. Bolouri. *Computation Modeling of Genetic and Biochemical Networks.* Cambridge, Mass.: MIT Press, 2001.

SYSTEM IDENTIFICATION AND ROBUSTNESS ANALYSIS OF THE CIRCADIAN REGULATORY NETWORK VIA MICROARRAY DATA IN ARABIDOPSIS THALIANA *

C. W. LI

Systems Biology Group, Automatic Control and signal Laboratory, Hsinchu, 300, Taiwan
Email:d925908@oz.nthu.edu.tw

W. C. CHANG

Systems Biology Group, Automatic Control and signal Laboratory, Hsinchu, 300, Taiwan
Email:cwli@moti.ee.nthu.edu.tw

B. S. CHEN †

Department of Electrical Engineering, National Tsing Hua University, Hsinchu, 300, Taiwan
Email:bschen@ee.nthu.edu.tw

The circadian regulatory network is one of the main topics of plant investigations. The intracellular interactions among genes in response to the environmental stimuli of light are related to the foundation of functional genomics in plant. However, the sensitivity analysis of the circadian system has not analyzed by perturbed stochastic dynamic model via microarray data in plant. In this study, the circadian network is constructed for *Arabidopsis thaliana* using a stochastic dynamic model with sigmoid interaction, activation delay, and regulation of input light taken into consideration. The describing function method in nonlinear control theory about nonlinear limit cycle (oscillation) is employed to interpret the oscillations of the circadian regulatory networks from the viewpoint that nonlinear network will continue to oscillate if its feedback loop gain is equal to 1 to support the oscillation of circadian network. Based on the dynamic model via microarray data, the system sensitivity analysis is performed to assess the robustness of circadian regulatory network via biological perturbations. We found that the circadian network is more sensitive to the perturbation of the *trans*-expression threshold, is more sensitive to the activation level of steady state, rather than the *trans*-sensitivity rate.

1 Introduction

Biological phenomena at different organismic levels have implicitly revealed some sophisticated systematic architectures of cellular and physiological activities. These architectures were built upon the biochemical processes before the emergence of proteome and transcriptome [1,2]; and most biological phenomena such as metabolism, stress response [3], and cell cycle are directly or indirectly influenced by genes and have been well studied on the molecular basis. Thus, the identification of a signal transduction pathway could be traced back to the genetic regulatory level. The rapid advances of

* This work is supported by National Science Council, Taiwan.
† Work partially supported by grant NSC 93-3112-B-007-003 of the National Science Council, Taiwan.

genome sequencing and DNA microarray technology make possible the quantitative analysis of signaling regulatory network besides the qualitative analysis [4].

In this study, The ARX dynamic system approach is applied to the circadian regulatory pathway of *Arabidopsis thaliana* with microarray data sets publicly available on the net [5]. According to the synchronously dynamic evolution of microarray data, we have successively identified the core signaling transduction from light receptors of phytochromes [6] and crytochromes [7] to the endogenous biological clock [8], which is coupled to control the correlatively physiological activity with paces on a daily basis. With the dynamic system approach, not only the regulatory abilities, but also the oscillatory frequency and the delays of regulatory activity were specified. Moreover, we design several simulation assays with the biological senses to mimic the biological experiments.

2 Dynamic System Description of Circadian Regulatory Model

We can consider any gene expression profile as a system response or output stimulated by some inputs from other gene expressions and environmental stimuli. According to this description, let $X_i(k)$ denote the expression profile of the i-th gene at time point k. Then the following general form of ARX difference equation is proposed to model the expression level of the i -th gene as the synthesis of n upstream genetic signals $\widetilde{X_i}$, $i=1,2,\cdots,n$ and an external input signal u under their τ delays, (see figure 1)

$$
\begin{aligned}
X_i(k) = {} & d_{1,i1}\widetilde{X_1}(k-\tau_i)+d_{1,i2}\widetilde{X_2}(k-\tau_i)+\cdots+d_{1,ii}X_i(k-\tau_i)+\cdots+d_{1,in}\widetilde{X_n}(k-\tau_i)+ \\
& d_{2,i1}\widetilde{X_1}(k-2\tau_i)+d_{2,i2}\widetilde{X_2}(k-2\tau_i)+\cdots+d_{2,ii}X_i(k-2\tau_i)+\cdots+d_{2,in}\widetilde{X_n}(k-2\tau_i)+ \\
& \quad \vdots \qquad\qquad\qquad \vdots \qquad\qquad\qquad \vdots \qquad\qquad\qquad \vdots \\
& d_{q,i1}\widetilde{X_1}(k-Q\tau_i)+d_{q,i2}\widetilde{X_2}(k-Q\tau_i)+\cdots+d_{q,ii}X_i(k-Q\tau_i)+\cdots+d_{q,in}\widetilde{X_n}(k-Q\tau_i)+ \\
& b_i\cdot u(k-\tau_{iu})+\varepsilon_i(k) \quad ,i=1,2,\cdots,n
\end{aligned} \tag{1}
$$

where $\tilde{X}_j(k-q\tau_i)$, j=1,2,$\cdots$,n; $q=1,2,\cdots,Q$ is the upstream genetic signal transformed by $X_j(k)$ with the q-th order of τ_i delay and through a sigmoid activation function to denote the binding of transcription factor $X_j(k)$ on gene i, and the genetic kinetic parameter $d_{q,ij}$ denotes the regulation abilities of transcription factor $\tilde{X}_j(k)$ on gene i. Meanwhile, $u(k-\tau_{iu})$, which denotes the external input light with a delay τ_{iu} affecting $X_i(k)$, correlates with the output genetic expression $X_i(k)$ with the input kinetic parameters b_i . $\varepsilon_i(k)$ is the stochastic noise of current microarray data or the residue of the model. Here τ_i and τ_{iu}, which are essential to the activation-time estimation, should be determined previously and will be discussed later. The ARX model (AutoRegressive with eXternal input), which admit a reformation to the linear regression model, is the special case of the ARMAX model (autoregressive moving average with exogenous input). Moreover, an oscillation will exist in circadian regulatory network by the feedbacks through other genes if these feedbacks are limited by sigmoid functions to avoid their unstable propagations, which will be discussed by describing function method [9] in the sequel.

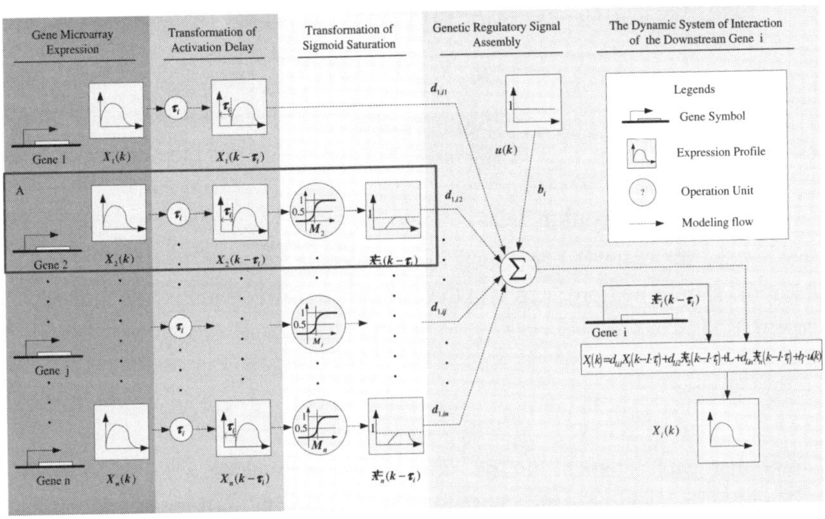

Figure 1. Illustration of the dynamic system scheme using the ARX(1) model. Block A represents the transformation of the genetic regulatory signal, $\tilde{X}_j(k-q\tau_i)$, for j=2 and q=1.

For the limited influence expression of $\tilde{X}_j(k-q\tau_i)$ (see Block A in Figure 1), the sigmoid function is chosen to express the nonlinear 'on' and 'off' activities of physically genetic interactions with parameters $\theta_j = \{\gamma, M_j, \tau_i\}$ as follows,

$$\widetilde{X_j}(k-q\cdot\tau_i) = \frac{1}{1+e^{-\gamma\left(X_j(k-q\cdot\tau_i)-M_j\right)}} \tag{2}$$

where γ is the trans-sensitivity rate, and M_j is the trans-expression threshold derived from the mean of the j-tth gene's profile. γ could determine the transition time of activation between the states of 'off' and 'on' from X_j to X_i, for which a larger γ is with a less transition time, to mimic the transient state of the genetic interaction on the *trans* level. M_j can determine the threshold of the half activation level of X_j to X_i, for which a larger M_j is with a less activating ability, to mimic the steady state of the genetic interaction on the trans level. For the biological reason of small activation delay on mRNA level and less modeling complexity, we can reduce the order of the ARX model to no more than 2, Q=1 (i.e. ARX(1)) or Q=2 (i.e. ARX(2)) in Eq. (1). We will determine an adequate order for our interesting system later. And now we take the second order ARX model for illustration as follows,

$$\begin{aligned} X_i(k) = {} & d_{1,i1}\widetilde{X_1}(k-\tau_i) + d_{1,i2}\widetilde{X_2}(k-\tau_i) + \cdots + d_{1,ii}X_i(k-\tau_i) + \cdots + d_{1,in}\widetilde{X_n}(k-\tau_i) + \\ & d_{2,i1}\widetilde{X_1}(k-2\tau_i) + d_{2,i2}\widetilde{X_2}(k-2\tau_i) + \cdots + d_{2,ii}X_i(k-2\tau_i) + \cdots + d_{2,in}\widetilde{X_n}(k-2\tau_i) + \\ & b_i \cdot u(k) + \varepsilon_i(k) \quad , i = 1, 2, \cdots, n \end{aligned} \tag{3}$$

Consequently, the vector difference form underlined in this equation is applied to m time points in order.

$$\underline{X_i} = d_{1,i1}\underline{\widetilde{X_{1,\tau_i}}} + \cdots + d_{1,ii}\underline{X_{i,\tau_i}} + \cdots + d_{1,in}\underline{\widetilde{X_{n,\tau_i}}} +$$
$$d_{2,i1}\underline{\widetilde{X_{1,2\tau_i}}} + \cdots + d_{2,ii}\underline{X_{i,2\tau_i}} + \cdots + d_{2,in}\underline{\widetilde{X_{n,2\tau_i}}} + b_i \cdot \underline{u} + \underline{\varepsilon_i} \tag{4}$$

where $\underline{X}_i = \begin{bmatrix} X_i(k_1) \\ X_i(k_2) \\ \vdots \\ X_i(k_m) \end{bmatrix}, \underline{X}_{n,q} = \begin{bmatrix} X_n(k_1-q) \\ X_n(k_2-q) \\ \vdots \\ X_n(k_m-q) \end{bmatrix}, \underline{\tilde{X}}_{n,q} = \begin{bmatrix} \tilde{X}_n(k_1-q) \\ \tilde{X}_n(k_2-q) \\ \vdots \\ \tilde{X}_n(k_m-q) \end{bmatrix}, \underline{u}_{\tau_{iu}} = \begin{bmatrix} u(k_1) \\ u(k_2) \\ \vdots \\ u(k_m) \end{bmatrix}, \underline{\varepsilon}_i = \begin{bmatrix} \varepsilon_i(k_1) \\ \varepsilon_i(k_2) \\ \vdots \\ \varepsilon_i(k_m) \end{bmatrix}$, and m denotes the number of time points. τ_i is the specific activation delay.

In the next step, to estimate the kinetic parameters $d_{q,in}$, $q=1,2$; $n=1,2,\cdots$, and b_i , the formula Eq. (4) should be translated into the difference matrix equation as follows,

$$Y_i = A_i\Omega_i + E_i \tag{5}$$

where $Y_i = \underline{X}_i$, $\Omega_i = \begin{bmatrix} d_{1,i1} & \cdots & d_{1,in} & d_{2,i1} & \cdots & d_{2,in} & b_i \end{bmatrix}^T$, and $E_i = \underline{\varepsilon}_i$ are in vector forms, while $A = \begin{bmatrix} \underline{\tilde{X}}_{1,\tau_i} & \cdots & \underline{\tilde{X}}_{n,\tau_i} & \underline{\tilde{X}}_{1,2\tau_i} & \cdots & \underline{\tilde{X}}_{n,2\tau_i} & \underline{u}_{\tau_{iu}} \end{bmatrix}$ is a matrix.

We assume that each element in the stochastic noise vector, $\varepsilon_i(k_l)$, $i=\{1,\cdots,m\}$,is an independent random variable with a normal distribution with zero mean and variance σ^2, which is unknown and needs to be estimated. Thus, we will estimate the parameter $\hat{\Omega}_i$ using the maximum likelihood method.

The maximum likelihood estimate of σ^2 is the estimate of noise covariance. Substituting Eq. (8) into Eq. (7) yields,

$$L(\Omega_i, \sigma^2) = -\frac{m}{2}\ln\left[2\pi\sigma^2\right] - \frac{m}{2} \tag{6}$$

where $\sigma^2 = \frac{1}{m}\sum_{l=1}^{m}[Y_i - A_i\Omega_i]^T[Y_i - A_i\Omega_i]$

Therefore, we can find the maximum likelihood estimation of Ω_i by minimizing the value of σ^2. From Eq. (6) best choice of parameter vector Ω_i to minimize σ^2 using the least-squares method is obtained as follows [10],

$$\widehat{\Omega}_i = \left(A_i^T A_i\right)^{-1} A_i^T Y_i \tag{7}$$

After the parameter estimation in Eq. (7), substituting Ω_i in Eq. (7) into stochastic model in Eq. (3) lead to the estimated circadian regulatory network equations.

3. Assay of the Model

3.1. Assay of ARX System Model

The assays of the ARX system model are divided into four categories. The first is the confirmation of the oscillation frequency of circadian regulatory network by the oscillatory characteristics of the dynamic circadian regulatory model; the second is the input stimulus changes; the third is the trans disturbance; and the last is about the cis perturbation. For each pair of gene expressions from both the biological assay and the simulation, we calculated the Pearson correlation coefficient between the genes' mRNA expression profiles of $X_i(k)$ *in vivo* and $\widehat{X}_i(k)$ *in silico* at all time points $k = k_1, k_2, \cdots, k_m$ as follows.

$$r\left(X_i, \widehat{X}_i\right) = \frac{\sum_{l=1}^{m} X_i(k_l)\widehat{X}_i(k_l) - \left(\frac{\sum_{l=1}^{m} X_i(k_l) \sum_{l=1}^{m} \widehat{X}_i(k_l)}{m}\right)}{\sqrt{\left[\sum_{l=1}^{m} X_i^2(k_l) - \frac{\left[\sum_{l=1}^{m} X_i(k_l)\right]^2}{m}\right]\left[\sum_{l=1}^{m} \widehat{X}_i^2(k_l) - \frac{\left[\sum_{l=1}^{m} \widehat{X}_i(k_l)\right]^2}{m}\right]}} \tag{8}$$

To measure the period of the time-course expression profile, the power spectrum, which has different magnitudes in different frequencies (the reciprocal of periods), is employed to detect which frequency has the largest magnitude. First, we should take the Discrete Fourier Transform of $X_i(k)$ for $k = k_1, k_2, \cdots, k_m$ as follows,

$$X_i(w) = \sum_{l=1}^{m} X_i(k_l) e^{-jwk} \tag{9}$$

where w is the radian frequency.

Then we can detect the frequency with the maximum magnitude,

$$\omega_i = \arg\max_{\omega} |X_i(\omega)| = \frac{2\pi}{T_i} \tag{10}$$

where T_i is the period of $X_i(k)$ and can be determined from the reciprocal of the detected frequency ω_i. Furthermore, the measure of mean expression of $X_i(k)$ is important for distinguishing the deviation of expression profile under different assays as follows,

$$M_i = \frac{1}{m} \sum_{l=1}^{m} X_i(k_l) \tag{11}$$

3.1.1. Determination of system order

In this study, the formulated ARX model should be first assigned with a proper modeling order and an activation delay to analyze the experimental expression data of microarray. According to Eq. (1), we compared the first-order (Q=1) ARX model (i.e. ARX(1)) and the second-order (Q=2) ARX model (i.e. ARX(2)) with different activation delays τ as shown in Fig. 2a. We exploited the mean similarity between the raw expression and the simulation of all 16 system genes, which is measured by Pearson correlations, to evaluate the performance of the network model. Owing to the least difference at 0.5-hr delay between ARX(1) and ARX(2), we would prefer the more flexible ARX(1) model with a 0.5-hr activation delay as the system model for the circadian regulatory network. Consequently, the simulation expressions of the derived circadian network model are shown in Fig. 2b, $\tilde{X}_j(k - q\tau_i)$, for j=2 and q=1. The detection of the static structural characteristics will help reconstruct their hidden significance of cis connectivity as in the signaling transduction network of Fig. 3.

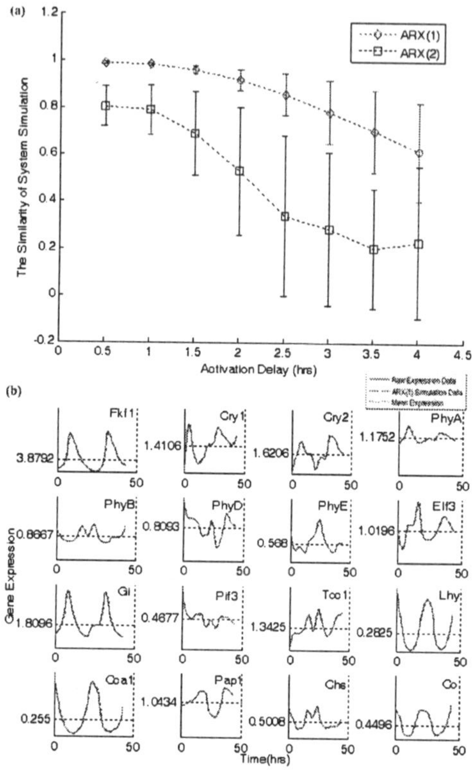

Figure 2. ARX system modeling with determination of system modeling order and activation delay. (a) The average similarity (measured by Pearson correlation) of all system genes under different activation delays. (b) The dynamic data fitting of 16 genes in the circadian network with ARX(1) model and 0.5-hr activation delay.

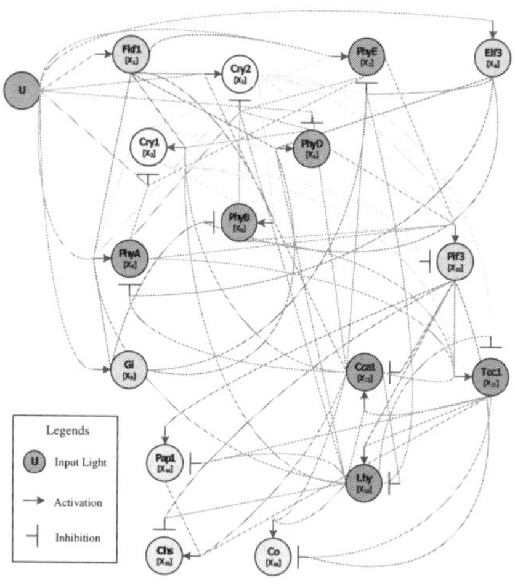

Figure 3. Signaling transduction network of system genes and input light in the circadian network of *Arabidopsis*. The colored circles indicate the system genes with their names and notations of $X_1 \sim X_{16}$.

3.2. Sensitivity Analysis of Circadian System

The sensitivity measure of the circadian system for the analysis of robustness can also be derived from the system model. For illustration, we would rearrange Eq. (3) into the following difference matrix equation,

$$\underline{Y}(k) = D_1\tilde{\underline{Y}}(k-\tau) + D_2\underline{Y}(k-\tau) + Bu(k) \tag{12}$$

where $D_1 = \begin{bmatrix} 0 & d_{1,12} & \cdots & \cdots & d_{1,1n} \\ d_{1,21} & 0 & d_{1,23} & \cdots & \vdots \\ \vdots & d_{1,32} & \ddots & \ddots & \vdots \\ \vdots & \vdots & \ddots & \ddots & d_{1,(n-1)n} \\ d_{1,n1} & \cdots & \cdots & d_{1,n(n-1)} & 0 \end{bmatrix}$, $D_2 = \begin{bmatrix} d_{2,11} & 0 & \cdots & \cdots & 0 \\ 0 & d_{2,22} & 0 & \cdots & \vdots \\ \vdots & 0 & \ddots & \ddots & \vdots \\ \vdots & \vdots & \ddots & \ddots & 0 \\ 0 & \cdots & \cdots & 0 & d_{2,nn} \end{bmatrix}$ $\underline{Y}(k) = \begin{bmatrix} X_1(k) \\ X_2(k) \\ \vdots \\ X_n(k) \end{bmatrix}$, $\underline{Y}(k-\tau) = \begin{bmatrix} X_1(k-\tau_1) \\ X_2(k-\tau_2) \\ \vdots \\ X_n(k-\tau_n) \end{bmatrix}$ $\tilde{\underline{Y}}(k-\tau) = \begin{bmatrix} \tilde{X}_1(k-\tau_1) \\ \tilde{X}_2(k-\tau_2) \\ \vdots \\ \tilde{X}_n(k-\tau_n) \end{bmatrix}$

$B = \begin{bmatrix} b_1 \\ b_2 \\ \vdots \\ b_n \end{bmatrix}$, $\underline{U} = \begin{bmatrix} u(k) \\ u(k) \\ \vdots \\ u(k) \end{bmatrix}$ and n is the number of genes.

3.2.1. Circadian clock frequency assay

While we obtain the oscillation frequencies w_i of circadian network by the intersection in Eq. (10), we will compare with the oscillation frequencies calculated by Eq. (9) and (10) to validate the accuracy of the proposed dynamic model in the sequel.

A dynamic system with saturation (or sigmoid function) nonlinear feedback will lead to oscillation (limit cycle) [9]. This oscillation phenomenon can be interpreted by the theory of the describing function, which will be used to describe the circadian regulatory network of *Arabidopsis thaliana*. According to Eq. (12), we get

$$\underline{Y}(k) = (I - z^{-\tau}D_2)^{-1}D_1\tilde{\underline{Y}}(k-\tau) + (I - z^{-\tau}D_2)^{-1}Bu(k) \tag{13}$$

where $z^{-\tau} = \begin{bmatrix} z^{-\tau_1} & 0 & \cdots & 0 \\ 0 & z^{-\tau_2} & \ddots & \vdots \\ \vdots & \ddots & \ddots & 0 \\ 0 & \cdots & 0 & z^{-\tau_n} \end{bmatrix}$,and $z^{-\tau_i}$ denotes delay operator of τ_i .

If the oscillation (limit cycle) occurs in circadian network, then the sigmoid function $\tilde{Y}(k)$ in Eq. (2) can be approximated by the describing function N(A) as [9]

$$\tilde{\underline{Y}}(k) = \begin{bmatrix} \tilde{Y}_1(k) \\ \tilde{Y}_2(k) \\ \vdots \\ \tilde{Y}_n(k) \end{bmatrix} \approx N(A)\underline{Y}(k) \tag{14}$$

where the describing function matrix $N(A) = \begin{bmatrix} z^{-\tau_1} & 0 & \cdots & 0 \\ 0 & z^{-\tau_2} & \ddots & \vdots \\ \vdots & \ddots & \ddots & 0 \\ 0 & \cdots & 0 & z^{-\tau_n} \end{bmatrix}$, and $N_i(A_i)$ denotes the describing function of the i-th gene of oscillation and A_i denotes the amplitude of oscillation of the i-th gene of gene j is free of oscillation, then the corresponding $N_j(A_j) = 0$. From Eq. (13) and (14), we can approximate the circadian network as

$$\underline{Y}(k) = (I - z^{-\tau}D_2)^{-1}D_1N(A)z^{-\tau}\underline{Y}(k) + (I - z^{-\tau}D_2)^{-1}Bu(k) \tag{15}$$

There are two rhythms, one is circadian rhythm and another is diurnal rhythm. The first term with gain equal to 1 on the right hand side of Eq. (16) is the response for circadian rhythm; and the second term for diurnal rhythm, which is controlled by diurnal cycling of light and dark u(k) and some photoreceptor genes are of this case. Since the oscillation exists in the circadian network, by control theory, the closed loop gain should be lossless in order to support the oscillation, i.e.

$$D_1 N(A) = z^{\tau} - D_2 \tag{16}$$

At frequency domain, we can get

$$e^{jw_i\tau_i} - d_{2,ii} = \sum_{j \neq i}^{n} d_{1,ij} N_j(A_j) \tag{17}$$

where $e^{jw\tau} = \begin{bmatrix} e^{-jw_1\tau_1} & 0 & \cdots & 0 \\ 0 & e^{-jw_2\tau_2} & \ddots & \vdots \\ \vdots & \ddots & \ddots & 0 \\ 0 & \cdots & 0 & e^{-jw_n\tau_n} \end{bmatrix}$

For example, for gene PhyE, $e^{jw_7\tau_7} - d_{2,77} = -0.1579 - 0.1226i$ and $\sum_{j \neq 7}^{N} d_{1,7j} N_j(A_j) = -0.1339 - 0.1253i$,which matches the oscillation condition in Eq. (17). For gene Lhy, $e^{jw_{12}\tau_{12}} - d_{2,1212} = -0.2181 - 0.1045i$ and, $\sum_{j \neq 12}^{N} d_{1,12j} N_j(A_j) = -0.2144 - 0.0589i$ which roughly matches the oscillation condition of describing function in the nonlinear circadian system. By describing analysis of nonlinear oscillation [9], the intersection of $e^{jw_i\tau_i} - d_{2,ii}$ and $\sum_{j \neq i}^{N} d_{ij} N(A_i)$ in Eq. (17) implies the occurrence of oscillation and the A_i and w_i at the interaction point are the oscillation amplitude and oscillation frequency.

3.2.2. Trans-perturbation assay

As in the description of Eq. (2), γ is the trans-sensitivity rate which is related to the transition time of trans-activaton and M_j is the trans-expression threshold that determines the saturating transformation level of expression. We also induce the corresponding sensitivity in the following,

$$\frac{\Delta \underline{Y}}{\Delta \underline{F}} \approx D_I \frac{\Delta \underline{\tilde{Y}}}{\Delta \underline{F}} \tag{18}$$

where $\underline{F} = \{\underline{\gamma}, \underline{M}\}$ and $\frac{\Delta \underline{\tilde{Y}}}{\Delta \underline{\gamma}} = \frac{-\underline{Y}e^{-\underline{\gamma}(\underline{Y}-\underline{M})}}{(1+e^{-\underline{\gamma}(\underline{Y}-\underline{M})})^2}$. Hence we could discuss the sensitivity on the trans level like the input sensitivity.

3.2.2.1. Trans-sensitivity rate γ simulation of gene

In a similar way as in input perturbation, we changed γ from 100% to 0% (-100%) and 200% (+100%) of system genes in pathway to compare with their sensitivities to γ, as shown in Table 1A. We also average the three measure indexes of each gene, which are shown in Fig. 4.

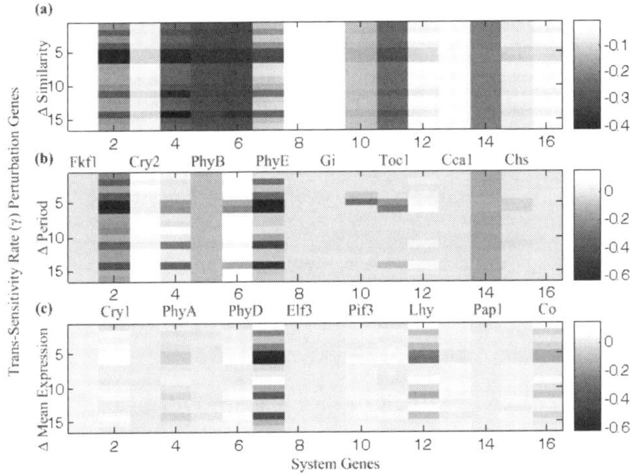

Figure 4. Deviation representations of the system genes under the perturbation of trans-sensitivity rate γ. The perturbation is performed in the vertical axis and the responses of 16 system genes are shown in the horizontal axis, and the colored bars of degree are on the right-hand side of each inset. (a) Δ Similarity (measured by the Pearson correlation), (b) Δ Period, and (c) Δ Mean expression.

3.2.2. Trans-expression threshold M_j simulation of gene

We varied M_j to 100% lower (-100%) and higher (+100%) than the original mean expression of the j-th gene respectively and compared with their sensitivities of M_j, which are shown in Table 1B; and their average measure indexes are shown in Fig. 5.

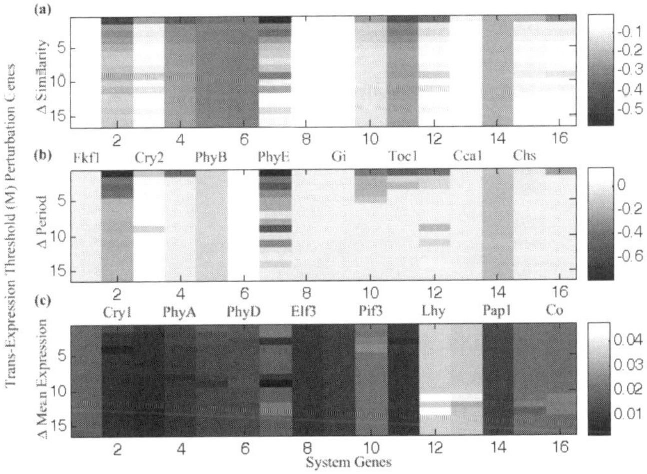

Figure 5. Deviation representations of the system genes under the perturbation of trans-expression threshold M. The perturbation is performed in the vertical axis and the responses of 16 system genes are shown in the horizontal axis, and the colored bars of degree are on the right-hand side of each inset. (a) Δ Similarity (measured by the Pearson correlation), (b) Δ Period, and (c) Δ Mean expression.

Table 1. The sensitivities of the system genes in the circadian regulatory network under different perturbations due to the *input* light, *trans* level, and *cis* level. The bold values represent significant sensitivities (less robustness). In general, the fact that sensitivities are not large implies that the circadian regulatory network is robust enough.

Perturbation	Type of Sensitivity	Specification	The Sensitivites of the system genes in the circadian regulatory network															
			Fkfl	Cry1	Cry2	PhyA	PhyB	PhyD	PhyE	Elf3	Gi	Pif3	Toc1	Lhy	Cca1	Pap1	Chs	Co
Trans level	A. Trans-sensitivity rate (γ)	+0	0.0112	**0.017**	0.0119	0.008	0.0104	**0.0129**	-0.008	0.0112	0.0112	0.0113	**0.013**	0.0059	0.0106	0.0116	0.01	0.0069
		-100%	0.0112	**0.017**	0.011	0.0096	0.0112	**0.0128**	-0.008	0.0112	0.0112	0.0115	**0.013**	0.0055	0.011	0.0115	0.0096	0.0062
		+100%	0.0112	**0.017**	0.0121	0.0073	0.0102	**0.013**	-0.008	0.0112	0.0112	0.0114	**0.013**	0.0066	0.0106	0.0117	0.0105	0.0076
		Mean	0.0112	**0.017**	0.0117	0.0083	0.0106	**0.0129**	-0.008	0.0112	0.0112	0.0114	**0.013**	0.006	0.0107	0.0116	0.01	0.0069
	B. Trans-expression level (M)	+0	0.0112	-0.068	0.0088	0.0474	0.026	0.00403	**0.267**	0.0112	0.0112	0.0236	-0.031	**0.122**	0.0116	0.0074	0.0228	**0.105**
		-100%	0.0112	0.01	-0.007	0.06	0.0301	-0.0003	**0.272**	0.0112	0.0112	-0.008	-0.031	**0.11**	0.0011	0.0104	-0.002	**0.095**
		+100%	0.0112	-0.047	-0.043	0.033	0.0211	0.00812	**0.233**	0.0112	0.0112	0.0156	0.0002	**0.124**	-0.002	0.0106	0.0139	**0.107**
		Mean	0.0112	-0.035	-0.014	0.0468	0.0257	0.00394	**0.257**	0.0112	0.0112	0.0103	-0.021	**0.119**	0.0037	0.0095	0.0116	**0.102**

4. Results

In the perturbation of *trans*-sensitivity rate (γ), we will discuss whether the transition rate, which determines the transition time of one gene binding to or interacting with another one, affects the system gene's expression in this model system. It seems that the similarity (Fig. 4a) remains unchanged for most system genes except Cry1 [X_2], PhyA [X_4], PhyD [X_6], and PhyE [X_7] [11]. If we consider the periodic variation in Fig. 4b, Cry2's [X_3] period is lengthened about 10%, whilst that of Cry1 [X_2] and Pap1 [X_{14}] are shortened about 20%, respectively. The diversity and sensitivity of period due to perturbation of the transition time are evident as in Fig. 4c. The mean expressions of system genes are almost unaltered but PhyE [X_7] [10] is reduced. Because the largest difference in the mean sensitivity of each gene of Table 1A is about 0.025, we would conclude that the *trans*-sensitivity rate, which determines the transition time indicating the transient state of *trans* activation, has less influence on the circadian system.

In another perturbation of *trans*-expression threshold M, there are five genes of Cry1 [X_2], Cry2 [X_3], PhyD [X_6], Pif3 [X_{10}] and Toc1 [X_{11}] with perceptible variations, which have the same behavior in the measures of similarity and period (see Figs. 5a and 5b). Owing the largest difference in the mean sensitivity of each gene of Table 1B being close to 0.29, the circadian network is more sensitive to the perturbation of the *trans*-expression threshold M, is more sensitive to the activation level of steady state, rather than the *trans*-sensitivity rate γ.

5. Discussion

In our dynamic system approach applied to the circadian network using ARX, we not only can identify the regulatory abilities via ARX(1) with activation delays, but also indicate the regulatory strength from the input-light signal. The greatest importance of the proposed dynamic model is the convenience of the consequent system analysis, for example, sensitivity analysis, to gain more insight about the circadian regulatory network. There are some shortcomings in our study. First, although the time-course microarray data are available, its lower samplings will distort the real changes of gene expressions, especially for fast dynamic evolution. A more sampling experiment with respect to the intrinsic turnover rate is expected for a more precise analysis. Second, we formulate our

ARX circadian network model using the biological knowledge of the correlations between the circadian genes. In the circadian regulatory network, it is enough to reconstruct the system because of its simulation similarity approaching 0.99 in Fig. 2a.

In the near future, as the system modeling algorithms are further developed, we expect this dynamic system approach to have immense impact in elucidating the underlying molecular mechanisms of network in a variety of organisms besides the circadian network in Arabidopsis thaliana, especially after the maturation of the protein chips.

Acknowledgments

We thank the National Science Council, Taiwan. for grants NSC 93-3112-B-007-003.

References

1. J. R. Kettman, J. R. Frey and I. Lefkovits. Proteome, transcriptome and genome: top down or bottom up analysis. *Biomol. Eng.*, 18:207-212, 2001.
2. J. Scheel, M.C. Von Brevern, A. Horlein, A. Fischer, A. Schneider and A. Bach. Yellow pages to the transcriptome. *Pharmacogenomics*, 3:791-807, 2002.
3. S. Motaki, K. Ayako, Y. S. Kazuko and S. Kazuo. Molecular response to drought, salinity and frost: common and different paths for plant protection. *Cur. Opn. Biotechnol.,* 14:194-199, 2003.
4. T. S. Hughes, B. Abou-Khalil, P. J. Lavin, T. Fakhoury, B. Blumenkopf and S. P. Donahue. Visual field defects after temporal lobe resection: a prospective quantitative analysis. *Neurology.*, 53:167-72, 1999.
5. S. L. Harmer, J. B. Hogenesch, M. Straume, H. S. Chang, B. Han, T. Zhu, X. Wang, J. A. Kreps and S. A. Kay. Orchestrated transcription of key pathways in *Arabidopsis* by the circadian clock. *Science*, 290:2110-2113, 2000.
6. J. J. Casal. Phytochromes, cryptochromes, phototropin: photoreceptor interactions in plants. *Photochem. Photobiol.*, 71:1-11, 2000.
7. C. Fankhauser and D. Staiger. Photoreceptors in Arabidopsis thaliana: light perception, signal transduction and entrainment of the endogenous clock. *Planta.*, 216:1-16, 2002.
8. D. Alabadi, T. Oyama, M. J. Yanovsky, F. G. Harmon, P. Mas and S. A. Kay. Reciprocal regulation between TOC1 and LHY/CCA1 within the *Arabidopsis* circadian clock. *Science*, 293:880-883, 2001.
9. J. J. E. Slotine and W Li. Applied Nonlinear Control, Prentice Hall, Englewood Cliffs, NJ, 1991.
10. R. Johansson. *System Modeling and Identification*, PrenticeHall, Englewood Cliffs, NJ, 1993.
11. M. J. Aukerman, M. Hirschfeld, L. Wester, M. Weaver, T. Clack. R. M. Amasino and R. A. Sharrock. A deletion in the *PHYD* gene of the Arabidopsis Wassilewskija ecotype defines a role for phytochrome D in red/far-red light sensing. *Plant Cell.*, 9:1317-1326, 1997.

PROTEIN SUBCELLULAR LOCALIZATION PREDICTION WITH WOLF PSORT

PAUL HORTON [A], KEUN-JOON PARK [A,C], TAKESHI OBAYASHI [B,D,E], AND KENTA NAKAI [C]

[a] *Computational Biology Research Center, AIST, Tokyo, Japan*
E-mail: horton-p@aist.go.jp

[c] *Human Genome Center, Institute of Medical Science, University of Tokyo*
E-mail: park-kj@hgc.jp, knakai@ims.u-tokyo.ac.jp

[b] *Graduate School of Bioscience and Biotechnology, Tokyo Institute of Technology*
E-mail: toobayas@bio.titech.ac.jp

[d] *Graduate School of Pharmaceutical Sciences, Chiba University*

[e] *Core Research for Evolutional Science and Technology*

We present a new program for predicting protein subcellular localization from amino acid sequence. WoLF PSORT is a major update to the PSORTII program, based on new sequence data and incorporating new features with a feature selection procedure. Following SWISS-PROT, we divided eukaryotes into three groups: fungi, plant, and animal. For the 2113 fungi proteins divided into 14 categories; we found that, combined with BLAST, WoLF PSORT yields a cross-validated accuracy of 83%, eliminating about 1/3 of the errors made when using BLAST alone. For 12771 animal proteins a combined accuracy of 95.6% is obtained, eliminating 1/4 of BLAST alone errors, and for 2333 plant proteins the accuracy can be improved to 86% from 84%.

1. Introduction

Protein localization is a central issue in understanding cells. More than 20 papers[6] have been published in major international journals describing programs for predicting localization from amino acid sequence in eukaryotic cells. Some of these works such as PSORT,[9] PSORTII,[7] the SignalP[11,10] family of programs and others[4] use sorting signal information for prediction. However the majority of prediction programs developed use amino acid content in some form. These methods exploit the longstanding observation[12] that amino acid content correlates strongly with localization site.

In this paper we present WoLF PSORT, a major extension to PSORTII which combines (mostly signal based) features from PSORT and iPSORT[2] with amino acid content. Prediction accuracy is increased by using feature selection while retaining the simple k nearest neighbor classifier used by PSORTII. The dataset constructed and web service are available at `wolfpsort.org`

2. Methods

2.1. *Dataset*

We prepared the dataset primarily from Swiss-Prot[3] Release 45.0 annotation, ignoring entries with weakening qualifiers such as "by similarity". In addition several hundred *Arabidopsis* entries were added from the Gene Ontology[1] web site (up to the 2004/12/4 release). Entries with evidence codes {TAS, IDA, IMP} were included, with revisions by hand in a few cases.

2.1.1. *Site Definition*

The distribution of localization sites in our database is shown in Table 1. The sites reflect common usage found in localization labeled Swiss-Prot entries. Table 2 shows the corresponding Gene Ontology numbers for the Gene Ontology derived sequences.

Table 1. The distribution of localization sites for each category of organisms.

localization	animal	plant	fungi	localization	animal	plant	fungi
*nucl*ear	2682	433	667	cyto_nucl	245	9	91
*plas*ma membrane	3195	160	220	cyto_mito	18	3	8
*extr*acellular	3130	113	140	cyto_pero	10	0	2
*cyto*sol	1555	452	383	cyts_plas	5	0	0
*mito*chondria	938	200	389	cyto_plas	4	0	0
*chlo*roplast	N/A	744	N/A	cyto_golg	4	0	0
E.R.	425	65	66	E.R._mito	18	0	4
*pero*xisome	217	47	77	E.R._golg	9	0	0
*lyso*some	132	N/A	N/A	extr_plas	19	0	0
*golg*i body	100	25	38	mito_pero	15	0	0
*vacu*ole	16	72	23	mito_nucl	2	0	0
*cyto*skeletal	32	10	5	**sum**	**12771**	**2333**	**2113**

Italics indicate the abbreviated name. Localization names joining two abbreviated names with "_", such as "cyto_nucl", indicate dual localization.

2.2. *WoLF PSORT system*

WoLF is a feature selection program, (the name "WoLF" is loosely inspired by the words "Learning", and "Weighted Features"). WoLF PSORT is the combination of WoLF with a version of PSORTII slightly extended for this purpose. The extended version outputs amino acid features and some iPSORT[2] features as well as the PSORT features. An overview of the system is shown in Figure 1.

Table 2. The correspondence between the localization sites used in our study and the GO numbers for the entries derived from GO annotation.

description	GO numbers	depth	WoLF PSORT site
cytoskeleton	GO:0005856	2	cyts
cytosol	GO:0005829	0	cyto
endoplasmic reticulum	GO:0005783	0	E.R.
extracellular	GO:0005576	0	extr
cell wall	GO:0005618	0	extr
Golgi apparatus	GO:0005794	1	golg
mitochondrion	GO:0005739	0	mito
nucleus	GO:0005634	0	nucl
plasma membrane	GO:0005886	0	plas
peroxisome	GO:0005777	2	pero
vacuolar membrane	GO:0005774	2	vacu
chloroplast	GO:0009507	0	chlo
thylakoid lumen	GO:0009543	0	chlo

Depth indicates the number of levels of "part_of" descendents included with the GO number. For example GO:0005856 and all of its "part_of" children and grandchildren were included in cyts. Note that "extr" and "chlo" are the union of two lines from this table.

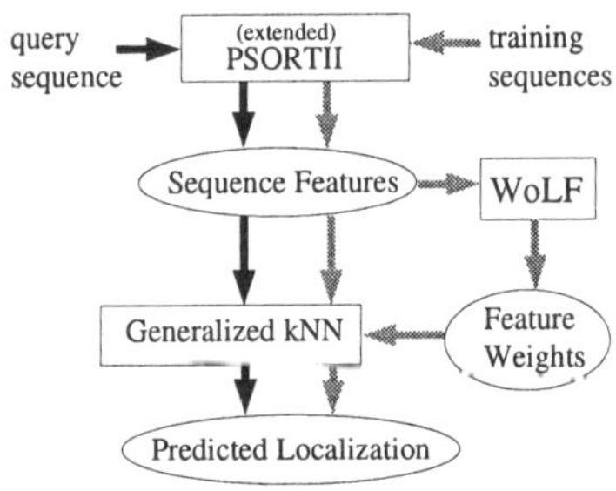

Figure 1. A schematic of the WoLF PSORT system is shown. Rectangles represent programs or procedures and ovals represent computed quantities. The black arrows denote information derived from the query sequence while gray arrows denote information from the training sequences.

2.3. *Classification*

2.3.1. *Candidate Features*

We used fours kinds of features: PSORT[9] features, iPSORT[2] features, amino acid content, and sequence length. Each feature is a deterministic mapping from amino acid sequence to the reals. Since the numerical range of the raw features are not homogeneous, we normalized each feature to its percentile value in the training data. Values observed in test data may not appear in the training data – in which case our programs use linear interpolation (extrapolation) to obtain a pseudo-percentile value.

2.3.2. *Classification Algorithm*

We adopted a weighted version of the *k*NN (*k*-Nearest Neighbors)[5] algorithm for classification. As in standard *k*NN, our method classifies based on the k nearest instances in the dataset. However we slightly extended the distance calculation. In our variation two weights, w_{1i} and w_{2i}, are associated with each feature i. More formally, let F_{ji} and F_{ki} denote the values of feature i in protein instance j and k respectively. The distance $d_W(j,k)$ between j and k is defined as:

$$d_W(j,k) = \sum_i w_{1i}|F_{ji} - F_{ki}| + \sqrt{\sum_i w_{2i}(F_{ji} - F_{ki})^2} \tag{1}$$

combining elements of Euclidean and Manhattan (city block) distances.

2.3.3. *Extensions for Dual Localization Prediction*

The dataset contains some dually localized proteins. We gave partial credit for partially correct predictions as shown Table 3.

Table 3. Example Utility Values.

label	prediction	utility	label	prediction	utility
nucl	nucl	1	nucl_cyto	nucl_mito	0.333
nucl_cyto	nucl_cyto	1	nucl	cyto	0
nucl_cyto	nucl	0.5	mito	nucl_cyto	0
nucl	nucl_cyto	0.5	mito	nucl	0

Examples of the utilities used in this study are shown. "label" denotes the localization site according to the dataset annotation. Note that we chose to use utilities that are symmetric with regard to the label and prediction.

2.3.4. *Feature Selection and Weighting*

We developed a C++ program, WoLF, for selecting the weights in Equation 1. Given a set of candidate features, WoLF selects non-negative integer weights for each feature (a weight of zero is equivalent to exclusion of the feature). WoLF uses a greedy, neighborhood search algorithm to find a locally optimal set of weights. The program uses the jackknife (leave one out cross-validation) utility on the training data to evaluate weight vectors. In the case of ties the simpler weight vector is chosen.

2.3.5. *Reducing Over-reliance on Sequence Similarity*

When training WoLF PSORT we employed a *taboo list* which disallows the use of highly similar sequences (with identical localization sites) as a neighbor when classifying a given instance. We determined the threshold by inspection of the correlation between best hit eValue and co-localization. The thresholds used were -33, -63, -33.4 $\log_2$ eValue for fungi, plant, and animal respectively.

2.3.6. *Evaluation of WoLF PSORT Accuracy*

5-fold cross-validation was used to estimate the accuracy. For each partition, feature and k value selection were performed using a jackknife test on the training partition.

3. Results

3.1. *Effect of Feature Weighting*

The results of various feature weight vectors, including those selected by WoLF PSORT are shown in Table 4. For those tables the value of k was optimized separately for the taboo and no taboo list cases. The WoLF PSORT weight vectors (one per partition) however were only trained using the taboo list. The confusion matrices for the WoLF PSORT cross-validation for yeast (which had the highest fraction of dual localization annotations) is shown in Table 6.

In addition to cross-validation studies, we also ran the WoLF feature weighting procedure on the complete datasets. The selected features (trained with the taboo list) are shown in table 5.

3.2. *WoLF PSORT Combined with BLAST*

We calculated the utility of combining WoLF PSORT with BLAST in a trivial way; namely using the WoLF PSORT prediction for queries whose best BLAST hit eValue exceeds a given threshold, while predicting the localization of the best BLAST hit otherwise. Ties for the best BLAST hit (especially with eValue=0) were fairly common, in which case we voted amongst the best hits (breaking ties using the overall proportion of each localization in the given dataset). In the rare cases in which no BLAST hit was obtained, the majority classifier was used in lieu of BLAST. The results of this hybrid predictor for the three datasets are shown in Figure 2.

3.3. *WoLF PSORT Server*

The WoLF PSORT server is freely available at `wolfpsort.org`. Detailed information about the features of the query sequence and its k nearest (by Equation 1) neighbors are given. These tables give the user a chance to examine the *evidence* behind the prediction. For example Figure 3 shows a partial screen shot of the detailed page (one click away from the summary page) when the protein AEP_YARLI is used as a query. From the first row in the displayed table one can see that the variable *gvh*, the signal peptide detecting weight matrix score of Gunnar von Heijne,[13] has a very high value (93 percentile), consistent with the prediction of

Table 4. Cross-validated utility with various feature weight vectors.

weight vector type	#weights	taboo	k	% utility
		Fungi Dataset		
psortEuclid	31	no	9.2(2.2)	64.7(2.8)
psortEuclid		yes	15.6(6.0)	**61.3(3.1)**
allEuclid	56	no	4.6(3.5)	73.9(2.2)
allEuclid		yes	26.8(11.3)	**69.3(2.9)**
allWeights	112	no	3.2(1.8)	74.3(1.7)
allWeights		yes	31.2(18.0)	**69.4(2.7)**
WoLF PSORT	22.2(4.3)	no	18.2(4.3)	72.6(0.7)
WoLF PSORT		yes	18.2(4.3)	**70.7(1.1)**
		Plant Dataset		
weight type	#weights	taboo	k	% utility
psortEuclid	31	no	4.8(1.9)	66.5(2.0)
psortEuclid		yes	41.2(24.3)	**53.6(2.0)**
allEuclid	56	no	1(0)	85.3(1.2)
allEuclid		yes	19(3.8)	**60.0(3.8)**
allWeights	112	no	1(0)	85.2(1.2)
allWeights		yes	20.6(3.2)	**60.0(3.7)**
WoLF PSORT	21.2(1.3)	no	3(2.1)	76.7(2.4)
WoLF PSORT		yes	13.8(7.0)	**65.1(2.6)**
		Animal Dataset		
weight type	#weights	taboo	k	% utility
psortEuclid	31	no	1(0)	79.7(0.5)
psortEuclid		yes	36.0(9.3)	**72.2(0.8)**
allEuclid	56	no	1(0)	92.3(0.5)
allEuclid		yes	34.6(6.5)	**77.8(0.6)**
allWeights	112	no	1(0)	93.1(0.6)
allWeights		yes	30.2(7.2)	**79.0(0.7)**
WoLF PSORT	25.8(4.2)	no	39.4(7.2)	83.2(1.2)
WoLF PSORT		yes	39.4(7.2)	**79.7(1.0)**

Utility is given as percent of the maximum possible. The number of (non-zero) weights is omitted when it is the same as the row above. Numerical entries represent averages over 5-fold cross-validation with standard deviations given in parenthesis. The "psortEuclid" weight vector has weight 1 for the quadratic term of each PSORT feature, "allEuclid" has weight 1 for the quadratic term of all features, "allWeights" has weight 1 for all possible terms, and WoLF PSORT is the weight vector selected by WoLF PSORT.

extracellular. One can also see that the value of the *mit* feature is very different between the query and two of its neighbors (PEPF_ASPFU and PEPA_ASPOR). *mit* is a variable designed to discriminate between mitochondrial and non-mitochondrial proteins[9] so this does not seem to weaken the evidence in this case.

4. Discussion

4.1. *Interpretable Results*

WoLF PSORT alone achieves accuracy estimates (with sequence similarity reduction used for training but not evaluation) of 73%, 77%, 83%, on the yeast, plant

Table 5. Selected Features.

feature type	feature name	fungi	plant	animal
iPSORT	net charge(1, 25)	1	0	1^2
	max negative charge(1,20)	0	1	0
	max hydropathy(1,30)	2	1	$1,1^2$
PSORT signal peptide	gvh	1	0	1^2
	psg	0	2	0
PSORT targeting signal	mip	0	2	1^2
	mit	1	2	1
PSORT membrane protein related	alm	1	0	3^2
	m1a	1^2	1	0
	m1b	1	3	0
	m3a	0	1	0
	m3b	1	1	0
	mNt	1	0	0
	tms	2	1^2	2
PSORT sorting motif	erl	1^2	1^2	0
	leu	1	0	1
	nuc	1	1	1^2
	pox	1^2	1^2	0
	tyr	2	1	1^2
	yqr	1	0	0
	vac	0	0	1
PSORT non-sorting motif	dna	1^2	1^2	$1,4^2$
	rib	2^2	1^2	1^2
	myr	1^2	0	$1,1^2$
	rnp	1	1	1^2
	act	0	0	3^2
miscellaneous	length	0	1	1^2

The selected features are shown with their weights. Features with 0 weight for all datasets were omitted. "1^2" and "2^2" indicate the quadratic term rather was selected with a weight of 1 or 2 respectively. The amino acid content features (by one letter code) selected for the three datasets were "ARNDQEGIKMFSWV", "ACQIIILSV", and "CIKS" respectively. In each of those cases the weight was 1, the weight type was always linear for the fungi and plant dataset and always quadratic for the animal dataset. "sorting" motifs refer to motifs such as the E.R. retention signal "erl" with a direct causal relationship to localization. Descriptions of the variables can be found on the WoLF PSORT server. For the PSORT variables useful documentation can also be found on the PSORT help page `www.psort.nibb.ac.jp`

and animal datasets, with a small number of features and the trivially simple k nearest neighbors classifier. We do not claim that this will meet the accuracy of sophisticated classifiers such as the popular support vector machine. However the template based nature of the kNN classifier makes individual classifications easier to interpret. The web server provides a tabular display to facilitate this process.

4.2. *Evaluation in the Presence of Similar Sequences*

In this study we included many similar sequences. This makes achieving a high accuracy easy. However by comparing with BLAST we were able to show that our method can be effective even when sequence similarity is low. For example the

Table 6. Confusion Matrix for the Fungi Dataset.

site	nucl	cyto nucl	cyto	cyto mito	mito	mito nucl	plas	extr	cyto pero	pero	E.R.	golg	vacu	cyts	sum
nucl	**556**	33	37	0	27	2	6	0	0	6	0	0	0	0	667
cyto_nucl	49	**10**	24	0	6	0	0	0	0	2	0	0	0	0	91
cyto	88	16	**232**	3	30	0	1	5	0	7	1	0	0	0	383
cyto_mito	1	0	0	**0**	7	0	0	0	0	0	0	0	0	0	8
mito	38	3	22	0	**299**	3	9	6	0	8	1	0	0	0	389
mito_nucl	1	0	1	0	2	**0**	0	0	0	0	0	0	0	0	4
plas	6	0	3	0	2	0	**205**	2	0	1	1	0	0	0	220
extr	2	2	4	0	2	0	0	**129**	0	0	1	0	0	0	140
cyto_pero	1	0	1	0	0	0	0	0	**0**	0	0	0	0	0	2
pero	16	3	23	0	5	0	6	2	1	**20**	1	0	0	0	77
E.R.	4	2	5	0	2	0	30	11	0	4	**7**	1	0	0	66
golg	10	1	1	0	1	0	10	8	0	0	2	**5**	0	0	38
vacu	2	0	2	0	2	0	10	6	0	0	1	0	**0**	0	23
cyts	4	0	1	0	0	0	0	0	0	0	0	0	0	**0**	5
sum	778	70	356	3	385	5	277	169	1	48	15	6	0	0	2113

The rows represent the dataset labels. The columns represent predictions made by WoLF PSORT

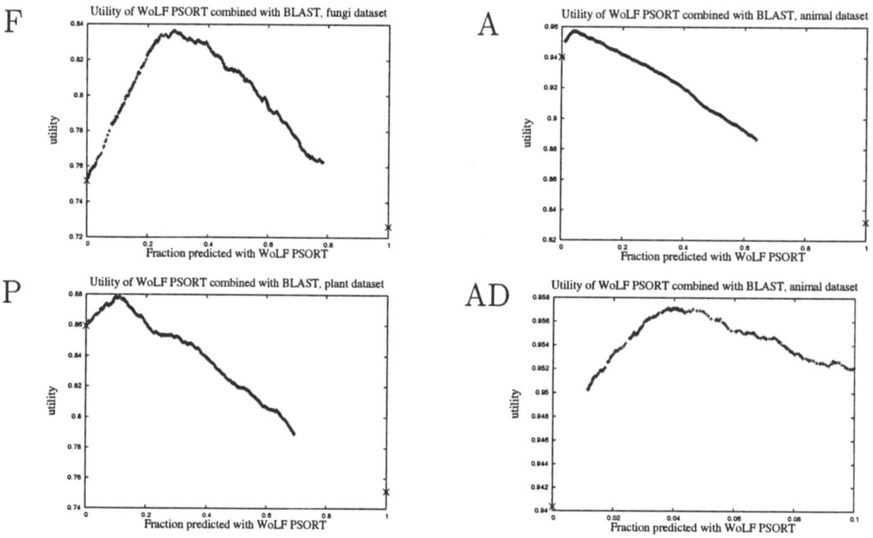

Figure 2. The utility obtained by combining WoLF PSORT and BLAST with a simple eValue threshold is shown for the fungi (F), plant (P), and animal (A) datasets. A detailed view of the animal data (AD) is shown as well. The point at $x = 1$ represents using WoLF PSORT alone. The lack of points between 0.8 and 1 for the fungi data is due to the fact that approximately 20% (30% for plant, 36% for animal) of the proteins have hits with an eValue of "0", a similar (but smaller) gap occurs at the other end due to proteins with no blast hits.

number of errors produced by using BLAST alone on the animal dataset can be reduced by about a quarter (accuracy of $\approx 95.6\%$ *vs* $\approx 94.0\%$, yielding 179 errors) by using WoLF PSORT for proteins without good BLAST hits.

		Normalized F											
id	site	iPSORT		PSORT Featu									
		-1 25	MxHv1 30	alm	dna	erl	gvh	leu	m1a	m1b	m3b	mNt	mit
AEP_YARLI	extr?	55	92	21	46	50	93	47	50	49	48	49	58
AEP_YARLI	extr	55	92	21	46	50	93	47	50	49	48	49	58
PEPF_ASPFU	extr	55	96	17	46	50	96	47	50	49	48	49	83
CAR1_CANPA	extr	55	93	20	46	50	96	47	50	49	48	49	55
PEPA_ASPOR	extr	55	93	21	46	50	93	47	50	49	48	49	94

Figure 3. A screen shot of the server showing the feature table displayed for query and nearest neighbors is shown. The query id, predicted localization, and features are showed aligned with the id, localization, and features of its nearest neighbors. Color is used to attract attention to large differences between the query and nearest neighbors. In the example shown the query is "AEP_YARLI", which also appears on the next line as the nearest neighbor in the dataset.

4.3. *Predicting Dual Localization*

WoLF PSORT was designed with dual localization in mind. The only dual localization category for which SWISS-PROT currently contains a significant number of entries is dual localization to the nucleus and cytosol. This is an important category of proteins, for example some transcription factors are regulated by conditional localization to the nucleus[8].

The prediction results seen in the confusion matrix (Table 6) are mixed. Unfortunately most of the 91 proteins labeled "cyto_nucl" are misclassified as either nuclear or cytosol. On the other hand, perhaps this mistake should not be looked at too harshly – as a "half-right" prediction of a dually localized protein is the *best possible* prediction for existing prediction methods which do not consider multiple localization at all. It seems likely that the current annotation in SWISS-PROT is conservative relative to multiple localization. For example in a recent large scale experiment using GFP fusion proteins to determine localization in yeast, approximately 20% out of a total of over 4000 measured proteins were found to dually localize to the nucleus and cytoplasm.

5. Conclusion

WoLF PSORT achieves a dramatic improvement in prediction accuracy over PSORTII while maintaining the simple, easily understood classifier which has been one reason for PSORTII's widespread use. Indeed by applying a feature selection algorithm, WoLF PSORT is actually *simpler* than PSORTII in the sense that it uses fewer features for classification. WoLF PSORT is also one of the most serious attempts to date to incorporate dually localized proteins into a prediction scheme for eukaryotic cells.

6. Acknowledgement

C.J. Collier designed the initial server task handler. H. Harada helps maintain the current server. Dr. H. Ohta gave valuable advice in preparing the entries extracted from GO. This work was partly supported by MEXT.

References

1. M. Ashburner, C. A. Ball, J. A. Blake, D. Botstein, H. Butler, J. M. Cherry, A. P. Davis, K. Dolinski, S. S. Dwight, J. T. Eppig, M. A. Harris, D. P. Hill, L. Issel-Tarver, A. Kasarskis, S. Lewis, J. C. Matese, J. E. Richardson, M. Ringwald, G. M. Rubin, and G. Sherlock. Gene ontology: tool for the unification of biology. *Nature Genetics*, 25(1):25–29, 2000.
2. Hideo Bannai, Yoshinori Tamada, Osamu Maruyama, Kenta Nakai, and Satoru Miyano. Extensive feature detection of N-terminal protein sorting signals. *Bioinformatics*, 18:298–305, 2002.
3. B. Boeckmann, A. Bairoch, R. Apweiler, M. Blatter, A. Estreicher, E. Gasteiger, M. J. Martin, K. Michoud, C. O'Donovan, I. Phan, S. Pilbout, and M. Schneider. The SWISS-PROT protein knowledgebase and its supplement TrEMBL in 2003. *Nucleic Acids Research*, 31:365–370, 2003.
4. M.G. Claros and P. Vincens. Computational method to predict mitochondrially imported proteins and their targeting sequences. *European Journal of Biochemistry*, 241:779–786, 1996.
5. Richard O. Duda and Peter E. Hart. *Pattern Classification and Scene Analysis*. John Wiley & Sons, New York, 1973.
6. Paul Horton, Yuri Mukai, and Kenta Nakai. Protein localization prediction. In Limsoon Wong, editor, *The Practical Bioinformatician*, pages 193–215. World Scientific, 5 Toh Tuck Link, Singapore 596224, 2004.
7. Paul Horton and Kenta Nakai. Better prediction of protein cellular localization sites with the k nearest neighbors classifier. In *Proceeding of the Fifth International Conference on Intelligent Systems for Molecular Biology*, pages 147–152, Menlo Park, 1997. AAAI Press.
8. José M. Mingot, Eduardo A. Espeso, Eliecer Díez, and Miguel A. Peñalva. Ambient pH signaling regulates nuclear localization of the *aspergillus nidulans* PacC transcription factor. *Molecular and Cellular Biology*, 21(5):1688–1699, March 2001.
9. Kenta Nakai and Minoru Kanehisa. A knowledge base for predicting protein localization sites in eukaryotic cells. *Genomics*, 14:897–911, 1992.
10. Henrik Nielsen. Machine learning approaches for the prediction of signal peptides and other protein sorting signals. *Protein Engineering*, 12(1):3–9, 1999.
11. Henrik Nielsen, Jacob Engelbrecht, Søren Brunak, and Gunnar von Heijne. Identification of prokaryotic and eukaryotic signal peptides and prediction of their cleavage sites. *Protein Engineering*, 10:1–6, 1997.
12. K. Nishikawa and T. Ooi. Correlation of the amino acid composition of a protein to its structural and biological characters. *J. Biochem.*, 91(5):1821–1824, 1982.
13. Gunnar von Heijne. A new method for predicting signal sequence cleavage sites. *Nucleic Acids Research*, 14:4683–4690, 1986.

PREDICTING RANKED SCOP DOMAINS BY MINING ASSOCIATIONS OF VISUAL CONTENTS IN DISTANCE MATRICES

PIN-HAO CHI AND CHI-REN SHYU

Medical and Biological Digital Library Research Lab,
Department of Computer Science, University of Missouri, Columbia, MO 65211, USA
E-mail: pinhao@diglib1.cecs.missouri.edu, ShyuC@missouri.edu

Protein tertiary structures are known to have significant correlations with their biological functions. To understand the information of the protein structures, Structural Classification of Protein (SCOP) Database, which is manually constructed by human experts, classifies similar protein folds in the same domain hierarchy. Even though this approach is believed to be more reliable than applying traditional alignment methods in structural classifications, it is labor intensive. In this paper, we build a non-parametric classifier to predict possible SCOP domains for unknown protein structures. With supervised learning, the algorithm first maps tertiary structures of training proteins into two-dimensional distance matrices, and then extracts signatures from visual contents of matrices. A knowledge discovery and data mining (KDD) process further discovers relevant patterns in training signatures of each SCOP domain by mining association rules. Finally, the quantity of rules whose patterns match signatures of unknown proteins determines predicted domains in a ranked order. We select 7,702 protein chains from 150 domains of SCOP database 1.67 release as labelled data using 10 fold cross validation. Experimental results show that the prediction accuracy is 91.27% for the top ranked domain and 99.22% for the top 5 ranked domains. The average response time takes 6.34 seconds, exhibiting reasonably high prediction accuracy and efficiency.

1. Introduction

Protein structure information is known to be more conserved than amino acid sequences and serves as ideal references to study protein structure-to-function relationships. Similar protein folds may suggest similar biochemical functions.[27] In our knowledge, the most reliable structural comparison method is to manually inspect similar protein structures such as SCOP.[17] Proteins with high structural similarity will be classified into the same hierarchical SCOP domain. Even though manual inspection provides more accurate structural classification, it is labor intensive for a large number of protein tertiary structures. Automated structural comparison methods such as Distance Alignment (DALI)[12] and Combinatorial Extension (CE)[22] algorithms globally find a structural alignment between two polypeptide chains such that superimposed segments of amino acids can have a good structural match within a small Root Mean Square Deviation (RMSD) threshold. Due to the huge combination of possible alignments, exhaustively searching a local optimal solution is known to be computationally expensive, proving a complexity of NP-Hard.[9] Therefore, life science researchers and biologists have a great demand on efficient and accurate protein structure classification systems.

Several well-known structural classification databases have been studied in computational molecular biology literatures. Secondary Structure Alignment Program (SSAP) utilizing double phases dynamic programming techniques for optimal structural alignment of two proteins becomes a framework to construct CATH database.[18] The DALI algorithm that applies Monte Carlo heuristics to compare structural similarities from distance matrices is used to conduct structural classifications in FSSP database.[13] Applying specific heuristics for reducing computational complexity, these classical structural alignment algorithms may return variant classification results from the same protein set. To avoid suffering from drawbacks of subjective heuristics, recent classification works[3,6] that maintain higher accuracies than applying each individual method by intersecting multiple intermediate results of existing structural alignment algorithms. Even though structural alignment methods present satisfactory classification accuracies, the process of performing multiple pairwise alignments between an unknown protein and known proteins in databases is still incapable of providing fast predictions.

With the advent of x-ray diffraction and high-resolution nuclear magnetic resonance (NMR) techniques, the amount of newly discovered proteins has grown rapidly in recent years. As July 5th, 2005, Protein Data Bank (PDB), announces 32,107 protein structures and 8,070 of them have not been classified in the latest SCOP release (1.67). This noticeable gap is well-recognized and continues to grow. Hence, there is an urgent need to develop an efficient domain classification method with sufficiently high accuracy to streamline the labor-intensive classification process. It is noteworthy to mention that, instead of replacing human inputs from this classification process, a more realistic approach is to suggest a handful set of top ranked domains for further studies.

In this work, we extend our recent research results in a real-time tertiary structure retrieval system called ProteinDBS[7,14,24] and develop a series of knowledge discovery and data mining techniques to perform fast SCOP domain predictions with reasonably high accuracy. This paper is organized as follows. Section 2 introduces our unique model to cast protein backbone structures into high-dimensional feature vectors. Section 3 describes the algorithm to transform feature values into a set of feature intervals and illustrates the association rule mining using a supervised learning technique. Experimental results of prediction accuracy and efficiency are reported in Section 4. Finally, we conclude this paper and discuss possible future works in Section 5.

2. Preliminaries

Protein Tertiary Structure refers to a single polypeptide chain that is constructed by a long amino acid string. For a protein chain k with n amino acids, its backbone is represented by a n-dimensional vector $\{\vec{C_\alpha^{k,1}}, \vec{C_\alpha^{k,2}}, ..., \vec{C_\alpha^{k,n}}\}$, where the element, $\vec{C_\alpha^{k,i}}$, is the three-dimensional coordinate of the i-th C_α atom. The distance matrix of k is defined as a $n \times n$ symmetric real matrix whose element at i-th column and j-th row is the Euclidean distance between $\vec{C_\alpha^{k,i}}$ and $\vec{C_\alpha^{k,j}}$. A distance matrix is generally sufficient to recover the original three-dimensional backbone structure in polynomial time using distance geometry methods.[11] Several literatures[12,16,26] study comparing similar distance matrices as a equivalent problem to protein tertiary structure comparisons. Our assumption is based on the fact

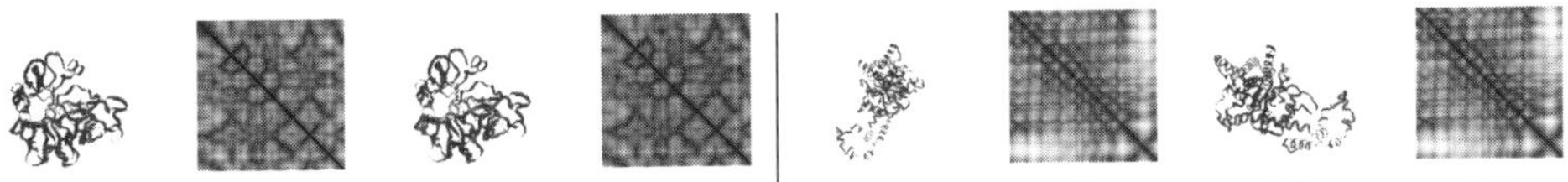

(a) (b) (c) (d) (e) (f) (g) (h)

Figure 1. The three-dimensional backbone structures and distance matrices from protein chains selected from the SCOP domain *Carbonic anhydrase*: (a-b)$1am6$, (c-d)$1bic$, and the SCOP domain *D-xylose isomerase*: (e-f)$9xim\_D$, (g-h)$1xlb\_A$

that similar protein folds should have distance matrices with similar visual contents. We also expect that proteins in the same SCOP domain should present high similarities in distance matrices. To pictorially explain our assumption, Figure 1 shows that protein chains from SCOP *Carbonic anhydrase* and *D-xylose isomerase* domains present high similarities in both three-dimensional tertiary structures and two-dimensional distance matrices. Even though similar visual patterns can be identified by manual inspections, it is still a challenging research topic to mimic distance matrix comparisons automatically using computational techniques. Fortunately, there exists a rich body of literatures in the area of content-based image retrieval (CBIR) since early 80's.[5,21,23] The concept of CBIR is to retrieve visually similar images from databases for a query image. This is a perfect fit to the protein distance matrix comparisons. To effectively retrieve similar candidates in a large population of distance matrices, extracting relevant features becomes an important issue to study. In our previous works,[7,24] the distance matrix is divided into six band regions, parallel to its diagonal. In each band, four local features are computed by histograms of four bins of distance ranges: [0-5], [6-10], [11-15], and [16-∞]. We also have extracted nine global features from visual patterns of distance matrices using a suite of standard computer vision algorithms.[10,20,19] After features are extracted, each protein backbone can be transformed into a high-dimensional feature vector and clustered in the feature space. Readers are referred to our previous publications[7,24] for the details of the feature extraction algorithms applied in this work.

The distribution of feature values is expected to have significant correlation to protein domains in SCOP. A set of features with certain ranges could best describe structural patterns of proteins in a specific SCOP domain. Figure 2 depicts a simplified example using three features, namely the 8th localized histogram (The 4th gray-scale level in the 2nd partitioned band region of distance matrix), the 5th texture[10] ($Homogenity$), and the 9th texture ($Cluster\_Tendency$). For proteins in SCOP domains *Carbonic anhydrase* (D_1), *D-xylose isomerase* (D_2) and *Calmodulin* (D_3), these three features are partially overlapped in multiple intervals. From the top range line of Figure 2, it is clear that all database protein structures from D_1 and D_2 mix in the same "Histogram 8" feature interval. Similarly, the "Texture 5" feature is unable to separate proteins in D_2 from those in D_3. Adding association information among feature intervals, the algorithm is able to predict an unknown protein structure to $D_1 : \{f_{Histogram8} \in [0.040,0.045)$ and $f_{Texture9} \in [0.005,0.010)\}$, $D_2 : \{f_{Histogram8} \in [0.040,0.045)$ and $f_{Texture5} \in [0.085,0.090)\}$, or $D_3 : \{f_{Texture5} \in [0.085,0.090)$ and $f_{Texture9} \in [0.005,0.010)\}$.

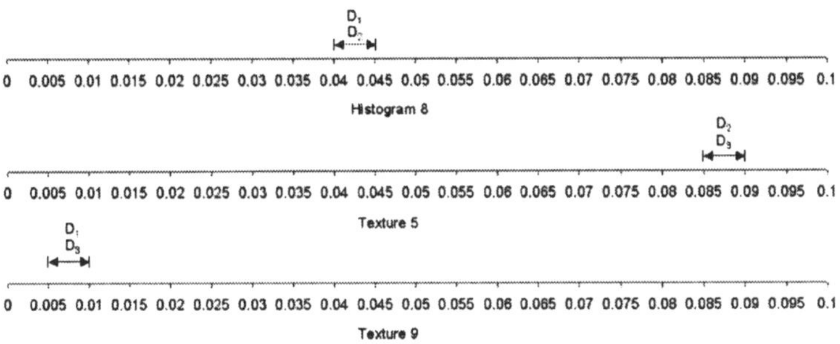

Figure 2. An example of feature intervals for SCOP domains, D_1:*Carbonic anhydrase*, D_2: *D-xylose isomerase* and D_3: *Calmodulin*

Knowledge discovery and data mining techniques have been widely studied in high-throughput data analysis of various aspects such as classification,[15] mining in web usage, spatial data, document indexing,[8] and biological domains.[26] Among data mining techniques, association rule (*AR*) mining is able to retrieve hidden patterns and discover meaningful information from the data. Given a protein chain p_1, it will be preprocessed into an m-dimensional feature vector $\{f_1^{p_1}, f_2^{p_1}, f_3^{p_1}, ..., f_m^{p_1}\}$, where $f_i^{p_1}$ has been normalized in $R[0,1]$ and $1 \leq i \leq m$. Then, the algorithm partitions $R[0,1]$ space of each individual feature of proteins into a set of disjoint intervals $\{[0,\eta_1], (\eta_1,\eta_2], ..., (\eta_n, 1]\}$, where $0 < \eta_1 < \eta_2 < ... < \eta_n < 1$. To discuss data mining algorithms used in this work, each feature interval $(\eta_i, \eta_{i+1}]$ is defined as an *item*. For example, there exist three feature intervals (items) generated from a partition of R[0,1] that are associated with the jth feature of all database proteins such as $I_1 = [0.0, 0.2]$, $I_2 = (0.2, 0.75]$, and $I_3 = R(0.75, 1.0]$. For a protein p_1, the jth feature value, $f_j^{p_1} = 0.5$, will be transformed into *item* I_2. Applying the same item mapping process for m features, each backbone structure is then represented by a set of m items ($m = 33$ in our work). This collection of items forms a *transaction* for mining item associations. In addition, a database D that includes n proteins can be described by n transactions. With a set of items, I, an association rule is defined as an implication rule composed of items with a form $\{X \Rightarrow Y\}$, where $X, Y \subseteq I$ and $X \cap Y = \emptyset$. *Itemsets* X and Y are called *Antecedent* and *Consequent*, respectively. For an association rule represented by $\{X \Rightarrow Y\}$, the *support* of the rule is the percentage of all *transactions* in D that include $\{X \cup Y\}$ items. The *confidence* of the rule is a ratio of the total amount of *transactions* that contain $\{X \cup Y\}$ to *transactions* with $\{X\}$ items. The association rule mining generates relevant rules in the database with the *support* and *confidence* that can pass *minimal_support* and *minimal_confidence* thresholds, respectively.

3. Method

To precisely predict an unknown protein structure among hundreds or even thousands of SCOP domains, it is critically important to identify appropriate feature intervals, as well as associations among these relevant intervals within each SCOP domain. The way to formulate a partition of a real space $R[0,1]$ has vital impact on determining relevant items.

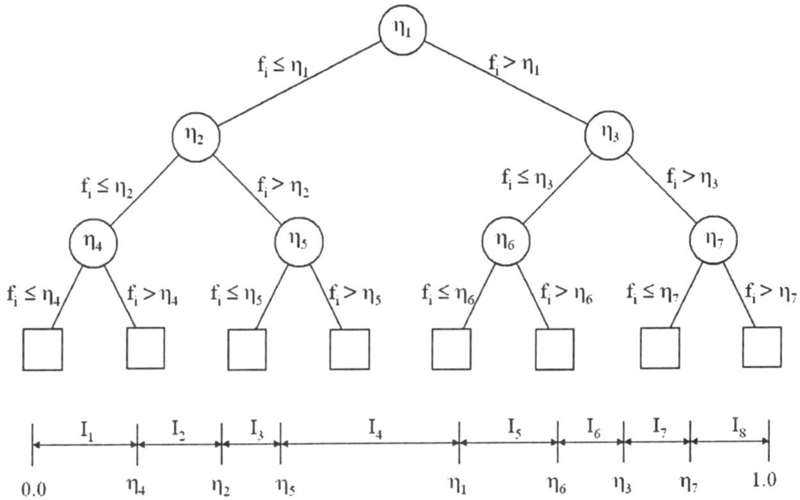

Figure 3. A binary decision tree to determine thresholds for a space partition of feature f_i

Partitioning a real space too finely will generate many tiny intervals within one domain, resulting in huge amount of association rules. A coarse partition of space will create intervals that mix multiple domains without enough discriminatory power. Instead of randomly or evenly partitioning the real space into intervals, we apply C4.5 decision tree[25] to find relevant intervals for each feature among all database domains.

3.1. *Space Partition Algorithm Using C4.5 Decision Tree*

For each individual feature of all m-dimensional feature vectors, the algorithm constructs a C4.5 decision tree. In total, there are 33 trees for all features used in this work. The splitting criterium to grow the decision tree is based on the minimization of entropy. Let D^t be the set of protein features at a certain node t. The entropy, $H(D^t)$, of node t and the weighted entropy, $H(D^{t'})$, of its child nodes t_l and t_r are computed as follows:

$$H(D^t) = -\sum_{j=1}^{r} p^t_{d_j} \times log(p^t_{d_j}), H(D^{t'}) = \alpha \times H(D^{t_l}) + (1-\alpha) \times H(D^{t_r}) \qquad (1)$$

where $p^t_{d_j}$ denotes the ratio of proteins in domain d_j to the total number of proteins that exist in node t. To compute $H(D^{t'})$, α represents the percentage of protein chains that have been dispatched from a parent node to the left child by the threshold η, which is an optimal threshold and selected based on the maximization of $H(D^t) - H(D^{t'})$. With a top-down iterative node splitting, the algorithm collects sorted thresholds of k internal nodes using in-order traversal, and the space R[0,1] will be partitioned into $k+1$ intervals as a set of *items*. For example, Figure 3 shows that eight *items*, $I_1 = R[0.0, \eta_4]$, $I_2 = R(\eta_4, \eta_2]$, ..., $I_8 = R(\eta_7, 1.0]$, are partitioned by seven threshold values $\{\eta_4, \eta_2, \eta_5, \eta_1, \eta_6, \eta_3, \eta_7\}$. Each protein is then mapped into a 33-item transaction for mining item associations using the intervals selected by the decision trees.

Items	Features	Intervals
I_1	Histogram(8)	R(0.035,0.04]
I_2	Histogram(8)	R(0.04,0.045]
I_3	Texture(5)	R(0.39,0.395]
I_4	Texture(5)	R(0.495,0.5]
I_5	Texture(9)	R(0.005,0.01]
I_6	Texture(9)	R(0.03,0.035]

$\{I_1, I_3\} \rightarrow$ *Carbonic anhydrase*
$\{I_1, I_5\} \rightarrow$ *Carbonic anhydrase*
$\{I_3, I_5\} \rightarrow$ *Carbonic anhydrase*
$\{I_1, I_3, I_5\} \rightarrow$ *Carbonic anhydrase*

$\{I_2, I_4\} \rightarrow$ *D-xylose isomerase*
$\{I_2, I_6\} \rightarrow$ *D-xylose isomerase*
$\{I_4, I_6\} \rightarrow$ *D-xylose isomerase*
$\{I_2, I_4, I_6\} \rightarrow$ *D-xylose isomerase*

Figure 4. Association Rules generating from partitioned feature intervals using Apriori algorithm

3.2. *Mining Training Data and Prediction Model*

After transforming three-dimensional protein backbones into the form of *transactions*, the system then mines associations of the items from training data by applying the Apriori algorithm.[2] The main concept of Apriori algorithm is to generate association rules from frequent itemsets whose *support* is greater than the *minimal_support* threshold. Since any subset of a large transaction is still a frequent itemset, the algorithm finds candidates of frequent itemsets with n_i items from frequent itemsets with $n_i - 1$ items, where $n_i \geq 1$. In Apriori algorithm, *minimal_support* is an important criterium to determine the quantity of association rules. Due to the non-uniformly distributed proteins among all domains, it is inappropriate to mine rules from the entire database using a single *minimal_support*. Therefore, for each domain d, we perform Apriori algorithm and each frequent itemset, I, refers to an association rule $I \Rightarrow d$. For instance, itemsets $\{I_1, I_3, I_5\}$ and $\{I_2, I_5, I_6\}$ are frequent for SCOP domain *Carbonic anhydrase* and *D-xylose isomerase*, respectively. Examples of association rules for domain predictions are shown in Figure 4. After obtaining rules from all SCOP domains, a small portion of rules (2.81%) shared by multiple domains has been pruned out prior to the prediction stage. Our current setting of the *minimal_support* is 90% within each domain. Mining training proteins of 150 SCOP domains populates 2,354 association rules. Discovered rules has been efficiently organized and loaded into main memory for fast predictions.

The next task is to design a scoring function that suggests possible SCOP domains in a ranked order. For an unknown protein t, a complete itemset, $I^t = \{I_1^t, I_2^t, ..., I_m^t\}$, is formed by mapping features into item intervals as discussed in Section 2, where m is the total number of features ($m = 33$ in our work). Given k association rules in domain d, each rule can be represented by $\{I_1^i, I_2^i, ..., I_n^i\} \Rightarrow d$, where $m \geq n \geq 2$ and $k \geq i \geq 1$. Among these rules, we group them into two sets: *matched rules* R_c^d and *mismatched rules* R_m^d, where $|R_c^d| + |R_m^d| = k$. The i-th rule is categorized as *matched rules* when the condition, $\{I_1^i, I_2^i, ..., I_n^i\} \subseteq I^t$, is satisfied. Contrarily, a mismatched rule has at least one item in its antecedent that is not included in I^t for the unknown protein t. The scoring function rewards matched rules and penalizes mismatched rules in each domain. For the i-th matched rule, the scoring function further considers the degree of reward N_i, which is the size of its antecedent. To gauge the degree of penalty for mismatched rules, we use a *discrete distance* measurement, which is demonstrated as follows. Let r_m:$\{I_1^m, I_2^m, ..., I_n^m\} \Rightarrow d$ be a mismatched rule, $f_{ea}(I_i^m)$ be a function that returns which feature maps item I_i^m, and $i_{dx}(I_i^m)$ be a function to return the index value of item I_i^m in integer. As an example, a decision tree for the 3rd feature generates 10 items $\{I_1^{'}, I_2^{'}, ..., I_{10}^{'}\}$, which are sequen-

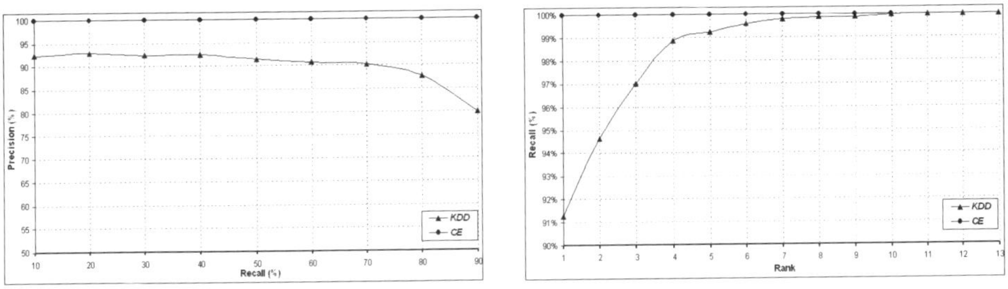

(a) (b)

Figure 5. (a) A precision-to-recall chart for 10 rounds of experiments (b)An accumulated recall chart for top 13 predicted domains

tially stored in an array of position $\{65, 66, ..., 74\}$. Since item $I_1^{'}$ is partitioned from the 3rd feature, $f_{ea}(I_1^{'})$ is equal to 3 and $i_{dx}(I_1^{'})$ returns 65. For any two items x and y, we define $g(x, y) = 1$ when $f_{ea}(x) = f_{ea}(y)$ and $g(x, y) = 0$ if $f_{ea}(x) \neq f_{ea}(y)$. The *discrete distance* between a mismatched rule r_m and an unknown protein t is defined as: $\sum_{x \in \delta_{r_m}} \sum_{n=1}^{m} |i_{dx}(x) - i_{dx}(I_n^t)|^2 \times g(x, I_n^t)$, where δ_{r_m} is the set of mismatched items in r_m. From the same decision tree, items in the neighborhood of partitioned feature intervals are expected to have structural similarities, resulting in a small *discrete distance*. This penalty is then normalized by M_d, the total number of mismatched items from R_m^d.

$$Score(d) = \frac{\sum_{i=1}^{|R_c^d|} N_i}{(\sum_{j=1}^{|R_m^d|} \sum_{x \in \delta_j} \sum_{n=1}^{m} |i_{dx}(x) - i_{dx}(I_n^t)|^2 \times g(x, I_n^t))/M_d} \tag{2}$$

Taking both reward and penalty into consideration, the scoring function for each domain is defined in Eq. (2). To predict ranked domains of an unknown protein, the algorithm computes and ranks scores for all domains.

4. Experiment

We evaluate the performance in accuracy and efficiency for predicting SCOP domains. Experiments are conducted using 10 fold cross validation on a large-scale dataset. With 7,702 protein chains from 150 SCOP domains, 10% of proteins from each domain are randomly selected for blind test. To evaluate the prediction accuracy, we use *Precision and Recall* in the context of machine learning.[4] Given n_r possible SCOP domains, let N_P^d be the number of testing proteins that are predicted to the domain d, N_{TP}^d be the number of testing proteins whose predicted domain d matches its true SCOP domain and N_T^d be the number of testing proteins that are from domain d, where $1 \leq d \leq n_r$. The performance metrics are defined as follows:

$$Precision = \frac{1}{n_r} \sum_{d=1}^{n_r} \frac{N_{TP}^d}{N_P^d}, \; Recall = \frac{1}{n_r} \sum_{d=1}^{n_r} \frac{N_{TP}^d}{N_T^d} \tag{3}$$

Figure 5(a) presents a plot of *Precisions* against *Recall* ranging from 10% to 90%. The ideal case occurs when all testing proteins are predicted correctly, achieving 100%

precision at any recall rate. Our KDD algorithm exhibits 92.42% precision with a 10% recall, 91.35% precision recalling half of them, and 79.77% precision recalling 90% of the entire testing protein set. Normally, the precision will drop by increasing the recall rate. A more practical goal for domain prediction is to suggest a small set of candidate domains to streamline the manual process. To demonstrate the usefulness of our prediction model, we also measure the recall rate by accumulating *True Positives* from the top predicted SCOP domains in the ranked results. In Figure 5(b), our KDD method delivers 91.27% recall rate from the top predicted domain and 99.22% from the top 5 predicted domains. 100% recall rate is achieved by top 13 predictions. What this means is that a human domain classifier only needs to examine 5 domains to guarantee 99.22% coverage of the true domain and 13 domains for 100% coverage.

To evaluate the efficiency of predictions, we measure the average response time. Our system is hosted on a standard Linux Redhat platform with Dual Xeon IV 2.4GHz processors and 2GB RAM. Figure 6(a) shows that the response time of prediction, including feature extractions, itemset generations, and the ranked scores computation. When the protein size increases, it demands more computational resource to extract features on larger distance matrices. This reflects the gaps between two curves in Figure 6(a), where the top curve reports the response time with feature extraction and the bottom curve depicts the response time for computing scores and ranking domains. On the average, predicting an unknown protein to a SCOP domain takes 6.34 seconds. Comparing to a well-recognized structural alignment algorithm, CE,[22] on the same testing data, we conduct pairwise structural alignments for 1 against 7,701 proteins using the *Leave-One-Out* strategy. The SCOP domain of protein with the highest score is specified as the predicted result. We find that CE predicts SCOP domains of all testing proteins correctly. However, pairwise alignments using CE take 15,461.29 seconds. Sacrificing supportable accuracy, our algorithm runs near 2,439 times faster than the CE algorithm. Even though computer algorithms present high prediction accuracy in empirical results, classification by human experts is still believed to be more reliable. Instead of replacing manual classifications, our proposed method assists human experts to make the task of SCOP domain classification achievable and efficient.

In addition, our method is able to predict the SCOP fold of an unknown protein structure from the predicted domain by referencing the known mapping information between domain and fold. For the fold level, our approach exhibits 94.47% prediction accuracy, which is higher than the accuracy of SCOP domain predictions, 91.27%. Due to *one-to-many* relationship between fold and domains, it has a chance to conclude correct folds from incorrectly predicted domains. Therefore, SCOP domain predictions are more challenging than predictions in fold level. For instance, a SCOP fold f_1 contains three domains, such as d_1, d_2, and d_3, respectively. Even though the algorithm predicts a testing protein of SCOP domain d_1 as d_2, the fold is still mapped to f_1. Since the standard testbed of SCOP fold predictions is not available at this moment, we briefly compare to a recent approach in terms of data size, precision, and response time. A prominent work called 3-step scheme(PA+CP+DALI)[1] reports 98.8% accuracy in fold prediction and the average response time is 24,501 seconds. It is noteworthy to mention that their experiments are

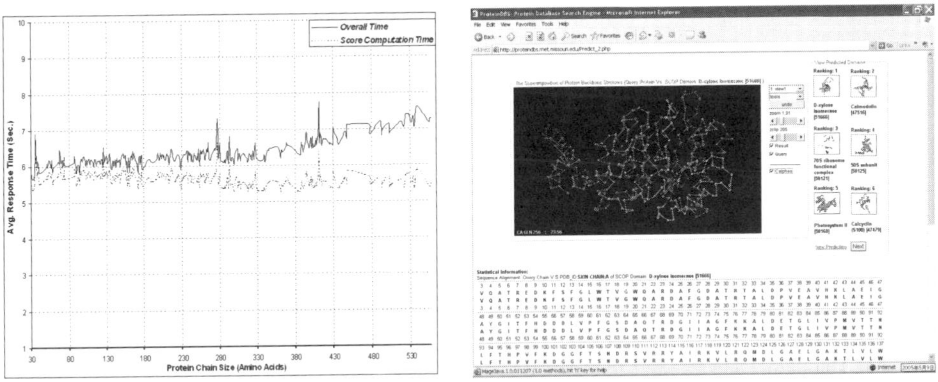

(a) (b)

Figure 6. (a)Average response times to predict SCOP domains with various protein chain sizes (b) The publicly available domain prediction system based on this our prediction model.

conducted on a comparably small testing set (600 proteins) from 15 SCOP folds.

5. Conclusion

Our automatic SCOP domain ranking and prediction algorithm accelerates the processes of structural recognition for newly discovered proteins. In this paper, we introduce an advanced algorithm to convert high-level features of distance matrices into itemsets for rule mining. The advantage of this KDD approach is to effectively reveal the hidden knowledge from similar protein tertiary structures for ranking and predicting possible SCOP domains. Although a multi-variate decision tree might be able to give comparable performance in classification and response time, the tree approach normally could not provide reasonable ranking results that are more valuable in the real world setting, as discussed previously. From the experimental results, our method can achieve reasonably high prediction performance in both accuracy and efficiency. To extend the scope of SCOP domain predictions, one possible direction is to computationally analyze text-based gene annotations, especially the passages related to gene functions, from structurally similar proteins.

To provide a tool for the research community, we have implemented a web-based interface to predict possible SCOP domains for unknown protein structures. Users are allowed to upload a protein file that follows PDB ATOM format. In Figure 6(b), the superimposition view shows that the query protein is structurally similar to a protein $5xin\_A$ from the top ranked SCOP domain *D-xylose isomerase*. Our system is publicly accessible at http://ProteinDBS.rnet.missouri.edu/Predict.php.

References

1. Z. Aung and K.-L. Tan. Clasifying Protein Folds using Multi-level Information of Protein Strutures. *The Third Asia Pacific Bioinformatics Conference SIG-Structure Meeting*, 2005.
2. R. Agrawal, T. Imielinski, and A. Swami. Database mining: a performance perspective. *IEEE Transactions on knowledge and data engineering*, 5(6):914–925, 1993.

3. T. Can, O. Camoglu, A.K. Singh, and Y.F. Wang. Automated Protein Classification Using Consensus Decision. *Proc. of the 3rd Int. IEEE Comput. Soc. Comput. Syst. Bioinformatics Conference*, 224–235, 2004.
4. R. Caruana and A. Niculescu-Mizil. Data mining in metric space: an empirical analysis of supervised learning performance criteria. *Proc. of the ACM SIGKDD Int. conference on Knowledge discovery and data mining*, 69–78, 2004.
5. S.K. Chang and T.L. Kunii. Pictorial dataBase systems. *IEEE Computer*, 14:13–21, 1981.
6. S. Cheek, Y. Qi, S.S. Krishna, L.N. Kinch, and N.V. Grishin. SCOPmap: Automated assignment of protein structures to evolutionary superfamilies. *BMC Bioinformatics*, 5(1):197–197, 2004.
7. P.H. Chi, G. Scott, and C.R Shyu. A fast protein structure retrieval system using image-based distance matrices and multidimensional index. *Int. J. of Softw. Eng. and Know. Eng.*, 15(3),527–545, 2005.
8. M.H. Dunham. Data Mining: Introductory and Advanced Topics. Prentice Hall, New Jersey, USA, 164–192, 2003.
9. A. Godzik. The structural alignment between two proteins: Is there a unique answer?*Protein Sci.*, 5:1325–1338, 1996.
10. R.M. Haralick, K. Shanmugam, and I. Dinstein. Textural features for image classification. *IEEE Trans. on Syst., Man, and Cybernetics*, SMC-3:610–621, 1973.
11. T.F. Havel, I.D. Kuntz and G.M. Crippen. The theorey and practice of geometry. *Bull. Math. Biol.*, 45:665–720, 1983.
12. L. Holm and C. Sander. Protein structure comparison by alignment of distance matrices. *J. Mol. Biol.*, 233:123–138, 1993.
13. L. Holm and C. Sander. Mapping the protein universe. *Science*, 273:595–602, 1996.
14. M. Leslie. Protein Matchmaking. *Science*, 305:1381, 2004.
15. B. Liu, W. Hsu, Y. Ma. Integrating Classification and Association Rule Mining. *Proc. of the Fourth Int. Conference on Knowledge Discovery and Data Mining*, 80–86, 1998.
16. R. Kolodny and N. Linial. Approximate protein structural alignment in polynomial time. *Proc. Natl. Acad. Sci.*, DOI:10.1073/pnas.0404383101, 12201–12206, 2004.
17. A.G. Murzin, S.E. Brenner, T. Hubbard, and C. Chothia. SCOP:a structural classification of proteins database for the investigation of sequences and structures. *J. Mol. Biol.*,247:536–540,1995.
18. C.A. Orengo, F.M.G. Pearl, J.E. Bray, A.E. Todd, A.C. Martin, L. Lo Conte, and J.M. Thornton. The CATH Database provides insights into protein structure/function relationships. *Nucl. Acids. Res.*, 27(1):275–279, 1999.
19. N. Otsu. A threshold selection method from gray-level histogram. *IEEE Trans. on Syst., Man, and Cybernetics*, SMC-9:62–66, 1979.
20. A. Rosenfeld and A.C. Kak. Digital picture processing. Academic Press, New York, 1982.
21. A.W.M. Smeulders, M. Worring, S. Santini, A. Gupta and R. Jain. Content-based image retrieval at the end of the early years. *IEEE Trans. on Pattern and Machine Intell.*, 2:1349–1380,2000.
22. H.N. Shindyalov and P.E. Bourne. Protein structure alignment by incremental combinatorial extension (CE) of the optimal path. *Protein Eng.*, 9:739–747, 1998.
23. A.W.M. Smeulders, T.S. Huang, T. Gevers. Special issue on content-based image retrieval. *Int. J. Computer Vision*, 56:5–6, 2004.
24. C.R. Shyu, P.H. Chi, G. Scott, and D. Xu. ProteinDBS - A content-based retrieval system for protein structure databases. *Nucl. Acids. Res.*, 32:w572–575, 2004.
25. J.R Quinlan. C4.5 Programs for Machine Learning. *Morgan Kauffman*, 1993.
26. M.J. Zaki, S. Jin, C. Bystroff. Mining Residue Contacts in Proteins Using Local Structure Predictions. *IEEE Trans. on Syst., Man, and Cybernetics*, 33(5):789–801, 2003.
27. T.I. Zarembinski, L.W. Hung, H.J. Mueller-Dieckmann, K.K. Kim, H. Yokota, R. Kim and S.H. Kim. Structure-based assignment of the biochemical function of a hypothetical protein: A test case of structural genomics. *Proc. Natl. Sci. USA*, 95:15189–15193, 1998.

RECOMP: A PARSIMONY-BASED METHOD FOR DETECTING RECOMBINATION

DEREK RUTHS   LUAY NAKHLEH

Department of Computer Science, Rice University, Houston, Texas 77005, USA.
{druths,nakhleh}@cs.rice.edu

The central role phylogeny plays in biology and its pervasiveness in comparative genomics studies have led researchers to develop a plethora of methods for its accurate reconstruction. Most phylogeny reconstruction methods, though, assume a single tree underlying a given sequence alignment. While a good first approximation in many cases, a tree may not always model the evolutionary history of a set of organisms. When events such as interspecific recombination occur, different regions in the alignment may have different underlying trees. Accurate reconstruction of the evolutionary history of a set of sequences requires recombination detection, followed by separate analyses of the non-recombining regions. Besides aiding accurate phylogenetic analyses, detecting recombination helps in understanding one of the main mechanisms of bacterial genome diversification. In this paper, we introduce RECOMP, an accurate and fast method for detecting recombination events in a sequence alignment. The method slides a fixed-width window across the alignment and determines the presence of recombination events based on a combination of topology and parsimony score differences in neighboring windows. On several synthetic and biological datasets, our method performs much faster than existing tools with accuracy comparable to the best available method.

1. Introduction

Phylogeny, i.e., the evolutionary history of a set of organisms, plays a major role in representing and understanding relationships among the organisms. The rapidly-growing host of applications of comparative genomics has moved phylogeny to the forefront, rendering it an indispensable tool for analyzing and understanding the structure and function of genomes and genomic regions. Further, understanding evolutionary change and its mechanisms also bears direct impact on unraveling the genome structure and understanding phenotypic variations. One such mechanism of evolutionary change is *interspecific recombination*—the exchange of genetic material among different organisms across species boundaries.

Accurate detection of recombination is important for at least two major reasons. Studies have shown that the presence of recombination events has negative effects on the quality of the reconstructed phylogenetic tree.[9,12] Therefore, accurate reconstruction of the evolutionary history of a set of sequences that contains recombination events necessitates first detection of recombination events and then individual analyses of the non-recombined regions. Further, recombination plays a significant role in bacterial genome diversification. Whereas eukaryotes evolve mainly though lineal descent and mutations, bacteria obtain a large proportion of their genetic diversity through the acquisition of sequences from distantly related organisms, via horizontal gene transfer (HGT) or recombination.[6] Further, recombination is one of the processes by which bacteria develop resistance to antibiotics.[1,7]

In light of their effects on the accuracy of phylogenetic methods and their significance as a central evolutionary mechanism, developing accurate methods for detecting recombination is imperative. Many methods have been proposed for this problem (for example, Posada studied the performance of 14 different recombination detection methods[8]). Recombination detection methods fall into various categories, depending on the strategies they employ.[10] Among those categories, phylogeny-based detection methods are currently the most commonly used.[10] Recombination events result in different phylogenetic trees underlying different regions of the sequence alignment, and it is this observation that forms the basis for phylogeny-based recombination detection methods. The most recent methods include PLATO (Partial Likelihood Assessed through Tree Optimization),[2] DSS (Difference of Sum of Squares),[5] and PDM (Probabilistic Divergence Measure).[3,4] Central to all these methods is the idea of sliding a window along the alignment of sequences, fitting data in each window to a phylogeny, and comparing phylogenies in neighboring windows.

Ruths and Nakhleh addressed the limitations of these methods, and introduced preliminary measures for recombination detection.[12] In this paper, we extend our previous work by considering both the topologies of trees and their parsimony scores across adjacent windows of the alignment. We introduce a new phylogeny-based framework, RECOMP (RECOMbination detection using Parsimony), that uses parsimony-based tree reconstruction and evaluation, coupled with measurement of topological differences. We have implemented and studied the performance of four different measures (within the RECOMP framework) on synthetic as well as biological datasets. Our results show that RECOMP's accuracy is comparable to the most accurate existing methods, and is much faster.

The rest of the paper is organized as follows. In Section 2 we briefly describe interspecific recombination and review the most recent phylogeny-based methods for its detection. In Section 3, we describe our new method, RECOMP. We describe our experimental settings and results in Section 4, and conclude in Section 5 with final remarks and directions for future research.

2. Phylogeny-based Recombination Detection

Interspecific (or inter-species) recombination is a process by which genetic material is exchanged between different species lineages. When interspecific recombination events occur, different regions in the sequence alignment may have different underlying trees, as illustrated in Figures 1 and 2. The sequence alignment depicted in Figure 1 has three non-recombining regions I, II, and III, defined by a recombination event that involves the exchange of region II sequences between organisms B and D. The phylogenetic tree shown in Figure 2(a) models the evolutionary history of regions I and III of the alignment, whereas the phylogenetic tree in Figure 2(b) models the evolutionary history of region II of the alignment.

The scenario depicted in these two figures illustrates that recombination events may result in different phylogenetic trees underlying different regions; this phenomenon is the basis for phylogeny-based recombination detection methods. Three of the most recent and

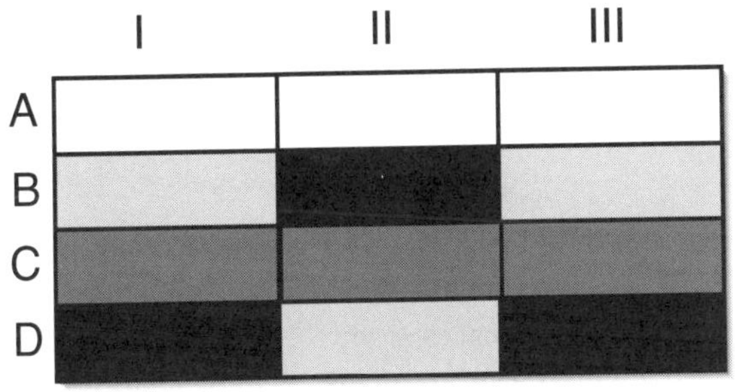

Figure 1. An alignment of four sequences whose evolutionary history contains a recombination event that involves the exchange of sequences in region II between organisms B and D.

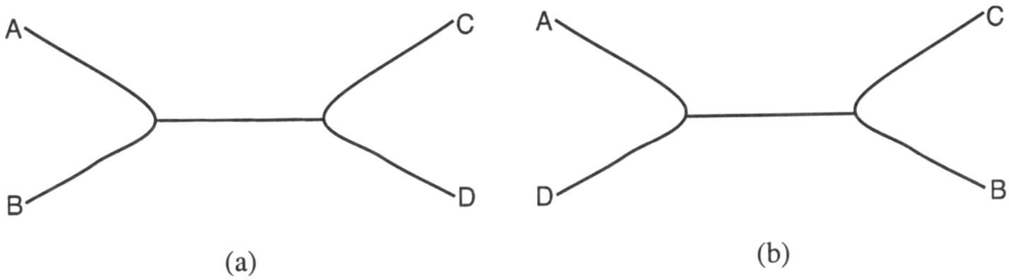

Figure 2. (a) The phylogenetic tree underlying regions I and III of the alignment in Figure 1. (b) The phylogenetic tree underlying region II of the alignment in Figure 1.

accurate phylogeny-based recombination detection methods are PLATO (Partial Likelihood Assessed through Tree Optimization),[2] DSS (Difference of Sum of Squares),[5] and PDM (Probabilistic Divergence Measure).[3,4] Central to all these methods is the idea of sliding a window along the alignment of sequences, fitting data in each window to a phylogeny, and comparing phylogenies in neighboring windows.

PLATO computes the likelihood of various regions of the sequence alignment from a single reference tree. The idea is that recombination regions will have a low likelihood score. The main problem with this approach is that the reference tree may be inaccurate since it is estimated from the whole sequence alignment.

DSS improves upon PLATO by sliding a window along the alignment, computing a tree on the first half of the window, and estimating the fit of the second half of the window to that tree (using a distance-based measure). The main problem with this approach is that it uses distance-based methods; such methods are inaccurate, especially given short sequences (which is the case when using DSS).

PDM addresses the shortcomings of DSS by (1) considering a likelihood approach for fitting the data to a tree, (2) using a distribution over trees, rather than a single tree (to capture the uncertainty of tree estimation from short sequences), and (3) comparing trees based on changes to their topologies. Later, Husmeier and Wright further improved the performance of PDM by incorporating sophisticated tree clustering techniques.[4] Since

PDM uses a probabilistic approach, it is very slow in practice. Further, since the tree space has very high dimensionality, clustering trees may be problematic.

3. RECOMP

Our proposed method is similar to PDM in principle, yet much simpler and faster, and comparable in accuracy. We slide a window of width w along the alignment, obtaining a set $\mathcal{T}_i$ of trees on $\mathcal{S}_i$, the set of sequences in the ith window, using a maximum parsimony heuristic (heuristic search with branch swapping, as implemented in PAUP*[13]), and comparing the sets $\mathcal{T}_i$ and $\mathcal{T}_{i+1}$ of trees. The MP heuristic we use returns a set of trees, sorted by their parsimony scores: some trees may have an identical parsimony score. We denote the set of *all* j^{th} ($j = 1, 2, \ldots$) best parsimony trees (with respect to their scores, sometimes called the jth *level*) by LVL^j, and the set of trees in the top k levels by $OPT(k)$ ($k \geq 1$), formally the set $\cup_{1 \leq \ell \leq k} LVL^{\ell}$. In the experimental study of our method, we considered $\mathcal{T}_i = OPT(k)$, and studied the performance of the method as a function of the k value (we used $k = 1, 2, 3, 4$).

Let $\mathcal{T}$ be a set of trees. We define the *center* of the set, $c(\mathcal{T})$, to be the strict consensus[a] of all trees in the set, and the *radius*, $r(\mathcal{T}) = \max\{RF(c(\mathcal{T}), T) : T \in \mathcal{T}\}$, where RF denotes the Robinson-Foulds distance between a pair of trees.[11] Further, we define $d_{min}(T, \mathcal{T}) = \min\{RF(T, T') : T' \in \mathcal{T}\}$ and $d_{max}(T, \mathcal{T}) = \max\{RF(T, T') : T' \in \mathcal{T}\}$. We write $P(S, T)$ to denote the parsimony score of tree T leaf-labeled by set S of sequences, and $P(S, \mathcal{T})$ to denote $\min_{T \in \mathcal{T}} P(S, T)$. We investigated four functions for comparing the sequences in two adjacent windows W_i and W_{i+1}:

- **Intersection**(W_i,W_{i+1}) = $\frac{|\{T : T \in \mathcal{T}_{i+1} \text{ and } RF(T, c(\mathcal{T}_i)) \leq r(\mathcal{T}_i)\}|}{|\mathcal{T}_{i+1}|} + \frac{|\{T : T \in \mathcal{T}_i \text{ and } RF(T, c(\mathcal{T}_{i+1})) \leq r(\mathcal{T}_{i+1})\}|}{|\mathcal{T}_i|}$.
- **AvgMin**(W_i,W_{i+1}) = $\frac{\sum_{T \in \mathcal{T}_i} d_{min}(T, \mathcal{T}_{i+1})}{|\mathcal{T}_i|} + \frac{\sum_{T \in \mathcal{T}_{i+1}} d_{min}(T, \mathcal{T}_i)}{|\mathcal{T}_{i+1}|}$.
- **AvgMax**(W_i,W_{i+1}) = $\frac{\sum_{T \in \mathcal{T}_i} d_{max}(T, \mathcal{T}_{i+1})}{|\mathcal{T}_i|} + \frac{\sum_{T \in \mathcal{T}_{i+1}} d_{max}(T, \mathcal{T}_i)}{|\mathcal{T}_{i+1}|}$.
- **ParsDiff**(W_i,W_{i+1}) = $|P(\mathcal{S}_{i+1}, \mathcal{T}_{i+1}) - P(\mathcal{S}_i, \mathcal{T}_{i+1})|$.

Further, we normalized the values computed by each of the four functions as follows. Let m and n be the minimum and maximum values, respectively, obtained by a function across all windows for a given sequence alignment. We normalize each value x computed by the function on the alignment by

$$\frac{x - m}{n - m}.$$

Therefore, the four functions return values in the range $[0, 1]$. The rationale behind the functions is as follows. Given an alignment of sequences, each of length L, let i be a site falling at a recombination breakpoint. Further, assume that the window we consider is of

[a]The strict consensus of a set of trees is the maximally resolved tree (i.e., the tree that has a maximum number of edges) in which every edge is also an edge of every tree in the set.

width w. Then, the tree T on which sites $(i-w)\ldots(i-1)$ is different from tree T' on which sites $i\ldots i+(w-1)$ evolved. Due to the inaccuracy of phylogeny reconstruction methods, and the potential errors in evolutionary assumptions made, T and T' may be unattainable; hence the need for considering sets of trees, rather than a single tree. When sets $\mathcal{T}_i$ and $\mathcal{T}_{i+1}$ correspond to sequence regions that fall on different sides of a recombination breakpoint, we expect the trees to differ between the two sets, which implies a lower **Intersection** value, and higher **AvgMin**, **AvgMax**, and **ParDiff** values. When the two sets of tree correspond to sequence regions that fall on the same side of any recombination event, we expect a higher **Intersection** value, and lower **AvgMin**, **AvgMax**, and **ParDiff** values. For consistency purposes, we always report $1-I$, where I is the value computed by the comparison function.

The outline of the RECOMP method is as follows:

RECOMP(S,w,t**)**

> **for** $i=0$ **to** $L-w$
>
> > $X_i = f(W_i, W_{i+1})$;
> >
> > $i = i + t$;
>
> **Plot** X.

The sequence alignment is denoted by S, the window size by w, and the step size by t. The parameter L denotes the length of the sequences in S, f can be any of the aforementioned four functions, and W_i denotes the sequence alignment in window i. The output of RECOMP is a graphical representation of the output of the functions. Choosing a threshold that distinguishes the recombination sites can be determined by inspecting the graphical output of RECOMP (as is the case with all phylogeny-based methods that have graphical output). Further, such a threshold can be automatically computed by a careful training of the method on datasets with characteristics similar to those of the dataset under investigation.

4. Empirical Performance

4.1. *Data*

To test our method, we applied it to the three synthetic and one biological datasets used in another paper.[4] For the three synthetic datasets $SD1$, $SD2$, and $SD3$, the evolution of three DNA sequence alignments, each of 5500 nucleotides, was simulated down trees with 8 leaves. Each of the two datasets $SD1$ and $SD2$ contains two recombination events: an ancient event affecting the region between sites 1000 and 1500, and a recent event affecting the region between sites 2500 and 3000. Further, they both contain a mutational hot spot between sites 4000 and 4500 (sites were evolved at an increased nucleotide substitution rate) to test whether the detection method can successfully distinguish between recombination and rate variation. The average branch lengths of the phylogenetic trees underlying datasets $SD1$ and $SD2$ are 0.1 and 0.01, respectively. The third synthetic dataset, $SD3$, contains two recombination events: an ancient event affecting the region between sites 1000 and

2000, and a recent recombination event between sites 3000 and 4000. The branch lengths of the phylogenetic tree underlying dataset $SD3$ were drawn from a uniform distribution on the interval $[0.003, 0.005]$.

The biological dataset, HD, consists of 10 Hepatitis B Virus sequences each of 3049 nucleotides, with evidence for recombination events (the dataset contained two recombinant strains and eight nonrecombinant strains). For more details on the datasets, the reader is referred to the original paper.[4]

4.2. *Results*

We ran RECOMP with all four functions on the four datasets. We considered four different values of k (1,2,3, and 4) for sets $OPT(k)$ of trees, two window sizes 300 and 500, and step size 100. We describe our results of the four functions on all datasets when using $OPT(3)$ for window size 500, which produced the best results among all parameter settings. These results are shown in Figures 3–5 for the three synthetic datasets, and in Figure 6 for the biological dataset.

In the case of the $SD1$ dataset, our method detected the four recombination breakpoints (at sites 1000, 1500, 2500, and 3000) based on all four functions (Figure 3). There are clear threshold values that could be used as cutoff values between recombination/non-recombination regions: 0.8 for the **Intersection** function, 0.4 for the **AvgMin** and **AvgMax** functions, and 0.1 for the **ParsDiff** function. Clearly, the signal for a recombination breakpoint at sites 2500 and 3000 is stronger than that at sites 1000 and 1500. The reason for this is that the recombination event involving the region between sites 2500 and 3000 occurred between more distantly related taxa, which results in larger topological differences and parsimony score differences among trees across recombination breakpoints. Observe that the **ParsDiff** function is very robust, in this case, to the mutational hotspots: it correctly predicts no recombination in the mutational hotspot region between sites 4000 and 4500. The **Intersection** function has the strongest signal of recombination at all four recombination breakpoints (sites 1000, 1500, 2500, and 3000); however, the function is sensitive to mutational hotspots, and exhibits large fluctuations.

Similar behavior was obtained by the four functions on the dataset $SD2$ (Figure 4). However, in the case of this dataset, the **AvgMin** and **AvgMax** functions showed a weak signal for the ancient recombination event between sites 1000 and 1500. The **Intersection** and **ParsDiff** function still showed clear signal for recombination at all four breakpoints. Once again, the **ParsDiff** outperformed all other three functions in robustness with respect to mutational hotspots. The $SD2$ dataset was evolved with a lower rate of evolution than that of $SD1$ and hence was harder to analyze (which is the case for the other existing methods[4]).

The $SD3$ dataset was evolved down the tree with the lowest rate of evolution among all three synthetic datasets, and hence was the hardest for the methods to analyze (which is the case for the other existing methods[4]). As with the other two datasets, detecting the recent recombination event is easier, which is shown in the performance of all four functions in Figure 5. In particular, all four functions had a weak signal of recombination at site 2000.

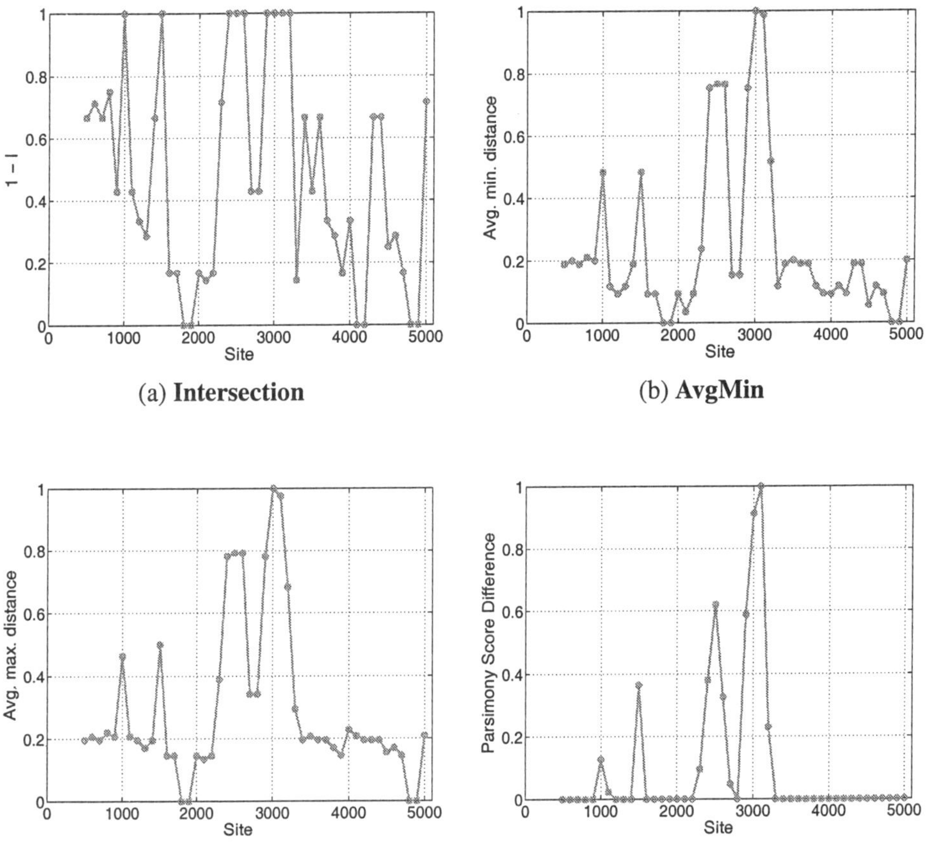

(a) **Intersection** (b) **AvgMin**

(c) **AvgMax** (d) **ParsDiff**

Figure 3. Results of the four functions on $SD1$.

Yet again, most of the sites in this alignment were synonymous, which made it hard for all methods to detect recombination.

On the Hepatitis B dataset, both the DSS and PDM methods detected three breakpoints around sites 600, 1700, and 2200. Our method shows peaks at these three points, based upon the four functions we used (Figure 6). Nevertheless, the **Intersection** and **ParsDiff** functions gave the clearest signal among the two.

The performance of PLATO, DSS, and PDM on the same datasets is provided by Husmeier and Wright.[4] The performance of our method is comparable to that of PDM, which performed best among those three methods. Further, since our method uses a fast MP heuristic, calculates parsimony scores (which is computable in polynomial time), and computes simple functions, it is much faster (orders of magnitude) than PDM, which uses compute-intensive Bayesian analysis techniques.

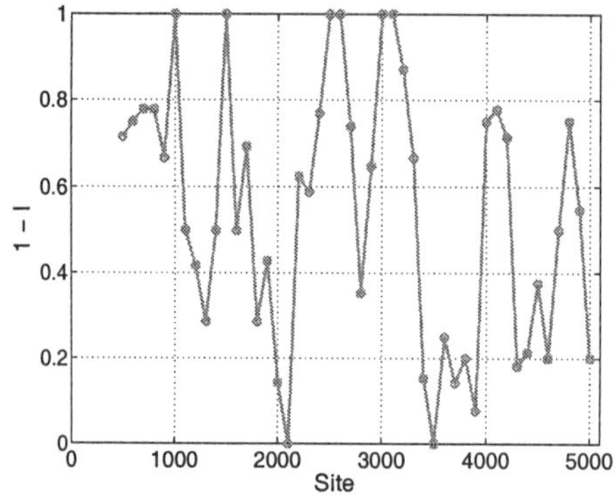

(a) **Intersection**

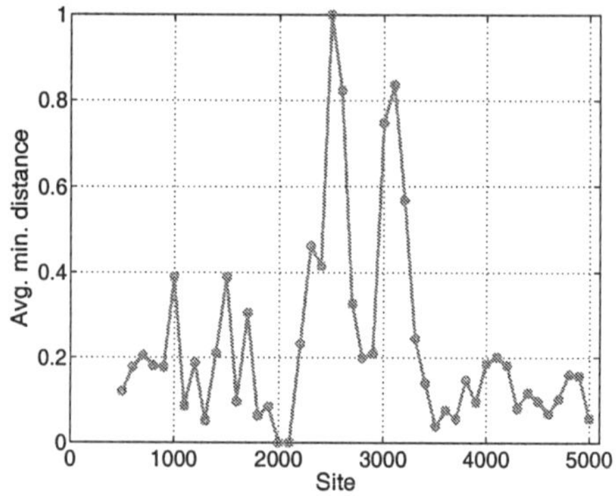

(b) **AvgMin**

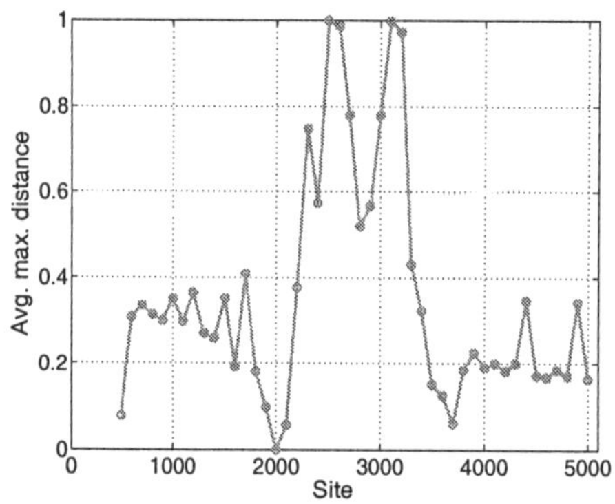

(c) **AvgMax**

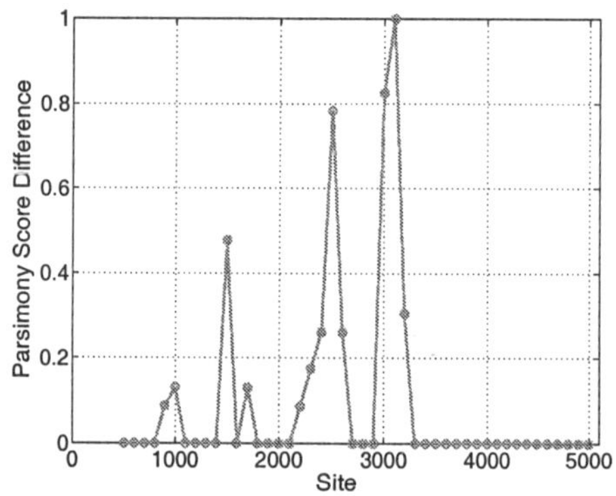

(d) **ParsDiff**

Figure 4. Results of the four functions on $SD2$.

5. Conclusions and Future Work

In this paper, we introduced a simple, effective and fast parsimony-based method for detecting recombination. In experimental studies involving both synthetic and biological datasets, our method produced very good results—comparable to those of the best known methods (and ran orders of magnitude much faster). Our future work includes exploring ways to improve the performance of our method in the presence of mutational hot spots. Further, we are interested in devising methods for detecting the locations of the recombination events on the organismal tree. An open-source, stand-alone implementation of RECOMP is currently available for download and use. It is implemented in the Sequoia software suite as both a command-line tool as well as a Java library which allows its incorporation into larger programs.

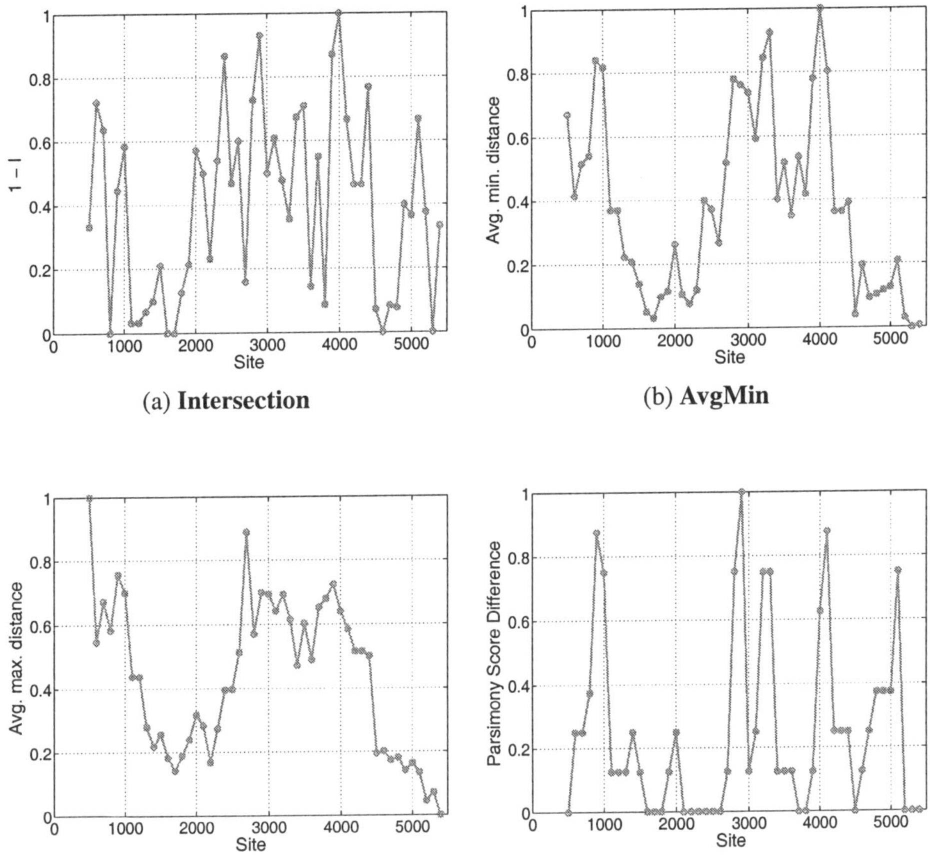

(a) **Intersection** (b) **AvgMin**

(c) **AvgMax** (d) **ParsDiff**

Figure 5. Results of the four functions on $SD3$.

References

1. M.C. Enright, D.A. Robinson, G. Randle, E.J. Feil, H. Grundmann, and B.G. Spratt. The evolutionary history of methicillin-resistant Staphylococcus aureus (MRSA). *Proc. Natl. Acad. Sci. USA*, 99(11):7687–7692, 2002.
2. N.C. Grassly and E.C. Holmes. A likelihood method for the detection of selection and recombination using nucleotide sequences. *Molecular Biology and Evolution*, 14:239–247, 1997.
3. D. Husmeier and F. Wright. Probabilistic divergence measures for detecting interspecies recombination. *Bioinformatics*, 17:S123–S131, 2001.
4. D. Husmeier and F. Wright. Detecting interspecific recombination with a pruned probabilistic divergenec measure. *Unpublished manuscript*, 2004.
5. G. McGuire, F. Wright, and M.J. Prentice. A graphical method for detecting recombination in phylogenetic data sets. *Molecular Biology and Evolution*, 14:1125–1131, 1997.
6. H. Ochman, J.G. Lawrence, and E.A. Groisman. Lateral gene transfer and the nature of bacterial innovation. *Nature*, 405(6784):299–304, 2000.
7. I.T. Paulsen *et al.* Role of mobile DNA in the evolution of Vacomycin-resistant Enterococcus

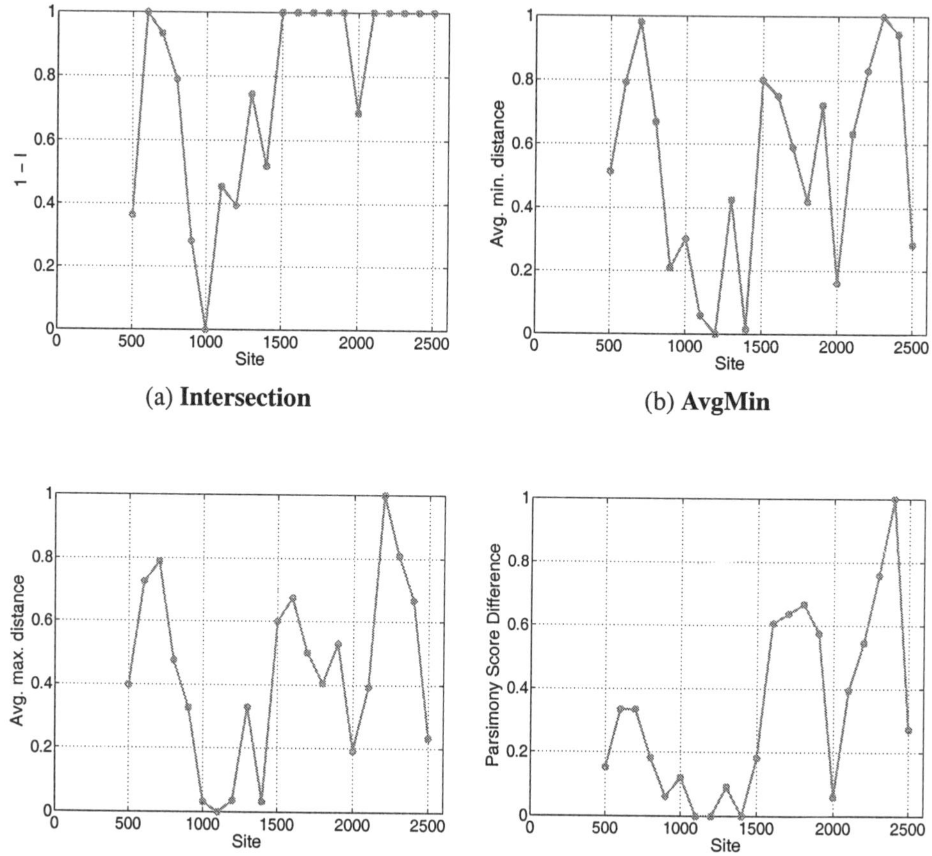

(a) **Intersection** (b) **AvgMin**

(c) **AvgMax** (d) **ParsDiff**

Figure 6. Results of the four functions on HD.

faecalis. *Science*, 299(5615):2071–2074, 2003.

8. D. Posada. Evaluation of methods for detecting recombination from DNA sequences: empirical data. *Molecular biology and evolution*, 19:708–717, 2002.
9. D. Posada and K.A. Crandall. The effect of recombination on the accuracy of phylogeny estimation. *Journal of Molecular Evolution*, 54:396–402, 2002.
10. D. Posada, K.A. Crandall, and E.C. Holmes. Recombination in evolutionary genomics. *Annual Review of Genetics*, 36:75–97, 2002.
11. D. Robinson and L. Foulds. Comparison of phylogenetic trees. *Mathematical Biosciences*, 53:131–147, 1981.
12. D. Ruths and L. Nakhleh. Recombination and Phylogeny: Effects and Detection. *The International Journal of Bioinformatics Research and Applications*, In press, 2005.
13. D. L. Swofford. PAUP*: Phylogenetic analysis using parsimony (and other methods), 1996. Sinauer Associates, Underland, Massachusetts, Version 4.0.

ALIGNSCOPE : A VISUAL MINING TOOL FOR GENE TEAM FINDING WITH WHOLE GENOME ALIGNMENT

HEE-JEONG JIN,[1] HYE-JUNG KIM,[1] JEONG-HYEON CHOI[2] AND HWAN-GUE CHO[1]

[1]*Dept.of Computer Science and Engineering, Pusan National University, South Korea*
E-mail : {hjjin,hjkim,hgcho}@pusan.ac.kr
[2]*School of Informatics, Indiana University, Bloomington, IN 47404, USA*
E-mail : jeochoi@indiana.edu

One of the main issues in comparative genomics is the study of chromosomal gene order in one or more related species. Recently identifying sets of *orthologous* genes in several genomes has become getting important, since a cluster of similar genes helps us to predict the function of unknown genes. For this purpose, the whole genome alignment is usually used to determine horizontal gene transfer, gene duplication, and gene loss between two related genomes. Also it is well known that a novel visualization tool of the whole genome alignment would be very useful for understanding genome organization and the evolutionary process. In this paper, we propose a method for identifying and visualizing the alignment of the whole genome alignment, especially for detecting *gene clusters* between two aligned genomes. Since the current rigorous algorithm for finding gene clusters has strong and artificial constraints, they are not useful for coping with "noisy" alignments. We developed the system *AlignScope* to provide a simplified structure for genome alignment at any level, and also to help us to find gene clusters. In this experiment, we have tested AlignScope on several microbial genomes.

1. Introduction

Alignment is a procedure that compares two or more sequences by searching for a series of individual characters or character patterns that are in the same order in the sequences. This procedure assists in designating the functions of unknown proteins, determining the relatedness of organisms, and identifying structurally and functionally important elements and other useful functions.[9,12] Many widely divergent organisms are descended from a common ancestor through a process called evolution. The inheritance patterns and diversities of these organisms have significant information regarding the nature of small and large-scale evolutionary events.

The complexity and the size of the genome make it difficult to analyze. Because the large amount of biological noises is present when visualizing genomes, it is not enough to simply draw the aligned pairs of various genomes. Therefore an alignment visualization tool needs to provide a method for viewing the global structure of whole genome alignment in a simplified form at any level of detail. Figure-1 clearly illustrates this problem. In Figure-1, the resolution of the snapshot is 800 by 600 pixels, so one pixel corresponds about 6000 bases of a given genome sequence.

Currently there are several systems for visualizing the alignment of genomes. The NCBI Map Viewer[14] provides graphical displays of biological features on NCBI's assembly of human genomic sequence data. GeneViTo[5] is a JAVA-based computer application that serves as a workbench for genome-wide analysis through visual interaction. GenomePixelizer[1] generates custom images of the physical or genetic positions of specified sets of genes in whole genomes or parts of genomes. This method assists in comparing of specified genes between genomes. However, the available genome viewers do not conveniently display the global structure between two genomes since these viewers only draw sets of alignment or gene pairs (Figure-1) and genome alignment has a lot of aligned pairs. So, we propose a novel visualization system to establish meaningful features, especially *gene clusters*, in whole genome alignment using "zoned hierarchical clustering" method. Zoned hierarchical clustering method clusters all alignment pairs to illustrate a simplified view.

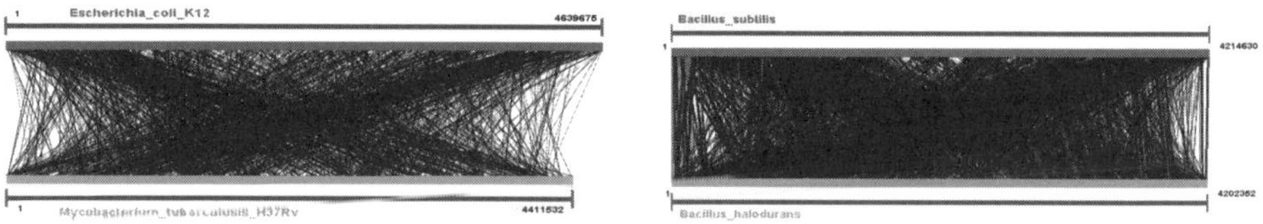

(a) (b)

Figure 1. A snapshot of whole genome alignment using AlignScope. The snapshot (a) shows the alignment between *E. coli K12 (4.6Mbp)* and *M. tuberculosis H37Rv (4.4Mbp)*. The snapshot (b) shows the alignment between *Bacillus subtilis (4.2Mbp)* and *Bacillus halodurans (4.2Mbp)*.

A fundamental question in genomics is the relationship between chromosomal gene order and function. Current evidence suggests that genes are not randomly distributed on the chromosomes and that genes which are physically close to each other tend to represent groups of genes with a functional relationship even if they are not contiguous. These groups of genes are called "gene clusters" or "gene teams" in two or more genomes.[3,6,7] Identifying conserved gene clusters is important for many biological problems, such as genome comparative mapping, studying transcriptional neighborhoods, predicting gene functions and other problems. For this purpose, Steffen[7] and Takeaki[13] proposed that a common interval in a sequence be used, but in practice, they also considered assumptions. Corteel[3] introduced the concept of a "gene team", which is a set of orthologous genes that appear in two or more species, possibly in a different order yet with the distance between the genes on the team for each chromosome is always within a certain threshold. For this model to function properly each gene can have at most one orthologous partner in the other chromosome. Xin[6] removes this constraint in the original model, and thus allows the analysis of complex prokaryotic or eukaryotic genomes with extensive paralogs. Kim[8] searches for gene clusters with and/or without physical proximity constraint. Other work identifies functional modules from genomic association of genes using protein interaction networks.[11]

Currently the rigorous algorithms for finding gene clusters require strong and artificial constraints. For example, distances between adjacent genes within a cluster are smaller than given threshold. Since they allow a few "noise" genes which don't included in a cluster, they are not useful for coping with the "noisy" genome that occurs in practice. Our method enables not only to provide a simplified structure of genome alignment but also to assist in detecting of gene clusters.

2. Clustering for Gene Pairs

Since the previous visualization tools for genome alignment mainly consider genetic information such as ORF and gene prediction and annotation, they are not useful in the study of the relationships.Our AlignScope visualizes relationships at any simplified level and also predicts gene clusters using a zoned hierarchical clustering algorithm.

2.1. *Preliminary*

In this paper, we only consider a pairwise whole genome alignment. Many local alignment tools such as BLAST, FASTA, MUMer and GAME can be used to obtain the alignment. Without loss of generality, we denote an upper genome by U and a lower genome by L.

- $u_i = [v_i,\ w_i]$ ($l_j = [p_j, q_j]$) is a subsequence of U (L) where v_i and w_i are the start and end positions of u_i at U and p_j and q_j are those of l_j at L.
- A geometry center $M(u_i) = \frac{v_i+w_i}{2}$ and $M(l_j) = \frac{p_j+q_j}{2}$.
- An alignment pair $a_i = (u_i, l_j)$ where u_i is the opponent of l_j in terms of an aligned pair.
- The geometry center $M(a_i) = (M(u_i), M(l_j))$ where $a_i = (u_i, l_j)$.
- The distance $\Delta(a_i, a_j)$ between two alignment pairs, $a_i = (u_{i_1}, l_{i_2})$ and $a_j = (u_{j_1}, l_{j_2})$ is defined by

$$\begin{aligned}\Delta(a_i, a_j) &= \Delta(M(a_i), M(a_j)) = |M(a_i) - M(a_j)| \\ &= |M(u_{i_1}) - M(u_{j_1})| + |M(l_{i_2}) - M(l_{j_2})| = \frac{|v_{i_1} - v_{j_1} + w_{i_1} - w_{j_1}| + |p_{i_2} - p_{j_2} + q_{i_2} - q_{j_2}|}{2}\end{aligned}$$

- A cluster $c_x = \{a_i \mid \forall a_p, a_q \in c_x \text{ and } \forall a_k \notin c_x, \Delta(a_p, a_q) < \Delta(a_p,\ a_k) \text{ and } \Delta(a_p,\ a_q) < \Delta(a_q,\ a_k)\}$.
- $|c_x|$ denotes the cardinality of a cluster c_x.
- An interval $I(c_x)$ = ($[v_x,\ w_x]$, $[p_x,\ q_x]$) where v_x and w_x are the minimum and maximum positions of c_x at U, and p_x and q_x are those of c_x at L.
- A distance between two clusters, c_x and c_y,

$$\Lambda(c_x,\ c_y) = \frac{\sum_{i,j} \Delta(\ a_i,\ a_j)}{(|c_x| + |c_y|)},\quad \text{for } a_i \in c_x, a_j \in c_y.$$

Figure-2 shows an example of our notations.

- In genome U, $u_1 = [1,8]$, $u_2 = [11,15]$, $u_3 = [21,26]$.
- In genome L, $l_1 = [3,7]$, $l_2 = [14,19]$, $l_3 = [23,30]$.
- $a_1 = (u_1, l_3)$, $a_2 = (u_2, l_1)$, $a_3 = (u_3, l_2)$.
- $I(a_1) = ([1,8],[23,30])$, $I(a_2) = ([11,15],[3,7])$, $I(a_3) = ([21,26],[14,19])$

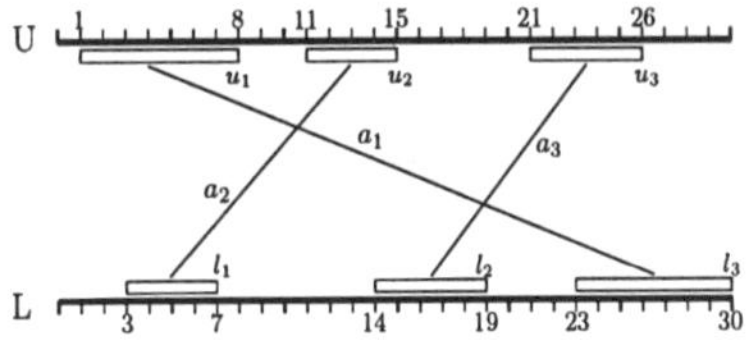

Figure 2. The plot shows an example of genome alignment between two genomes U and L. In genome U, $u_1 = [1,8]$, $u_2 = [11,15]$, $u_3 = [21,26]$. In genome L, $l_1 = [3,7]$, $l_2 = [14,19]$, $l_3 = [23,30]$.

2.2. *Zoned Hierarchical Clustering*

Hierarchical clustering is a statistical method for finding relatively homogeneous clusters and determines clusters of similar data points in multi-dimensional spaces based on empirical data and a certain kind of distance measure.[10] It starts with each case in a separate cluster and then combines the clusters sequentially, reducing the number of clusters at each step until only one cluster is left. When there are N input data, this involves $N-1$ clustering steps, or fusions.

Now we will cluster all alignment pairs to illustrate a simplified view from the whole genome alignment in order to construct a simplified structure of alignment pairs. We use the concept of hierarchical clustering modified, "*zoned hierarchical clustering*", for clustering of alignment pairs. There are 6 stages for zoned hierarchical clustering:

(1) Sort all alignment pairs by their geometric center on the basis of upper genome.
(2) Assign each alignment pair to a base cluster, i.e., $c_x = \{a_x\}$. If we have n alignment pairs initially, we will have n clusters at the start.
(3) For each cluster, we must compute local "effecting zone" which is dependent on the structure of the on-going cluster. (This procedure will be explained later.)
(4) Find the nearest pair of clusters using $\Delta(c_x,\ c_y)$. In this procedure, we consider only "effecting zone" of each cluster for search area. This will prevent searching all candidate clusters.
(5) Merge the pair of the nearest clusters. This is the basic procedure of hierarchical clustering and procedure produced the clustering tree.
(6) Repeat steps 3 and 5 until all clusters are merged.

Effecting Zone of each cluster

Typical hierarchical clustering techniques find the nearest pair of each pair in the alignment pairs. In the case of single-link clustering, this process takes very long time, $O(N^2)$. For fast clustering, we need to consider a small local zone to find the nearest clusters. So our method takes $O(k \cdot N)$, where the number of data in a local zone is k.

2.3. *Two Density Measures for Level-of-Detail and Detection of Gene Clusters*

The zoned hierarchical clustering process generates a tree or a dendrogram, where each step in the clustering process is illustrated by merging two clusters. Each internal node in clustering tree represents a cluster. We define two *density* measures for each cluster to view the simplified structure of whole genome alignment as level of detail. One is *"area density* (σ_a)*"* and the other is *"line density* (σ_l)*"*. Area density is the number of alignment pairs divided by the sum of the two interval distances at U and L of the geometric polygon (in fact a trapezoid) that covers alignment pairs in c_x (Figure-4). Line density is the number of *properly* contained alignments divided by all alignments on either side of c_x in U and L. The formal definition of σ_a and σ_l is given in the following:

- Given $I(c_x) = ([a,b],[c,d])$, $\sigma_a(c_x) = |c_x|/((b-a)+(d-c)+2)$.
- Let the number of alignment pairs contained in $I(c_x)$ be t, then $\sigma_l(c_x) = 2*|c_x|/t$.

In Figure-4, $\sigma_a(c_x)$ is $4/((U_e-U_s)+(L_e-L_s)+2)$ and $\sigma_l(c_x)$ is 8/12. AlignScope can visualize at any simplification level to select clusters with a certain σ_a or σ_l in a tree. And since σ_l means a noise rate of a cluster, we can detect gene clusters according to the noise rate.

2.4. *Spliced Gene Clusters*

In addition to providing a simplified structure for detecting gene clusters, AlignScope can find gene clusters. Figure-3 shows the clusters detected by our method and common interval method. Note that several alignment pairs have been mixed at *Genome_1* while most of them are pure at *Genome_2*. The clusters consist of these alignment pairs, they shall be called "*spliced clusters*". The spliced clusters are determined by zoned hierarchical clustering. We speculate that the spliced cluster between the genomes is related to evolution.

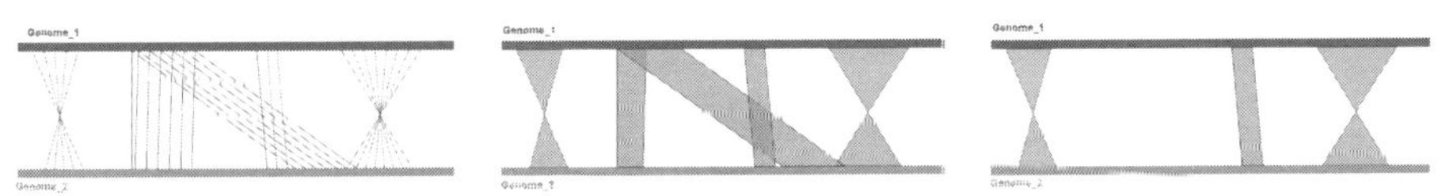

(a) (b) (c)

Figure 3. Comparison of zoned hierarchical clustering and common interval method. The plot (a) shows test data for alignment pairs. The plot (b) shows clusters obtained by zoned hierarchical clustering. The plot (c) shows clusters found by common interval method.

3. Visualization of Gene Pairs

Data Filtering

AlignScope is able to display the features of the whole genome in real-time. Since it is sometimes hard to understand genes in specific region or with certain alignment score, AlignScope supports various filtering options to analyze these data. It supports the following filtering functions:

- Filtering by alignment score: AlignScope uses a large number of alignment pairs from alignment programs such as BLAST[2] and GAME[4] with specific alignment score.
- Filtering by physical position: Since images drawn by AlignScope fit typical computer monitors without scrolling due to the large size of genome, it is difficult to view information on individual genes or alignment pairs. So, AlignScope extracts data by selected regions of interest.
- Filtering by $|c_x|$s within a cluster: Zoned hierarchical clustering generates clusters with a various $|c_i|$. AlignScope can extract clusters by a $|c_i|$.

Shape and Color of Alignment pair

There are two kinds of gene clusters, *parallel* clusters, and *reverse* clusters which mean that the order of the aligned genes is reversed. The colors and shapes of alignment pairs represent the degree that they are parallel or reverse. Let $Color(c_x)$ be a color assigned to cluster c_x, N be the number of crossing alignment pairs in c_x, and $|c_x| = M$. In Figure-4, the total number of crossing alignment pairs is 8. And two shapes, which are sand clock and rectangle (Figure-3), are used to represent clusters according to the number of crossing alignment pairs in c_x. If the number of crossing alignment pairs in c_x is larger than $\frac{M(M-1)}{4}$, the shape of the cluster is a sand clock, and in other case, it is a rectangle.

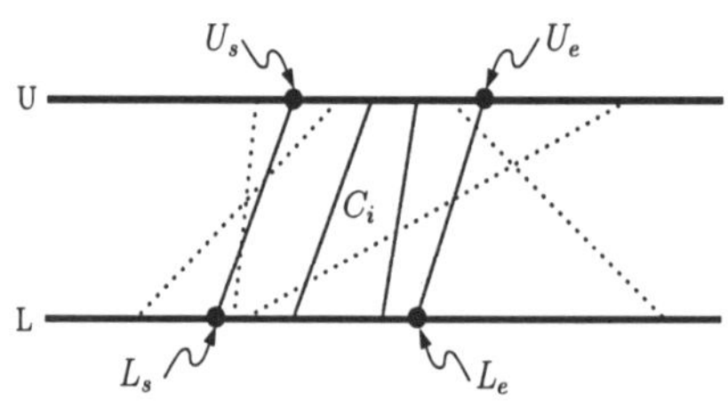

Figure 4. The plot shows an example of a cluster c_x where all solid lines are aligned pairs and contained in c_x. The line density $\sigma_l(c_x) = 8/12$, since $|c_x| = 4$ and the number of aligned pairs contained in $I(c_x)$ at U and L, respectively, is 6.

- $Color(c_x) = rgb(0,\ 255,\ 0)$, if $N = 0$.
- $Color(c_x) = rgb(255,\ 0,\ 0)$, if $N = M(M-1)/2$.
- $Color(c_x) = rgb(255 \cdot 2k/M(M-1),\ 255 \cdot (1-2k/M(M-1),\ 0)$, for any.

4. Experiment Result

We have tested the performance of AlignScope on two data sets. AlignScope uses a large number of alignment pairs from alignment programs such as BLAST and GAME, or gene pairs between two genomes. The first data set is a group of three prokaryote genomes, *A. fulgidus*, *P. abyssi*, and *M. thermautotrophicus*. We generated by COGs database alignment pairs from *A. fulgidus* vs. *P. abyssi* and *A. fulgidus* vs. *M. thermautotrophicus*, respectively, i.e., an alignment pair is made if a pair of genes in each genome is belonged to the same

Table 1. The longest cluster produced by AlignScope with σ_l = 0.9 for a pair of *A. fulgidus* vs. *M. thermautotrophicus*.

Index	Our	Kim[8]	COG	gene of A. fulgidus	gene of M. thermautotrophicus	description
1	O	O	COG0201	secY	secY	protein translocase, subunit SEC61 alpha (secY)
.	.	.	.	.	.	.
16	O	O	COG1588	-	-	RNAse P protein subunit P29
17	O	X	COG0092	rps3P	rps3P	SSU ribosomal protein S3P (rps3P)
18	O	X	COG0091	rpl22P	rpl22P	LSU ribosomal protein L22P (rpl22P)
19	O	X	COG0185	rps19P	rps19P	SSU ribosomal protein S19P (rps19P)
20	O	X	COG0090	rpl2P	rpl2P	LSU ribosomal protein L2P (rpl2P)
21	O	X	COG0089	rpl23P	rpl23P	LSU ribosomal protein L23P (rpl23P)
22	O	X	COG0088	rpl4P	rpl4P	LSU ribosomal protein L4P (rpl4P)
23	O	X	COG0087	rpl3P	rpl3P	LSU ribosomal protein L3P (rpl3P)

COG. Figure-5 shows an example of the steps to detect gene cluster using AlignScope. Table-1 shows the longest cluster produced by AlignScope with $\sigma_l \geq 0.9$. While Kim et al.[8] detected 16 genes, AlignScope detected 23 genes. As the result of comparing

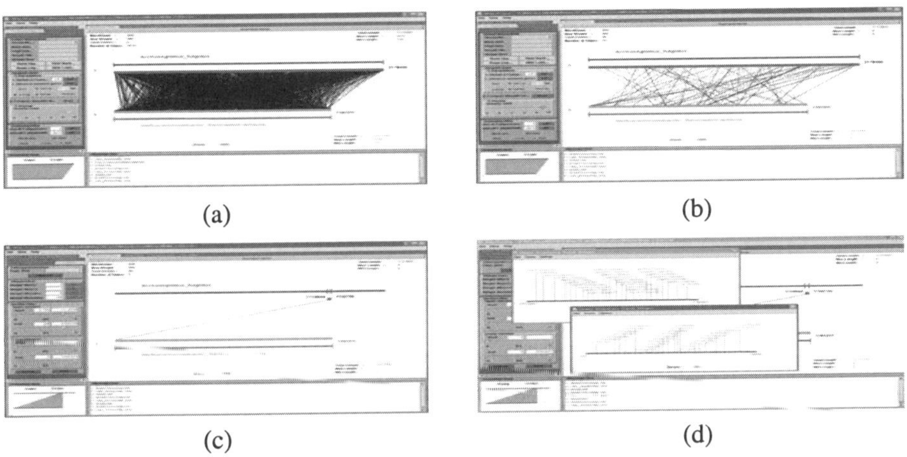

Figure 5. Example of the steps to detect a gene cluster which is shown at Table-1 using AlignScope. The plot (a) shows a snapshot of all alignment pairs of *A. fulgidus* and *M. thermautotrophicus*. The plot (b) shows a snapshot of detected gene clusters with $\sigma_l \geq 0.9$. The plot (c) shows a snapshot after filtering by physical position [1710044, 1742720] at *A. fulgidus*. The plot (d) shows a snapshot of genes information in a cluster of (c).

A. fulgidus with *M. thermautotrophicus*, the longest cluster found by AlignScope has the same genes except COG0255 as shown in Table 2. It is interesting that all genes except COG0255 in *A. fulgidus* are parallel with those in *M. thermautotrophicus*, but all genes in *A. fulgidus* cross completely to those in *P. abyssi*. Note that AlignScope represents the rate of crossings of genes in a cluster to its color and shape (See Section 3).

The second data-set is a pair of *E.coli K12* vs. *B.subtilis*. Alignment pairs are made in the case that a pair of genes in each genome has the same gene name. AlignScope enables to adjust the simplified level of clusters (See Section 2.4). Figure-6 shows global structures

Table 2. Example of a conserved gene cluster within three archaebacteria, *A. fulgidus*, *M. thermautotrophicus* and *P. abyssi*.

	1	2	3	4	5	6	7	8	9
A. fulgidus	COG0201	COG0200	COG1841	COG0098	COG0256	COG2147	COG1717	COG0097	COG0096
M. thermautotrophicus	COG0201	COG0200	COG1841	COG0098	COG0256	COG2147	COG1717	COG0097	COG0096
P. abyssi	COG0087	COG0088	COG0089	COG0090	COG0185	COG0091	COG0092	COG0255	COG1588
	10	11	12	13	14	15	16	17	18
A. fulgidus	COG0199	COG0094	COG1471	COG0198	COG0093	COG0186	COG1588	COG0255	COG0092
M. thermautotrophicus	COG0199	COG0094	COG1471	COG0198	COG0093	COG0186	COG1588		COG0092
P. abyssi	COG0186	COG0093	COG0198	COG1471	COG0194	COG0199	COG0096	COG0097	COG1717
	19	20	21	22	23	24			
A. fulgidus	COG0091	COG0185	COG0090	COG0089	COG0088	COG0087			
M. thermautotrophicus	COG0091	COG0185	COG0090	COG0089	COG0088	COG0087			
P. abyssi	COG2147	COG0256	COG0098	COG1841	COG0200	COG0201			

of clusters at several simplified level, $Sim(level)$; 0, 0.4, 0.8, and 1. It is worth notifying that the global structure of clusters becomes clear and vivid. Table-3 shows the result of

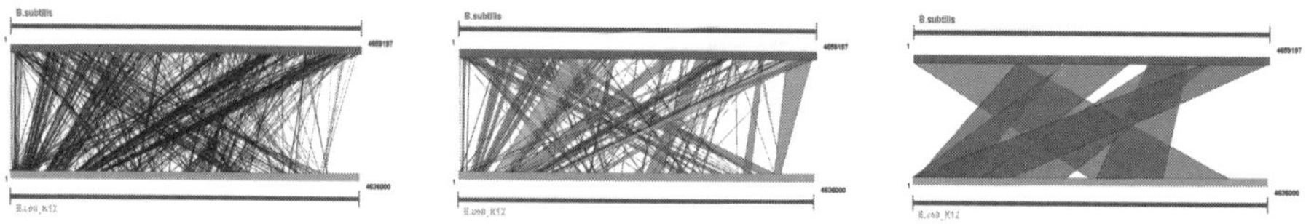

(a) $Sim(0)$ (b) $Sim(0.4)$ (c) $Sim(0.8)$

Figure 6. The simplified structure of gene clusters for alignment pairs between *E.coli K12* and *B.subtilis*. The plot (a) shows input alignment pairs. The plot (b) shows clusters which the smallest size is 10 and the largest size is 50. The plot (c) shows clusters which the smallest size is 45 and the largest size is 120.

comparing the spliced clusters described Section 2.5 with Xin's.[6] We checked some clusters and compared to the result of previous work. For the selected gene clusters, AlignScope detected most of sets of gene clusters found by Xin.[6] Interestingly, there were gene clusters of candidate clusters determined by AlignScope that Xin did not detect. To the best of our knowledge, the concept of a "spliced" gene ordering or alignment has not been reported before. Now we are investigating the biological meaning of the "spliced genes" found by AlignScope. Figure-7 (b) shows a set of spliced clusters in Table-3.

5. Conclusion

In this paper, we proposed a new method for visualization of whole genome alignment. A novel visualization tool for the whole genome alignment would be very useful for understanding genome organization and the evolutionary process. Several genome viewers already exist, each one serves a different need and research interest. AlignScope is easy to use in a computer environment and helps understanding the relationship and gene clusters between two genomes. Our system is freely available on http://jade.cs.pusan.ac.kr/~alignscope. The main features of AlignScope are as follows:

Table 3. Example of a set of spliced clusters. Since AlignScope does not consider "noisy" alignment pairs at clustering procedure, we can obtain the fine spliced clusters.

OUR	Xin[6]	COG	gene	product
O	X	COG0789	ycgE	putative transcriptional regulator
		-	ycgJ	orf, hypothetical protein
		-	ycgK	orf, hypothetical protein
		COG3100	ycgL	orf, hypothetical protein
		COG2983	ycgN	orf, hypothetical protein
		COG2719	ycgB	putative sporulation protein
		-	ycgR	orf, hypothetical protein
		COG1607	yciA	Acyl-CoA hydrolase
		COG2917	yciB	Intracellular septation protein A
		-	yciC	orf, hypothetical protein
O	O	COG2894	minD	cell division inhibitor, a membrane ATPase, activates minC
		COG0850	minC	cell division inhibitor, inhibits ftsZ ring formation
O	O	COG0243	narG	nitrate reductase 1, alpha subunit
		COG1140	narH	nitrate reductase 1, beta subunit
		COG2180	narJ	nitrate reductase 1, delta subunit, assembly function
		COG2181	narI	nitrate reductase 1, cytochrome b(NR), gamma subunit
O	O	COG0747	oppA	oligopeptide transport; periplasmic binding protein
		COG0601	oppB	oligopeptide transport permease protein
		COG1173	oppC	homolog of Salmonella oligopeptide transport permease protein
		COG0444	oppD	homolog of Salmonella ATP-binding protein of oligopeptide ABC transport system
		COG1124	oppF	homolog of Salmonella ATP-binding protein of oligopeptide ABC transport system

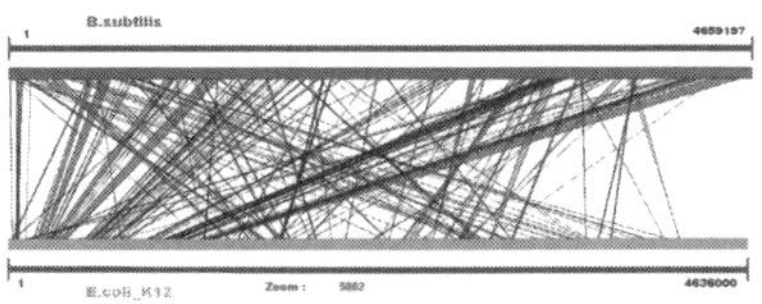

(a) all clusters between *E.coli K12* and *B.subtilis*

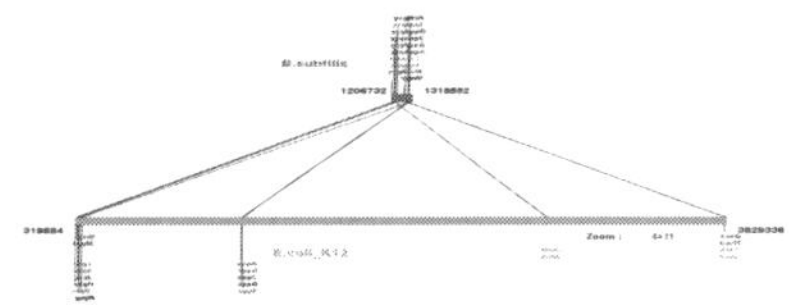

(b) a set of spliced clusters in Table-3

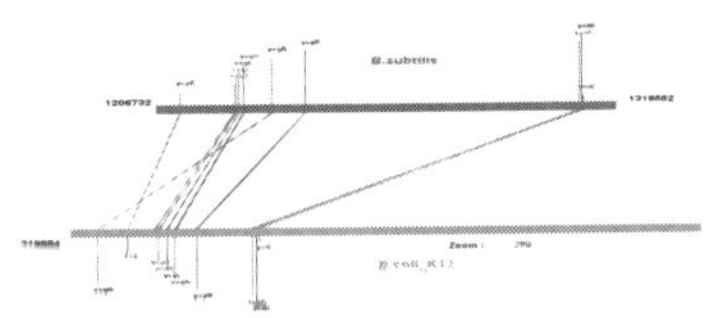

(c) information of first cluster in (b)

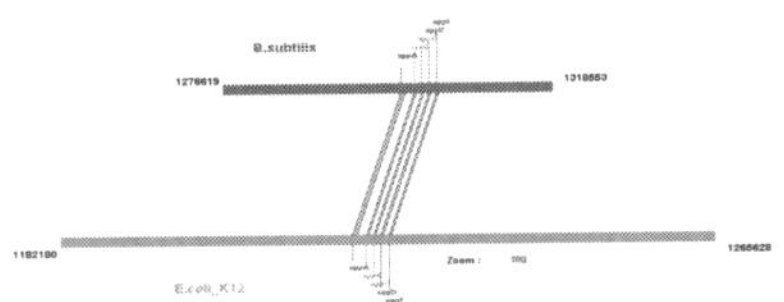

(d) information of second cluster in (b)

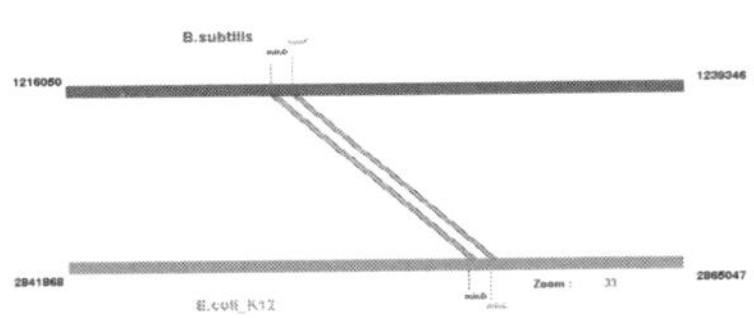

(e) information of third cluster in (b)

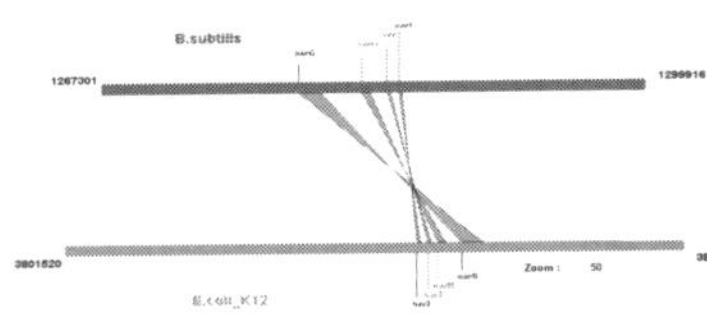

(f) information of fourth cluster in (b)

Figure 7. A set of "spliced clusters" found by AlignScope. The plot (a) shows detected spliced clusters between *E.coli K12* and *B.subtilis*. The plot (b) shows whole spliced clusters in Table-3 and the plot (c)∼(f) show the information of each cluster in (b).

- AlignScope provides an intuitive controls for the visualization of whole genome alignment at any simplified level.

- AlignScope is very-fast since we only consider a few "effecting zones" in each alignment and not the whole region of a genome. This improves the intractability between biologist and bioinformatics software.
- By using AlignScope, the candidate sets of gene clusters in whole genome can be easily found. In addition, AlignScope can detect the interesting "spliced" gene transfer, which was not possible in the previous approaches based on single string algorithms.

Acknowledgments

This work was supported by the Korea Research Foundation Grant(F01-2004-000-10016-0). We gratefully credit the thoughtful reviewers, who provided substantial constructive criticism on an earlier version of this note.

References

1. Kozik A, Kochetkova E, and Michelmore R. Genomepixelizer–a visualization program for comparative genomics within and between species. *Bioinformatics*, pages 335–336, 2002.
2. S.F. Altschul, W. Gish, W. Miller, E.W. Myers, and D.J. Lipman. Basic local alignment search tool. *Journal of Molecular Biology*, pages 403–410, 1990.
3. A. Bergeron, S. Corteel, and M. Raffinot. The algorithmic of gene teams. *In Proc. Second Annual Workshop on Algorithms in Bioinformatics*, pages 464 – 476, 2002.
4. Jeong-Hyeon Choi, Hwan-Gue Cho, and Sun Kim. Multiple genome alignment by clustering pairwise matches. *RECOMB Comparative Genomics Satellite Workshop, Lecture Notes in Bioinformatics*, pages 30–41, 2004.
5. Vernikos GS, Gkogkas CG, Promponas VJ, and Hamodrakas SJ. Genevito: visualizing gene-product functional and structural features in genomic datasets. *BMC Bioinformatics*, page 53, 2003.
6. Xin He and Michael H. Goldwasser. Identifying conserved gene clusters in the presence of orthologous groups. *In Proc. Research in Computational Molecular Biology*, pages 272–280, 2004.
7. Steffen Heber and Jens Stoye. Finding all common intervals of k permutations. *Proceedings of CPM 01,volume 2089 of Lecture Notes in Computer Science*, 2001.
8. Sun Kim, Jeong-Hyeon Choi, and Jiyoung Yang. Gene teams with relaxed proximity constraint. *IEEE Computational Systems Bioinformatics (CSB'05)*, 2005.
9. Needleman S.B. and C.D. Wunsch. A general method applicable to the search for similarities in the amino acid sequences of two proteins. *Journal of Moleculer Biology*, pages 443–453, 1970.
10. R. Sibson. Slink: An optimally efficient algorithm for the single-link cluster method. *Comput. J*, pages 93–95, 1973.
11. Berend Snel, Peer Bork, and Martijn A. Huynen. The identification of functional modules from the genomic association of genes. *Proceedings of the National Academy of Sciences*, pages 5890–5895, 2002.
12. Smith T.F and Waterman M.S. Identification of common molecular subsequences. *Journal of Moleculer Biology*, pages 195–197, 1981.
13. Takeaki Uno and Mutsunori Yagiura. Fast algorithms to enumerate all common intervals of two permutations. *Algorithmica*, pages 290 – 309, 2000.
14. NCBI Map Viewer. http://www.ncbi.nih.gov/mapview.

AN EFFICIENT ALGORITHM FOR STRING MOTIF DISCOVERY*

FRANCIS Y.L. CHIN AND HENRY C.M. LEUNG†

Department of Computer Science, The University of Hong Kong, Pokfulam Hong Kong, China

Finding common patterns, motifs, in a set of DNA sequences is an important problem in bioinformatics. One common representation of motifs is a string with symbols A, C, G, T and N where N stands for the wildcard symbol. In this paper, we introduce a more general motif discovery problem without any weaknesses of the Planted (l,d)-Motif Problem and also a set of control sequences as an additional input. The existing algorithms using brute force approach for solving similar problem take $O(n(t+f)l5^l)$ times where t and f are the number of input sequences and control sequences respectively, n is the length of each sequence and l is the length of the motif. We propose an efficient algorithm, called VAS, which has an expected running time $O(nfl(nt)^k(4^{k-1}+1/4^{k-1})^l)$ using $O((nt)^k(4^{k-1}+1/4^{k-1})^l)$ space for any integer k. In particular when $k = 3$, the time and space complexities are $O(nlf(nt)^3(1.0625)^l)$ and $O((nt)^3(1.0625)^l)$ respectively. This algorithm makes use of voting and graph representation for better time and space complexities. This technique can also be used to improve the performances of some existing algorithms.

1 Introduction

A *genome* is a DNA sequence consisting of four types of nucleotides (symbols): A, C, G and T. During the *gene expression* process, some substrings of the genome, called *genes*, are decoded to produce proteins. In order to start the gene expression process, a molecule called *transcription factor* binds to a *binding site*, represented by a short substring, in the *promoter region* of the gene. Genes seldom work alone. One kind of transcription factor may bind to the binding sites of several genes, allowing the genes to be decoded together. Such binding sites should then have the same length and similar patterns. Finding the common pattern, motif, of the binding sites from a set of sequences representing the promoter regions is an important problem for understanding how gene expression works.

A motif is usually represented by a string [3,4,7,8,12,14-23] or a matrix [1,2,5,6,9-11,13]. When a motif is represented by a $4 \times l$ probability matrix M, where l is the length of the binding sites, the i-th column of M represents the occurrence probabilities of A, C, G and T at the i-th position of a binding site. Although many real biological motifs can be better represented by a matrix, most existing algorithms [1,2,5,6,9,10,13] cannot guarantee finding the optimal matrix-represented motif from a given set of sequences and those algorithms that can may take a prohibitively long time to do so when l is large [11].

When a length l string P is used to represent the motif, all binding sites are length-l strings similar to P with at most d point substitutions. In other words, the Hamming distance between the motif and each binding site is at most d. Since there are a finite

---

* The research was supported in parts by the RGC grant HKU 7135/04E

† email: {chin, cmleung2}@cs.hku.hk

number of length-l strings (4^l possible strings), many algorithms [4,8,12,15,17-19,21-23] can guarantee finding the best string motif. The drawback is that some real biological motifs cannot be represented by strings.

Pevzner and Sze [17] define a precise version of motif discovery problem using string representation which has been considered in [4,12,14,15,18,19].

> **Planted (*l*,*d*)-Motif Problem:** Suppose there is a fixed but unknown nucleotide sequence P (the motif) of length l. Given t length-n nucleotide sequences and each sequence contains a planted variant of P, we want to determine P without knowing the positions of the planted variants. A variant is a substring derivable from P with at most d point substitutions.

Many algorithms have been developed to solve this problem [4,14,15,18]. However, this problem makes various assumptions and has the following weaknesses.

1. Because of experimental error and noise, some input sequences may not contain any variants of P. On the other hand, some promoter regions may contain more than one binding site [5,10,11,13,20].
2. Although the binding sites of a transcription factor may be different from each other, in real biological data, there are some conserved positions where all binding sites have the same nucleotides. The Planted (l,d)-Motif Problem does not exploit this property, making the defined problem more difficult than the actual problem [11,23].
3. It is difficult for biologists to determine the parameter d without any knowledge about the motif and the binding sites.
4. There are some patterns which are not motifs, but occur frequently in some parts of the genome, inside and outside the promoter regions. Algorithms for solving the Planted (l,d)-Motif Problem may mistakenly find these patterns as motifs [2].

Many algorithms [12,18,20-22] have been studied to solve the motif discovery problem without these weaknesses. Sinha and Tompa [21] modified the planted (l,d)-motif problem to overcome the first three weaknesses. In their model, a sequence may contain zero or more variants of the motif which is represented by a length-l sequence, called *pattern*, consisting of symbols {A, C, G, T, N}. A *variant* of a pattern P is a substring exactly the same as P except that each wildcard symbol N is replaced by A, C, G, or T where d is the number of wildcard symbols N in P. Patterns with different d values are compared by their z-scores (the number of standard deviations by which the number of variants of a pattern in the input sequences exceeds its expected number). Patterns with higher z-scores are more likely to be the correct motifs and the optimal motif is the pattern with the highest z-score.

During experiments, biologists usually get a set of sequences (control set) that do not contain many binding sites of the transcription factors as a by-product [2]. Takusagaw and Gifford [2,22] also considered motifs with wildcard symbols but with a set of control sequences as an additional input to overcome the last weakness. Patterns with relatively more variants in the input sequences than in the control set are likely to be the correct motifs.

There are a number of algorithms [8,20] which can find motifs with wildcard symbols, a similar problem model as given in [21,22]. However, all these algorithms find the motif by brute force. Since there are 5^l possible motifs (patterns), the running times of these algorithms increase exponentially with l.

In this paper, we give the first algorithm which solves the motif discovery problem without any of the above weaknesses, an extension of the motif discovery problem stated in [22]. Our algorithm called VAS (stands for Voting Algorithm from sets of Substrings) uses a new technique based on voting to find the maximum clique of a graph constructed from the set of sequences. Algorithm VAS can effectively find the optimal motif, in a few seconds/minutes instead of hours/days before. This new technique, when applied to some existing algorithms, should be able to greatly improve their performances too. In [17], a graph is constructed for solving the Planted (l,d)-Motif Problem. Each length-l substring in the input sequence is represented by a vertex, and an edge between two vertices exists if the two corresponding substrings differ by no more than $2d$ point substitutions. The Planted (l,d)-Motif Problem can be reduced to the finding of the maximum clique of the constructed graph, which takes $O((nt)^{(t-1)+2.376})$ time when d is large and $O((nt)^2)$ space. In [4], a voting algorithm was introduced for finding the motif. This algorithm, through some heuristics, can solve the Planted (l,d)-Motif Problem for large l and d with a high probability. However, its time and space complexity are $O(nt(3l)^d)$ and $(O(n(3l)^d)$ if we want to find the motifs with 100% certainty.

Algorithm VAS, based on voting from k similar substrings (when $k = 2$, the two similar substrings form an edge), has the merits of voting and graph representation and has better expected time and space complexities. For example, when $k = 2$, the expected time complexity and space complexity of VAS are $O(nlf(nt)^2(1.25)^l)$ and $O((nt)^2(1.25)^l)$ respectively; when $k = 3$, VAS takes $O(nlf(nt)^3(1.0625)^l)$ time and $O((nt)^3(1.0625)^l)$ space, where n is the length of the input sequences, t is the number of the input sequences, f is the number of the sequences in the control set and k can be any positive integer. Experimental results show that VAS has good performances on both simulated data and real biological data.

This paper is organized as follows. In Section 2, we define the extended motif discovery problem with control set. In Section 3, we describe VAS for solving this extended motif discovery problem. Experimental results on both simulated data and real biological data are shown in Section 4 followed by a conclusion in Section 5.

2 The Motif Discovery Problem

In order to address the weaknesses of the Planted (l,d)-Motif Problem, we define the motif discovery problem without any of these weaknesses as long as *P has relatively more variants in the set of input sequence T than in the set of control sequences F*. Before proceeding further, we have to formally define the meaning of "*P* has relatively more variants in *T* than in *F* " in the above problem definition.

Let t and f be the number of length-n sequences in T and F respectively. Barash et al. [2] determined whether a pattern P is the motif by considering the "random selection null hypothesis" that sequences in T are randomly selected from all the $t + f$ sequences. P is the motif if this hypothesis is false. However, they assumed that each sequence contains only one binding site and did not consider sequences with zero or multiple binding sites. Similar approach has also been used in [22]. In this paper, we verify whether P has relatively more variants in T than in F using a similar hypothesis as Barash et al. with the extra assumption that each sequence may contain zero or more variants of the motif.

Similar to [3,10], we break down the sequences in T and F into $\alpha = t(n - l + 1)$ and $\beta = f(n - l + 1)$ length-l substrings. Assume that T and F contain k_t and k_f variants of a motif with pattern P, consider the null hypothesis that the sequences in T are constructed by combining α substrings randomly selected from the $\alpha + \beta$ substrings without replacement (we may not be able to combine α substrings to construct t length-n sequences). Given a pattern P with $k_t + k_f$ variants in set T and F respectively, the probability that k_t of them are in set T is

$$P_{hyper}(k_t \mid k_t + k_f, \alpha, \beta) = \frac{\binom{\alpha}{k_t}\binom{\beta}{k_f}}{\binom{\alpha+\beta}{k_t + k_f}}$$

followed from the hyper-geometric probability distribution. The p-value that the null hypothesis is true can be calculated by summing up the tail of the probability distribution for $k_t' > k_t$.

$$p-value = P(k_t, k_f, \alpha, \beta) = \sum_{k_t'=k_t}^{\min\{M, k_t+k_f\}} P_{hyper}(k_t' \mid k_t + k_f, \alpha, \beta)$$

A pattern P with a small p-value means the null hypothesis is unlikely, i.e. P is likely to be the motif. Based on what we have discussed, we give the formal definition of the extended motif discovery problem without any of the above weaknesses as follows:

> **Extended Motif Problem with Control Set:** Suppose there is a fixed but unknown pattern P (the motif) of length l with symbols A, C, G, T and N. Given k_t variants of P in the t length-n nucleotide sequences in T and k_f variants of P in the f length-n nucleotide sequences in F, where $k_t/t >> k_f/f$ in the sense that P has a small p-value,we want to determine P with knowledge of the motif length l only.

In practice, there might be a few patterns with small p-values. Our algorithm will find the optimal motif which is the pattern with the smallest p-values. Note that the input of d is not necessary in the above problem definition because the correct pattern P should include the knowledge of d. Our algorithm will exhaust all values of d to find the pattern P with the smallest p-value.

3 Algorithm

Since there are 5^l possible length-l patterns and checking which pattern has the smallest p-value by brute force takes $O(5^l nl(t + f))$ time which can be extremely long for large l. Existing algorithms for solving the planted motif problem, like WINNOVER [17], PROJECTION [3] and SPELLER [19], cannot be extended to solve the extended motif discovery problem easily because they either do not guarantee finding the motifs or need a long running time when d is large.

Algorithm 1: VAS when $k = 1$.

```
Create a hash table V with zero at each entry        {V stores the number votes for each pattern}
min_p ← 1                                            {min_p is the minimum p-value}
For each length-l substring S in T
    For d ← 0 to l
        For each pattern P with exactly d symbol Ns such that S is a variant of P
            V(H(P)) ←V(H(P)) + 1                     {H(P) is the hash value of P}
sort the patterns in V in non-increasing order of the number of votes V(H(P))
For each pattern P
    k_t ←V(H(P))
    If P(k_t, 0, α, β) < min_p
        count the number of variants k_f of P in F
        If P(k_t, k_f, α, β) < min_p
            min_p ←P(k_t, k_f, α, β)
            motif ← P
    Else output motif
```

We might apply the basic idea of the Voting algorithm [4] to solve the extended motif discovery problem (Algorithm 1). For each length-l substring S in the input sequence, one vote is given to patterns P such that S is a variant of P. Note that all patterns with different d values, $0 \leq d \leq l$, have been considered in the algorithm. Since there are $t(n - l + 1)$ length-l substrings in T, and a length-l substring can be a variant of $\binom{l}{d}$ possible patterns with exactly d symbol Ns, the time needed for the algorithm is

$$O\left(n(t-l+1)l\sum_{d=0}^{l}\binom{l}{d}\right) = O(ntl2^l)$$

Since exactly $t(n - l + 1)2^l$ votes will be issued, there will be at most $n(t - l + 1)2^l$ entries in the hash table V. The time needed to count the number of variants of a pattern in F is nlf. Therefore the time needed for verifying each entry in V is at most $nlft(n - l + 1)2^l = O(nlf(nt)2^l)$. The total running time of the algorithm is $O(ntl2^l + nlf(nt)2^l) = O(nlf(nt)2^l)$. The memory needed for storing the hash table V is $O(nt2^l)$.

Although the base number of the exponent is reduced from 5 to 2, the time and space complexities of this direct voting algorithm are still very large. The space complexity remains impractical for large l.

The planted motif problem can also be viewed as a maximum clique problem [17]. Even though the maximal clique problem is NP-complete, this approach has the advantage that the space complexity is at most $O((nt)^2)$ as there are $t(n - l + 1)$ substrings (vertices) in T. In order to reduce the time and space complexities, we modify the Voting algorithm to vote patterns by a set of substrings instead of a single substring.

Normally the hidden motif should have at least two variants in T. One vote will be given to those patterns P such that length-l substrings S and S' in T are variants of P. Thus, a pattern with k variants in T should get exactly $\binom{k}{2}$ votes. Intuitively, the time and space complexities can be reduced because the hash table V does not needed to handle those patterns with only one variant in T. The expected time complexity and space complexity can be calculated as follows.

Assume the occurrence probabilities of A, C, G and T are 0.25. The probability that S differs from S' in i positions is $\binom{l}{i}0.25^{l-i}0.75^i$ and the number of patterns P such that both S and S' are variants of P is

$$\sum_{d=i}^{l}\binom{l-i}{d-i}=\sum_{j=0}^{l-i}\binom{l-i}{j}=2^{l-i}$$

So the expected number of patterns voted by each pair of substrings is

$$\sum_{i=0}^{l}\binom{l}{i}\left(\frac{1}{4}\right)^{l-i}\left(\frac{3}{4}\right)^{i}2^{l-i}=\sum_{i=0}^{l}\binom{l}{i}\left(\frac{2}{4}\right)^{l-i}\left(\frac{3}{4}\right)^{i}=\left(\frac{5}{4}\right)^{l}$$

Since there are $O((nt)^2)$ pairs of substrings in T, the time complexity of the algorithm (including checking the patterns in F) is

$$O\left(l(nt)^2\left(\frac{5}{4}\right)^{l}+nlf(nt)^2\left(\frac{5}{4}\right)^{l}\right)=O\left(nlf(nt)^2\left(\frac{5}{4}\right)^{l}\right)$$

And the space complexity of the algorithm is

$$O\left((nt)^2\left(\frac{5}{4}\right)^{l}\right)$$

In order not to miss motifs with only one variant in T, we check whether each length-l substring in T can be the motif. This checking step takes $O(nlf(nt))$ time and $O(1)$ space which do not affect the time and space complexities of VAS.

With this approach, the time and the space complexities are reduced by a factor of

$$nt2^l/(nt)^2\left(\frac{5}{4}\right)^{l}=\frac{1}{nt}\left(\frac{8}{5}\right)^{l}$$

if we vote on patterns by pairs of substrings instead of single substrings. The algorithm can be speeded up if and only if $nt < (8/5)^l$. Therefore, voting by pairs of substrings is beneficial when the size of the input sequences is small or the length of the motif is long. Similar improvement can be performed by giving votes to patterns of k substrings. The expected time complexity and space complexity for voting from k substrings are $O(nlf(nt)^k(4^{k-1}+1/4^{k-1})^l)$ and $O((nt)^k(4^{k-1}+1/4^{k-1})^l)$ respectively. In practice, VAS has the best time complexity when $k = 2$ or 3 depending on the size of the input sequences and the length of the motif.

4 Experimental Results

We have implemented VAS in C++ and used it to find motifs in both simulated and real biological data. In this section, we describe the performance of VAS and compare it with some existing motif discovery algorithms. All experiments were taken on a 2.4GHz P4 CPU with 1 GB memory.

Table 1. Successful rate and running time of VAS.

l	d	$b = 10$		$b = 20$		$b = 30$		$b = 40$	
		Success rate	Time	Success rate	Time	Success rate	Time	Success rate	Time
8	1	100%	13.7s	100%	13.7s	100%	13.8s	100%	13.8s
	2	74%	13.8s	76%	13.7s	100%	13.7s	100%	13.8s
10	3	100%	17.1s	100%	17.0s	100%	16.7s	100%	16.8s
	4	62%	17.1s	54%	16.7s	100%	16.9s	100%	16.7s
12	5	100%	23.1s	100%	23.3s	100%	23.3s	100%	23.4s
	6	44%	23.0s	84%	23.1s	100%	23.2s	100%	23.3s
14	7	100%	37.8s	100%	37.6s	100%	37.7s	100%	37.6s
	8	68%	37.7s	96%	37.7s	100%	37.7s	100%	37.7s
16	9	100%	67.1s	100%	67.1s	100%	67.1s	100%	67.1s
	10	100%	67.0s	82%	67.2s	100%	67.2s	100%	67.0s
18	11	100%	132s	100%	132s	100%	131s	100%	132s
	12	100%	132s	100%	132s	100%	131s	100%	132s
20	13	100%	256s	100%	256s	100%	255s	100%	256s
	14	100%	255s	100%	256s	100%	256s	100%	256s

4.1. *Simulated Data*

The simulated data were generated as follows. All input instances contain $t = 20$ length-600 sequences in T and $f = 20$ length-600 sequences in F. Each nucleotide of these sequences was generated independently with the same occurrence probability 0.25. Then a length-l motif M with d Ns was picked randomly from all possible patterns and b variants of M were planted in the sequences in T at random positions. The motif length l and the sequences in T and F were inputted to VAS for finding the motifs. For each set of parameter l, d and b, we ran 50 test cases. Table 1 shows the successful rate and the average running time of VAS when $k = 2$ (votes are given by pairs of substrings in T).

Since the number of votes given by each pair of substrings in T is almost independent of the number of planted variants in T and the number of Ns in the motif pattern, the running time of VAS is independent of these factors as shown in Table 1. Algorithm VAS may not find the motif when d, the number of Ns in the motifs, is relatively large (e.g. (8,2), (10,4), (12,6)) and the number of planted variants in T is small ($b = 10$ or $b = 20$).

It is because random patterns P might have more variants in T and less variants in F than the motif M in these cases. Since VAS cannot distinguish M from these random patterns P, VAS fails to find the motif. However, when the number of non-N symbols in M is reasonably large (> 6), VAS can find the motif M successfully with high probability.

Common motif discovery algorithms like PROJECTION [3] and VOTING [4] are developed for solving planted motif problem without control set. In order to compare the performance of these algorithms with VAS, we reduce the values of d for these algorithms such that they can theoretically find the motif [3] and plant exactly one variant in each sequence in T. Table 2 shows the results of these algorithms.

Table 2. Successful rate and running time of brute force algorithm, PROJECTION, VOTING And VAS

l	d	Brute Force		PROJECTION		VOTING		VAS	
		Success rate	Time	Success rate	Time	Success rate	Time	Success rate	Time
8	1	100%	268s	94%	18s	100%	<1s	100%	13.8s
10	2	100%	72min	98%	77s	100%	0.7s	100%	16.7s
12	3	-	-	88%	371s	100%	28.4s	100%	23.4s
14	4	-	-	76%	650s	100%	412s	100%	37.7s
16	5	-	-	82%	20min	100%	31min	100%	67.0s
18	6	-	-	88%	34min	-	-	100%	132s
20	7	-	-	86%	48min	-	-	100%	256s

Although brute force algorithms can find motif when l is small, they fail to find the motif when $l > 10$. Voting algorithm [4] (we use the basic voting algorithm without heuristic search) has a better performance than the brute force algorithms because its running time increases exponentially with d instead of l. The running time of PROJECTION does not increase sharply with l because it performs heuristic search for finding the motifs. However, it does not guarantee that the motifs can be found all the time and has a success rate less than 100%. When compared with these algorithms, VAS has the best performance in both accuracy and running time.

4.2. *Real Biological Data*

SCPD [24] contains different transcription factors for yeast. For each set of genes regulated by the same transcription factor, we chose the 600 base pairs in the upstream of the genes as the input sequences T. 100 random sequences in the upstream of yeast's genes were picked randomly as the set of control sequences F. The lengths of the motifs were same as those of the published motifs. For PROJECTION and the Voting algorithm, we tested all possible d from 0 to l. Experimental results are showed in Table 3. Some motifs with many wildcard symbols (e.g. GAL4) cannot be represented properly by the planted motif problem and can be found by VAS only. Since PROJECTION and the voting algorithm do not consider the set of control sequences, they fail to find motifs when relatively less variants are in T (e.g. ACE2). On the other hand, VAS can find the motifs in these cases with the help of the control sequences. Note that we have not shown all the experimental results because PROJECTION, the Voting algorithm and VAS have the same performance on the rest transcription factors in the SCPD.

Table 3. Experimental results on real biological data

Name	Published Pattern	PROJECTION	VOTING	VAS
CuRE	TTTGCTC	TTTGCTC	TTTGCTC	TTTGCTC
GATA	CTTATC	CTTATC	CTTATC	TTATCG
ACE2	GCTGGT	-	-	GCTGGT
AP1	TTANTAA	-	TTACTAA	TTANTAA
GAL4	$CGGN_{11}CCG$	-	-	CGGNGNNCTNTNGNCCG
ROX	YYNATTGTTY	-	-	TCCATTGTTC

Symbol Y means C or T. N_{11} means 11 Ns.

5 Discussion

In this paper, we have introduced VAS for solving the extended motif discovery problem with control set using $O(nlf(nt)^k(4^{k-1}+1/4^{k-1})^l)$ time and $O((nt)^k(4^{k-1}+1/4^{k-1})^l)$ space for any positive integer k. Not only VAS can solve the motif discovery problem with least assumptions, experimental results show that VAS has the best performance than existing algorithms in both speed and accuracy. Since VAS can find the number of variants of every length-l patterns in T in short running time, the new technique used in VAS can also be applied to find string motifs for other motif discovery algorithms for those problems without control set F [12] or based on other hypotheses [20]. For example, if the input does not contain any control sequences, we cannot use the hyper-geometric distribution for the evaluation of p-values by. In this case, we may have to evaluate the p-values based on the background occurrence probabilities of the nucleotides. The extension of this work will have similar performance as VAS and will be included in the full paper.

VAS works well on the extended motif discovery problem because it is easy to find the set of patterns to be voted by a substring in T. This task may become difficult when the definition of variants is changed. In the future, we will investigate how to use VAS to solve motif discovery problems with other definitions of variants, for example, motif with IUPAC symbols.

6 References

1 T. Bailey and C. Elkan. Unsupervised learning of multiple motifs in biopolymers using expectation maximization. *Machine Learning*, 21:51-80, 1995

2 Y. Barash, G. Bejerano and N. Friedman. A simple hyper-geometric approach for discovering putative transcription factor binding sites. *WABI*, p278-293, 2001.

3 J. Buhler and M. Tompa. Finding motifs using random projections. *RECOMB*, p69-76, 2001.

4 F. Chin and H. Leung. Voting Algorithms for Discovering Long Motifs. *APBC*, p261-271, 2005.

5 F. Chin, H. Leung, S.M. Yiu, T.W. Lam, R. Rosenfeld, W.W. Tsang, D. Smith and Y. Jiang. Finding Motifs for Insufficient Number of Sequences with Strong Binding to Transcription Factor. *RECOMB*, p125-132, 2004

6 G. Z. Hertz and G. D. Stormo. Identification of consensus patterns in unaligned dna and protein sequences: a large-deviation statistical basis for penalizing gaps. *In*

Proceedings of the 3rd International Conference on Bioinformatics and Genome Research, p201-216, 1995

7 U. Keich and P. Pevzner. Finding motifs in the twilight zone. *RECOMB*, p195-204, 2002

8 S. Kielbasa, J. Korbel, D. Beule, J. Schuchhardt and H. Herzel. Combining frequency and positional information to predict transcription factor binding sites. *Bioinformatics*, 17:1019-1026, 2001

9 C. Lawrence, S. Altschul, M. Boguski, J. Liu, A. Neuwald and J. Wootton. Detecting subtule sequence signals: a Gibbs sampling strategy for multiple alignment. *Science*, 262:208-214, 1993

10 C. Lawrence and A. Reilly. An expectation maximization (em) algorithm for the identification and characterization of common sites in unaligned biopolymer sequences. *Proteins: Structure, Function and Genetics*, 7:41-51, 1990

11 H. Leung and F. Chin. Finding Exact Optimal Motif in Matrix Representation by Partitioning. *Bioinformatics*, 21(supp 2):ii86-ii92, 2005

12 H. Leung and F. Chin. Generalized Planted (*l,d*)-Motif Problem with Negative Set. *WABI*, p264-275, 2005

13 H. Leung, F. Chin, S.M. Yiu, R. Rosenfeld and W.W. Tsang. Finding Motifs with Insufficient Number of Strong Binding Sites. *Jour. Comp. Biol.*, 2005 (will appear)

14 M. Li, B. Ma, and L. Wang. Finding similar regions in many strings. *Journal of Computer and System Sciences*, 65:73-96,2002

15 S. Liang. cWINNOWER Algorithm for Finding Fuzzy DNA Motifs. *Computer Society Bioinformatics Conference*, p260-265, 2003

16 G. Pesole, N. Prunella, S. Liuni, M. Attimonelli, and C. Saccone. Wordup: an efficient algorithm for discovering statistically significant patterns in dna sequences. *Nucl. Acids Res.*, 20(11):2871-2875,1992.

17 P. Pevzner and S.H. Sze. Combinatorial approaches to finding subtle signals in dna sequences. *In Proc. of the Eighth International Conference on Intelligent Systems for Molecular Biology*, p269-278, 2000.

18 S. Rajasekaran, S. Balla and C.H. Huang. Exact algorithms for planted motif challenge problem. *APBC*, p249-259, 2005.

19 M.F. Sagot. Spelling approximate repeated or common motifs using a suffix tree. *In C.L. Lucchesi and A.V. Moura editors, Latin'98: Theoretical Informatics, volume 1380 of Lecture Notes in Computer Science*, p111-127, 1998.

20 S. Sinha. Discriminative motifs. *In Proc. of the Sixth Annual International Conference on Computational Biology*, p291-298, 2002

21 S. Sinha and M. Tompa. A statistical method for finding transcription factor binding sites. *In Proc. of the Eighth International Conference on Intelligent Systems for Molecular Biology*, p344-354, 2000.

22 K.T. Takusagawa and D.K. Gifford. Negative information for motif discovery. *PSB*, p360-371, 2004

23 M. Tompa. An exact method for finding short motifs in sequences with application to the ribosome binding site problem. *In Proc. of the 7th International Conference on Intelligent Systems for Molecular Biology*, p262-271, 1999.

24 J. Zhu and M. Zhang. SCPD: a promoter database of the yeast Saccharomyces cerevisiae. *Bioinformatics* 15:563-577, 1999. http://cgsigma.cshl.org/jian/

DISCRIMINATIVE DETECTION OF CIS-ACTING REGULATORY VARIATION FROM LOCATION DATA

YUJI KAWADA AND YASUBUMI SAKAKIBARA

Department of Biosciences and Informatics, Keio University
3-14-1 Hiyoshi, Kohoku-ku, Yokohama, 223-8522, Japan
yuji@dna.bio.keio.ac.jp, yasu@bio.keio.ac.jp

The interaction between transcription factors and their DNA binding sites plays a key role for understanding gene regulation mechanisms. Recent studies revealed the presence of "functional polymorphism" in genes that is defined as regulatory variation measured in transcription levels due to the *cis*-acting sequence differences. These regulatory variants are assumed to contribute to modulating gene functions. However, computational identifications of such functional *cis*-regulatory variants is a much greater challenge than just identifying consensus sequences, because *cis*-regulatory variants differ by only a few bases from the main consensus sequences, while they have important consequences for organismal phenotype. None of the previous studies have directly addressed this problem. We propose a novel discriminative detection method for precisely identifying transcription factor binding sites and their functional variants from both positive and negative samples (sets of upstream sequences of both bound and unbound genes by a transcription factor) based on the genome-wide location data. Our goal is to find such discriminative substrings that best explain the location data in the sense that the substrings precisely discriminate the positive samples from the negative ones rather than finding the substrings that are simply over-represented among the positive ones. Our method consists of two steps: First, we apply a decision tree learning method to discover discriminative substrings and a hierarchical relationship among them. Second, we extract a main motif and further a second motif as a *cis*-regulatory variant by utilizing functional annotations. Our genome-wide experimental results on yeast *Saccharomyces cerevisiae* show that our method presented significantly better performances for detecting experimentally verified consensus sequences than current motif detecting methods. In addition, our method has successfully discovered second motifs of putative functional *cis*-regulatory variants which are associated with genes of different functional annotations, and the correctness of those variants have been verified by expression profile analyses.

1. Introduction

Transcription factors (TFs) are DNA-binding proteins at the terminals of signal transduction networks and, in genomic sequences, a TF binding site (motif) is a set of *cis*-regulatory elements that preserve a certain nucleotide composition, playing a key role in transcriptional regulations. Each transcription factor recognizes a specific binding site composed of similar substrings, referred to as *cis*-regulatory variants. Recently, such subtle variations were hypothesized to also play a key role in transcription control. [1,5] It is generally assumed that *cis*-regulatory variants are hard to be detected only by sequence analyses but rather require extensive experimental studies. [1]

While a number of methods have been proposed previously, computational identification of TF binding sites is still a challenging and unsolved problem. Most existing methods for detecting motifs examine only the upstream sequences of clustered, and presumably co-regulated, groups of genes or bound genes by the same TF, and search for statistically over-represented motifs among them. Such well-known motif detecting algorithms include AlignACE, Multiple EM for Motif Elicitation (MEME), Yeast Motif Finder (YMF), and MDScan. [4] Since biological signals are subject to mutations and usually do not appear exactly, they typically use probability weight matrix (PWM) to represent motifs. On the other hand, genome-wide location analyses, referred to as chromatin immunoprecipitation (ChIP) microarray experiments, recently elucidated *in vivo* physical interactions between TFs and their chromosomal targets on the genome. [2,3] The ChIP microarray technique can be thought to provide reliable and useful information about direct binding of a specific protein complex to DNA. In other words, the ChIP data provide us the explicit interaction information about not only TF-DNA "binding" but also TF-DNA "unbinding".

Our fundamental idea for detecting motifs is that the true motif appears only in the upstream sequences of the target genes controlled and bound by the TF and does NOT appear in those of the unbound ones. This idea leads us to a discriminative approach to find true motifs that distinguish the upstream sequences between bound and unbound genes.

Compared with most existing methods, our new strategy has three distinct features. First, our method takes unbound upstream sequences into account as negative samples as well as bound sequences as positive ones. Several approaches using ChIP data have been proposed previously, [4] but they still focus on the positive samples alone. Second, we define motifs as "discriminative" substrings that correctly distinguish the upstream sequences of positive samples from those of negative ones instead of statistically over-represented patterns or well-conserved ones. Even if using statistical criteria, methods that only focus on over-represented patterns suffer from numerous spurious random similarities. Third, we use a discriminative machine learning technique for detecting motifs, and we search for motifs using an exact-match, which is the opposite of the current probabilistic search strategies. Existing methods try to represent a motif by one single model allowing biological noises (mismatches, insertions and deletions) to some extent. Yet their obtained model is characterized by one specific substring, referred to as *consensus*. If one single consensus sequence characterizes the positive samples, it must be more precisely detected by using an exact-match search when negative samples are taken into account. In addition, by allowing ambiguity, current methods can not distinguish between the consensus sequences and their functional variants. As a result, they fail to detect the subtle differences of motifs that lead to important consequences for organismal phenotype. In contrast with most existing methods, we search for main motifs and their functional variants by focusing on the subtle differences among substrings rather than allowing and unifying them.

To search for the most discriminative substrings, we employ the decision tree learning method. Decision trees are used for classification tasks whose concepts are defined in terms of a set of attribute-value pairs. A text-classification tree classifies an input text (sequence) into one category according to several tests whether the input sequence contains some specific substrings. The inductive learning problem of decision trees is to construct such a text-classification tree from already classified sequences. In this paper, we use the

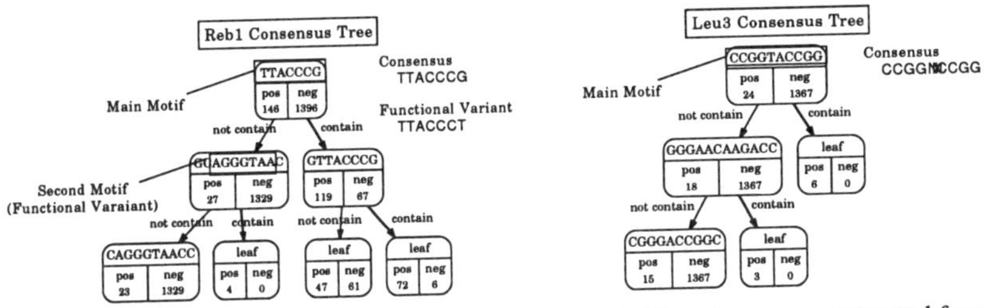

Figure 1. Motif detection by a decision tree learning method. These trees are constructed from both positive and negative samples of Reb1 and Leu3. The number of samples is shown in each node. The correctly identified consensus sequence and its previously inferred functional variant (only for Reb1) are shown inside the rectangles.

decision tree learning method for extracting sequence motifs given positive and negative samples. As a result of learning, substrings that are the most important and predictive for distinguishing the upstream sequences between positive and negative samples are extracted and are assigned to each internal node of the learned tree, which we call here a *consensus tree*. Figure 1 demonstrates the effectiveness of our method using the consensus tree. Our method correctly identified the consensus sequences for Reb1 and Leu3. As for the case of Reb1, a previous computational study [5] based on phylogenetic analysis only inferred the presence of Reb1 consensus variant. Our method succeeded to identify this variant and presented the relationships among them as a hierarchical tree structure. Further, our method inferred a number of *cis*-regulatory variants that have not previously been detected for many TFs through genome-wide experiments on S. cerevisiae.

2. Methods

Our method consists of two steps: (i) build a consensus tree by decision tree learning method, and (ii) search for highly functionally enriched motifs from the extracted substrings that are assigned in the internal nodes of the consensus tree.

In the preprocessing step, we select highly ChIP-array-enriched genes (binding P-value ≤ 0.001) as positive samples and low ChIP-array-enriched genes (binding P-value ≥ 0.80) as negative ones. The genome-wide location analyses assign P-value (confidence value) to each interaction between a TF and an intergenic region. It is reported that the empirical rate of false positives at a stringent P-value threshold ($P \leq 0.001$) is $6 - 10\%$ in the data of Ref. 3 and 4% in the data of Ref. 2. Since we assume that true motifs appear only in the upstream sequences of positive samples and not in those of negative ones, the use of a high confidence P-value threshold is required.

2.1. *Consensus Tree Construction*

We define motifs as informative substrings that can correctly classify genes into proper classes ('positive'/'negative') based on their upstream sequences. Thus, given a specific TF's positive and negative samples, our aim is to search for the most informative and hence discriminative substrings from the positive samples.

To accomplish this task, we use the decision tree learning method. We denote a se-

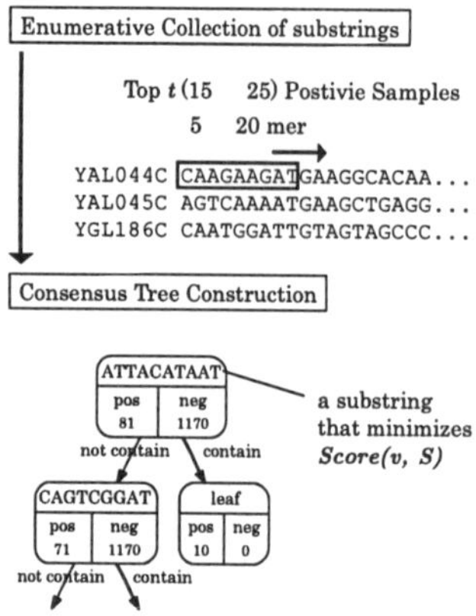

Figure 2. Consensus Tree Construction.

$\underline{BTLEARN(S, prnsrt, nsrt, keyl1, keyl2)}$:

(1) Collect all the substrings.
$Keywords = \{v \mid keyl1 \leq |v| \leq keyl2\}$
(2) Output a consensus tree T.
$T = BTFIND(S,\ Keywords,\ prnsrt,\ nsrt)$

$\underline{BTFIND(S, Keywords, prnsrt, nsrt)}$:

(1) If $(|S| - Occur(S, c_i))/|S| \leq nsrt$ is satisfied, return a subtree $T = c_i$.
(2) If $|S| \leq prnsrt$ is satisfied and the major class label with S is c_i, return a subtree $T = c_i$.
(3) If a substring v_g that minimizes $Score(v_g, S)$ is found from $v \in Keywords$, return v_g as an informative substring of the current node, a left-sided subtree T_0 and a right-sided subtree T_1.
$T_0 = BTFIND(S_0^v,\ Keywords - v,\ prnsrt,\ nsrt)$
$T_1 = BTFIND(S_1^v,\ Keywords - v,\ prnsrt,\ nsrt)$

Figure 3. Decision Tree Learning Algorithm.

quence by w, a substring by v, class labels ('positive'/'negative') by c and c_i, samples by S, and by S_0^v, S_1^v, $Occur$ as follows: $S_0^v = \{(w, c) \in S \mid w \text{ does not contain } v\}$, $S_1^v = \{(w, c) \in S \mid w \text{ contains } v\}$, $Occur(S, c_i) = |\{(w, c) \mid c = c_i\}|$. And v is "informative" if and only if $S_0^v \neq \emptyset$ and $S_1^v \neq \emptyset$.

If we have two classes ('positive' and 'negative') and denote their class labels by c_1 and c_2 respectively, the objective function is defined in Equation 1.

$$
\begin{aligned}
I(S) &= -\sum_{i=1}^{2} \frac{Occur(S, c_i)}{|S|} \log_2 \frac{Occur(S, c_i)}{|S|} \\
Loss(v, S) &= \frac{|S_0^v|}{|S|} I(S_0^v) + \frac{|S_1^v|}{|S|} I(S_1^v) \\
Score(v, S) &= Loss(v, S) + \tau \frac{1}{l} \log(p(v))
\end{aligned}
\qquad (1)
$$

where l is the length of v, $p(v)$ is the probability of generating v from a third-order Markov background model estimated from all the intergenic regions. τ is a free parameter and is chosen empirically. $Loss$ function indicates a weighted sum of the entropies of two sets that are divided by the presence of one specific substring. With the minimum entropy criterion, the most discriminative substring is the one that minimizes the $Score$ function.

The procedure of constructing a consensus tree by our decision tree learning method is shown in Figure 2. We begin by collecting every nonredundant w-mer in both strands of the top t $(15 - 25)$ positive samples, and then recursively search for the substring that minimizes the objective function with the current positive and negative samples from the collection of substrings, and divide both samples by the presence of it. The algorithm of decision tree learning is outlined in Figure 3. Given samples (S), two values for condition precedent ($prnsrt$ and $nsrt$) and lower and upper bounds of the length of the substring ($keyl1$ and $keyl2$), *BTLEARN* returns a learned consensus tree. By examining three TFs, we set $prnsrt = 10$, $nsrt = 0.01$, $keyl1 = 5$, and $keyl2 = 20$. We normalized the log likelihood of the background model, and set $\tau = 0.05$.

As a result of learning, substrings that are the most important and predictive for discrimination are extracted and are assigned to each internal node of the learned tree. Our decision tree learning method recursively split the search space, which is equivalent to clustering genes recursively by the presence of specific substrings. Therefore, we apply the following strategy for extracting the consensus sequence and their second variants: In a hierarchical structure of the learned tree, the main consensus sequence is extracted from the root node, and its significant second variants are extracted from the left children and the left descendants of the root node (Fig. 2). Since we assume that the number of significant functional variants is not large, we set the maximum depth of consensus trees to three.

2.2. *Extractions of cis-regulatory elements based on functional annotations*

After constructing the consensus tree, we search for a highly functionally enriched motif from an extracted substring in each internal node of the learned tree. Highly functionally enriched motif, which we call here a *functional consensus*, is the one whose target genes are highly associated with a same functional annotation. Target genes of a motif mean the genes which are included in the positive samples of the TF and whose upstream sequences contain a perfect-match to the motif. We assume that motifs are composed of several functional consensuses each of which regulates a specific set of genes. Since it is not usually possible to predict which nucleotide changes in motifs might affect expression, we search for main motifs and their variants by utilizing functional annotations.

We slide a window of length more than six along a discriminative substring in the node, and evaluate a motif in the window at each position by measuring its functional enrichment. For each window position, we calculate the hypergeometric P-value of independence between genes which are targets of the motif in the window and genes with the same GO biological process category, adjusted by Bonferroni correction for multiple testing. We collect the most functionally enriched motif as a functional consensus from every node.

The hypergeometric P-value is given by Equation 2.

$$P\text{-value} = \sum_{i=I}^{T} \frac{\binom{B}{i}\binom{G-B}{T-i}}{\binom{G}{T}} \tag{2}$$

where G is the total number of genes, B is the total number of genes in a particular biological process category, T is the number of target genes of the motif, and I is the number of genes which are targets of the motif and are in the particular biological process.

From the information-theoretic point of view, the most discriminative substrings are not necessarily be functionally enriched. Intuitively, they are too "informative" in the following sense. Since the ratio of nucleotide distribution in S. cerevisiae is approximately given by: A : T : G : C = 32 : 32 : 18 : 18, the average information content of one nucleotide is: $I_{ave} = -\sum_{i\in\{A,T,G,C\}} p_i \log_2(p_i) \approx 1.94$ bits, where p_i is the frequency of occurrence of nucleotide i. The amount of information required to identify γ sites out of a possible Γ is given by: $I_\gamma = -\log_2 \frac{\gamma}{\Gamma}$ bits. Thus, if a motif is six base long and it occurs exactly once in every 1000 bases and may be placed in either of the two DNA strands in n sequences, the average information required to identify a motif is then: $I_{actual} = -\log_2 \frac{n}{(n\times(1000-6+1)\times 2)} \approx 10.96$ bits. Therefore, $I_{actual}/I_{ave} \approx 5.64$ nucleotides are required to identify a motif from positive samples alone. However, in the discriminative framework, we search for a motif which appears only in the positive samples and must not appear in the negative ones. If we have p positive samples and q negative ones, the average information required to identify such a motif is: $I_{req} = -\log_2 \frac{p}{((p\times q)\times(1000-10+1)\times 2)}$ bits. In the case of $p = 50$ and $q = 1200$, $I_{req}/I_{ave} \approx 10.91$ nucleotides are required to identify such a discriminative motif.

The discussion stated above is just a rough approximation. In the discriminative framework, however, the required information tends to become high. Thus, to correctly identify functional consensuses, we need to "decompose" them by utilizing functional annotations. From the discussion stated above, we set the minimum length of a sliding window to six.

3. Experimental Results

3.1. *Data*

We collected the sequences of 1000 bp upstream of the translation start sites for 6270 genes on S.cerevisiae from SGD and SCPD, and two published genome-wide location data. [2,3] To search for functional consensuses and to assess the reliability of discovered *cis*-regulatory variants, we also collected various types of functional annotations, such as GO annotations for S.cerevisiae (process, component, and function), MIPS categories for S.cerevisiae (function, complex, motif, protein class, and phenotype), and a compendium of 827 gene expression profiles from 29 different publications. For evaluating obtained motifs, we collected all the 20 experimentally verified consensus sequences from TRANSFAC database and 25 from the literature that were reported in at least two papers. The average of the length of the collected motifs was 7.20 and the standard deviation of that was 2.27.

The total numbers of the location data that we used was 148. The number of positive samples ranged from 1 to 282 and that of negative ones ranged from 552 to 2084, with an average of 63 positive samples and 1177 negative ones per a TF. Due to the page limitation, we will only show typical experimental results for several TFs. The full results are available at our web site (`http://www.dna.bio.keio.ac.jp/reg_motifs`).

3.2. *Detection of Known Motifs*

We compared the motif detecting performance of our method with four other programs including AlignACE, MEME, YMF and MDScan. [4] AlignACE and MEME employ a heuristic local search approach, YMF employs an enumerative one, and MDScan employs a hybrid of enumerative and heuristic ones. Each program was run with default parameters. Note that, since the published consensus sequences are obtained empirically, they may not be the most functionally enriched and they are slightly different from literature to literature. Therefore, a discovered substring was considered to be consistent with the published consensus sequence if it contained at most one mismatch, insertion, or deletion.

When we only evaluated the top scoring motifs, that is, substrings that were assigned to the root nodes in the learned trees, our method correctly identified 38 out of 45 published

Table 1. Comparison of Discovered Motifs.

TF name	Consensus	Our Method	AlignACE	MEME	YMF	MDScan
Abf1	TCAYTNTNNACG	TCACTATATACG			CACTWNAYACG	
Ace2	GCTGGT	GGGCGGGTG			CACACNCACAC	
Arg80	TTAAGTG	GCCGTTAAGT			CCGCGNCCGAC	
Bas1	TGACTC	CTGACTCCG			CACACNCACAC	
Cad1	TTASTAA	ATTAGTCAGC			CACACNCACAC	
Cbf1	TCACGTG	GTCACGTG			RTCACGTGAY	
Fkh1	GGTMAACAA	AAGGTAAACAA			AGGGGCGGGG	
Fkh2	GTAAACAA	TTGTTTACCTTT			CACACNCACAC	
Gcn4	TGACTCA	TGACTCA			AYATANATAYA	
Gln3	GATAAG	GATAAGATAAG			AYATANATAYA	
Hsf1	TTCNNGAA	TTCTAGAAG			CCCGTCTAGC	
Ino2	TTTCACATG	TTTTCACATGC			CGACCNCCGSG	
Ino4	TTTCACATG	TTTCACATG			ACGTANGTACG	
Leu3	CCGGNNCCGG	CCGGTACCGG			CCGGTNCCGGC	
Mbp1	ACGCGT	GACGCGTT			ACGCGWCGCG	
Mcm1	TCCYAATTNGG	CCAAATTAGG			CCGGGNCGTGC	
Msn2	MAGGGG	GCAGGGGCG			CTSCCNCATCC	
Msn4	CCCCT	CGAGAGCCCCA			CGAGGGCGCC	
Ndd1	TTGTTTAC	TTGTTTACCTTT			CACACNCACAC	
Pho4	CACGTG	CCACGTGC			CTACCCGGAG	
Rap1	ACCCATACA	ACCCATACA			CAYCCNTACAY	
Rcs1	TGCACCC	ACTGCACCC			ATATANRTATA	
Reb1	TTACCCG	TTACCCG			TSCGGGTAAY	
Ste12	TGAAACA	ATTTGAAACA			CCTGANTCAGG	
Sum1	GTGACNC	CTGACACCTG			ATATANATATA	
Swi4	CNCGAAA	GACGCGAAA			CACACNCACAC	
Swi6 (SCB)	CRCGAAA	TTTCGCGTC	none		none	none
Swi6 (MCB)	ACGCGT	CGCGTCG			ACGCGWCGCG	
Yap1	TTACTAA	TTAGTCAGCAT			CGACGNCGACG	

Table 2. Most Associated Functional Category.

Database		Motif	Associated Category	*P*value
GO	Process	TGACTC	amino acid metabolism	1.03 E-33
		GACTAA	nitrogen compound metabolism	5.782 E-19
	Function	TGACTC	molecular_function	5.89 E-10
		GACTAA	catalytic activity	1.07 E-04
	Component	TGACTC	cellular_component	2.57 E-10
		GACTAA	cellular_component	9.01 E-05
MIPS	function	TGACTC	amino acid metabolism	4.00 E-28
		GACTAA	amino acid metabolism	4.78 E-16
	complexes	TGACTC	Complexes by Systematic Analysis	2.31 E-16
		GACTAA	Complexes by Systematic Analysis	1.45 E-04
	protein class	TGACTC	Cys6 cysteine-zinc cluster	9.58 E-11
		GACTAA	Cys6 cysteine-zinc cluster	1.29 E-04
	phenotype	TGACTC	Auxotrophies, carbon and nitrogen utilization defects	1.43 E-05
		GACTAA	Methionine auxotrophy	1.41 E-03

Table 3. Differences of Expression Profiles.

Motif		Motif	t-test *P* value	Source
TGACTC	vs	GACTAA	1.27 E-07	Ploidy

Table 4. GO Terms for Gcn4.

GO Terms (assigned by YPD)
Amino acid biosynthesis
Cellular response to glucose starvation
Cellular response to nitrogen starvation
Cellular response to starvation
Response to stress
Nucleotide biosynthesis
Nucleobase, nucleoside, nucleotide and nucleic acid metabolism
Transcription from RNA polymerase II promoter
Regulation of transcription from RNA polymerase II promoter

Note: Terms that were associated with the main motif and the second motif are underlined.

consensus sequences. AlignACE identified 12, MEME identified 16, YMF identified 17, and MDScan identified 17 among 45 published consensus sequences. Within seven consensus sequences that our method failed to identify, four consensus sequences were discovered in other nodes of the learned trees. When we used random sequences generated from a third-order Markov background model as negative samples, our method identified 25. Table 1 shows 28 examples of 45 TFs used in our experiments, and shows discriminative substrings discovered by our method and discovered motifs with other four programs. In Table 1, motifs that were consistent with the published consensus sequences are underlined for our method and YMF and are shown with the mark of rectangles for AlignACE, MEME, and MDScan.

Our method clearly outperformed other programs, because all the existing methods only focus on the positive samples even if some of them were designed to utilize the location data. [4] In addition, our approach of using negative samples based on the location data was quite effective compared with using a random background model for negative samples. We assume that this result also contributed to the motif detecting performance of our method.

3.3. *Putative Cis-Regulatory Variants*

By performing the genome-wide search with our method on S. cerevisiae, we discovered putative functional variants for 17 TFs in total that were verified by both functional data analyses and expression profile analyses.

To assess the difference of expression profiles of two groups of targets, we used the paired t-test among all the Pearson correlations between every pair of genes within one group and those between every pair of genes each of which belongs to the different group. In other words, we assessed the difference between intra-cluster expression similarities and inter-cluster expression similarities. To select a meaningful threshold for both a hypergeometric P-value (functional enrichment) and a t-test P-value (expression difference), we calculated the average P-value of 1000 randomly selected motifs' targets for 10 times respectively, and we set a hypergeometric threshold to 0.1 and a t-test threshold to 0.01.

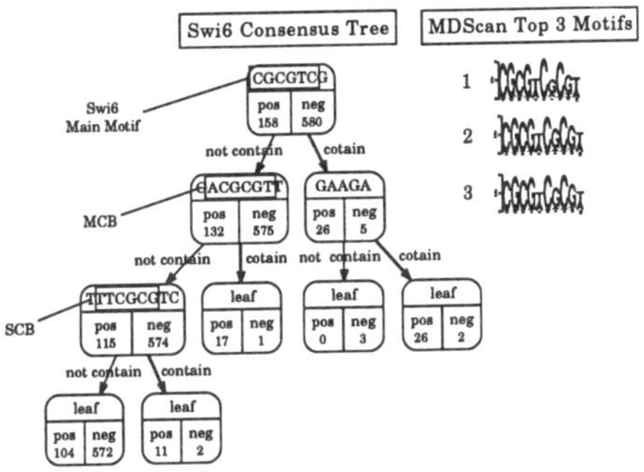

Figure 4. Relationship among different motifs induced by different complexes formed from the same non-DNA-binding cofactor, Swi6. (Swi6 forms two different complexes with different TFs, and each complex recognizes a specific motif)

Table 5. Most Associated GO Category.

Motif	Category	*P* value
CGCGTC	cell cycle	4.54 E-07
ACGCGT	cell cycle	1.10 E-06
TTCGCG	G1/S transition of mitotic cell cycle	6.82 E-07

Table 6. Top Two Associated TFs.

Motif	TFs	Overlaps	*P* value
ACGCGT	Swi6	46	1.58 E-56
	Mbp1	42	2.53 E-56
TTCGCG	Swi4	55	1.76 E-58
	Swi6	36	5.62 E-30

Table 7. Differences of Expression Profiles.

Motif		Motif	t-test *P* value	Source
CGCGTC	vs	ACGCGT	1.00 E-03	Stress Response
CGCGTC	vs	TTCGCG	1.32 E-10	Mitotic Cell Cycle
ACGCGT	vs	TTCGCG	7.81 E-06	Cell Cycle

Due to the page limitation, we only pick up Gcn4 as an example. Gcn4 regulates general control in response to amino acid or purine starvation. It involves in induction of genes required for utilization of poor nitrogen sources.

The discriminative substrings discovered in the root node and in the left children were TGACTCA (Table 1) and GATGACTAAC respectively. The discovered functional consensuses from them were TGACTC and GACTAA. Table 2-4 show the most associated functional categories, the difference of the expression profiles between those two motifs' targets, and the GO Terms for Gcn4 respectively. Table 2 and 4 indicate that targets of the main motif, TGACTC, primarily involve in the amino acid metabolism, while those of the second variant, GACTAA, involve in the nitrogen compound metabolism. Note that both target genes were predicted to be bound by the same TF from the location data, and targets of GACTAA had any significant overlaps with those of other TF's main motif. However, the expression profile analyses for them (Table 3) showed targets of GACTAA had a distinct biological property compared with those of the main motif of Gcn4 (TGACTC). Therefore, we concluded that GACTAA is a putative functional *cis*-regulatory variant of Gcn4.

3.4. *Detection of Multiple Motifs of Non-DNA-Binding Cofactors*

The representation of motifs as a hierarchical tree structure can be used for analyzing a relationship among multiple motifs induced by different complexes formed from the same cofactor. Our method correctly identified those relationships among motifs. To illustrate this, we pick up Swi6 as an example. (shown in Figure 4)

Swi6 is a non-DNA-binding cofactor of Mbp1 and Swi4. Swi6 and Mbp1 form MBF and Swi6 and Swi4 form SBF, both heterodimers are active during G1/S phase. Although Swi6 involves in both complexes, each complex recognizes a specific motif. MBF binds MCB (consensus:ACGCGT) and SBF binds SCB (consensus:CGCGAAA). Our method successfully identified both MCB and SCB from the positive and negative samples of Swi6, while MDScan failed to detect SCB. Further, our method presented the relationships between MCB and SCB as a hierarchical tree structure.

The functional consensuses of each internal node of the learned tree (Fig. 4) were CGCGTC, ACGCGT, and TTCGCG respectively. Table 5 shows the most associated GO biological process category for each motif's targets. Although these targets were predicted to be bound by Swi6 from the location data, targets of TTCGCG showed a distinct biological property. Their hypergeometric P-value associated with "cell cycle" category was just 0.00147. Table 6 shows the top two associated TFs with each motif. To determine the most associated TFs, we calculated the hypergeometric P-value of independence between targets of each motif and those of each TF's main motif, adjusted by Bonferroni correction (CGCGTC was excluded, since it was the main motif of Swi6). ACGCGT was highly associated with Mbp1, and TTCGCG was highly associated with Swi4. Table 7 shows the differences of expression profiles among each motif's targets. Targets of TTCGCG showed different expression profiles compared with others.

Table 5-7 clearly show the multimodality of Swi6. We assumed that the signal of MCB was stronger than that of SCB, since MCB-like motifs (CGCGTC, and ACGCGT) were discovered twice by our method and MDScan could only detect MCB. The consensus tree is thus able to reveal a relationship among multiple motifs of the same cofactor as a hierarchical tree structure.

4. Conclusion

We present a novel discriminative motif detection method based on the location data. Our method significantly outperformed other motif detecting methods. Further, our method successfully detected putative functional *cis*-regulatory variants and also revealed the relationships among multiple motifs of the same cofactor for several TFs. Since our motifs obtained in this paper are just substrings, ongoing efforts for combining this method with methodologies of profile hidden Markov models will be published soon.

With the progress of genome-wide location analyses, we hope that our method can provide a useful platform for analyzing the regulatory functions of motifs including functional variants, and hence present more detailed analyses for transcriptional regulations.

References

1. C. Cowles, J. Hirschhorn, D. Altshuler, and E. Lander. Detection of regulatory variation in mouse genes. *Nature Genetics*, 32(3):432–437, 2002.
2. C. Harbison, D. Gordon, T. Lee, N. Rinaldi, K. Macisaac, N. Hannett T. Danford, J. Tagne, D. Reynolds, J. Yoo, E. Jennings, et al. Transcriptional regulatory code of a eukaryotic genome. *Nature*, 431:99–104, 2004.
3. T. Lee, N. Rinaldi, F. Robert, D. Odom, Z. Bar-Joseph, G. Gerber, N. Hannett, C. Harbison, C. Thompson, I. Simon, et al. Transcriptional Regulatory Networks in *Saccharomyces cerevisiae*. *Science*, 298:799–804, 2002.
4. X. Liu, D. Brutlag, and J. Liu. An algorithm for finding protein-DNA binding sites with applications to chromatin-immunoprecipitation microarray experiments. *Nature Biotechnology*, 20(8):835–839, 2002.
5. A. Tanay, I. Gat-Viks, and R. Shamir. A Global View of the Selection Forces in the Evolution of Yeast *Cis*-Regulation. *Genome Research*, 14(5):829–834, 2004.

ON THE COMPLEXITY OF FINDING CONTROL STRATEGIES FOR BOOLEAN NETWORKS

TATSUYA AKUTSU * MORIHIRO HAYASHIDA

Bioinformatics Center, Institute for Chemical Research, Kyoto University
Uji-city, Kyoto 611-0011, Japan
E-mail: {takutsu, morihiro}@kuicr.kyoto-u.ac.jp

WAI-KI CHING †

Department of Mathematics, The University of Hong Kong
Pokfulam Road, Hong Kong, China
E-mail: wkc@maths.hku.hk

MICHAEL K. NG ‡

Department of Mathematics, Hong Kong Baptist University
Kowloon Tong, Hong Kong, China
E-mail: mng@math.hkbu.edu.hk

This paper considers a problem of finding control strategies for Boolean networks, where Boolean networks have been used as a model of genetic networks. This paper shows that finding a control strategy leading to the desired global state is NP-hard even if there is only one control node in the network. This result justifies existing exponential time algorithms for finding control strategies for probabilistic Boolean networks. On the other hand, this paper shows that the problem can be solved in polynomial time if the network has a tree structure.

1. Introduction

One of the important future directions of bioinformatics and systems biology is to develop a control theory for complex biological systems. For example, Kitano[1,2] mentions that identification of a set of perturbations that induces desired changes in cellular behaviors may be useful for systems-based drug discovery and cancer treatment. Though many attempts have been done based on control theory, existing theories and technologies are not satisfactory. Many important results in control theory are based on linear algebra, but it seems that biological systems contain many non-linear subsystems. Therefore, it is required to develop a control theory for complex biological systems.

*Work partially supported by Grant Nos. #17017019 and #16300092 from MEXT, Japan.
†Work partially supported by RGC Grant Nos. HKU 7126/02P, and HKU CRGC Grant Nos. 10203919,10204437
‡Work partially supported by RGC Grant Nos. HKU 7130/02P, 7046/03P, 7035/04P, 7035/05P.

Various mathematical models have been proposed for modeling complex and non-linear biological systems. Among them, the *Boolean network* (BN)[3] has been well-studied.[3,4,5,6,7,8] BN is a very simple model: each node (e.g., gene) takes either 0 (inactive) or 1 (active) and the states of nodes change synchronously. Though Boolean networks can not model detailed behaviors of biological systems, it may provide good approximations to the nonlinear functions appearing in many biological systems.[6] For example, Harris *et al.*[7] analyzed published data for over 150 regulated transcription systems and discussed relations between real transcription networks and Boolean networks. Therefore, it is reasonable to seek for a control theory for BNs. Even if a control theory for BNs is not practical, it may provide a new theoretical insight for systems biology.

Many studies have been done for understanding dynamical properties of BNs. For example, distribution of attractors,[3,5] relationship between network topology and chaotic behavior,[6] and inference of BNs from gene expression data[4,8] have been extensively studied. However, not much attention has been paid for finding control strategies on BNs. Recently, Datta *et al.*[9,10,11] proposed methods for finding a control strategy for probabilistic Boolean networks (PBNs), where a PBN[12] is an extension of a BN (therefore, a BN is a special case of a PBN). In their approach, it is assumed that states of some nodes can be externally controlled and the objective is to find a sequence of control actions with the minimum cost that leads to the desirable state of a network. Since BNs are special cases of PBNs, their methods can also be applied to finding a control strategy for BNs. However, their methods require high computational costs: it is required to handle exponential size matrices. Thus, their methods can only be applied to small biological systems. Therefore, it is reasonable to ask how difficult it is to find control strategies for BNs.

In this paper, we show that the control problem on BNs is NP-hard in general. This result justifies the use of exponential time algorithms for general BNs (and PBNs) as done by Datta *et al.* We further show that the control problem remains NP-hard even for some restricted cases of BNs. On the other hand, we show that the control problem can be solved in polynomial time if a BN has a tree topology. We finally discuss biological implications of the theoretical results.

2. Boolean Network and Its Control

First, we briefly review BN.[3] A BN is represented by a set of *nodes* and a set of regulation rules for nodes, where each node corresponds to a gene if BN is treated as a model of a genetic network. Each node takes either 0 or 1 at each discrete time t, a regulation rule for each node is given by a Boolean function, and the states of nodes change synchronously. An example is given in Fig. 1. In this case, the state of node v_1 at time $t+1$ is determined by the logical AND of the states of nodes v_2 and v_3 at time t. Dynamics of a BN is well-described by a state transition table shown in Fig. 1. The first row means that if the state of BN is $[0,1,1]$ at time t then the state will be $[1,0,0]$ at time $t+1$. PBN[12] is an extension of BN, in which multiple Boolean functions are assigned to each node and one function is selected at each time t according to a given probability distribution. Therefore, BN is a special case of PBN in which the same function is always selected for each node.

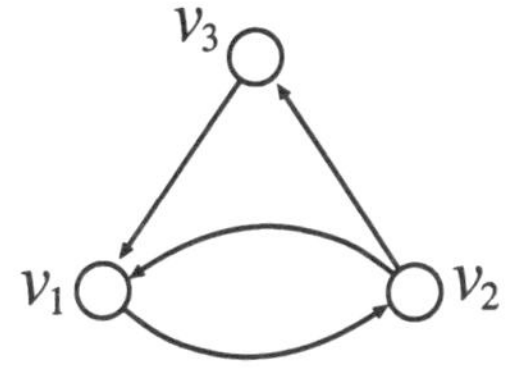

$v_1(t+1) = v_2(t)$ AND $v_3(t)$
$v_2(t+1) = v_1(t)$
$v_3(t+1) =$ NOT $v_2(t)$

time t			time t+1		
v_1	v_2	v_3	v_1	v_2	v_3
0	0	0	0	0	1
0	0	1	0	0	1
0	1	0	0	0	0
0	1	1	1	0	0
1	0	0	0	1	1
1	0	1	0	1	1
1	1	0	0	1	0
1	1	1	1	1	0

Figure 1. Example of a Boolean network (BN). Dynamics of BN (left) is well-described by a state transition table (right). For example, if the state of BN is $[0, 1, 1]$ at time t, the state will be $[1, 0, 0]$ at time $t + 1$.

In order to consider the control problem, we add *external control nodes* to a BN (original nodes are called *internal nodes*). The states of external nodes are not determined by Boolean functions. Instead, these are given externally.

Now, we formally define the control problem. A BN with external control is represented by a set V of $n + m$ nodes $V = \{v_1, \ldots, v_n, v_{n+1}, \ldots, v_{n+m}\}$, where $v_1, \ldots, v_n$ are internal nodes (corresponding to genes) and $v_{n+1}, \ldots, v_{n+m}$ are external control nodes. We also use x_i to denote an external node v_{n+i} when it is convenient to distinguish external nodes from internal nodes. Each node takes either 0 or 1 at each discrete time t, and the state of node v_i at time t is denoted by $v_i(t)$. The value of each v_i $(i = 1, \ldots, n)$ is directly controlled by k_i other nodes. Let $IN(v_i) = \{v_{i_1}, \ldots, v_{i_{k_i}}\}$ be the set of controlling elements of v_i, where $1 \leq i_j \leq n + m$. We assign to each v_i a Boolean function $f_i(v_{i_1}, \ldots, v_{i_{k_i}})$. Then the dynamics of the system is given by

$$v_i(t+1) = f_i(v_{i_1}(t), \ldots, v_{i_{k_i}}(t)).$$

We define the set of edges E by $E = \{(v_{i_j}, v_i) | v_{i_j} \in IN(v_i)\}$. Then, $G(V, E)$ is a directed graph representing network topology of a BN. We let $\mathbf{v}(t) = [v_1(t), \ldots, v_n(t)]$ and $\mathbf{x}(t) = [x_1(t), \ldots, x_m(t)]$. Note that a node without incoming edges is either an external node or a *constant node*, where a constant node is a node with a constant state.

Definition 2.1. (BN-CONTROL)
Suppose that for a BN, we are given an initial state of the network (for internal nodes) $\mathbf{v}^0$ and the desired state of the network $\mathbf{v}^M$ at the M-th time step. Then, the problem (BN-CONTROL) is to find a sequence of 0-1 vectors $\langle \mathbf{x}(0), \ldots, \mathbf{x}(M) \rangle$ such that $\mathbf{v}(0) = \mathbf{v}^0$ and $\mathbf{v}(M) = \mathbf{v}^M$. If there does not exist such a sequence, "No" should be the output.

In this paper, a *control strategy* denotes a sequence of states of control nodes $\langle \mathbf{x}(0), \mathbf{x}(1), \ldots, \mathbf{x}(M) \rangle$. Fig. 2 illustrates BN-CONTROL. The left part is a BN, where v_1, v_2, v_3 are internal nodes, and x_1, x_2 are external nodes. We are also given initial and desired states as in the right top part of Fig. 2. If the control sequence is given as in the shaded region of Fig. 2, the state of BN will change as in the right bottom part and we will have the desired state at time $t = 3$.

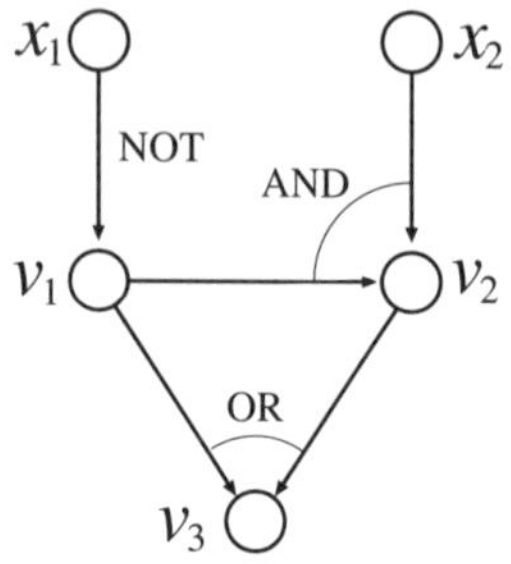

initial (t=0)	0	0	0
desired (t=3)	0	1	1

⇩

t	v_1	v_2	v_3	x_1	x_2
0	0	0	0	0	1
1	1	0	0	0	1
2	1	1	1	1	1
3	0	1	1	0	0

Figure 2. Example of BN-CONTROL. In this problem, given initial and desired states of internal nodes (v_1, v_2, v_3), it is required to compute a sequence of states of external nodes (x_1, x_2) that leads to the desired state.

The desired states of all nodes are specified in the above. However, it may not be required to specify states of all the nodes because we may be interested only in controlling several important nodes (a set of these nodes is denoted by V' in this paper). We call this case *partial BN-CONTROL.*

In this paper, we assume that the number of input variables for each Boolean function is bounded by a constant. Otherwise, it is computationally difficult to find a control strategy even for one Boolean function (for example, one can consider a function representing a SAT formula). Due to this assumption, we can assume that enumeration of satisfying assignments can be done in constant time per Boolean function.

3. Hardness of Finding Control Strategies

As mentioned before, Datta *et al.*[9,10,11] proposed algorithms for finding control strategies for PBN based on Markov chains and dynamic programming. However, their algorithms are not efficient because it is required to consider all possible states of PBN (or BN) at all time steps between the initial and final time steps. For example, we need to consider state transition matrices of size $O(2^n \times 2^n)$ because there are $O(2^n)$ possible states and transitions among them must be also considered. We show here that the control problem is NP-hard in general, which implies that the approach by Datta *et al.* is reasonable.

Theorem 3.1. *BN-CONTROL is NP-hard.*

Proof. We present a simple polynomial time reduction from 3SAT[13] to BN-CONTROL (see Fig. 3), where a similar reduction was used in a study on Bayesian networks.[14]

Let $y_1, \ldots, y_N$ be Boolean variables (i.e., 0-1 variables). Let $c_1, \ldots, c_L$ be a set of clauses over $y_1, \ldots, y_N$, where each clause is a logical OR of at most three literals. It should be noted that a literal is a variable or its negation (logical NOT). Then, 3SAT is a problem of asking whether or not there exists an assignment of 0-1 values to $y_1, \ldots, y_N$ which satisfies all the clauses (i.e., the values of all clauses are 1).

From an instance of 3SAT, we construct an instance of BN-CONTROL as follows. We let the set of nodes $V = \{v_1, \ldots, v_L, x_1, \ldots, x_N\}$ where each v_i corresponds to c_i and

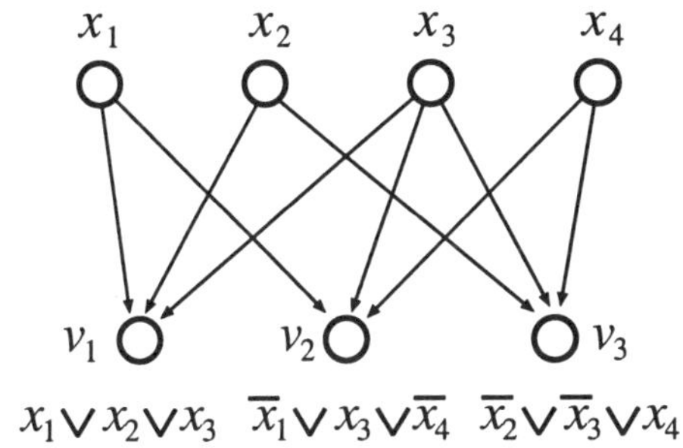

Figure 3. Reduction from 3SAT to BN-CONTROL. An instance of 3SAT $\{y_1 \vee y_2 \vee y_3, \overline{y_1} \vee y_3 \vee \overline{y_4}, \overline{y_2} \vee \overline{y_3} \vee y_4\}$ is transformed into an instance of BN-CONTROL in a simple way that external nodes correspond to variables in 3SAT, internal nodes correspond to clauses, and all the nodes must have value 1 at the desired state.

each x_j corresponds to y_j. Suppose that $f_i(y_{i_1}, \ldots, y_{i_3})$ is a Boolean function assigned to c_i in 3SAT. Then, we assign $f_i(x_{i_1}, \ldots, x_{i_3})$ to v_i in BN-CONTROL. Finally, we let $M = 1$, $\mathbf{v}^0 = [0, 0, \ldots, 0]$ and $\mathbf{v}^M = [1, 1, \ldots, 1]$.

Then, there exists a sequence $\langle \mathbf{x}(0), \mathbf{x}(1) \rangle$ which makes $\mathbf{v}(1) = [1, 1, \ldots, 1]$ if and only if there exists an assignment which satisfies all the clauses (see Fig. 3). Actually, a satisfying assignment for 3SAT corresponds to $\mathbf{x}(0)$. Since the above reduction can be done in linear time, BN-CONTROL is NP-hard. □

Since BN-CONTROL is a special case of partial BN-CONTROL, NP-hardness of partial BN-CONTROL directly follows from the above result. We can still prove that partial BN-CONTROL is NP-hard even if the desired state of only one node is specified. For that purpose, we simply add an internal node v_{L+1} to the BN in the above proof. Then, we let f_{L+1} be the conjunction of $v_1, \ldots, v_L$, and let $M = 2$, $v_{L+1}^0 = 0$ and $v_{L+1}^M = 1$.

Corollary 3.1. *Partial BN-CONTROL is NP-hard.*

Datta *et al.*[9] considered general cost functions C_k and C_M. We can consider a special case where $C_k = 0$ and C_M is the Hamming distance between the specified desired state and the final state given by a control strategy. Then, BN-CONTROL corresponds to the problem of asking whether or not the minimum cost is 0. Since BNs are special cases of PBNs, it follows that finding an optimal control strategy for PBN is NP-hard.

Corollary 3.2. *Finding an optimal control strategy for PBN is NP-hard.*

It is also possible to show that approximation of the Hamming distance is quite hard. For that purpose, we modify the network in the proof of Corollary 3.1. We add h nodes v_{L+1+i} $(i = 1, \ldots, h)$ with regulation rules $v_{L+1+i}(t+1) = v_{L+1}(t)$. Then, we let $V' = \{v_{L+2}, \ldots, v_{L+1+h}\}$, $M = 3$, $v_i^0 = 0$ and $v_i^M = 1$ for all $v_i \in V'$. Then, the cost is either 0 or h, which implies that obtaining approximate solutions (within a factor of $O(n)$ if we let $h = O(n)$) is still NP-hard.

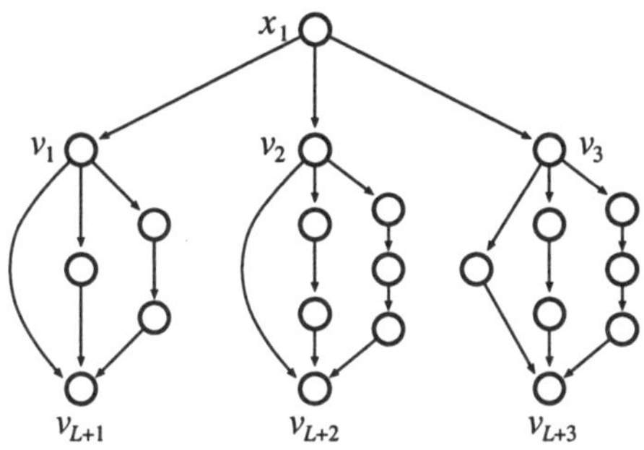

Figure 4. The network constructed (in the proof of Thm. 3.2) from the same 3SAT instance as in Fig. 3.

In the above, we used many control nodes. However, it is not plausible that we can control many genes. Thus, it is worthy to consider the following special case.

Theorem 3.2. *BN-CONTROL and partial BN-CONTROL are NP-hard even if there exists only one control node and the network structure is an almost tree of bounded degree.*

Proof. We give a proof for the partial control problem. Modification of the proof for BN-CONTROL is omitted in this version. As in Thm. 3.1, we use a reduction from 3SAT (see also Fig. 4). We construct an instance of the partial control problem so that the sequence of values of the single control node x_1 constitutes the satisfying assignment. For each clause c_i, we construct two special nodes v_i and v_{L+i}. Suppose that variables $y_{i_1}, y_{i_2}, y_{i_3}$ appear in clause c_i in 3SAT. Then, we create 3 paths from v_i to v_{L+i}, where the lengths of paths are i_1, i_2 and i_3, respectively. The identify function is assigned to each gene (except v_{L+i}) in the paths, and a function corresponding to c_i is assigned for v_{L+i}. Then, we let $V' = \{v_{L+1}, \ldots, v_{2L}\}$, $M = N + 1$, $v_i^0 = 0$ and $v_i^M = 1$ for $v_i \in V'$.

Then, the state $x_1(N - i)$ corresponds to an assignment of 0-1 value to y_i. From this, there exists a sequence $\langle \mathbf{x}(0), \mathbf{x}(1), \ldots, \mathbf{x}(N + 1) \rangle$ which makes $v_i(N + 1) = 1$ for all $v_i \in V'$ if and only if there exists an assignment which satisfies all the clauses. Therefore, partial BN-CONTROL is NP-hard even if there is only one control input.

Note that the above network structure belongs to the class of *almost trees*, where an undirected graph is called an almost tree if the number of edges in each bi-connected component is at most the number of nodes in the component plus some constant. Though the degree of x_1 can be high, it can be reduced to 3 by using a substructure like binary tree. □

4. Algorithms for Trees

In this section, we present polynomial time algorithms for special cases of the control problem. First, we consider the case where the network has a rooted tree structure (all paths are directed from leaves to the root). In order to compute a control strategy, we employ dynamic programming. Though dynamic programming is also employed in exponential time algorithms[9,10] for PBNs, it is used here in a significantly different way.

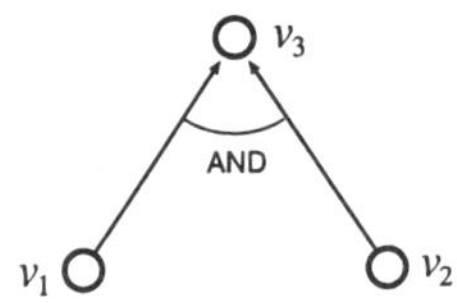

Figure 5. Computation of $S[v_3, t, 1]$. In this case, $S[v_3, t+1, 1] = 1$ if and only if $S[v_1, t, 1] = 1$ and $S[v_2, t, 1] = 1$. $S[v_3, t+1, 0] = 1$ if and only if $S[v_1, t, 0] = 1$ or $S[v_2, t, 0] = 1$.

In order to apply dynamic programming, we define $S[v_i, t, b]$ as below, where v_i is a node, t is a time step and b is a Boolean value (i.e., 0 or 1). Here $S[v_i, t, b]$ is 1 if there exists a control sequence (up to time t) that makes $v_i(t) = b$ (see also Fig. 5).

$$S[v_i, t, 1] = \begin{cases} 1, \text{ if there exists } \langle \mathbf{x}(0), \ldots, \mathbf{x}(t) \rangle \text{ such that } v_i(t) = 1, \\ 0, \text{ otherwise.} \end{cases}$$

$$S[v_i, t, 0] = \begin{cases} 1, \text{ if there exists } \langle \mathbf{x}(0), \ldots, \mathbf{x}(t) \rangle \text{ such that } v_i(t) = 0, \\ 0, \text{ otherwise.} \end{cases}$$

Then, $S[v_i, t, 1]$ can be computed by the following dynamic programming procedure.

$$S[v_i, t+1, 1] = \begin{cases} 1, \text{ if there exists } [b_{i_1}, \ldots, b_{i_k}] \text{ such that } f_i(b_{i_1}, \ldots, b_{i_k}) = 1 \text{ holds and} \\ \quad S[v_{i_j}, t, b_{i_j}] = 1 \text{ holds for all } j = 1, \ldots, k, \\ 0, \text{ otherwise.} \end{cases}$$

$S[v_i, t, 0]$ can be computed in a similar way. It should be noted that each leaf is either a constant node or an external node. For a constant node, either $S[v_i, t, 1] = 1$ and $S[v_i, t, 0] = 0$ hold for all t, or $S[v_i, t, 1] = 0$ and $S[v_i, t, 0] = 1$ hold for all t. For an external node, $S[v_i, t, 1] = 1$ and $S[v_i, t, 0] = 1$ hold for all t.

In the control problems, states of some (or all) internal nodes at the M-th step (more generally, at the t-th step) may be specified. Let $C[v_i, t, b] = 1$ denotes the constraint that the state of v_i at the t-th step can be b ($b \in \{0, 1\}$), otherwise $C[v_i, t, b] = 0$. For example, if $v_i(M) = 1$ must hold, we let $C[v_i, M, 1] = 1$ and $C[v_i, M, 0] = 0$. Then, we can modify the recurrence in dynamic programming as:

$$S[v_i, t+1, 1] = \begin{cases} 1, \text{ if } C[v_i, t+1, 1] = 1 \text{ and there exists } [b_{i_1}, \ldots, b_{i_k}] \text{ such that} \\ \quad f_i(b_{i_1}, \ldots, b_{i_k}) = 1 \text{ holds and } S[v_{i_j}, t, b_{i_j}] = 1 \text{ holds for all} \\ \quad j = 1, \ldots, k, \\ 0, \text{ otherwise.} \end{cases}$$

Then, we can decide whether or not there exists a control sequence by checking whether $S[v, M, 1] = 1$ or $S[v, M, 0] = 1$ holds for each node v. The required control sequence can be obtained by using the well-known traceback technique.[15]

Based on the above algorithm, we have the following theorem where the proof is omitted in this version.

Theorem 4.1. *If a BN has a rooted tree structure, both BN-CONTROL and partial BN-CONTROL can be solved in* $O((n+m)M)$ *time.*

We can generalize Thm. 4.1 for the case of unrooted trees. We call v_i a *branching node* if v_i has at least two outgoing edges. We call v_i an *outmost branching node* if either v_i is the only one branching node, or all paths from v_i to other branching nodes must pass the same branching node v_j. We denote such v_j by $nb(v_i)$.

Then, we can determine $S_O[v_i, t, b]$'s by repeatedly removing outmost branching nodes (see also Fig. 6 and Fig. 7), where we use $S_O[v_i, t, b]$ to denote the required table. For an outmost branching node v, we let

$$\Gamma^+(v) = \{w|(v,w) \in E\} - \{u\} \quad \text{and} \quad \Gamma^-(v) = \{w|(w,v) \in E\} - \{u\},$$

where u is the node adjacent to v and lying between v and $nb(v)$. If there is only one branching node, u can be empty. For each adjacent node w (except u) of v, we let $T_{v,w}$ be the subtree induced by $\{v,w\} \cup \{z|dist(v,z) < dist(nb(v),z)\}$, where $dist(v,z)$ denotes the number of edges of the path connecting v and z (without considering directions of edges). If $(u,v) \in E$, T_v is the subtree induced by v, u and the nodes in $\cup_{w \in \Gamma^-} T_{v,w}$. Otherwise (i.e., $(v,u) \in E$ or u is empty), T_v is the subtree induced by v and the nodes in $\cup_{w \in \Gamma^-} T_{v,w}$. It is worthy to note that $T_{v,w}$ is always a rooted tree and thus the algorithm for rooted trees can be used as a subroutine. Using the following procedure, we can determine $S_O[v, t, b]$.

Procedure BN-CONTROL-TREE
  for all v, t and $b \in \{0,1\}$ **do** $S_O[v,t,b] \leftarrow 1$; $C[v,t,b] \leftarrow 1$
  while there exists a branching node **do**
    Select an arbitrary outmost and non-processed branching node v
    for all $w \in \Gamma^+(v)$ **do**
      for all t_O and b_O **do**
        if there does not exist a control strategy for $T_{v,w}$ such that $S[v, t_O, b_O] = 1$
        then $S_O[v, t_O, b_O] \leftarrow 0$
      Delete nodes in $T_{v,w}$ (except v)
    for all t and b **do** $C[v,t,b] \leftarrow S_O[v,t,b] \wedge C[v,t,b]$
    if $(u,v) \in E$ **then**
      for all t_O and b_O **do**
        if there does not exist a control strategy for T_v such that $S[u, t_O, b_O] = 1$
        then $S_O[u, t_O, b_O] \leftarrow 0$
      for all t and b **do** $C[u,t,b] \leftarrow S_O[u,t,b] \wedge C[u,t,b]$
      Delete nodes in T_v (including v)
    else
      for all t_O and b_O **do**
        if there does not exist a control strategy for T_v such that $S[v, t_O, b_O] = 1$
        then $S_O[v, t_O, b_O] \leftarrow 0$
      for all t and b **do** $C[v,t,b] \leftarrow S_O[v,t,b] \wedge C[v,t,b]$
      Delete nodes in T_v (except v)

Based on the above procedure, we have the following where the proof is omitted here.

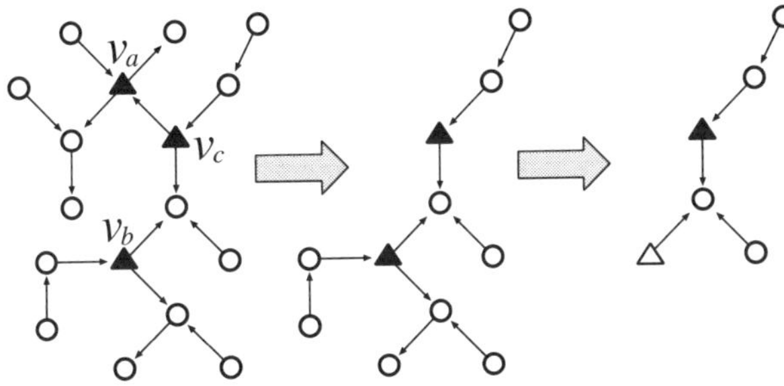

Figure 6. Illustration of the procedure for unrooted trees, where v_a, v_b and v_c are branching nodes. At the beginning, v_a and v_b are outmost branching nodes and $nb(v_a) = nb(v_b) = v_c$.

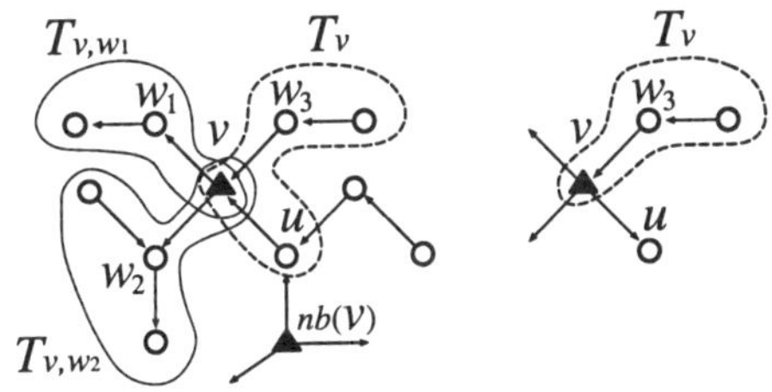

Figure 7. Example of $T_{v,w}$ and T_v. It should be noted that T_v includes u if $(u, v) \in E$ (left), whereas T_v does not include u if $(v, u) \in E$ (right). In both cases, $\Gamma^+(v) = \{w_1, w_2\}$ and $\Gamma^- = \{w_3\}$.

Theorem 4.2. *If a BN has a tree structure, both BN-CONTROL and partial BN-CONTROL can be solved in $O((n + m)M^2)$ time.*

The above algorithm may also be useful even if the network has a few loops. Suppose that the network becomes a forest if H nodes are removed. Though it is difficult to find the minimum H, a greedy approach may work well to find an appropriate H. Then, we examine all possible time series for these H nodes and apply the algorithm in Thm. 4.2. This tree-based method takes $O(2^{HM}(m + n)M^2)$ time. On the other hand, we can use the algorithm by Datta *et al.*[9] to solve BN-CONTROL and partial BN-CONTROL. Then, it will take $O(2^{2n+m}M)$ time. However, it is very time consuming even for small n (e.g., $n = 10$). Therefore, the tree-based method may be much more useful for BN-CONTROL and partial BN-CONTROL than the algorithm by Datta *et al.* when HM is small enough.

It should also be noted that the algorithm for trees can be extended for other discrete and finite domains. For that purpose, we modify $S[v_i, t, b]$ so that b takes values in the target domain and we replace Boolean functions with discrete functions for the domain.

5. Concluding Remarks

We have shown that finding a control strategy for Boolean networks is computationally very hard. Hardness results still hold for other models of biological systems if those can represent Boolean formula for 3SAT using control variables. Since close relationships between biological systems and Boolean circuits are suggested,[7,16,17] it seems difficult to find control strategies efficiently for all types of biological networks.

However, many biological sub-networks have special features. For example, Kitano[1,2] suggested that negative feedback loops play an important role in biological systems: these contribute to keeping robustness of biological systems. Such sub-networks are considered to be significantly different from the networks constructed in this paper because it seems impossible to describe negative and robust feedback loops using Boolean functions. Therefore, one of important future studies is to develop an efficient algorithm for finding control strategies for such robust sub-networks.

References

1. H. Kitano. Computational systems biology. *Nature*, 420:206–210, 2002.
2. H. Kitano. Cancer as a robust system: implications for anticancer therapy. *Nature Reviews Cancer*, 4:227–235, 2004.
3. S. A. Kauffman. *The Origins of Order: Self-organization and Selection in Evolution.* Oxford Univ. Press, 1993.
4. T. Akutsu, S. Miyano and S. Kuhara. Inferring qualitative relations in genetic networks and metabolic pathways. *Bioinformatics*, 16:727–734, 2000.
5. R. Albert and A-L. Barabási. Dynamics of complex systems: scaling laws for the period of Boolean networks. *Physical Review Letters*, 84:5660-5663, 2000.
6. L. A. Amaral, A. Diaz-Guilera, A. A. Moreira, A. K. Goldberger and L. A. Lipsitz. Emergence of complex dynamics in a simple model of signaling networks. *Proc. National Academy of Sciences USA*, 101:15551-15555, 2004.
7. S. E. Harris, B. K. Sawhill, A. Wuensche and S. Kauffman. A model of transcriptional regulatory networks based on biases in the observed regulation rules. *Complexity*, 7:23-40, 2002.
8. S. Liang, S. Fuhrman and R. Somogyi. REVEAL, a general reverse engineering algorithm for inference of genetic network architectures. *Proc. Pacific Symposium on Biocomputing*, 3:18–29, 1998.
9. A. Datta, A. Choudhary, M. L. Bittner and E. R. Dougherty. External control in Markovian genetic regulatory networks. *Machine Learning*, 52:169-191, 2003.
10. A. Datta, A. Choudhary, M. L. Bittner and E. R. Dougherty. External control in Markovian genetic regulatory networks: the imperfect information case. *Bioinformatics*, 20:924–930, 2004.
11. A. Datta, A. Choudhary, M. L. Bittner and E. R. Dougherty. Intervention in context-sensitive probabilistic Boolean networks. *Bioinformatics*, 21:1211-1218, 2005.
12. I. Shmulevich, E. R. Dougherty, S. Kim and W. Zhang. Probabilistic Boolean networks: a rule-based uncertainty model for gene regulatory networks. *Bioinformatics*, 18:261–274, 2002.
13. M. R. Garey and D. S. Johnson. *Computers and Intractability. A Guide to the Theory of NP-Completeness.* W.H. Freeman and Co., 1979.
14. G. Cooper. The computational complexity of probabilistic inference using Bayesian belief networks. *Artificial Intelligence*, 42:393-405, 1990.
15. P. Clote and R. Backofen. *Computational Molecular Biology: An Introduction.* John Wiley and Sons Ltd., 2000.
16. H. H. McAdams and L. Shapiro. Circuit simulation of genetic networks. *Science*, 269:650–656, 1995.
17. C-H. Yuh, H. Bolouri and E. H. Davidson. Genomic Cis-regulatory logic: experimental and computational analysis of a sea urchin gene. *Science*, 279:1896–1902, 1998.

CHARACTERIZATION OF MULTI-CHARGE MASS SPECTRA FOR PEPTIDE SEQUENCING

KET FAH CHONG, KANG NING,[†] HON WAI LEONG

Department of Computer Science, National University of Singapore, 3 Science Drive 2, Singapore 117543

PAVEL PEVZNER

Department of Computer Science & Engineering, University of California, San Diego, La Jolla, CA 92093-0114

Sequencing of peptide sequences using tandem mass spectrometry data is an important and challenging problem in proteomics. In this paper, we address the problem of peptide sequencing for multi-charge spectra. Most peptide sequencing algorithms currently handle spectra of charge 1 or 2 and have not been designed to handle higher-charge spectra. We give a characterization of multi-charge spectra by generalizing existing models. Using these new models, we have analyzed spectra with charges 1-5 from the GPM [8] datasets. Our analysis shows that higher charge peaks are present and they contribute significantly to prediction of the complete peptide. They also help to explain why existing algorithms do not perform well on multi-charge spectra. We also propose a new *de novo* algorithm for dealing with multi-charge spectra based on the new models. Experimental results show that it performs well on all spectra, especially so for multi-charge spectra.

1 Introduction

Proteomics is the large-scale study of proteins, particularly their sequences, structures and functions. In proteomics, the identification of the protein sequences is very important, and peptide sequencing is essential to the identification of the proteins. Currently, peptide sequencing is largely done by tandem mass spectrometry. The analysis of the spectrum data is a non-trivial problem. This is in part because the spectrum obtained from MS/MS usually contains lots of noise, which do not belong to the peptide, but introduced because of the impurity of the peptide, and the inaccuracy of the machines. The problem becomes more difficult since for one peptide sequence, not all of its subsequences have the corresponding ions in the spectrum.

Deducing peptide sequences from raw MS/MS data is slow and tedious when done manually. Instead, the most popular approach is to do a *database search* of known peptide sequences with the un-interpreted experimental MS/MS data. A number of such database search algorithms have been described, the most popular being Mascot [1] and Sequest [2]. These methods are effective but often give false positives or incorrect identifications. Searching databases with masses and partial sequences (sequence tags) derived from MS/MS data give more reliable results [3]. For unknown peptides, *de novo* sequencing [4-7] is used in order to predict sequences or partial sequences. However, the

[†] Contact: ningkang@comp.nus.edu.sg.

prediction of peptide sequences from MS/MS spectra is dependent on the quality of the data, and this result in good predicted sequences only for very high quality data.

This paper focuses on the important issue of the *amount of charge* on the ions in the spectra, particularly multi-charge spectra (charges 3 to 5). In the case of an ESI/MALDI source, the parent ion and many fragments may have multiple charge units assigned to them. Multi-charged spectra (with charges up to 5) are available from the GPM [8] web-site. Current *de novo* methods work well on good quality spectra of charges 1 and 2. However, they do not work well on spectra with charges 3 to 5 since they do not explicitly handle multi-charge ions (one notable exception is PEAKS [6] which does conversion of multi-charge peaks to their singly-charged equivalent before sequencing). Lutefisk [7] works with singly-charged ion only, while Sherenga [4] and PepNovo [5] works with singly- and doubly-charged ions. Therefore, it is not surprising that some of the higher charged peaks are mis-annotated by these methods leading to lower accuracy.

In this paper, we propose *a generalized model* that better describes multi-charge spectra (multi-charge to mean charge $\geq$ 3) and *quality measures* for multi-charge spectra based on the new model. Our evaluation of multi-charged spectra from GPM with the new model shows that the theoretically attainable accuracy increases as we consider higher charge ions meaning that multi-charge ions are significant. In addition, we show that *any* algorithm that considers *only* charge 1 or 2 ions will suffer from low prediction accuracy. Our experiments show that the accuracy[1] of these methods on multi-charge spectra is very low (less that 35%), and this accuracy decrease as the charge of the spectra increases (for charge 4 spectra, the accuracy of Lutefisk is less than 7%).

We also proposed a *simple de novo* sequencing algorithm called *GBST* (greedy best strong tag) that considers higher charge ions based on our new model. Experimental results on GPM spectra show that *GBST* outperforms many of the other *de novo* algorithms on spectrum data with charge of 3 or more.

2 Modeling of Multi-Charge Spectra

Consider an experimental mass spectrum $S = \{p_1, p_2, \ldots p_n\}$ of maximum charge α that is produced by an MS/MS experiment on a peptide $\rho = (a_1 a_2 \ldots a_l)$, where a_j is the jth amino acid in the sequence. The parent mass of the peptide ρ is given by $M = m(\rho) = \sum_{j=1}^{l} m(a_j)$. Consider a peptide prefix fragment $\rho_k = (a_1 a_2 \ldots a_k)$, for $k \leq n$, that has mass $m(\rho_k) = \sum_{j=1}^{k} m(a_j)$. Suffix masses are defined similarly. Then, the set of all possible prefixes and suffixes of a peptide forms the "full ladder" of the peptide. Let $TS_0(\rho) = \{m(\rho_1), m(\rho_2), \ldots, m(\rho_n)\}$ to be the set of all possible (*uncharged*) prefix fragment masses of the peptide ρ. A peak in the experimental spectrum S then corresponds to the detection of some charged prefix or suffix peptide fragment that results from peptide fragmentation in the mass spectrometer. Each peak p_i in the experimental spectrum S is described by its *intensity*(p_i) and *mass-to-charge ratio* $mz(p_i)$.

---

[1] The accuracy measure we use is defined in Section 3.3.

However fragmentation is usually not very clean and other types of fragments occur. Noise and contaminants can also cause a peak in the experimental spectrum. In peptide sequencing, we are given an experimental spectrum with true peaks and noise and the problem is to try to determine the original peptide ρ that produced the spectrum.

The Theoretical Spectrum for a Known Peptide: To theoretically characterize a multi-charge spectrum of a known peptide ρ, we consider the set of all *possible* true peaks that correspond to prefix fragments (N-terminal ions) and suffix fragments (C-terminal ions). Each peak p can be characterized by the ion-type, that is specified by $(z, t, h) \in (\Delta_z \times \Delta_t \times \Delta_h)$, where z is the charge of the ion, t is the basic ion-type, and h is the neutral loss incurred by the ion. In this paper, we restrict our attention to the set of ion-types $\Delta = (\Delta_z \times \Delta_t \times \Delta_h)$, where $\Delta_z = \{1,2,\ldots, \alpha\}$, $\Delta_t = \{a\text{-ion}, b\text{-ion}, y\text{-ion}\}$ and $\Delta_h = \{\varnothing, -H_2O, -NH_3\}$.[2] The (z, t, h)-ion of the peptide fragment q (prefix or suffix fragment) will produced an observed peak p_i in the experimental spectrum S that has a mass-to-charge ratio of $mz(p)$, that can be computed using a shifting function, *Shift*, defined as follows:

$$m(q) = Shift(p_i, (z,t,h)) = mz(p_i) \cdot z + (\delta(t) + \delta(h)) - (z-1) \qquad (1)$$

where $\delta(t)$ and $\delta(h)$ are the mass differences associated with the ion-type t and the neutral-loss h, respectively. We say that peak p_i is a *support peak* for the fragment q and *has ion-type* (z, t, h) and we say that the fragment q is *explained* by the peak p_i.

We define the *theoretical spectrum* $TS_\alpha^\alpha(\rho)$ *for* ρ *for maximum charge* α to be the set of all *possible* observed peaks that may be present in an experimental spectrum for the peptide ρ with maximum charge α. More precisely, $TS_\alpha^\alpha(\rho) = \{ p : p$ is an observed peak for the (z, t, h)-ion of peptide prefix fragment ρ_k, for all $(z, t, h) \in \Delta$ and $k=1,\ldots,n\}$.

Extended Spectrum: Conversely, the *real* peaks in an experimental spectrum $S = \{p_1, p_2, \ldots p_n\}$ of maximum charge α, may have come from different ion-type of different fragments (may be prefix or suffix fragment, depending on the ion-type). We do not know, a priori, the ion-type $(z, t, h) \in \Delta$ of each peak p_i. Therefore, we "extend" each peak p_i by generating a set of $|\Delta|$ pseudo-peaks (or guesses), one for each of the different ion-types $(z, t, h) \in \Delta$. More precisely, in the extended spectrum S_α^α, for each peak $p_i \in S$ and ion-type $(z, t, h) \in \Delta$, we generate a pseudo-peak, denoted by $(p_i, (z, t, h))$, with an "assumed" (uncharged) fragment mass computed using the *Shift* function (1). At most one of these pseudo-peaks is a real peak, while the others are "introduced" noise.

We always express a fragment mass in experimental spectrum using its *PRM* (prefix residue mass) representation, which is the mass of the prefix fragment. For suffix fragments (y-ions), we use its corresponding prefix fragment. Mathematically, for a fragment q with mass $m(q)$, we define $PRM(q) = m(q)$ if q is a prefix fragment ($\{b\text{-ion}\}$); and we define $PRM(q) = M - m(q)$ if q is a suffix fragment ($\{y\text{-ion}\}$). By calculating the *PRM* for all fragments, we can treat all fragments masses uniformly.

---

[2] The definitions and results in this paper also apply to any set of ion-types considered.

We illustrate the extended spectrum with an example shown in Figure 1. For simplicity, we only consider ion-types Δ_t = {*b*-ions, *y*-ions} and Δ_h={Ø}. Given a peptide ρ = GAPWN, with parent mass $M = m(\rho) = 525.2$, and an experimental spectrum S = {113.6, 412.2, 487.2} with maximum charge 2. The first peak "113.6" is a (2, *b*-ion, Ø)-ion of the prefix fragment GAP; the peak 412.2 is a (1, *b*-ion, Ø))-ion of the prefix fragment GAPW; and "487.2" is a (1, *y*-ion, Ø)-ion for the fragment G. In Figure 1(a), only charge 1 is considered and S_1^2 = {112, 430, 411, 132, 486, 57}. The entries in the table are the *PRM* values. For example, the possible fragment masses of 112 and 430 correspond to the extension of the first peak for ion-types (1, *b*-ion, Ø) and (1, *y*-ion, Ø), respectively. However, if charge 2 is also considered, then S_2^2 = {112, 430, 225, 31, 411, 132, 486, 57} as shown in Figure 1(b).

Duality between Extended Spectrum and Theoretical Spectrum: We now describe a duality relationship between the extended spectrum S_β^α and the theoretical spectrum $TS_\beta^\alpha(\rho)$. Given an experimental spectrum S of a known peptide ρ, the set $RP_\alpha^\alpha(S,\rho)$ of *real peaks* in the spectrum S is given by:

$$RP_\alpha^\alpha(S,\rho) = TS_\alpha^\alpha(\rho) \cap S \tag{2}$$

The set $EF_\alpha^\alpha(S,\rho)$ of *explained fragments* in the peptide ρ, namely fragments that can be "explained" by the presence of support peak or pseudo-peak in S_α^α, is given by:

$$EF_\alpha^\alpha(S,\rho) = TS_0(\rho) \cap PRM(S_\alpha^\alpha)\,. \tag{3}$$

In the set $RP_\alpha^\alpha(S,\rho)$, there may be several real peaks that are support peaks for the same fragment . Similarly, in the set $EF_\alpha^\alpha(S,\rho)$, there may be multiple pseudo-peaks in S, that helps to "explain" the same fragment. Indeed, we have the following duality theorem:

Duality Theorem: Given an experimental spectrum S of a known peptide ρ, we have

$$EF_\alpha^\alpha(S,\rho) = PRM(Shift(RP_\alpha^\alpha(S,\rho))) \tag{4}$$

Modelling Current Algorithms: To take into account the fact that some algorithms consider only ion-types of charge up to β (usually β = 2), we extend the definition to $TS_\beta^\alpha(\rho)$ which is defined to be the subset of $TS_\alpha^\alpha(\rho)$ for which the charge $z \in \{1,2,\ldots,\beta\}$. The case β=1 reflects the assumption that all peaks are of charge 1, and makes use of the extended spectrum S_1^α. Algorithms such as PepNovo and Lutefisk works with a subset of the extended spectrum S_2^α, even for spectra with charge $\alpha > 2$. In general, $TS_\beta^\alpha(\rho)$ does not account for peaks that correspond to ion-types with higher charges $z=\beta+1, \ldots, \alpha$. Of course, the more charge we take into account, the more accurate will be the accuracy that can be attained since $TS_1^\alpha(\rho) \subseteq TS_2^\alpha(\rho) \ldots \subseteq TS_\alpha^\alpha(\rho)$.

The Extended Spectrum Graph: We also introduce an extended spectrum graph, denoted by $G_d(S_\beta^\alpha)$, where d is the "connectivity". Each vertex v in this graph represents a pseudo-peak (p_i, (z, t, h)) in the extended spectrum S_β^α, namely, the (z, t, h)-

ions for the peak p_i. Thus $v = (p_i, (z, t, h))$. Each vertex represents a possible peptide fragment mass given by $PRM(Shift(p_j, (z, t, h)))$. Two special vertices are added – the start vertex v_0 corresponding to the empty fragment with mass 0 and the end vertex v_M corresponding to the parent mass M.

In the "standard" spectrum graph, we have a directed edge (u, v) from vertex u to vertex v if $PRM(v)$ is larger than $PRM(u)$ by the mass of a single amino acid. In the extended spectrum graph of connectivity d, $G_d(S_\beta^\alpha)$, we extend the edge definition to mean "*a directed path of no more than d amino acids*". Thus, we connect vertex u and vertex v by a directed edge (u, v) if the $PRM(v)$ is larger than $PRM(u)$ by the total mass of d' amino acids, where $d' \leq d$. In this case, we say that the edge (u, v) is connected by a path of length up to d amino acids. Note that the number of possible paths to be searched is 20^d and increased exponentially with d. We use d=2, unless otherwise stated.

z	$mz(p_1)$= 113.6		$mz(p_2)$= 412.2		$mz(p_3)$=487.2	
	B	*Y*	*B*	*Y*	*B*	*Y*
1	V_1 112.6	V_2 430.6	V_3 411.2	V_4 132	V_5 486.2	V_6 57

(a) The spectrum S_1^2 (only B and Y ions considered)

z	$mz(p_1)$= 113.6		$mz(p_2)$= 412.2		$mz(p_3)$=487.2	
	B	*Y*	*B*	*Y*	*B*	*Y*
2	V_7 225.2	V_8 318	- -	- -	- -	- -

(b) Extending the peaks for charge 2 ions.

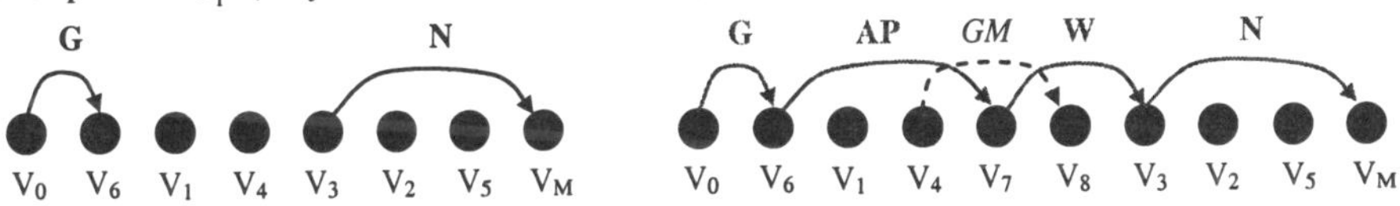

(c) The spectrum graph $G_2(S_1^2)$ (d) The extended spectrum graph $G_2(S_2^2)$

Figure 1. Example of extended spectrum graph for mass spectrum regenerated from peptide GAPWN.

Two extended spectrum graphs (with connectivity d=2) are shown in Figure 1. The spectrum graph $G_2(S_1^2)$ is shown in Figure 1(c). We can see that only the edges (v_0, v_6) for amino acid G and (v_3, v_M) for amino acid N can be obtained. The subsequence APW is longer than 2 amino acids long and so $G_2(S_1^2)$ is unable to elucidate this information. By considering S_2^2 (in (a) and (b)), we obtain the graph $G_2(S_2^2)$ shown in (d). New edges can be obtained, edge (v_6, v_7) for path AP of length 2 amino acids and (v_7, v_3) for amino acid W. This gives a full path from v_0 to v_M and the full peptide can now be elucidated. However we also note that in $G_2(S_2^2)$, fictitous edges may also be introduced due to the introduction of more noise. One example is shown in (d) using dashed line for the fictitious edge (v_4, v_8). Many such fictituous edges can result in fictituous paths from v_b to v_e, thus giving a higher rate of false positives.

2.1. *Quality Measures for Evaluating Mass Spectra*

We have extensively analyzed many multi-charge spectra using our new characterization. In this exercise, we are only analyzing the quality of the spectra, and we are not doing sequencing or prediction. We define two *quality measures* of a multi-charge spectra

- *Specificity*(α, β) $= |TS_\beta^\alpha(\rho) \cap S| \,/\, |S|$ $= |RP_\beta^\alpha(S,\rho)| \,/\, |S|$
- *Completeness*(α, β) $= |TS_0(\rho) \cap PRM(S_\alpha^\alpha)| \,/\, |\rho|$ $= |EF_\beta^\alpha(S,\rho)| \,/\, |\rho|$

Specificity measures the proportion of true peaks in the experimental spectrum *S*, and it can be also be consider the signal-to-noise ratio of *S*. However, for a given PRM, there may be multiple support peaks in $RP_\beta^\alpha(S,\rho)$, which lead to "double counting". The completeness measure avoids this by computing the proportion of the fragment masses that are explained by support peaks. Multiple support peaks for the same fragments are not double-counted.

2.2. *Experimental Data and Analysis*

The data being used for analysis and experimentation is the Amethyst data set from GPM (Global Proteome Machine) [8] (obtainable from ftp://ftp.thegpm.org/quartz). The GPM system is an open-source system for analyzing, storing, and validating proteomics information derived from tandem mass spectrometry. The database was designed to store the minimum amount of information necessary to search and retrieve data obtained from the publicly available data analysis servers. One feature of the Amethyst dataset is that there are lots of multi-charge spectra (up to charge 5). These data are MS/MS spectra obtained from QSTAR mass spectrometers. Both MALDI and ESI sources were included.

Using the $G_d(S_\beta^\alpha)$ extended spectrum graph model (with *d*=2), we have measured the average *Specificity*(α,β) and *Completeness*(α,β) on the enture Amethyst datasets from GPM using our extended spectra S_β^α for $1 \le \alpha \le 5$, and $1 \le \beta \le \alpha$. A mass tolerance of 0.5 Da is used for matching peak mass-to-charge ratios. All the data in the Amethyst dataset (12558 datasets in total, with 4000, 4561, 2483, 1175, 339 for charge 1, 2, 3, 4, 5, respectively) has been used for this purpose.

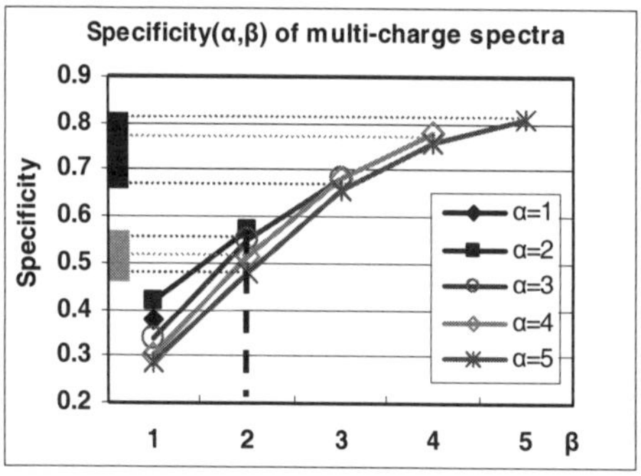

Figure 2. *Specificity*(α,β) of multi-charge spectra. Specificity increases as β increases. Most algorithms consider up to S_2^α (dashed black line). But considering S_α^α for spectra with $\alpha \ge 3$ improves the specificity (black line vs grey line).

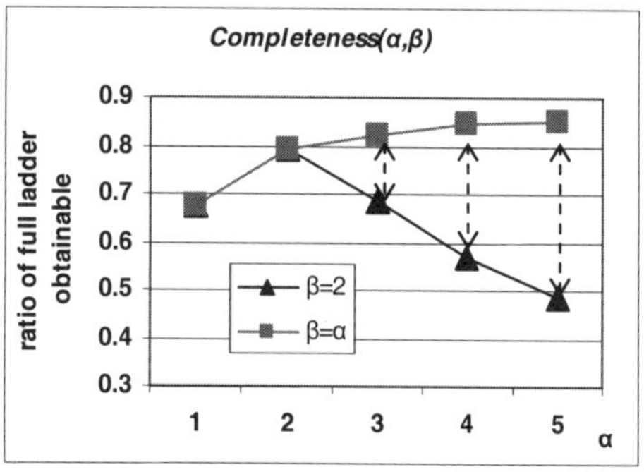

Figure 3. *Completeness*(α,β) of multi-charge spectra. We see that considering only S_2^α gives < 70% of the full ladder, which drops drastically as α gets bigger. On the other hand, considering S_α^α gives > 80% of full ladder.

The *Specificity*(α,β) results are showin in Figure 2. The results show that the GPM spectra contain an abundance of higher charged peaks in higher-charged spectra. For a fixed α, as β increases, the specificity increases – meaning that more true peaks are discovered. Furthermore, the increase is significant. For α=5, the specificity increases from 0.49 with β=2, to 0.81 when β=5. Algorithms that uses β =2 considering only charge 1 and 2 (like LuteFisk and PepNovo) are limited to specificity values of between 0.48 to 0.56, as indicated by the dashed vertical line at β=2.

The *Completeness*(α,β) results are showin in Figure 3. In this graph, we compare the *Completeness*(α,β) results for (a) using the full extended spectrum S_α^α versus (b) using only β=2, namely, S_2^α . Again, the results clearly show that significant improvement can be obtained by considering higher charge peaks. The disparity increases with α as seen from the widening gap indicated by the vertical arrows.

3 A Simple de Novo Algorithm for Multi-Charge Spectra

We now present a simple de novo peptide sequencing algorithm that takes into account multi-charged ion-types in the spectrum. Our main aim is to show that even with a simple algorithm, we can get improved results by considering multi-charged ions.

3.1 Strong Tags in the Multi-Charge Spectra

Tandem spectrum data analysis shows that peaks in many mass spectra can be grouped into closely-related sets, especially when the peptide is multi-charge. Within each set, the peaks can be interpreted as the same ion type (*b*-ions or *y*-ions), and the mass differences between "successive" peaks are such that they can form ladders (contiguous sequences). An example is shown in Figure 4, where we have computed the theoretical spectrum (the table) and the peaks from an experimental spectrum *S* are shown in bold. Several peaks are grouped together into contiguous sequences of *y*-ions and *b*-ions of charge 1.

This motivates us to call these contiguous sequences of strong ion-types (*b*-ions and *y*-ions of charge 1) "*strong tags*". More formally, they are defined as follows: Consider the extended spectrum graph, $G_1(S_1^\alpha)$, namely, only charge 1 ion-types. We define a *strong tag T of ion-type* (1, *t*, Ø) to be a maximal path (v_1, v_2, ..., v_r) in $G_1(S_1^\alpha)$ where each vertex $v_i \in T$ has the same ion-type (1, *t*, Ø) and each (v_i, v_{i+1}) is an edge in the graph , namely, their mass difference is the mass of one amino acid. (For our current algorithm, we consider only *b*-ions and *y*-ions, namely, *t* = *b*-ions or *y*-ions and strong tags must have at least 2 edges.)

Figure 5 shows the two strong tags obtained for the spectrum given in Figure 4.

To help the search for good strong tags, we define a *weight function* that is used to score vertices and strong tags. The weight of vertex $v_i \in G_1(S_\beta^\alpha)$ is defined as

$$w(v_i) = \frac{f_{support}(v_i) + f_{loss}(v_i) + f_{intensity}(v_i)}{f_{tolerance}(v_i)} \qquad (5)$$

- $f_{support\text{-}ion}(v_i)$ is a function of the number of v_j, with v_j having a different ion-type as v_i, but for the same subsequence
- $f_{loss}(v_i)$ is a function of the number of v_j, with $(PRM(v_i) - PRM(v_j))$=17 or 18,
- $f_{intensity}(v_i)$ is a function of ($\log_{10}$(intensity of the peak for which v_i represent)),
- $f_{tolerance}(v_i) = (1/N)\ (\sum \mid \mathrm{PRM}(v_j) - \mathrm{PRM}(v_i) - \mathrm{mass}(a_k) \mid)$, where N is the total number of incoming and outgoing edges for v_i, and a_k is the amino acid for each edge (v_i, v_j) or (v_j, v_i).

For a strong tag $T=(v_1, v_2, \ldots, v_r)$, the weight $W(T)$ of the strong tag T is just the sum of weight of the vertices in T, namely, $W(T) = \sum_{v_i \in T} w(v_i)$. Obviously, we are interested in finding a set of "best" strong tags, namely, tags that optimizes the weight $W(T)$. The spectrum graph $G_1(S_\beta^\alpha)$ is a DAG that may consist of several disjoint components. For each disjoint component C, we use a depth-first search (DFS) algorithm to compute a best strong tag for component C. We let *BST* denote the set of "best" strong tags from each of the components C in the spectrum graph.

bond	$^{+1}$y	$^{+1}$y*	$^{+1}$b	$^{+1}$b*
$^{S}1$	1807.0	1790.0	130.0	113.0
$^{I}2$	1693.9	1676.9	243.1	226.1
$^{R}3$	1537.8	1520.8	**399.2**	382.2
$^{V}4$	1438.8	1421.7	**498.3**	481.3
$^{T}5$	1337.7	1320.7	**599.3**	582.3
$^{Q}6$	1209.7	1192.6	**727.4**	710.4
$^{K}7$	1081.6	1064.5	**855.5**	838.5
$^{S}8$	994.5	977.5	942.5	925.5
$^{Y}9$	831.5	814.4	1105.6	1088.6
$^{K}10$	**703.4**	686.3	1233.7	1216.7
$^{V}11$	**604.3**	587.3	1332.8	1315.7
$^{S}12$	**517.3**	500.2	1419.8	1402.8
$^{T}13$	**416.2**	399.2	1520.8	1503.8
$^{S}14$	**329.2**	312.2	1607.9	1590.8
$^{G}15$	**272.2**	255.1	1664.9	1647.9
$^{P}16$	**175.1**	158.1	1761.9	1744.9

Figure 4. Theoretical spectrum for the peptide sequence "SIRVTQKSYKVSTSGPR", with parent mass of 1936.05 Da. "y" and "b" indicates y- and b-ions, "+1", "+2" indicates charge 1 and 2, and "*" indicates ammonia loss. Bold numbers are peaks present in experimental spectrum.

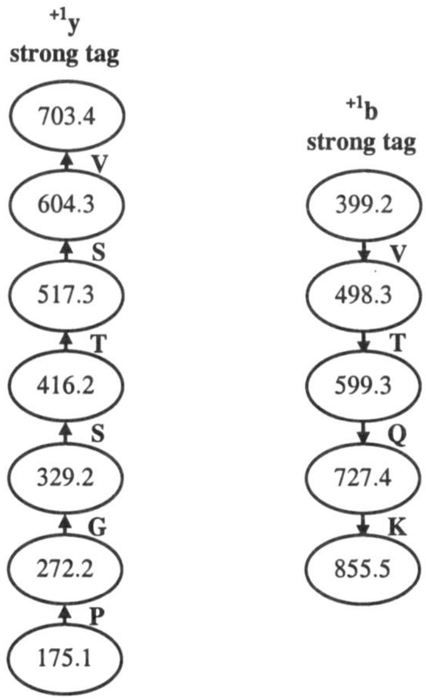

Figure 5. Example of strong tags in the spectrum graph for spectrum in Figure 4. There are 2 strong tags. Vertices (small ovals) represent fragment masses, and edges (arrows) represent amino acids whose mass are the same as the mass difference of the vertices.

3.2 *The GBST Algorithm*

We have developed a simple de novo peptide sequencing algorithm based on strong tag that we call the *Greedy Best Strong Tag (GBST)* algorithm which uses the strong tags in the spectrum graph. The *GBST* algorithm starts by computing the set *BST* of best strong tag as described in Section 3.1. After the BST is compute, the algorithm proceeds to find the best peptide sequence that can be obtained by "linking up" the strong tags in BST. We first build the strong tag graph $G_d(BST)$, where the vertices are the strong tags in *BST*, and we have an edge (u, v) from the *tail* vertex u of the tag T_u to the head vertex v of the tag T_v if $PRM(v)$ is larger than $PRM(u)$ by the total mass of d' amino acids, where $d' \leq d$. (We use d=2.) Compared to the spectrum graph G, the strong tag graph $G_d(BST)$ is very small – only $|BST|$ vertices and the number of edges is also small since we only connect strong tags in a head-to-tail manner. A path in $G_d(BST)$ is called a strong tag path since the vertices are strong tags. For a strong tag path $P = (T_1, T_2, \ldots, T_q)$, we define the weight $W(P)$ of the path P to be the sum of the weight of the strong tags in P, namely, $W(P) = \sum_{T_i \in P} W(T_i)$. The final step in the *GBST* algorithm is to use a DFS algorithm to compute the "best" strong tag path from v_0 to v_M in the graph $G_d(BST)$.

3.3 Experiments on Algorithms

The experimental data are selected from GPM spectrum datasets [8]. We have selected spectra data with different characteristics (average peak intensities, charges, etc.) for analysis. We have applied our algorithm on these spectrum data. For these spectrums, we have also compared our results with those of the Lutefisk [7] and PepNovo [5]. For comparison of prediction results, we have defined two accuracy measures:

- *Sensitivity* = #correct / $|\rho|$
- *Specificity* = #correct / $|P|$

where #correct is the "*number of correctly sequenced amino acids*". The number of correctly sequence amino acids is computed as the *longest common subsequence* (lcs) of the correct peptide sequence ρ and the sequencing result P. *Sensitivity* indicates the quality of the sequence with respect to the correct peptide sequence and a high sensitivity means that the algorithm recovers a large portion of the correct peptide. For fair comparison with algorithms like PepNovo that only outputs the highest scoring tags (subsequences) we also use the specificity measure.

Table 1: Results of GBST, compared with Lutefisk and PepNovo on GPM spectra. Results show that GBST is generally comparable and sometimes better, especially for multi-charge spectra. (*based on +1 and +2 spectra).

Charge	Number of spectrum	Lutefisk	PepNovo	GBST
1	756	0.261 / 0.258	0.322 / 0.186	***0.296 / 0.315***
2	874	0.243 / 0.241	0.316 / 0.215	***0.297 / 0.326***
3	454	0.111 / 0.113	-	***0.262 / 0.285***
4	207	0.065 / 0.063	-	***0.190 / 0.222***
5	37	0 / 0	-	***0.165 / 0.223***
All	**2328**	**0.203 / 0.202**	**0.319 / 0.202***	***0.278 / 0.304***

In the experiments, we have only run PepNovo on spectra with charge 1 and 2 (since it only handles charge 1 and 2), and compared the results with our algorithm. In Table 1, the accuracy values are represented in a (specificity/sensitivity) format.

Experiments results show that our algorithm generally perform comparable to or better than Lutefisk [7] and PepNovo [5]. This is obvious for multi-charge spectra. The relatively high specificity accuracy of our algorithms shows that our sequencing results have high signal-to-noise ratio, which are comparable with results of Lutefisk and PepNovo. The higher sensitivity accuracy shows that our algorithms can sequence more correct amino acids than Lutefisk and PepNovo.

4 Conclusion

Multi-charge spectra have not been adequately addressed by many *de novo* sequencing algorithms. In this paper, we give a characterization of multi-charge spectra and use it to analyze multi-charge spectra from GPM. Our results clearly show why existing algorithms do not perform well on multi-charged spectra. We also present a simple *de novo* sequencing algorithm (called *GBST* algorithm) which makes use of this model to predict sequences of such spectra. Our *de novo* algorithm not only works well for multi-charge spectra, but it still performs well on singly-charges spectra.

Acknowledgements

The authors would like to thank researchers at UCSD for providing us with experimental data and PepNovo program for our comparison. This work was partially supported by the National University of Singapore under grant R252-000-199-112.

References

1. D. N. Perkins, D. J. C. Pappin, D. M. Creasy and J. S. Cottrell. Probability-based protein identification by searching sequence databases using mass spectrometry data. *Electrophoresis*, 20:3551-3567, 1999.
2. J. K. Eng, A. L. McCormack and I. John R. Yates. An approach to correlate tandem mass spectral data of peptides with amino acid sequences in a protein database. *JASMS*, 5:976-989, 1994.
3. M. Mann and M. Wilm. Error-tolerant identification of peptides in sequence databases by peptide sequence tags. *Analytical Chemistry*, 66:4390-4399, 1994.
4. V. Dancik, T. Addona, K. Clauser, J. Vath and P. Pevzner. De novo protein sequencing via tandem mass-spectrometry. *J. Comp. Biol.*, 6:327-341, 1999.
5. A. Frank and P. Pevzner. PepNovo: De Novo Peptide Sequencing via Probabilistic Network Modeling. *Anal. Chem.*, 77:964 -973, 2005.
6. B. Ma, K. Zhang, C. Hendrie, C. Liang, M. Li, A. Doherty-Kirby and G. Lajoie. PEAKS: Powerful Software for Peptide De Novo Sequencing by MS/MS. *Rapid Communications in Mass Spectrometry*, 17:2337-2342, 2003.
7. J. A. Taylor and R. S. Johnson. Implementation and uses of automated de novo peptide sequencing by tandem mass spectrometry. *Anal Chem.*, 73:2594-2604, 2001.
8. R.Craig, J.P. Cortens and RC. Beavis. Open source system for analyzing, validating, and storing protein identification data. *J Proteome Res.*, 3:1234-1242, 2004.

EDAM: AN EFFICIENT CLIQUE DISCOVERY ALGORITHM WITH FREQUENCY TRANSFORMATION FOR FINDING MOTIFS

YIFEI MA GUOREN WANG
YONGGUANG LI AND YUEHAI ZHAO *

College of Information Science and Engineering
Northeastern University, Shenyang, China
E-mail: wanggr@mail.neu.edu.cn

Finding motifs in DNA sequences plays an important role in deciphering transcriptional regulatory mechanisms and drug target identification. In this paper, we propose an efficient algorithm, EDAM, for finding motifs based on frequency transformation and Minimum Bounding Rectangle (MBR) techniques. It works in three phases, *frequency transformation*, *MBR-clique searching* and *motif discovery*. In *frequency transformation*, EDAM divides the sample sequences into a set of substrings by sliding windows, then transforms them to frequency vectors which are stored in MBRs. In *MBR-clique searching*, based on the frequency distance theorems EDAM searches for MBR-cliques used for motif discovery. In *motif discovery*, EDAM discovers larger cliques by extending smaller cliques with their neighbors. To accelerate the clique discovery, we propose a range query facility to avoid unnecessary computations for clique extension. The experimental results illustrate that EDAM well solves the running time bottleneck of the motif discovery problem in large DNA database.

1. Introduction

In the process of gene expression, one or more proteins, called transcription factors have to bind to several specific regions named binding sites. These sites typically have a similar short DNA sequence pattern which is simply referred to *motif*. According to the traits of motif, the motif discovery problem is to find a pattern in sample sequences whose length is l, and in every sample sequence there is a pattern which has no more than d mismatches with this motif pattern [1]. The identification of short sequence motifs, such as transcription factor binding sites, is at the center of the transcriptional regulation understanding. The functional sites are constrained to contain motifs, since their changes will disrupt regulation, which is detrimental to the organism [2, 3].

Several motif-based methods have been proposed to count the total number of motifs rather than sequences, and construct a similar contingency table [4]. Some other methods including Consensus [5], Gibbs Sampler [6] and ANN-Spec [7] for multiple local alignment have been employed to resolve the identification of motifs problem. In many cases where motifs have been experimentally determined, these algorithms have been shown to yield the known motifs, indicating that such methods can discover unknown motifs from a

*This work was supported by the National Natural Science Foundation of China (Grant No. 60273079 and 60473074).

collection of sequences believed to be implanted motifs. Brazma et al. algorithms [8] find and analyze combinations of motif that occur in the upstream regions of genes in the yeast genome. These algorithms can identify all the motifs that satisfy given parameters with respect to a given sample sequences. However, they perform an exhaustive search through all 4^l l-letter patterns and find the high-scoring patterns, thus the algorithms become impractical for $l > 10$. Tompa raised the problem of Brazma, and improved this approach for longer patterns. One way around this problem is to limit the search spaces on the patterns appearing in the sample sequences [9–11].

WINNOWER is an outstanding algorithm for finding motifs in respect that it proposes a clique discovery approach to finding global optimal results [12]. WINNOWER indicates that the motif discovery problem is similar to the clique discovery problem. A clique is a set of nodes in a graph, each of which is connected to the others in this set. The sample sequences are divided into a set of substrings which are represented by nodes. If two substrings are similar, there will be an edge connecting them. Thereby, a motif can be taken as a clique in which different nodes are from different sample sequences. For a set of sample sequences $S = \{s_1, s_2, \ldots, s_q\}$, WINNOWER constructs a graph to find the cliques which represent the motifs in S. For each substring s_{ij} from position j to position $j + l - 1$ in sequence s_i , the algorithm constructs a node representing it. Two node s_{ij} and s_{pq} are connected by an edge, if s_{ij} and s_{pq} are similar ($i \neq p$). A q-clique in a graph is a q-nodes set, in which all the pair nodes are connected. Thereby, (l, d)-motif is a clique with size q in the graph. Since most of edges in the graph cannot make up a clique, called spurious edges, WINNOWER prunes some of these spurious edges to speed up searching. Suppose C is a clique, node n is a neighbor of C if and only if n connects to each node in C. If a clique has at least one neighbor, it is *extendable*. If an edge does not belong to any extendable clique of size q, it is *spurious*. WINNOWER prunes the spurious edges based on the observation that every edge in a q-clique belongs to at least $\binom{q}{k}$ extendable cliques of size k.

Although WINNOWER is a typical algorithm for motif discovery, it still has two main problems. (1) For the case that there are a few motifs in the sample sequences, so only a few cliques and edges in the graph. However, most of running time is spent to compute similarity of pairwise nodes during the construction of the graph. Therefore, most of similarity computations are unnecessary. (2) For the case that numerous motifs exist in the sample sequences, the graph will conclude numerous cliques and edges. In this case, WINNOWER needs huge spaces to record the edges. The space requirement of WINNOWER is often a bottleneck to find motifs in large sample sequences.

In this paper, we present an efficient clique discovery algorithm EDAM based on frequency transformation and MBRs. It works in three phases, *frequency transformation*, *MBR-clique searching* and *motif discovery*. In *frequency transformation*, EDAM divides the sample sequences into a set of substrings by sliding windows, then transforms them to frequency vectors which are stored in MBRs. In *MBR-clique searching*, based on the frequency distance theorems EDAM searches for MBR-cliques used for motif discovery. In *motif discovery*, EDAM discovers larger cliques by extending smaller cliques with their neighbors. To accelerate the clique discovery, we propose a range query facility to avoid unnecessary computations for clique extension. EDAM has the following advantages over WINNOWER. (1) EDAM avoids a lot of unnecessary similarity computations by MBR-cliques searching, since it only computes the similarity of nodes within the same MBR-clique. (2) Since EDAM uses MBRs to store similar substrings, it saves storage space compared with WINNOWER.

The rest of this paper is organized as follows. Section 2 formally defines the motif discovery problem. Section 3 describes the algorithm EDAM in detail. Section 4 gives an analysis of the time and space complexity of EDAM and WINNOWER. Section 5 shows the experimental results and compares the performance of EDAM with WINNOWER. Finally, Section 6 concludes this paper.

2. Problem Description

Known regulatory motifs are short, sometimes degenerate and appear frequently throughout the sample sequences. Additionally, Protein-binding DNA motifs often contain ambiguous nucleotides, which can have more than one equivalent nucleotide, so the problem is to discover the following motifs in a sample sequences [13].

Definition 1. *Motif discovery.* Given a sample sequences $S = \{s_1, s_2, \ldots, s_q\}$, the motif pattern length l and the maximum hamming distances between the motif occurrences d. Then the (l, d)-motif discovery problem is defined as finding such l-length pattern m.

$$(\forall s_i \in S)(\exists sub \in s_i)(Length(sub) = l \wedge hd(m, sub) \leq d) \tag{1}$$

Finding motifs, as WINNOWER demonstrated, is similar to the clique discovery problem. If we choose the hamming distance between a motif and any its occurrence is at most d, $2d$ is the longest acceptable distance between any two occurrences presenting a same motif. Therefore, a clique discovery problem corresponding to (l, d)-motif can be defined as follows.

Definition 2. *Clique discovery.* Given a sample sequences $S = \{s_1, s_2, \ldots, s_q\}$ and a (l,d)-motif discovery problem. Any l-length node set C is called a q-clique if and only if

(1) In C, different substrings come from different sample sequences.
(2) For any pair substrings s_i and s_j $(i \neq j)$ in C, $hd(s_i, s_j) \leq 2d$.

3. EDAM

EDAM is a different algorithm for finding motifs in sample sequences, and it has some advantages over WINNOWER. EDAM avoids a lot of unnecessary similarity computations by MBR-cliques searching, since it only computes the similarity of nodes within the same MBR-clique. Moreover, EDAM uses MBRs to store similar substrings, it saves storage space compared with WINNOWER.

3.1. *Frequency Transformation*

In Frequency transformation, EDAM divides the sample sequences into a series of substrings and transforms these substrings into frequency vectors that are stored in MBRs. Before we explain Frequency transformation, we first introduce frequency vector and MBR.

The frequency vector indicates the number of each kind of nucleotide in the DNA sequences. Since DNA sequences are composed of 4 different nucleotides, they always are treat as strings with the alphabet $\sum$ = {A,C,G,T}. EDAM transforms substrings divided from the sample sequences to a 4-dimensional vectors, and the value in every dimension indicates the number of one kind of nucleotide in the substring [14,15]. For example, given a substring $s = TAGCCGAA$, the frequency vector $f(s) = [3, 2, 2, 1]$.

Definition 3. *Frequency vector.* Given s be a substring and the alphabet $\sum = \{\alpha_1, \alpha_2, \ldots, \alpha_\sigma\}$, f_i indicates the number of ith nucleotide in $\sum$, then the frequency vector: $f(s) = [f_1, f_2, \ldots, f_\sigma]$

Minimum Bounding Rectangle(MBR) represents a subspace in the multidimensional space. Each dimension of MBR has a maximum and a minimum, which bound the subspace. The frequency vectors stored in the MBR are restricted in its subspace. In other words, for each frequency vector $f = [f_1, f_2, \ldots, f_\sigma]$ in a MBR $mbr = [(min_1, max_1), (min_2, max_2), \ldots, (min_\sigma, max_\sigma)]$, the value f_i of any dimension ($1 \leq i \leq \sigma$) must be in the interval $[min_i, max_i]$. In this way, the similar vectors representing similar substrings definitely are in an identical MBR or adjacent MBRs.

Frequency vector and MBR are two useful definitions for frequency transformation. In frequency transformation, EDAM reads only one sequence s_i of the sample $S = \{s_1, s_2, \ldots, s_q\}$ each time and sets up the MBRs for s_i. These MBRs divide the multidimensional space into different subspace (e.g. the multidimensional space is divided into subspaces by a grid using dichotomy). For each substring s_{ij} from position j to position $j + l - 1$ in sequence s_i, EDAM transforms it to the frequency vector $f(s_{ij})$ and stores $f(s_{ij})$ in the proper MBR.

3.2. *MBR-clique Searching*

Most of the frequency vectors in the MBRs cannot make up any clique, thus, how to avoid finding cliques in these frequency vectors is one of the foundational problems. In this section, we suggest using *MBR-clique searching* to resolve this problem base on the fact that the vectors in a clique are stored in the adjacent MBRs. The similarity of a pair substrings is generally measured by hamming distance, but hamming distance requires to count the number of mismatches, thus, it is difficult to calculate hamming distance by frequency vectors. Here, we suggest using frequency distance as the lower bourn of hamming distance.

Definition 4. *Frequency distance.* The summation of frequency differences (only positive) on every dimension in $\sum = \{\alpha_1, \alpha_2, \ldots, \alpha_\sigma\}$ of the given substrings s_1, s_2. $f_i(s_1)$ $f_i(s_2)$ denotes i^{th} dimension's value of s_1 and s_2 respectively. The frequency distance between s_1

and s_2 is defined as follow.

$$fd(s_1, s_2) = \sum_{i=1}^{\sigma} \begin{cases} f_i(s_1) - f_i(s_2) & \text{if } f_i(s_1) - f_i(s_2) \geq 0 \\ 0 & \text{else} \end{cases} \tag{2}$$

Suppose the hamming distance of a pair substrings s_1, s_2 is d, it means that if s_1 is transformed to s_2, based on that one mismatch needs one substitution, s_1 requires d substitution operations. According to the definition of frequency vector, d substitutions at most make d differences on frequency vectors.

Theorem 1. *Suppose s_1 and s_2 are two substrings. The frequency distance between s_1 and s_2 is a lower bound on their hamming distance.*

$$hd(s_1, s_2) \geq fd(s_1, s_2) \tag{3}$$

Since the clique in EDAM is a set of similar vectors, and these vectors are stored in adjacent MBRs, we estimate the distances between vectors by the distances between vectors and MBRs.

Theorem 2. *Suppose mbr is a MBR, v is a vector, not in mbr, then for any vector m in mbr, the frequency distance between m and v is no more than the minimum frequency distance between v and the bounding of mbr.*

$$fd(m, v) \geq fd(v, mbr) \tag{4}$$

For the vectors are stored in MBRs, we suggest using the MBR distance to estimate the distances between the vectors in them.

Definition 5. *MBR distance.* Suppose mbr_1 and mbr_2 are two MBRs, $min_i(mbr_j)$ and $max_i(mbr_j)$ are the minimum and the maximum of ith dimension in mbr_j. The frequency distance between mbr_1 and mbr_2 is the minimum frequency distance between the their bounds, it is defined as follow.

$$fd(mbr_1, mbr_2) = \sum_{i=1}^{\sigma} \begin{cases} min_i(mbr_1) - max_i(mbr_2) & \text{if } min_i(mbr_1) \geq max_i(mbr_2) \\ min_i(mbr_2) - max_i(mbr_1) & \text{if } min_i(mbr_2) \geq max_i(mbr_1) \\ 0 & \text{else} \end{cases} \tag{5}$$

Theorem 3. *Suppose mbr_1 and mbr_2 are two MBRs, v_1 and v_2 are frequency vectors that are stored in mbr_1 and mbr_2 respectively. The distance between mbr_1 and mbr_2 is the lower bound on the distance between v_1 and v_2.*

$$fd(v_1, v_2) \geq fd(mbr_1, mbr_2) \tag{6}$$

According to the clique definition and Theorem 3, we suggest using MBR-clique searching to record the MBRs which make up cliques, and then finding motifs in these MBR-cliques. A MBR-clique *MC* is a set of MBRs, the frequency distance between each pair of MBRs in MC does not excess the threshold.

Definition 6. *MBR-clique.* Given the sample sequences $S = \{s_1, s_2, \ldots, s_q\}$ and a (l,d)-motif discovery problem. A q-MBR set *MC* is called a MBR-clique if and only if

(1) In MC, different MBRs come from different sample sequences.

(2) For each pair of MBRs mbr_i and mbr_j $(i \neq j)$ in MC, $hd(mbr_i, mbr_j) \leq 2d$.

EDAM only searches for the cliques in MBR-cliques to speed up the discovery. The MBR-clique searching algorithm is illustrated in Algorithm 1. Before searching for MBR-cliques, (step 1) EDAM scans all the MBRs, (step 2 and 3) and filters out the MBRs which is empty. (step 4 and 5) EDAM searches for the MBRs that store the frequency vectors from the first sample sequence s_1, (step 6) initializes them as the 1-MBR-cliques, then extends these 1-MBR-cliques to q-MBR-cliques. (step 7) EDAM discovers motifs in the MBR-cliques.

Algorithm 1 MBRClique()

Input: the MBR set *smbr* that all the MBR in
Output: all the MBR-clique
1: FOR $\forall$ $mbr \in smbr$
2: IF *mbr* is empty
3: filter out *mbr* from *smbr*
4: FOR $\forall$ $mbr \in smbr$
5: IF $mbr.sequence = 1$
6: extending the 1-MBR-clique MC_1 to a q-MBR-clique C_q
7: searching for the motifs in C_q

For the motif pattern generally is short, the number of MBR is not large compared with the number of frequency vectors, and MBR-clique searching only takes a small part of the total running time for EDAM.

3.3. *Motif Discovery*

In this section, we illustrate the algorithm for finding motifs in the MBR-cliques found by MBR-clique searching. To discover the cliques representing the motifs, we employ a simple idea extending a known $(k-1)$-clique with its neighbor to a k-clique.

The motifs discovery problem implies us that for every sample sequence s_i, there is one and only one vector from s_i in the clique representing a motif. Following this clue, EDAM first finds a known clique $C = \{v_1, v_2, \ldots, v_k\}$ $(k \leq q)$, and every vector v_i $(1 \leq i \leq k)$ in C representing a substring from the sequence s_i, then searches for a neighbor v which is from the sequence s_{k+1} to extend C. Since any single vector makes up a 1-clique, in this way, EDAM can iteratively extend the 1-cliques made up of a vector from s_1 to q-cliques composed of vectors from every sample sequence.

Since the neighbor v must be similar to all the vectors in C, the extension has to calculate totally $e(k-1)$ times hamming distance (there are e neighbors). These calculations for extension cause a running time bottleneck for applications. To resolve this problem EDAM sets a signature on every neighbor ne of $C' = \{v_1, v_2, \ldots, v_{k-1}\}$, if $hd(ne, v_k) \leq d$, ne is also a neighbor of $C = \{v_1, v_2, \ldots, v_{k-1}, v_k\}$. EDAM can set the signature iteratively, because every vector is a neighbor of 0-clique.

Theorem 4. *Cliques combination property.* Given a hamming distance d, and two k-cliques $C_1 = \{v_1, v_2, \ldots, v_{k-1}, v'\}$ and $C_2 = \{v_1, v_2, \ldots, v_{k-1}, v''\}$

$$if \quad hd(v', v'') \leq d \quad then \quad C_3 = \{v_1, v_2, \ldots, v_{k-1}, v', v''\} \tag{7}$$

After the neighbors of the k-cliques have been found, we will extend the $(k+1)$-cliques to discovery larger motifs. For only a few of the new discovered $(k+1)$-cliques can be extended to q-cliques, it is necessary to prune the cliques named *spurious cliques* which can not be extended to q-cliques. According to the clique definition, if the $(k+1)$-clique $C_{k+1} = \{v_1, v_2, \ldots, v_{k+1}\}$ can be extended to q-clique $C_q = \{v_1, v_2, \ldots, v_q\}$, the neighbor v_{k+1} of C_k must be similar to every vector $v_i(k+2 \leq i \leq q)$. Thus, we use a range query based on the Theorem 2 to prune some spurious cliques. There are two important parameters v and r in range query, v is the query vector and r is the range radius. A range query $R(v, r)$ is to record the MBRs whose distances to v are within r. To prune some spurious cliques, we set the neighbor v_{k+1} as the query vector and the hamming distance $2d$ as the radius, then propose a rang query $R(v_{k+1}, 2d)$ in the MBR-clique. Based on Theorem 2, if any MBR in the MBR-clique is outside the range query, then C_{k+1} is a spurious clique, thus, EDAM prunes it to avoid unnecessary clique discoveries.

We describe the algorithm on the clique extension for finding motifs illustrated by Algorithm 2. Since every vector is a neighbor of 0-clique,EDAM initializes $vector_1$ from s_1 as the neighbor of 0-clique, and initializes the MBR-clique in which $vector_1$ stored as the query MBR-clique. (step 1-2) For every vector v in the query MBR-clique $mbrClique$, the algorithm calculates the hamming distance between v and the neighbor $vector_k$. If there is no vector in the known clique C_k that comes from same sequence as v does, moreover $hd(v, vector_k) \leq 2d$ and the signature on v has indicated v is a neighbor of C_{k-1}, then it is a neighbor of the clique C_k. Thus, (step 3) the algorithm resets a signature on v. (step 4) After every neighbor has been set signatures, EDAM extends C_k for finding q-cliques. (step 5) if v comes from the sequence next to $vector_k$ does, $C_{k+1} = C_k \cup \{v\}$ makes up a known $(k+1)$-clique. (step 6) If the C_{k+1} is a q-clique, (step 7) all the vectors in C_{k+1} that represent the occurrences of a motif are recorded. (step 8) If C_{k+1} is not spurious, (step 9) the algorithm extends C_{k+1} for further clique discovery. (step 10) After v is extended, if v has been set a signature, (step 11) the algorithm resets the signature on v.

Algorithm 2 searchMotifs()

Input: a known k-clique $C_k = \{vector_1, vector_2, \ldots, vector_k\}$; a neighbor $vector_k$; the query MBR-clique $mbrClique$;
Output: all the motifs

1: FOR $\forall\, v \in mbr$
2: IF $hd(v, vector_k) \leq 2d$ and $v.sequence > vector_k.sequence$ and $v.signature = vector_k.sequence - 1$
3: $v.signature = vector_k.sequence$
4: FOR $\forall\, v \in mbr$
5: IF $v.sequence = vector_k.sequence + 1$ and $v.signature = vector_k.sequence$
6: IF $\{vector_1, \ldots, vector_k, v\}$ has q vectors
7: record all the vectors $\{vector_1, vector_2, \ldots, v\}$, which represents a motif.
8: ELSE IF RangeQuery(v, mbr)=false
9: searchMotifs(v, mbr).
10: IF $v.signature = vector_k.sequence$
11: $v.signature = vector_k.sequence - 1$.

---

4. Analysis

In this section, we give an analysis of the time and space complexity of EDAM and WINNOWER.

4.1. *Space complexity*

For the sample sequences $S = \{s_1, s_2, \ldots, s_q\}$, there are about $N = \sum_{j=1}^{q} len_j$ subsequences. The spaces for WINNOWER are primarily composed by two parts: nodes and edges in the graph. For WINNOWER constructs a node for each valid subsequence in the sample sequences, it needs $O(N)$ nodes and $p_d O(N^2)$ edges, thereby, the WINNOWER's space complexity is $O(N^2)$. The spaces for EDAM are also composed by two parts: frequency vectors and MBRs. EDAM transforms subsequences divided for the sample sequences into the frequency vectors, thereby, there are $O(N)$ frequency vectors. If the MBR width is w, for every sequence $s_i \in S$, EDAM at most constructs $(l/w)^4$ MBRs. Since $l \ll N$, the EDAM's space complexity is approximately $O(N)$.

4.2. *Time complexity*

(a) The time complexity analysis of motif discovery computed by MATLAB.

(b) The practical performances of EDAM and WINNOWER in an identical sample.

Figure 1. The contrast between the time complexity analysis and the performance of EDAM of motif discovery in 15 4KB length sequences with increasing distance, fixed pattern length $l = 15$.

Given two l-length subsequences s_1 and s_2, the probability p_d that $hd(s_1, s_2) \leq d$ equals $\sum_{i=0}^{d}\binom{l}{i}\left(\frac{1}{4}\right)^{l-i}\left(\frac{3}{4}\right)^{i}$. If the similarity of the vectors in q-clique is completely independent ($d = l$), the probability c_q that q vectors make up a q-clique equals $p_d^{\binom{q}{2}}$. In contrast, the similarity of the vectors in q-clique is completely dependent ($d = 0$), c_q equals p_d^{q-1}. Suppose the length of the jth sample sequence is len_j, the expected number of q-cliques discovered in the sample sequences is in the interval $[p_d^{\binom{q}{2}} \prod_{j=1}^{q} len_j \, , p_d^{q-1} \prod_{j=1}^{q} len_j]$. The running time of WINNOWER is primarily spent in two parts: graph construction and clique discovery. For the graph construction has to compute the hamming distances of every pair nodes, it requires $0.5 \; l \; N^2$ calculations. Additionally, since each node in the graph has $p_d \; N$ edges, WINNOWER requires $\sum_{k=1}^{q}(c_k \; \prod_{j=1}^{k} len_j \; p_d \; N)$ calculations to find all the q-clique. In sum the time complexity of WINNOWER is $0.5 \, l \, N^2 + \sum_{k=1}^{q}(c_k \prod_{j=1}^{k} len_j \, p_d \, N)$.

Although EDAM works in three phases, the frequency transformation and MBR-clique searching only take small part of the total running time(primarily because the number of MBR is much smaller than the number of vectors). In motif discovery, each k-clique has potential $p_{d+w}\ len_j$ vectors for further extension. Consequently, the time complexity is $\sum_{k=1}^{q}(c_k \prod_{j=1}^{k} len_j\ p_{d+w}\ len_j\ l)$. Figure 4 illustrates the time complexity analysis and the practically performances of WINNOWER and EDAM with the same parameters.

5. Experimental Results

In this section, we illustrate EDAM's efficiency for discovering motifs in the sample sequences. The experiments were performed on a PC with 2.6GHz P4 CPU and 512MB memory, programmed in JAVA. The sample sequences originated from human gene sequences section (chr22) and the MBR width is 2. We do not present the experimental results in terms of some algorithms used the performance coefficients to the known motifs, which mainly measure the algorithm results accuracy. Because EDAM uses the same model (clique) as WINNOWER for finding motifs, and both of the two algorithms can find out all the global optimal results, thus, the performance coefficients of the two algorithms are the same. Beside above problem, running time is another big bottleneck, especially when we discover motifs in large DNA database, so in this paper we compare the running time of the two algorithms instead.

In Figure 1(b) WINNOWER took up a steady running time for all distances, for the running time was mostly for the graph construction. The performances of EDAM occurred a sharp rise as soon as the distance exceeded a percentage of pattern length. The sharp rise reveals that the techniques to avoid necessary computations in EDAM are efficient for short distances, but breaks for the liberal distances. In Figure 2(a) EDAM discovered $(2, 15)$-motifs in two different samples: original sample and synthetic sample. The original sample was from human gene sequences section. The synthetic sample sequences were implanted a serious of rational motifs with randomly distributed background, thus, the number of the results in synthetic sample emerged an outstanding increment over original sample. Due to the effect of the number of results, the running time in the synthetic sample was over in original sample for all the distances.Due to avoiding unnecessary computations, in Figure 2(b) EDAM's merit was obvious for the pattern length increment, whereas WINNOWER did not appear any distinct change. With the increment of the number of sequences in these sample, in Figure 2(c) the running time of WINNOWER for $(2, 15)$-motifs occurred a distinct rise, however, the running time of EDAM rose smoothly (we stopped the tests whose running time were more than one hour). Figure 2 imply that in the tests with a few number of the results or low distance-pattern ratio, EDAM performs better than WINNOWER. It is because EDAM well approximates hamming distance by frequency distance and avoids most of the unnecessary computations. On the other side, the merits are not so distinct.

6. Conclusions and Discussions

Although the motif discovery problem has a long history, it is still far away from being resolved. The well-known algorithm WINNOWER shows better performances than other

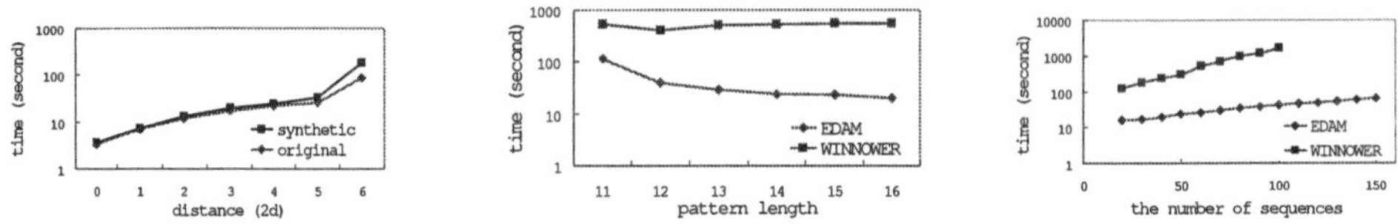

(a) EDAM discovered motifs in two different samples: original sample and synthetic sample, the figure illustrates the running time in original sample and synthetic sample.

(b) The performances of EDAM and WINNOWER in an identical sample with fix the distance d and different pattern length l from 11 to 16.

(c) The performances of EDAM and WINNOWER in a series of samples. These samples were composed of fix length sequences, and the number of sequences increased from 30 to 150.

Figure 2. The performances of WINNOWER and EDAM with different parameters.

algorithms, but it still has some shortages. In this paper, we suggest a unique algorithm EDAM using frequency transformation and MBR techniques to solve the running time problem of WINNOWER. The experimental results indicate that EDAM is more efficient than WINNOWER for motif discovery. Although EDAM shows excellent performances, further improvements are still necessary.

References

1. X. Dong, S. Y. Sung, W. Sung and C. L. Tan. Constraint based method for finding motifs in DNA sequences. *In proc. BIBE'04, 2004.*
2. M.Lapidot and Y. Pilpel. Comprehensive quantitative analyses of the effects of promoter sequence elements on mRNA transcription. *Nucleic Acids Research, Vol. 31, No.13,3824C3828,2003*
3. J. Shapiro and D. Brutlag. FoldMiner: Structural motif discovery using an improved superposition algorithm. *Protein Science, 13:278C294,2004.*
4. R .Sharan, I.Ovcharenko, A.Ben-Hur, and R.M. Karp. CREME: a frame work for identifying cis-regulatory modules in human-mouse conserved segments. *Bioinformatics, 19 (Suppl1), I283-I291,2003.*
5. G. Hertz, and G. Stormo. Identifying DNA and protein patterns with statistically significant alignments of multiple sequences. *Bioinformatics 14:563-577, 1999.*
6. M. Jerrum. Large cliques elude the Metropolisprocess. *Random Structures and Algorithms 3(4):347-359,1992*
7. C.T.Workman and G.D. Stormo. ANN-Spec:a method for discovering transcription factor binding sites with improved specificity. *Pac.Symp.Biocomput.,467-478,2000.*
8. A. Brazma, I. Jonassen, I. Eidhammer and D. Gilbert. Approaches to the automatic discovery of patterns in biosequences. *Journal of Computational Biology 5:278-305, 1998.*
9. M. Li, B.Ma and L. Wang. Finding similar regions in many strings. *In proceedings of the 31st ACM Annual Symposium on Theory of computing.473-482, 1999.*
10. X. Liu , D.L. Brutlag and J. S. Liu. BioProspector:discovering conserved DNA motifs in upstream regulatory regions of co-expressed genes. *Pac Symp Biocomput 127-38, 2001.*
11. X. Liu, D.L. Brutlag and J. S. Liu. An algorithm for finding protein-DNA binding sites with applications to chromatin-immunoprecipitation microarray experiments. *Nat biotechnol 835-9, 2002.*
12. P. A. Pevzner and S. H. Sze. Combinatorial approaches to finding subtle signals in DNA sequences. *In Proc. ISMB'00, 2000.*
13. R. V. Satya, A. Mukherjee and U. Ranga. A Pattern Matching Algorithm for Codon Optimization and CpG Motif engineering in DNA Expression Vectors. *In proc. CSB'03, 2003.*
14. M. Garofalakis and A. Kumar. Deterministic Wavelet Thresholding for Maximum-Error Metrics. *In proc. PODS'04, 2004.*
15. T. Kahveci and A. K. Singh. An Efficient Index Structure for String Databases. *In proc. VLDB'01 351-360, 2001.*

A RECURSIVE METHOD FOR SOLVING HAPLOTYPE FREQUENCIES IN MULTIPLE LOCI LINKAGE ANALYSIS

MICHAEL K. NG *

Department of Mathematics, Hong Kong Baptist University
Kowloon Tong, Hong Kong, China
E-mail: mng@math.hkbu.edu.hk

ERIC S. FUNG WAI-KI CHING † YIU-FAI LEE

Department of Mathematics, The University of Hong Kong
Pokfulam Road, Hong Kong, China
E-mails: ericfung@hkusua.hku,hk, wkc@maths.hku.hk, ahyiu@graduate.hku.hk

Multiple loci analysis has become popular with the advanced development in biological experiments. A lot of studies have been focused on the biological and the statistical properties of such multiple loci analysis. In this paper, we study one of the important computational problems: solving the probabilities of haplotype classes from a large linear system $Ax = b$ derived from the recombination events in multiple loci analysis. Since the size of the recombination matrix A increases exponentially with respect to the number of loci, fast solvers are required to deal with a large number of loci in the analysis. By exploiting the nice structure of the matrix A, we develop an efficient recursive algorithm for solving such structured linear systems. In particular, the complexity of the proposed algorithm is of $O(m \log m)$ operations and the memory requirement is of $O(m)$ locations where m is the size of the matrix A. Numerical examples are given to demonstrate the effectiveness of our efficient solver.

1. Introduction

Linkage analysis is an important tool for the mapping of genetic loci. With the availability of numerous DNA markers throughout the human genome, linkage analysis has demonstrated its usefulness in mapping the mutations responsible for hundreds of Mendelian diseases (Kruglyak and Lander, 1995). The genotype of an individual at loci X and Y is formed by the haplotypes of two gametes $X_f Y_f$ inherited from the father, and $X_m Y_m$ inherited from the mother. The haplotype of a gamete produced by the individual consists of a mixture of paternal and maternal alleles. A gamete contains two alleles from the same parental gamete (non-recombinant), i.e., $X_f Y_f$ or $X_m Y_m$, or one allele from each parental gamete (recombinant), i.e., $X_f Y_m$ or $X_m Y_f$. The recombination fraction between the two loci is defined as the probability that a gamete is recombinant.

* Work partially supported by RGC Grant Nos. 7130/02P, 7046/03P, 7035/04P, 7035/05P, and FRG04-05/II-51
† Work partially supported by RGC Grant No. HKU 7126/02P

When the number of loci is large, a haplotype almost certainly constitutes a new combination of alleles, different from the parental and the maternal haplotypes (Sham, 1998). If n loci are involved, there are $(n-1)$ intervals between adjacent loci, each of which can either have an even or an odd number of crossovers. This produces 2^{n-1} classes of gametic haplotypes, and therefore $(2^{n-1}-1)$ independent gametic frequencies (the 2^{n-1} classes of gametic frequencies must sum up to one). The frequency of a gametic haplotype is equal to the joint probability of the co-occurrence of a set of recombination events. Liberman and Karlin (1984) applied the concept of recombination values to establish the relationship between recombination fractions and haplotype frequencies for $n > 3$. The recombination value of a set of intervals (not necessary contiguous), is the probability of an odd number of crossovers occurring in the intervals. Since each of the $(n-1)$ intervals can be included or excluded in a set of intervals, there are $(2^{n-1}-1)$ sets of intervals and hence $(2^{n-1}-1)$ recombination values. There is a relationship between these $(2^{n-1}-1)$ recombination values and the $(2^{n-1}-1)$ haplotype frequencies as specified by a linear system

$$\Theta = \Gamma A_n$$

where A_n is the m-by-m matrix with $m = 2^{n-1}$ being equal to the number of haplotype classes, and Θ and Γ are m-vectors containing the recombination values and haplotype frequencies respectively, see Section 2 for details about the derivation of $\Theta = \Gamma A_n$.

When the number n of loci increases, the size of A_n increases exponentially and therefore the cost of solving $\Theta = \Gamma A_n$ is very expensive. Here we will first establish the structure of A_n and a recursive formula relating A_{n+1} and A_n. We then present a recursive solver based on the recursiev formula to solve $\Theta = \Gamma A_n$ efficiently.

The rest of this paper is organized as follows. In Section 2, we give some background and basic properties on the matrix A_n. In Section 3, we show that A_n is nonsingular and give the explicit form for its inverse. According to the explicit form of A_n^{-1}, we obtain the haplotype frequencies efficiently by using a recursive scheme. We also give a cost analysis for the proposed algorithm. Numerical examples are given to illustrate the effectiveness of the proposed method. Finally, concluding remarks are given in Section 4.

2. The Recombination Matrix A_n

In this section, we give some background of the recombination matrix A_n. In the multi-locus situation ($n \geq 3$), we denote a haplotype of n loci by a $(n-1)$ string of 0s and 1s with respect the ith digit representing the recombination status of the $(i+1)$th allele with respect to the first allele. This string of $(n-1)$ digits specifies the recombination status between all $n(n-1)/2$ pairs of loci. Here pairs of loci with different digits are recombinants while the others are non-recombinants. Such strings refer to different rows of the matrix A_n. To apply the concept of recombination values of a set of non-contiguous intervals, we let the inclusion or the exclusion of the intervals be denoted by a vector of 0s and 1s, where 0 represents exclusion and 1 represents inclusion. Such intervals refer to different columns of the matrix A_n. For each haplotype class and each set of intervals, we set the entry of A to 1 exactly when there is an odd number of crossovers for the intervals in the set.

For examples, in the case of four loci, W, X, Y and Z, there are eight possible haplotype classes, 000, 001, 010, 011, 100, 101, 110 and 111. Each represents a unique combination of recombination status between the six possible pairs of loci (WX, WY, WZ, XY, XZ and YZ). There are seven possible sets of intervals (001, 010, 011, 100, 101, 110, 111), excluding the set with no intervals. In this case, the relationship between the haplotype classes and the recombination values can be described as follows:

Haplotype classes	Interval sets						
WXYZ	001	010	011	100	101	110	111
000	0	0	0	0	0	0	0
001	1	0	1	0	1	0	1
010	1	1	0	0	1	1	0
011	0	1	1	0	0	1	1
100	0	1	1	1	1	0	0
101	1	1	0	1	0	0	1
110	1	0	1	1	0	1	0
111	0	0	0	1	1	1	1

- The gamete of the haplotype class "001" is the recombinant with respect to the loci W and Z, and is the non-recombinant with respect to the loci W, X and Y. Correspondingly, the crossover only occurs in the interval YZ, and therefore, we assign one in the sets of intervals (001, 011, 101 and 111) as these intervals including YZ contribute the frequencies to the haplotype class "001". By using the same arguments, the haplotype class "100" can be considered similarly.
- For the haplotype class "011", the gamete is the recombinant with respect to the loci W and Y and the loci W and Z, and is the non-recombinant with respect to the loci W and X. In this case, the crossover only occurs in the interval XY. The sets of intervals including XY contributing to the frequencies of the haplotype class "011" are 010, 011, 110 and 111. The haplotype class "110" can be considered similarly.
- The gamete of the haplotype class "010" is the recombinant with respect to the loci W and Y, and is the non-recombinant with respect to the loci W, X and Z. It also implies that such haplotype is also the the recombinant with respect to the loci X and Y, and also the loci Y and Z. Correspondingly, the crossover only occurs in the interval XY or YZ, and therefore, we assign one in the sets of intervals (001, 010, 101 and 111) as these intervals including XY or YZ contribute the frequencies to the haplotype class "010". We note that the sets of intervals (011 and 111) include both XY and YZ and therefore the value 0 is assigned to them since an odd number of crossovers occurring in the intervals is counted. By using the same arguments, the haplotype class "101" can be considered similarly.
- For the haplotype class "111", the gamete is the recombinant with respect to the loci W and X, the loci W and Y, and the loci W and Z. In this case, the crossover

only occurs in the interval WX. The sets of intervals contributing to the frequencies of the haplotype class "111" are 100, 101, 110 and 111.

Finally, we note that the sum of all haplotype frequencies should be equal to one. With the above table and the additional constraint, the matrix A_4 is given as follows:

$$\text{Haplotype classes}\quad \begin{array}{c} \\ \\ 000 \to \\ 001 \to \\ 010 \to \\ 011 \to \\ 100 \to \\ 101 \to \\ 110 \to \\ 111 \to \end{array} \overset{\begin{array}{c}\text{Interval sets}\\ \;\;\;001\;010\;011\;100\;101\;110\;111\end{array}}{\begin{pmatrix} 1 & 0 & 0 & 0 & 0 & 0 & 0 & 0 \\ 1 & 1 & 0 & 1 & 0 & 1 & 0 & 1 \\ 1 & 1 & 1 & 0 & 0 & 1 & 1 & 0 \\ 1 & 0 & 1 & 1 & 0 & 0 & 1 & 1 \\ 1 & 0 & 1 & 1 & 1 & 1 & 0 & 0 \\ 1 & 1 & 1 & 0 & 1 & 0 & 0 & 1 \\ 1 & 1 & 0 & 1 & 1 & 0 & 1 & 0 \\ 1 & 0 & 0 & 0 & 1 & 1 & 1 & 1 \end{pmatrix}}.$$

In the following discussion, the binary strings of haplotype classes and interval sets are represented in ascending order, and the properties of the recombination matrix A_n can be summarized as follows:

(1) All the entries in the first column of A_n are equal to 1.
(2) The first row of A_n is a unit row vector with the first entry being equal to 1.
(3) For the (i, j)th entry of A_n, we express the integers i and j in a binary system:

$$i = 1 + \sum_{k=0}^{n-2} a_k^{(i)} 2^k \quad \text{and} \quad j = 1 + \sum_{k=0}^{n-2} b_k^{(j)} 2^k.$$

The haplotype class is represented by $a_0^{(i)} a_1^{(i)} \cdots a_{n-2}^{(i)}$ and the set of intervals is represented by $b_1^{(i)} b_2^{(i)} \cdots b_{n-2}^{(i)}$. The value of the (i, j)th entry of A_n is determined by the following formula:

$$[A_n]_{i,j} = \left(a_0^{(i)} b_0^{(i)} + \sum_{k=1}^{n-2} (a_k^{(i)} - a_{k-1}^{(i)}) b_k^{(j)} \right) \pmod 2.$$

We note that when $a_k^{(i)}$ and $a_{k-1}^{(i)}$ are different, it refers to the case that the gamete is recombinant with respect to themselves, and hence such interval should be included in the interval set if $b_k^{(j)}$ is equal to 1. The $a_0^{(i)}$ already indicates whether the gamete is recombinant with respect to the first two loci. Finally, the value one is assigned to $[A_n]_{i,j}$ under the modulo arithmetic if the number of intervals included is an odd number.

According to the above properties of A_n, we can construct the recombination matrix and then solve the linear system $\Theta = \Gamma A_n$ to obtain the haplotype frequencies. Since the size of A_n increases exponentially with respect to the number of loci n, fast solvers are required in order to compute haplotype frequencies efficiently in linkage analysis of multiple loci. Next we present a recursive formula for A_{n+1} and A_n based on the nice structure of the matrix A_{n+1}.

Theorem 2.1. *For $n \geq 1$, the recombination matrix A_{n+1} is given recursively as follows:*

$$A_{n+1} = \left(\begin{array}{c|c} A_n & A_n - R \\ \hline PA_n & N - PA_n + R \end{array}\right)$$

where

$$P = \begin{pmatrix} 0 & 0 & \dots & 0 & 1 \\ 0 & \dots & 0 & 1 & 0 \\ \vdots & \vdots & \vdots & \vdots & \vdots \\ \vdots & \vdots & \vdots & \vdots & \vdots \\ 1 & 0 & 0 & \dots & 0 \end{pmatrix}, \qquad R = \begin{pmatrix} 1\,0\,0 & \dots & 0 \\ 1\,0\,0 & \dots & 0 \\ \vdots\,\vdots\,\vdots & \vdots & \vdots \\ \vdots\,\vdots\,\vdots & \vdots & \vdots \\ 1\,0\,0 & \dots & 0 \end{pmatrix} \qquad \text{and} \qquad N = \begin{pmatrix} 1\,1 & \dots & 1\,1 \\ 1\,1 & \dots & 1\,1 \\ \vdots\,\vdots & \dots & \vdots\,\vdots \\ \vdots\,\vdots & \dots & \vdots\,\vdots \\ 1\,1 & \dots & 1\,1 \end{pmatrix}.$$

Proof. First of all, we partition A_{n+1} into four blocks, i.e.,

$$A_{n+1} = \left(\begin{array}{c|c} M_1 & M_2 \\ \hline M_3 & M_4 \end{array}\right)$$

where M_i are 2^{n-1}-by-2^{n-1} matrices. We note that the binary strings of haplotype classes and interval sets are represented in ascending order. Therefore, for the matrix M_1 corresponding to the first 2^{n-1} rows and the first 2^{n-1} columns, the first digit of their corresponding haplotype classes and interval sets is equal to 0. It implies that M_1 is just the recombination matrix A_n for the n loci problem.

For the submatrix M_2, we note that the first digit of the interval sets corresponding the columns of A_{n+1} between $(2^{n-1}+1)$th to 2^nth, is equal to 1. Since the first digit of the corresponding haplotype classes is equal to 0, there is no contribution of such haplotype classes to the interval set "$100\cdots000$". We assign the zero entries for the first column of M_2, and the other entries are the same as the matrix M_1. Therefore the resulting matrix M_2 is equal to $(A_n - R)$.

For the submatrix M_3, the corresponding haplotype class "$i_1 i_2 \cdots i_n$" can be viewed as the same as the haplotype class "$(1-i_1)(1-i_2)\cdots(1-i_n)$". The contributions of the haplotype class "$i_1 i_2 \cdots i_n$" and the haplotype class "$(1-i_1)(1-i_2)\cdots(1-i_n)$" to the interval sets are the same. It means that the kth row of the matrix M_3 is equal to the $(2^{n-1}-k+1)$th row of the matrix M_1. Such permutation can be implemented as $M_3 = PM_1$.

For the submatrix M_4, by using the similar argument for the submatrix M_3, the kth row of the matrix M_4 is equal to the $(2^{n-1}-k+1)$th row of the matrix M_2. Since the first digit of all the haplotype classes and all the interval sets corresponding to the matrix M_4 is equal 1, all the entries of M_4 should increase by 1. We note that an odd number of crossovers occurring in the set intervals is counted in the recombination matrix. Therefore, the matrix M_4 is given by $N - P(A_n - R)$. Hence the result follows. $\square$

In the next section, we demonstrate that an efficient solver based on the recursive formula for A_{n+1} can be developed to solve the linear system $\Theta = \Gamma A_n$.

3. Recursive Solvers

Since the number n of loci increases, the size of A_n increases exponentially. Fast solvers are required in order to compute haplotype frequencies efficiently in linkage analysis of multiple loci. In this section, we show that A_n is nonsingular for $n \geq 2$, and study the structure of A_n^{-1}. We then present our recursive solvers.

Theorem 3.1. *For $n \geq 1$, A_{n+1} is nonsingular, and we have the following properties of A_{n+1}^{-1}:*

(a) The matrix A_{n+1}^{-1} is given by $\frac{1}{2}\begin{pmatrix} A_n^{-1}+G & (A_n^{-1}-G)P \\ A_n^{-1}-G-H & (H+G-A_n^{-1})P \end{pmatrix}$ where

$$G = \begin{pmatrix} 1 & 0 & \cdots & 0 \\ 0 & 0 & \cdots & 0 \\ \vdots & \vdots & \ddots & \vdots \\ 0 & 0 & \cdots & 0 \end{pmatrix} \quad \text{and} \quad H = \frac{1}{2^{n-2}} \begin{pmatrix} 1 & 1 & \cdots & 1 \\ 0 & 0 & \cdots & 0 \\ \vdots & \vdots & \ddots & \vdots \\ 0 & 0 & \cdots & 0 \end{pmatrix}$$

[Here we assume that $A_1 = 1$.]

(b) The first row of A_{n+1}^{-1} is a unit row vector with the first entry being equal to 1.

(c) The row sum of A_{n+1}^{-1} is equal to zero except for the first row of A_{n+1}^{-1}.

Proof. Here we use mathematical induction. Let $S(k)$ be a statement that A_k is invertible and A_k^{-1} satisfies above properties. To begin with, we notice that A_2 and its inverse are:

$$A_2 = \begin{pmatrix} 1 & 0 \\ 1 & 1 \end{pmatrix} \quad \text{and} \quad A_2^{-1} = \begin{pmatrix} 1 & 0 \\ -1 & 1 \end{pmatrix}.$$

For $k = 3$, A_3 and its inverse are given by

$$A_3 = \begin{pmatrix} 1 & 0 & 0 & 0 \\ 1 & 1 & 0 & 1 \\ 1 & 1 & 1 & 0 \\ 1 & 0 & 1 & 1 \end{pmatrix} \quad \text{and} \quad A_3^{-1} = \frac{1}{2} \begin{pmatrix} 2 & 0 & 0 & 0 \\ -1 & 1 & 1 & -1 \\ -1 & -1 & 1 & 1 \\ -1 & 1 & -1 & 1 \end{pmatrix}.$$

It is clear that the last two properties are satisfied. We note that

$$\begin{aligned}
&\frac{1}{2}\begin{pmatrix} A_2^{-1}+G & (A_2^{-1}-G)P \\ A_2^{-1}-G-H & (H+G-A_2^{-1})P \end{pmatrix} \\
&= \frac{1}{2}\begin{pmatrix} \begin{pmatrix} 1 & 0 \\ -1 & 1 \end{pmatrix} + \begin{pmatrix} 1 & 0 \\ 0 & 0 \end{pmatrix} & \left[\begin{pmatrix} 1 & 0 \\ -1 & 1 \end{pmatrix} - \begin{pmatrix} 1 & 0 \\ 0 & 0 \end{pmatrix}\right]\begin{pmatrix} 0 & 1 \\ 1 & 0 \end{pmatrix} \\ \begin{pmatrix} 1 & 0 \\ -1 & 1 \end{pmatrix} - \begin{pmatrix} 1 & 0 \\ 0 & 0 \end{pmatrix} - \frac{1}{2^0}\begin{pmatrix} 1 & 1 \\ 0 & 0 \end{pmatrix} & \left[\frac{1}{2^0}\begin{pmatrix} 1 & 1 \\ 0 & 0 \end{pmatrix} + \begin{pmatrix} 1 & 0 \\ 0 & 0 \end{pmatrix} - \begin{pmatrix} 1 & 0 \\ -1 & 1 \end{pmatrix}\right]\begin{pmatrix} 0 & 1 \\ 1 & 0 \end{pmatrix} \end{pmatrix} \\
&= \frac{1}{2}\begin{pmatrix} 2 & 0 & 0 & 0 \\ -1 & 1 & 1 & -1 \\ -1 & -1 & 1 & 1 \\ -1 & 1 & -1 & 1 \end{pmatrix} = A_3^{-1}.
\end{aligned}$$

The statement is true for $k = 2$ and $k = 3$.

Now we assume $S(k)$ is true. We are going to prove that $S(k+1)$ is true. By using Theorem 2.1, we have

$$A_{k+1} = \begin{pmatrix} A_k & A_k - R \\ PA_k & N - PA_k + R \end{pmatrix}$$

Let us consider the following matrix-matrix multiplication:

$$\frac{1}{2}\begin{pmatrix} A_k^{-1} + G & (A_k^{-1} - G)P \\ A_k^{-1} - G - H & (H + G - A_k^{-1})P \end{pmatrix} \begin{pmatrix} A_k & A_k - R \\ PA_k & N - PA_k + R \end{pmatrix} = \begin{pmatrix} C_{11} & C_{12} \\ C_{21} & C_{22} \end{pmatrix}$$

where $P^2 = I$. Our task here is to show that the above right-hand-side matrix is the identity matrix. We expand the product of the two matrices and we have

$$\begin{aligned} C_{11} &= \frac{1}{2}[(A_k^{-1} + G)A_k + (A_k^{-1} - G)PPA_k] = \frac{1}{2}[(A_k^{-1} + G)A_k + (A_k^{-1} - G)A_k] \\ &= \frac{1}{2}[I + GA_k + I - GA_k] = I \end{aligned}$$

and

$$\begin{aligned} C_{12} &= \frac{1}{2}[(A_k^{-1} + G)(A_k - R) + (A_k^{-1} - G)P(N - PA_k + R)] \\ &= \frac{1}{2}[(A_k^{-1} + G)(A_k - R) + (A_k^{-1} - G)PP(N - A_k + R)] \\ &= \frac{1}{2}[(A_k^{-1} + G)(A_k - R) + (A_k^{-1} - G)(N - A_k + R)] \\ &= \frac{1}{2}[I - A_k^{-1}R + GA_k - GR + A_k^{-1}N - I + A_k^{-1}R - GN + GA_k - GR] \\ &= \frac{1}{2}[A_k^{-1}N - GN] = 0. \end{aligned}$$

According to Theorem 2.1, the first row of A_k is a unit row vector with first entry being equal to 1, we obtain $GA_k = GR = (1, 0, \ldots, 0)$. By proposition 3.1, we obtain $A_k^{-1}N = GN = 2^{n-2}H$. Thus we have

$$\begin{aligned} C_{21} &= \frac{1}{2}[(A_k^{-1} - G - H)A_k + (H + G - A_k^{-1})PPA_k] \\ &= \frac{1}{2}[(A_k^{-1} - G - H)A_k + (H + G - A_k^{-1})A_k] \\ &= \frac{1}{2}[I - GA_k - HA_k + HA_k + GA_k - I] = 0 \end{aligned}$$

and

$$
\begin{aligned}
C_{22} &= \frac{1}{2}[(A_k^{-1} - G - H)(A_k - R) + (H + G - A_k^{-1})P(N - PA_k + R)] \\
&= \frac{1}{2}[(A_k^{-1} - G - H)(A_k - R) + (H + G - A_k^{-1})PP(N - A_k + R)] \\
&= \frac{1}{2}[(A_k^{-1} - G - H)(A_k - R) + (H + G - A_k^{-1})(N - A_k + R)] \\
&= \frac{1}{2}[I - GA_k - HA_k - A_k^{-1}R + GR + HR + HN - HA_k + HR + GN - GA_k \\
&\qquad + GR - A_k^{-1}N + I - A_k^{-1}R] \\
&= I + \frac{1}{2}[2GR + 2HR - 2GA_k - 2HA_k - 2A_k^{-1}R + HN + GN - A_k^{-1}N] = I.
\end{aligned}
$$

Hence (a) is proved.

By using the induction assumption, it is easy to show that each row sum of $(A_k^{-1} + G) + (A_k^{-1} - G)P$ is equal to zero except the first row. Also it is clear that the first row sum of $(A_k^{-1} + G) + (A_k^{-1} - G)P$ is equal to two. Moreover, we have

$$
\begin{aligned}
A_k^{-1} - G - H + (H + G - A_k^{-1})P &= A_k^{-1} - G - H + H + (G - A_k^{-1})P \\
&= (A_k^{-1} - G) + (G - A_k^{-1})P.
\end{aligned}
$$

Therefore we can show that each row sum of $(A_k^{-1} - G - H) + (H + G - A_k^{-1})P$ is equal to zero. Thus (b) and (c) are proved. □

By using Theorem 3.1, a recursive method can be developed to solve the linear system $\Theta = \Gamma A_n$. The next theorem states how to solve the linear system $\Theta = \Gamma A_n$ without storing A_n^{-1}.

Theorem 3.2. *The complexity for solving Γ in $\Theta = \Gamma A_n$ with n loci is of $O(n2^n)$.*

Proof. To begin with, let us consider the complexity for calculating $2^{n-1}\Theta A_{n+1}^{-1}$ given that the computational complexity of the inverse of $2^{n-2}XA_n^{-1}$ is $\psi(n)$, where X is a 1-by-2^{n-1} vector. By Theorem 3.1, we have

$$
\Theta A_{n+1}^{-1} = \frac{1}{2}\begin{pmatrix} \Theta_1 & \Theta_2 \end{pmatrix} \begin{pmatrix} A_n^{-1} + G & (A_n^{-1} - G)P \\ A_n^{-1} - G - H & (H + G - A_n^{-1})P \end{pmatrix}
$$

where $\Theta = (\Theta_1, \Theta_2)$. It implies that

$$
2^{n-1}\Theta A_{n+1}^{-1} = \begin{pmatrix} 2^{n-2}(\Theta_1 + \Theta_2)A_n^{-1} + 2^{n-2}(\Theta_1 - \Theta_2)G + -2^{n-2}\Theta_2 H \\ 2^{n-2}(\Theta_1 - \Theta_2)A_n^{-1}P - 2^{n-2}(\Theta_1 - \Theta_2)GP + 2^{n-2}\Theta_2 HP \end{pmatrix}^T.
$$

Firstly, we observe that the cost for $2^{n-2}G$ requires one operation and there is no computational cost for $2^{n-2}H$ as they are given by

$$
2^{n-2}G = \begin{pmatrix} 2^{n-2} & 0 & \cdots & 0 \\ 0 & 0 & \cdots & 0 \\ \vdots & \vdots & \ddots & \vdots \\ 0 & 0 & \cdots & 0 \end{pmatrix} \quad \text{and} \quad 2^{n-2}H = \begin{pmatrix} 1 & 1 & \cdots & 1 \\ 0 & 0 & \cdots & 0 \\ \vdots & \vdots & \ddots & \vdots \\ 0 & 0 & \cdots & 0 \end{pmatrix}.
$$

The computational cost for obtaining either $(\Theta_1+\Theta_2)$ or $(\Theta_1-\Theta_2)$ requires 2^{n-1} operations. The cost for $2^{n-2}(\Theta_1+\Theta_2)A_n^{-1}$ and $2^{n-2}(\Theta_1-\Theta_2)A_n^{-1}$ requires $2\psi(n)$ operations. The cost for $2^{n-2}(\Theta_1-\Theta_2)G$ requires one operation as $2^{n-2}G$ contains only one non-zero element in $2^{n-2}G$. Similarly, there is no cost involved for the computation of $2^{n-2}\Theta_2 H$. This is also true for the matrix multiplication of P as it is just a permutation. Thus, the total computational cost $\psi(n+1)$ of $2^{n-1}\Theta A_{n+1}^{-1}$ is equal to $2\psi(n)+5\cdot 2^{n-1}+4$. It is easy to deduce that

$$\psi(n+1)=3\cdot 2^{n-1}+5(n-1)2^{n-1}+4\cdot(2^{n-1}-1)=5n\cdot 2^{n-1}+(2^n-4)=O(n2^n).$$

Hence the result follows. □

Theorem 3.3. *The storage cost for solving* Γ *in* $\Theta=\Gamma A_n$ *with* n *loci is* $3\cdot 2^n-5$.

Proof. To begin with, let us denote the storage cost for computing $2^{n-2}\Theta A_n^{-1}$ by $\phi(n)$. According to Theorem 3.2, we need to store such components

$$2^{n-2}G,\quad 2^{n-2}H,\quad \Theta_1 \text{ and } \Theta_2.$$

Their corresponding storage cost are are 1, 2^{n-1}, 2^{n-1} and 2^{n-1} respectively. The computational procedure of solving Γ in $\Theta=\Gamma A_n$ is summarized as follows:

Procedure	Current Storage requirement
Start with $2^{n-2}A_n^{-1}$	$\phi(n)$
Load Θ_1,Θ_2	$\phi(n)+2^n$
Compute $\Theta_1+\Theta_2$, $\Theta_1-\Theta_2$	$\phi(n)+2^n+2^n$
Remove Θ_1	$\phi(n)+2^n+2^{n-1}$
Compute $X_1=2^{n-2}(\Theta_1+\Theta_2)A_n^{-1}$	$\phi(n)+2^n+2^{n-1}+2^{n-1}$
Remove $\Theta_1+\Theta_2$	$\phi(n)+2^n+2^{n-1}$
Compute $X_2=2^{n-2}(\Theta_1-\Theta_2)A_n^{-1}$	$\phi(n)+2^n+2^{n-1}+2^{n-1}$
Remove $2^{n-2}A_n^{-1}$	2^{n+1}
Compute $X_2=X_2P$	2^{n+1}
Create $2^{n-2}G$	$2^{n+1}+1$
Compute $Y=2^{n-2}(\Theta_1-\Theta_2)G$	$2^{n+1}+1+1$
Remove $\Theta_1-\Theta_2$, $2^{n-2}G$	$2^n+2^{n-1}+1$
Compute $2^{n-2}\Theta_2 H$	$2^n+2^{n-1}+1+2^{n-1}$
Remove Θ_2	$2^n+2^{n-1}+1$
Compute $X_1+Y-2^{n-2}\Theta_2 H$	$2^n+2^{n-1}+1+2^{n-1}$
Remove X_1	$2^n+2^{n-1}+1$
Compute $Y=YP$	$2^n+2^{n-1}+1$
Compute $X_2-Y+2^{n-2}\Theta_2 H$	$2^n+2^{n-1}+1+2^{n-1}$
Remove $X_2, Y, 2^{n-2}\Theta_2 H$	2^n

Table 1: The Storage of the Algorithm.

From the above procedure, the maximum storage requirement is either $\phi(n)+2^{n+1}$ or $2^{n+1}+1$. Since $\phi(n+1)=\phi(n)+2^{n+1}=\cdots=2^{n+2}-5+2^{n+1}$, the total storage requirement is $3\cdot 2^{n+1}-5$. □

3.1. *Computational Results*

In this subsection, we demonstrate the effectiveness of the proposed recursive solver for solving $\Theta = \Gamma A_n$. Here we perform our test in a MATLAB platform with CPU=AMP 1800+ and memory=512Mb. Table 2 shows the times (in seconds) required for computing ΘA_n^{-1} and the ratio between the computational times of ΘA_n^{-1} $\Theta' A_{n-1}^{-1}$. We remark that the complexity of the proposed recursive algorithm for the n loci problem is of $O((n-1)2^n)$. From Table 1, we find that the computational times only increase linearly with respect to n for our tested cases. It clearly shows that the proposed recursive method is highly efficient.

n	10	11	12	13	14	15	16	17
time (seconds)	0.05	0.11	0.22	0.33	0.77	1.43	2.86	5.65
ratio	–	2.20	2.00	1.50	2.33	1.86	2	1.98

n	18	19	20	21	22	23	24	25
time (seconds)	11.37	22.91	46.08	92.83	187.68	379.04	765.94	1812.82
ratio	2.01	2.01	2.01	2.01	2.02	2.02	2.02	2.37

Table 2: The Computational Times for different n.

4. Concluding Remarks

In this paper, we give a systematic formulation for the linkage analysis problem and an efficient recursive solver is also proposed for solving the haplotype frequencies in multiple loci linkage analysis. The complexity of our method is shown to be $O((n-1)2^n)$ for n loci problem. It is much more efficient when compared to $O(2^{3n})$ operations required by the classical Gaussian elimination method. Previous applications of the linkage analysis only consider small values of n, see for instance (Sham 1998 and Zhao 1990). With our formulation of the problem and also the fast recursive solver, practitioners can now consider larger n and we expect the method can be more popular.

References

1. D. Curtis, Another Procedure for the Preliminary Ordering of Loci on Two-Point Lod Scores, *Annals of Human Genetics*, 58, 65-75, 1994.
2. G. Golub and C. Van Loan, *Matrix Computations*, The John Hopkins University Press, Baltimore 1989.
3. L. Kruglyak and E. S. Lander, High-Resolution Genetic Mapping of Complex Traits, *American Journal of Human Genetics*, 56, 1212-23, 1995.
4. U. Liberman and S. Karlin, Theoretical Models of Genetic MapFunctions, *Theoretical Population Biology*, 25, 331-46, 1984.
5. P. Sham, *Statistics in Human Genetics*, Edward Arnold, 1998.
6. D. E. Weeks, G. M. Lathrop and J. Ott, Multipoint Mapping under Genetic Interference, *Human Heredity*, 43, 86-97, 1993.
7. L. P. Zhao, E. Thompson and R. Prentice, Joint Estimation of Recombination Fractions and Interference Coefficients in Multilocus Linkage Analysis, *American Journal of Human Genetics*, 47, 255-265, 1990.

TRENDS IN CODON AND AMINO ACID USAGE IN HUMAN PATHOGEN *TROPHERYMA WHIPPLEI*, THE ONLY KNOWN ACTINOBACTERIA WITH REDUCED GENOME

SABYASACHI DAS, SANDIP PAUL and CHITRA DUTTA

Bioinformatics Centre, Indian Institute of Chemical Biology, Kolkata –700032, India

The factors governing codon and amino acid usages in the predicted protein-coding sequences of *Tropheryma whipplei* TW08/27 and Twist genomes have been analyzed. Multivariate analysis identifies the replicational-transcriptional selection coupled with DNA strand-specific asymmetric mutational bias as a major driving force behind the significant inter-strand variations in synonymous codon usage patterns in *T. whipplei* genes, while a residual intra-strand synonymous codon bias is imparted by a selection force operating at the level of translation. The strand-specific mutational pressure has little influence on the amino acid usage, for which the mean hydropathy level and aromaticity are the major sources of variation, both having nearly equal impact. In spite of the intracellular life-style, the amino acid usage in highly expressed gene products of *T. whipplei* follows the cost-minimization hypothesis. Both the genomes under study are characterized by the presence of two distinct groups of membrane-associated genes, products of which exhibit significant differences in primary and potential secondary structures as well as in the propensity of protein disorder.

1. Introduction

Whipple's disease is a rare multisystemic bacterial infection caused by an intracellular pathogenic actinobacteria *Tropheryma whipplei*.[1] The complete genome sequences of two different strains of this human pathogen, *T. whipplei* TW08/27 and *T. whipplei* Twist, reveal several atypical characteristics of the organism.[1,2] First, their genome sizes are small (<1 Mb) and G+C-content (46% approx.) are low, as compared to other actinomycete species, which have, in general, genome sizes ranging from 1 million bp to 8 million bp and high G+C contents.[3] Second, the genomes bear all the traits of strictly host-adapted organisms, including pronounced deficiencies in energy metabolism and a lack of key biosynthetic pathways.[1,2] Third, both the genomes exhibit a great deal of genetic variability, mostly directed towards the changes in cell-surface proteins, indicating that immune evasion and host interaction play an important role in their lifestyle. Such atypical characteristics of *T. whipplei* reflect the possible existence of special selection forces operative at the genome and/or proteome levels, unveiling of which calls for an in-depth analysis of the trends in codon and amino acid usage in the organism.

Synonymous codon usage in most of the unicellular organisms are primarily governed by directional mutational bias and translational selection,[4,5] though several other factors such as context-dependence,[6] replicational-transcriptional selection,[7-9] protein hydropathy,[10] ecological niches [11] etc. may also have significant influences. The amino acid usage in microbial organisms may also be influenced by several factors like hydrophobicity, expressivity and aromaticity of the respective proteins,[12] cost minimization and conservation of GC-rich amino acids in highly expressed gene-products etc.[9,13,14] Multivariate analyses carried out in the present study indicate that the codon and amino acid usage in this human pathogen might be a consequence of a complex balance between replicational-transcriptional selection, translational control and other physico-chemical properties of the gene-products. The study, apart from providing an insight into the underlying selection pressures operative at the gene/protein level of *T. whipplei,* may also offer a better understanding of evolution of this host-adapted microorganism.

2. Methods and Materials

2.1 Sequence retrieval

All protein-coding sequences of *T. whipplei* TW08/27 and Twist genomes were extracted from NCBI GenBank. In order to reduce sampling errors, the annotated genes with less than 100 codons were excluded from the analysis. The presumed duplicates, transposes genes and the genes with internal stop codons, untranslatable codons were also excluded. Finally 729 sequences for *T. whipplei* TW08/27 and 734 sequences for *T. whipplei* Twist were selected for analysis.

2.2 Sequence analysis for identifying trends in codon and amino acid usage

The genes present in leading and lagging strand were isolated on the basis of reported location of oriC in *T. whipplei* Twist by Raoult et al. (2003).[2] Based on the change in AT-skew signal using Oriloc and the conservation of dnaA-dnaN-recF gene cluster, the oriC region in *T. whipplei* TW08/27 is assumed to be located at 0 KB. The terminus is assumed at the second inflexion in AT-skew and thus allowed us to locate each coding sequence that present either in the leading or lagging strands of replication.

To identify the major factors shaping variation in relative synonymous codon usage (RSCU) and relative amino acid usage (RAAU) among *T. whipplei* genes, we applied correspondence analyses (COA) using CODONW 1.4.2. GC_{1+2} (G+C content at first and second codon positions), GC_{3S} and GT_{3S} (the frequency of G+C and G+T respectively at synonymous third codon positions) were calculated for each coding sequences in both strains of *T. whipplei*. Parameters like total number of occurrence of each codon, RSCU,[4] codon adaptive index (CAI),[4] RAAU, average hydrophobicity (Gravy score),[15] aromaticity [12] and average size/complexity quotient [16] of encoded proteins were also calculated to find out the factors influencing codon and amino acid usage. To examine the nucleotide substitution patterns, we estimated pairwise synonymous divergences (d_S) as well as non-synonymous divergences (d_N) between the Orthologous genes of *T. whipplei* TW08/27 and *T. whipplei* Twist using the MEGA program (version 2.1), as described by Nei and Gojobori (1986).[17] In order to detect the differences between the two classes of genes, if any, codon and amino acid abundance were compared in 2 x 2 contingency tables. Linear regression analysis was used to find out the significance of association between the positions of sequences along major axes of COA and biological variables using STATISTICA (Version 6.0). The prediction of protein secondary structure was performed using GOR IV algorithm[18] from Expasy proteomics server and the disordered regions within proteins were predicted using GlobPlot (http://globplot.embl.de).[19]

3. Results and Discussion

3.1 Asymmetrical mutational bias, coupled with replicational-transcriptional selection on synonymous codon usage

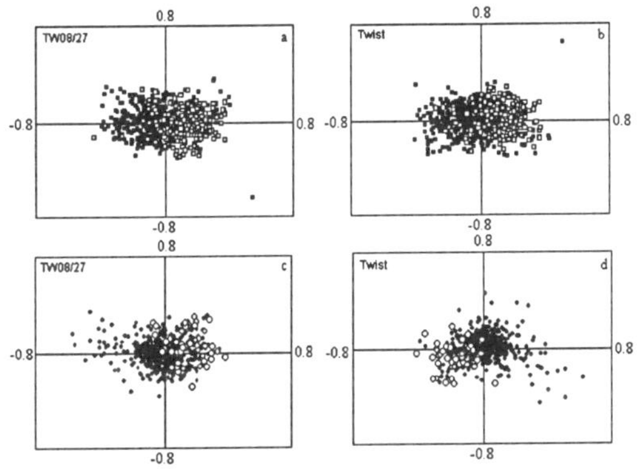

Figure 1. Position of genes along the first two principal axes generated by correspondence analysis (COA) on relative synonymous codon usage (RSCU) values from (a) *T. whipplei* TW08/27 (b) *T. whipplei* Twist genome and separately from leading strand of (c) TW08/27 and (d) Twist genome. The filled quadrangle and open quadrangle for figure a and b represent the genes transcribed in the leading and the lagging strands of replication respectively whereas for the figure c and d, the filled circles represent the genes transcribed only in the leading and the open circles represent the highly expressed genes of that strand.

Figure 1a and 1b show the position of the genes on the plane defined by the first and second major axes generated by COA on RSCU values of coding sequences in TW08/27 and *Twist* genomes respectively. The first principal axis accounts for 9.23% and 9.19%, while the second axis accounts for 5.15 % and 5.07 % of total variability for TW08/27 and *Twist* genomes respectively. In each plot, the genes transcribing in the leading and the lagging strands of replication are segregated in two discrete clusters with little overlap along the axis 1. Similar scatter plots with two distinct clusters of points were observed earlier in case of genomes with pronounced strand-specific mutational bias.[7-9] The chi square test on occurrences of different codons in two replicating strands showed that there are 21 G-ending/U-ending codons, usages of which significantly increase ($p<0.01$) in the leading strand genes, whereas usage of 24 and 25 C-ending/A-ending codons are significantly higher in the lagging strand of the TW08/27 and Twist genome respectively (data not shown). Thus, in both species under study, the primary cause of variation in synonymous codon usage is whether a gene is transcribed in the same direction as replication, or opposite to it. To our knowledge, this is for the first time the strong strand-specific biases is being reported to be the prime selection force behind the synonymous codon usage in any Actinobacteria. Furthermore, in both strains of *Tropheryma*, the number of predicted coding sequences is significantly higher in the leading strand (73.8 %

for TW08/27 and 74.1 % for Twist genome) than in the lagging strand and the distribution of highly expressed genes (*i.e.* ribosomal proteins, transcription and translation processing factors etc.) are also significantly skewed, most of the potential highly expressed genes (>70%) being transcribed from the leading strand. All these observations indicated that the replicational-transcriptional selection coupled with asymmetrical mutational bias is the primary cause of intra-genomic variations in codon usage pattern in *T. whipplei.*

3.2 Evidences for translational selection in T. whipplei

In order to examine the possible effect of translational selection, if any, on codon selection by the highly expressed genes of *T. whipplei*, we have performed a COA on RSCU values of the genes transcribed from the leading strand of replication only (as most of the highly expressed genes (>70%) are located in leading strand). Most of the potential highly expressed genes (*i.e.* genes encoding ribosomal proteins, transcription translation processing factors, heat-shock proteins etc.) are clustered at one extreme of the axis 1, which represents about 8% of total variance for both strains of *T. whipplei* (Fig. 1c,d). The first axis of COA on RSCU values of leading strand genes exhibits strong significant correlation ($r = 0.79$ and -0.77 at $p<0.0001$ for TW08/27 and Twist genome respectively) with the CAI values of genes. Therefore, the position of genes along the first axis of COA might be related to expressivity. In order to ascertain the influence of translation selection on highly expressed genes, we compared the estimated d_S values between 50 putative highly and 50 putative lowly expressed orthologous genes between TW08/27 and Twist genomes taken from the pooled data of the two extremes of axis1. The mean d_S value for high-expression genes (*i.e.* ribosomal proteins, transcription and translation processing factors, heat-shock proteins etc.) is 0.0049, which is significantly lower (t-test, $p<10^{-4}$) than that of low-expression genes (mean $d_S = 0.0058$) indicating that highly expressed genes have lower divergence at synonymous sites than lowly expressed genes. Therefore, it may be concluded that apart from the strand-specific mutational bias, a selection, putatively operating at the level of translation, also might have influenced the synonymous codon choice in genes of these two strains of *T. whipplei.*

3.3 Major sources of variation in proteome composition

To investigate whether the DNA strand specific mutational biases have any impact on amino acid usage in *T. whipplei* gene-products, we have performed COA on relative amino acid usage (RAAU) of the encoded proteins (Fig.2). There is no clear segregation of the proteins encoded by the leading and lagging strands genes. When the cumulative amino acid usage of encoded gene products in two strands are compared separately, only two amino acids encoded by G+U-rich codons (Phe and Val) are significantly over represented ($p<0.001$) in the leading strand and three residues encoded by A+C-rich codons (Lys, Asn and Thr) are more abundant ($p<0.001$) in the lagging strand.

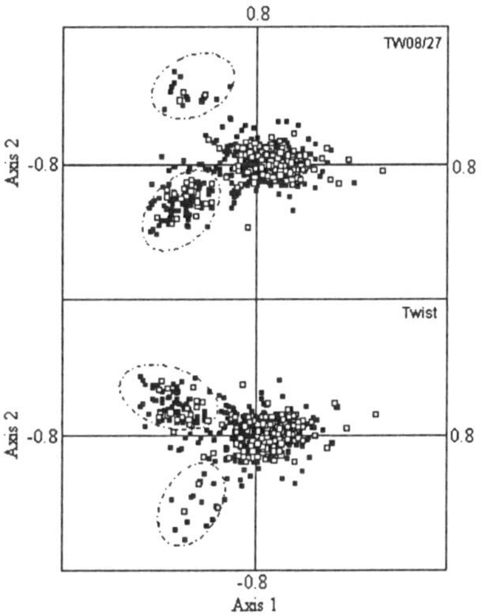

Figure 2. Position of each gene along the two major axes of variation generated by correspondence analysis on RAAU of encoded gene products for (a) *T. whipplei* TW08/27and (b) *T. whipplei* Twist. The filled quadrangle and open quadrangle represent the genes transcribed in the leading and lagging strands of replication respectively. Large and small dashed-line ovals represent the large and small clusters of genes encoding membrane-associated proteins.

In both strains of *T. whipplei*, the first three axes generated by COA on amino acid usage explain about 42% of total variability. The first and second major axes are strongly correlated both with hydrophobicity as well as with the aromaticity of encoded proteins (Table 1), implying hydropathy and aromaticity to be the major factors for amino acid variation in *T. whipplei* proteins. It is worth noting that there are two distinct clusters of proteins near the left end of the axis 1 (Fig. 2). A careful investigation reveals that the small cluster contained the genes for membrane associated proteins including WiSP family members and few hypothetical proteins, whereas genes coding for integral membrane proteins, several transporters, subunits of cytochrome C etc. are present in the large cluster. Although the genes present in both the clusters are mainly membrane-associated proteins, the amino acid usage profile and the propensities for formation of secondary structures are distinct from one another (Table 2). Proteins of these two clusters also differ in content of potential disordered structures. Disordered regions in proteins can be predicted by the lack of regular secondary structures whereas ordered regions (often termed globular) typically contain regular secondary structures packed into a compact globule.[19,20] As the probable coil forming regions are significantly higher in small cluster proteins (Table 2), disordered structures are more commonly found in the proteins of small cluster as compared to those comprising of the large cluster (Fig. 3). Recent investigations have indicated that disordered structures are usually more favored by proteins involved in regulatory functions and binding of various ligands.[20,21] Therefore, it may be presumed that the proteins in small cluster, which might play important roles in interactions with the host and/or immune evasion,[1,2] would be over represented by disordered structures. Members of the other cluster containing less disordered regions and exhibiting higher propensities for alpha-helical regions are primarily involved in transport and other membrane-associated processes. Compositions of the membrane-associated proteins of these two clusters are, therefore, influenced by evolutionary selective pressures resulting in a fine coordination between function, structure and stability.

Table 1. Non-parametric tests of association between the first three axes of COA on RAAU and multiple parameters of encoded proteins

	T. whipplei TW08/27			*T. whipplei* Twist		
	Variability explained(%)	Source of variation	Correlation coefficient* (r)	Variability explained(%)	Source of variation	Correlation coefficient* (r
1st Axis	18.9	Gravy Score Aromaticity	-0.83 -0.68	18.4	Gravy Score Aromaticity	-0.87 -0.68
2nd Axis	14.8	Gravy Score Aromaticity	-0.59 -0.45	14.1	Gravy Score Aromaticity	0.46 0.38
3rd Axis	8.8	Size/complexity GC_{12}	-0.79 0.65	8.8	Size/complexity GC_{12}	-0.74 0.66

* All correlations are significant at $p < 0.0001$.

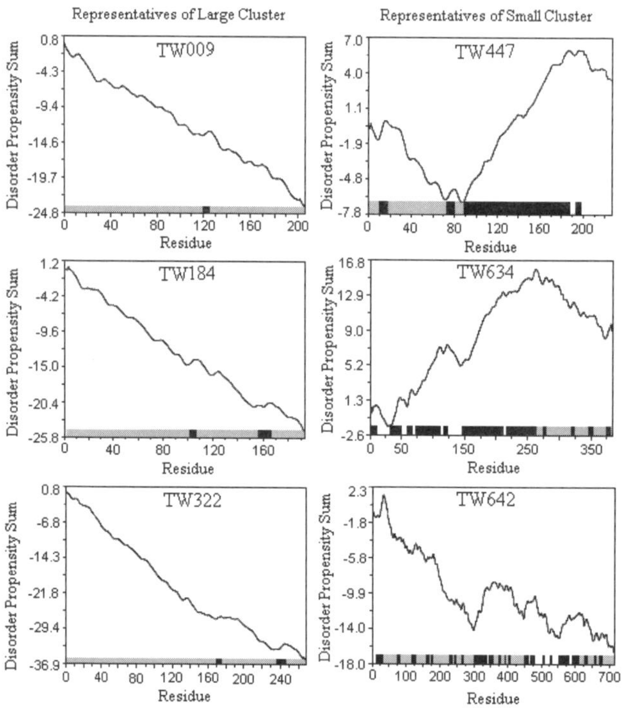

Figure 3. GlobPlot of the membrane associated proteins encoded by genes taken from large and small clusters generated by COA on RAAU for *T. whipplei* TW08/27 genome. The plots of the left and right panels are the representatives of large and small clusters respectively. Black colour indicates the disorder regions (lack of regular secondary structures) and gray colour indicates the ordered (globular) regions that typically contain regular secondary structures.

Another factor that largely influences the variations in amino acid usage is gene expressivity, as indicated by the presence of most of the potential highly expressed genes near the negative extreme end of the axis 3. A strong negative correlation of this axis with the average size/complexity quotient of the encoded proteins (Table 1) suggests that the highly expressed genes in *T. whipplei* have a tendency of avoiding the heavier residues including the aromatic ones in spite of their obligatory intracellular lifestyle. This supports the cost minimization hypothesis.[13] This is probably a genome-level adaptation to the host environment, as utilization of the less expensive and smaller residues by the highly expressed genes can minimize the energy exhaustion of the host and help them thereby to maintain the sustained infection, having least chance of elimination by the host.

Table 2. Amino acid usage and mean value of potential secondary structures of membrane associated proteins in the large and small clusters for *T. whipplei* TW08/27 and *T. whipplei* Twist genome

Amino acids and Predicted Secondary Structures	*T. whipplei* TW08/27		*T. whipplei* Twist	
	Large cluster (%)	Small cluster (%)	Large cluster (%)	Small cluster (%)
Phe	8.17*	3.17	7.87*	3.32
Leu	14.36*	8.24	14.46*	8.07
Ser	7.54	10.62+	7.66	10.27+
Tyr	2.97	4.70+	2.96	4.70+
Cys	1.07*	0.59	1.16*	0.59
Trp	1.81*	0.68	1.77*	0.79
Pro	4.27	8.07+	4.36	7.92+
His	1.33	1.73	1.39	1.47
Gln	1.83	3.20+	1.96	3.20+
Arg	4.56*	3.14	4.82*	3.18
Ile	8.76*	4.76	8.46*	4.77
Met	2.25*	0.50	2.18*	0.56
Thr	4.67	13.78+	4.67	13.93+
Asn	2.54	3.57+	2.57	3.44+
Lys	2.80	4.08+	2.83	3.97+
Val	9.29	8.74	9.25	8.99
Ala	9.43*	6.41	9.27*	6.43
Asp	2.33	3.67+	2.43	3.63+
Glu	1.97	2.10	2.08	2.16
Gly	7.71	8.09	7.49	8.46
Alpha helix	32.17*	9.44	31.92*	9.93
Beta sheet	24.29	29.95+	24.96	29.62+
Random coil	43.54	60.61+	43.12	60.45+

Values marked with *or + are significantly (p <0.001) more frequent in large or small cluster gene products respectively.

3.4 Relation between gene expression, protein conservation and amino acid usage

To understand the possible effect of gene expression on amino acid usage, we compared the average RAAU profiles of putative highly and lowly expressed gene-products of *T. whipplei* (Fig.4). It is interesting to note that except Pro, all other residues encoded by GC-rich codons (Gly, Ala and Arg) are significantly over represented ($p<0.001$) in the highly expressed gene-products, while usages of the residues encoded by all AU-rich codons (Phe, Tyr, Met, Ile and Asn) except Lys, are significantly higher ($p<0.001$) in lowly expressed gene-products. Increased usage of GC-rich or decreased usage of AU-rich codons in highly expressed genes of *T. whipplei* is supported by the significant positive correlation of GC_{1+2} with the coordinates of axis 3 of COA on RAAU (Table 1).

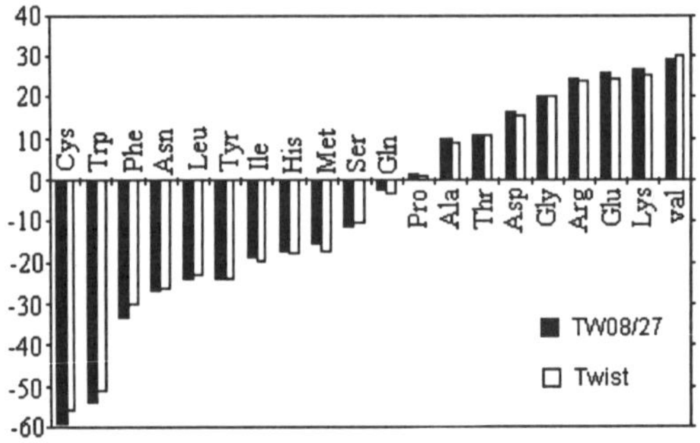

Figure 4. Difference in relative amino acid usage of highly and lowly expressed genes of *T. whipplei* TW08/27 (filled bars) and *T. whipplei* Twist (open bars). The differences were derived as R_{HL} = [{(Freq.H/Freq.L)-1} x 100].

Both the genomes under study have relatively low G+C-content (46% approx.), as compared to other actinomycete species and hence, a question arises: why do the highly expressed genes of these species exhibit higher usage of the residues encoded by G+C-rich codons? This might have happened for either of the following two reasons: (i) the highly expressed genes of *T. whipplei* are more conserved at the amino acid level than their lowly expressed counterparts and hence, they have retained a GC-richer composition which is closer to their putative ancestral state, or (ii) all genes are undergoing substitutions at a comparable rate irrespective of their level of expression, but due to some functional advantages, the highly expressed genes exhibit a positive selection in favour of the GC-rich residues. In order to examine which of these two possibilities is more likely to be true, we have compared the estimated pairwise non-synonymous divergences (d_N) between all orthologous of putative highly and lowly expressed genes. The mean $d_N = 0.021$ for highly expressed genes are significantly lower (t-test, $p<10^{-4}$) than that of lowly expressed genes ($d_N = 0.057$). It may be inferred therefore that in *T. whipplei*, the overall protein conservation in highly expressed genes is much higher and it also influences the discrimination in amino acid usage between highly and lowly expressed genes. The enhanced usage of the residues encoded by GC-rich codons in the highly expressed gene-products might have been possible due to more conservation at the amino acid level in highly expressed genes than the low-expression genes and hence, they retain GC-rich compositions which are closer to their putative ancestral state.

In summary, the present study reveals that the patterns of synonymous codon and amino acid usage in *T. whipplei* are the result of several factors. The replicational-transcriptional selection coupled with asymmetric mutational bias is the primary cause of intra-genomic variations in codon usage pattern. There is a residual intra-strand bias in synonymous codon usage by the highly expressed genes possibly due to the presence of translational selection. However, the influence of strand-specific mutational pressure is not so pronounced at the level of amino acid usage. The hydrophobicity and aromaticity seem to be the major sources of variation, both having nearly equal influence in amino acid usage. In spite of the intracellular life-style, the amino acid preferences in highly expressed gene products of *T. whipplei* follow the cost-minimization hypothesis. Another interesting finding is that the products of the highly expressed genes prefer to use the residues encoded by GC-rich codons, although the *T. whipplei* genomes, on an average, has only 46% G+C-content. The analysis presented here indicates that this might be due to greater conservation of a relatively GC-rich ancestral state in the highly expressed genes. Even the energetic cost of amino acid residues [22] might play a significant role in retaining the GC-rich residues in highly expressed genes. The study also shed lights on the diverse compositional and structural characteristics of two groups of membrane-associated proteins that might play distinct roles in host interactions.

References

1. S. D. Bentley, M. Maiwald, L.D. Murphy, et al. Sequencing and analysis of the genome of the Whipple's disease bacterium *Tropheryma whipplei. Lancet*, 361:637-644, 2003.
2. D. Raoult, H. Ogata, S. Audic, C. Robert, K. Suhre, M. Drancourt and J. M. Claverie. *Tropheryma whipplei* Twist: a human pathogenic Actinobacteria with a reduced genome. *Genome Res*, 13:1800-1809, 2003.
3. S. Casjens. The diverse and dynamic structure of bacterial genomes. *Annu Rev Genet*, 32:339-377, 1998.
4. P. M. Sharp and W. H. Li. The codon Adaptation Index--a measure of directional synonymous codon usage bias, and its potential applications. *Nucleic Acids Res*, 15:1281-1295, 1987.
5. A. Pan, C. Dutta, and J. DAS. Codon usage in highly expressed genes of *Haemophilus influenzae* and *Mycobacterium tuberculosis*: translational selection versus mutational bias. *Gene,* 215:405-413, 1998.
6. O. G. Berg and P. J. Silva. Codon bias in *Escherichia coli*: the influence of codon context on mutation and selection. *Nucleic Acids Res*, 25:1397-1404, 1997.
7. J.O. McInerney. Replicational and transcriptional selection on codon usage in *Borrelia burgdorferi. Proc. Natl. Acad. Sci. USA,* 95:10698-10703, 1998.
8. H. Romero, A. Zavala, H. Musto. Codon usage in *Chlamydia trachomatis* is the result of strand-specific mutational biases and a complex pattern of selective forces. *Nucleic Acids Res.,* 28:2084-2090, 2000.
9. S. Das, S. Paul, S. Chatterjee and C. Dutta. Codon and Amino Acid Usage in Two Major Human Pathogens of Genus *Bartonella* – Optimization Between Replication-Transcriptional Selection, Translational Control and Cost Minimization. *DNA Res.,* 12:91–102, 2005a.
10. A. B. de Miranda, F. Alvarez-Valin, K. Jabbari, W. M. Degrave and G. Bernardi. Gene expression, amino acid conservation, and hydrophobicity are the main factors shaping codon preferences in *Mycobacterium tuberculosis* and *Mycobacterium leprae*. *J. Mol. Evol.,* 50:45-55, 2000.

11. G. A. Singer and D. A. Hickey. Thermophilic prokaryotes have characteristic patterns of codon usage, amino acid composition and nucleotide content. *Gene,* 317:39-47, 2003.
12. J. R. Lobry and C. Gautier. Hydrophobicity, expressivity and aromaticity are the major trends of amino-acid usage in 999 Escherichia coli chromosome-encoded genes. *Nucleic Acids Res*, 22:3174-3180, 1994.
13. H. Seligmann. Cost-minimization of amino acid usage. *J Mol Evol*, 56:151-161, 2003.
14. S. Das, A. Pan, S. Paul, and C. Dutta. Comparative Analyses of Codon and Amino Acid Usage in Symbiotic Island and Core Genome in Nitrogen-Fixing Symbiotic Bacterium Bradyrhizobium japonicum. *J. Biomol. Struct. Dyn.,* In Press., 2005b.
15. J. Kyte and R. F. Doolittle. A simple method for displaying the hydropathic character of a protein, *J. Mol. Biol.,* 157:105-132, 1982.
16. M. J. Dufton. Genetic code synonym quotas and amino acid complexity: cutting the cost of proteins? *J Theor Biol*, 187:165-173, 1997.
17. M. Nei and T. Gojobori. Simple methods for estimating the numbers of synonymous and nonsynonymous nucleotide substitutions. *Mol Biol Evol*, 3:418-426, 1986.
18. J. Garnier, J. F. Gibrat and B. Robson. GOR method for predicting protein secondary structure from amino acid sequence. *Methods Enzymol.,* 266:540-553, 1996.
19. R. Linding, R. B. Russell, V. Neduva and T. J. Gibson. GlobPlot: Exploring protein sequences for globularity and disorder. *Nucleic Acids Res*, 31:3701-3708, 2003.
20. M. Fuxreiter, I. Simon, P. Friedrich and P. Tompa. Preformed structural elements feature in partner recognition by intrinsically unstructured proteins. *J Mol Biol*, 338:1015-1026, 2004.
21. A. L. Fink. Natively unfolded proteins. Curr Opin Struct Biol, 15; 35-41, 2005.
22. H. Akashi and T. Gojobori. Metabolic efficiency and amino acid composition in the proteomes of *Escherichia coli* and *Bacillus subtilis*. *Proc Natl Acad Sci USA*, 99:3695-3700, 2002.

CONSEQUENCES OF MUTATION, SELECTION AND PHYSICO-CHEMICAL PROPERTIES OF ENCODED PROTEINS ON SYNONYMOUS CODON USAGE IN ADENOVIRUSES *

SANDIP PAUL, SABYASACHI DAS and CHITRA DUTTA

Bioinformatics Centre, Indian Institute of Chemical Biology, Kolkata –700032, India

Trends in synonymous codon usage in adenoviruses have been examined through the multivariate statistical analysis on the annotated protein-coding regions of 22 adenoviral species, for which complete genome sequences are available. One of the major determinants of such trends is the G + C content at third codon positions of the genes, the average value of which varied from one viral genome to other depending on the overall mutational bias of the species. G_{3S} and C_{3S} interacted synergistically along the first principal axis of Correspondence analysis on the Relative Synonymous Codon Usage of adenoviral genes, but antagonistically along the second principal axis. Other major determinants of the trends are the natural selection, putatively operative at the level of translation and quite interestingly, hydropathy of the encoded proteins. The trends in codon usage, though characterized by distinct virus-specific mutational bias, do not exhibit any sign of host-specificity. Significant variations are observed in synonymous codon choice in structural and nonstructural genes of adenoviruses.

1. Introduction

Genomes of adenoviruses are characterized by linear, double-stranded DNA, with inverted terminal repeats (ITR) ranging from 36 to over 200 bp in length depending on the serotype.[1,2] Genes inherited by all existing adenoviruses from their common ancestor (Genus-common genes) are located centrally in the genome and are involved in replication and packaging of viral DNA as well as in the formation of virion. The other genes (Genus -specific genes) are captured in each lineage and mostly located near the genome termini.[2] These genes are generally involved in interactions with the host and probably contribute to the survival of viruses in respective biological niches.[2]

In recent years, the focus of adenovirus research has shifted from basic biology to adenovirus-based vector technologies.[3] Adenoviruses are often efficient at gene delivery to specific cell types. Genetically engineered, replication-deficient recombinant adenoviruses are gradually becoming popular as gene delivery vehicles for their high capacity to transfer therapeutic genes *in vivo*.[4] One of the crucial issues for development of promising vectors for gene therapy are transient, but high level of expression of delivered genes within the host. But development of an efficient gene expression system needs a detailed knowledge of codon and nucleotide preferences in genomes concerned.

* This work was supported by the Council of Scientific and Industrial Research, Government of India (Project No. CMM 0017) and Department of Biotechnology, Government of India (Grant Number BT/BI/04/055-2001).

Keeping this in mind, the present study attempts to analyze the nucleotide and codon usage patterns in all adenoviral genomes sequenced so far.

It is well established that synonymous codon usage in various organisms, particularly in unicellular ones, often reflect a balance between the genomic G+C-bias and translational selection.[5,6] The strength and direction of these selection forces vary at the intra- and inter-genomic level.[7] Various other factors like codon-anticodon interaction,[8] physical location of each gene on the chromosome,[9] replicational- transcriptional selection,[10-12] ecological niches [13] etc. may influence the biased usage of synonymous codons. In viruses, however, little is known about the extent and origin of the synonymous codon bias. In human immunodeficiency virus (HIV), codon usage bias is the result of strong preference of adenine base.[14] Codon bias due to uneven base composition has also been described in nucleopolyhedroviruses [15] and pneumoviruses.[16] In papillomaviruses, specific codon usage pattern linked with variation in A+T-content within the genomes may increase the replicational fitness in mammalian epithelial cells.[17] In human RNA viruses, mutational bias is not the only determining factor, translational selection may also have influence in shaping codon usage bias.[18] In the present report, through a multivariate analysis, attempt has been made to delineate the trends in codon and nucleotide selection in adenoviral genes and also to identify the selection forces governing such trends. Such information not only can offer an insight into the evolution of codon usage patterns along adenovirus lineages, but also may help in increasing the efficiency of gene delivery/expression systems.

2. *Methods and Materials*

2.1 *Retrieval of sequences*

The 22 available complete genome sequences of adenoviruses (listed in Table1) have been downloaded from NCBI GenBank (Version 145.0). To minimize the sampling error we have taken only those genes, which are greater than or equal to 150 bp. We have also eliminated the partial coding sequences and those sequences, which have internal termination codons. Finally 616 coding sequences were selected for analysis.

2.2 *Sequence analysis*

Relative synonymous codon usage (RSCU) was used to examine the synonymous codon usage variation among the genes without any confounding influence of amino acid composition.[5] To find out the extent of base compositional bias GC_{1+2} (G+C content at first and second codon positions), GC_{3S} and N_{3S} (the frequency of G+C and base N respectively at synonymous third codon positions) were calculated for each gene under study. To measure the general non-uniformity of synonymous codon usage, the effective number of codons (Nc) of each gene was calculated.[19] The GRAVY score, which indicates the mean hydropathy index of the encoded amino acid residues and hence, is an estimate of overall hydrophobicity,[20] was computed for each gene product. The predictions of protein secondary structure were performed using GOR IV algorithm.[21]

Correspondence analysis (COA) on RSCU values was carried out using CODONW 1.4.2 to investigate the major trend in codon usage variation among genes. To see the extent of divergence in codon usage more precisely, a cluster analysis was carried out using simple D-squared statistic method. D-squared statistics is the sum of squares of the difference between codons of the two codon usage tables; i.e. D^2 = sum over all 64 codons of: $(\text{frequency}_{(\text{codon, Table 1})} - \text{frequency}_{(\text{codon, Table 2})})^2$. A matrix containing the D^2 value of each pair was used to produce clusters (dendrogram) by neighbor-joining method.[22] Linear regression analysis was used to find out the correlation between synonymous codon usage bias and various codon usage indices. To test the heterogeneity in codon usage, one-way ANOVA was performed using STATISTICA (Version 6.0).

3. *Results and Discussion*

3.1 Inter-and Intra-species variation in compositional constraints on codon usage

The codon usage bias in the coding regions of 22 completely sequenced adenoviruses of varying G+C content has been investigated (Table1). The average values of the effective numbers of codons (Nc) in different adenoviruses varied from 38.97 (in *Porcine adenovirus A*) to 54.67 (in *Canine adenovirus*). The average GC_{3S} values for individual genomes varied from 22.78 (OAdV-A) to 79.61 (PAdV- A). In addition, there are marked intra-genomic variations in Nc (standard deviation > 3.5, except for BAdV-B) and GC_{3S} values (standard deviation > 5%). These observations indicate that there is a significant heterogeneity in compositional bias as well as in the codon usage pattern within and among the members of *Adenoviridae*. When the Nc values of each adenovirus gene are plotted against the corresponding GC_{3S}, only a small number of points lie on the expected curve and a large number of points lie well below the expected curve (not shown), suggesting that some additional selection pressure other than the species-specific mutational bias, acts on codon usage in adenovirus genomes.

3.2 Virus-specific synonymous codon usage patterns with no sign of host-specificity

Figure 1 depicts the position of each virus on the plane defined by the first (horizontal) and second (vertical) principal axes generated by COA on RSCU values of genes. The first and second principal axes account for 30.35 % and 7.6 % of total variability. The first principal axis exhibits strong correlation with GC_{3S}-content of the genes for all four genera of adenoviruses i.e. for atadenovirus, aviadenovirus, mastadenovirus and siadenovirus (Table 2). The viruses having highest GC_{3S} levels in their coding sequences display the most positive values along that axis (Fig. 1). When G_{3S} and C_{3S} are considered separately, the correlation coefficient exhibits by the positions of genes along the first axis with C_{3S} is significantly larger than that with G_{3S} (Table 2), indicating that the contribution of C_{3S} to the inter-species variation in overall GC_{3S}-content is greater than that of G_{3S}.

Table 1. Average codon usage bias (measured by effective number of codons) and base composition in 22 completely sequenced adenoviruses under study.

Virus	Abbreviation	Accession	GC %	Nc	A_{3S}	T_{3S}	G_{3S}	C_{3S}
ATADENOVIRUS								
Bovine adenovirus D	BAdV-D	NC_002685	35.2	44.46	32.93	39.05	16.53	11.49
Duck adenovirus A	DAdV-A	NC_001813	43.0	53.46	25.90	34.75	21.42	17.92
Ovine adenovirus D	OAdV-D	NC_004037	33.6	42.56	32.89	41.38	14.93	10.80
AVIADENOVIRUS								
Fowl adenovirus A	FAdV-A	NC_001720	54.3	52.36	16.58	21.66	28.05	33.71
Fowl adenovirus D	FAdV-D	NC_000899	53.8	51.01	17.11	20.43	28.05	34.41
MASTADENOVIRUS								
Bovine adenovirus A	BAdV-A	NC_006324	48.8	54.27	22.94	27.04	25.84	24.18
Bovine adenovirus B	BAdV-B	NC_001876	54.0	51.67	15.68	23.53	28.42	32.37
Canine adenovirus	CAdV	NC_001734	47.0	54.67	22.38	29.48	23.38	24.76
Human adenovirus A	HAdV-A	NC_001460	46.5	54.15	25.32	30.59	21.85	22.24
Human adenovirus B	HAdV-B	NC_004001	48.9	51.88	22.95	28.78	22.41	25.86
Human adenovirus C	HAdV-C	NC_001405	55.2	47.21	15.72	21.83	30.46	31.99
Human adenovirus D	HAdV-D	NC_002067	56.6	46.40	15.53	19.73	29.67	35.07
Human adenovirus E	HAdV-E	NC_003266	57.7	44.47	12.39	18.17	31.10	38.34
Human adenovirus F	HAdV-F	NC_001454	51.2	51.99	18.73	27.36	25.8	28.11
Murine adenovirus A	MAdV-A	NC_000942	47.8	53.09	22.35	31.36	24.28	22.01
Ovine adenovirus A	OAdV-A	NC_002513	43.6	49.73	26.94	31.55	20.63	20.88
Porcine adenovirus A	PAdV-A	NC_005869	63.8	38.97	7.63	12.66	35.99	43.72
Porcine adenovirus C	PAdV-C	NC_002702	50.5	50.64	18.48	23.55	27.02	30.95
Simian adenovirus A	SAdV-A	NC_006144	55.3	47.36	15.81	22.62	29.20	32.37
Tree shrew adenovirus	TSAdV	NC_004453	50.0	50.42	19.95	26.87	25.87	27.31
SIADENOVIRUS								
Frog adenovirus	FrAdV	NC_002501	37.9	44.81	33.98	35.10	16.76	14.16
Turkey adenovirus A	TAdV-A	NC_001958	34.9	44.09	30.53	41.34	16.49	11.64

The separation of one viral genome from another is found to be significant by one-way analysis of variance (ANOVA) on the first principle axis ($F_{21,\,594} = 57.805$, $p < 10^{-7}$), as that axis explains the major variation in codon usage. This result indicates that some virus-specific selection pressure might have influenced synonymous codon usage in different adenoviruses. However, no sign of host-specificity can be observed in the trends in codon usages. Viruses infecting the same host appear, in most cases, at distinct positions along axis1 and/or axis2 generated by COA on RSCU values of genes (Fig. 1). For example, the viruses infecting the bovine host (BAdV-A, BAdV-B and BAdV-D) are placed in three different positions significant distances apart from one another. Therefore, it seems that synonymous codon usage patterns in adenoviruses do not follow, in general, any host specific trend.

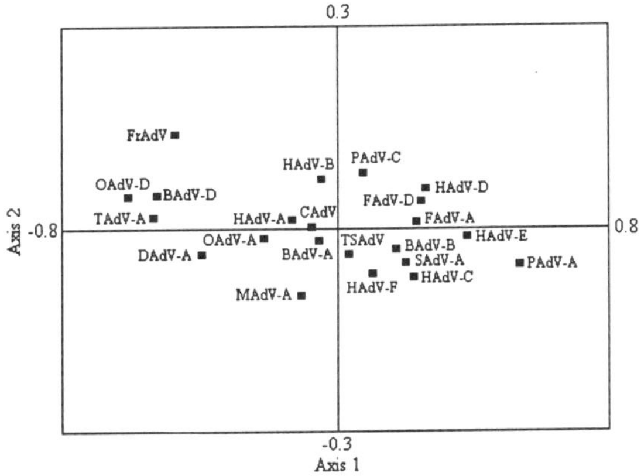

Figure 1: Positions of adenoviruses on the plane defined by first and second principle axes generated from Correspondence analysis of Relative Synonymous Codon Usage (RSCU) of corresponding genes. GC-poor species are on the left, while GC-rich on the right.

The members of different genera of adenoviruses (atadenovirus, aviadenovirus, mastadenovirus, siadenovirus) exhibit non-uniform distribution along the axis 1 in Fig. 1. The position of atadenoviruses and siadenoviruses are at the extreme left of axis1, whereas aviadenoviruses exhibit the positive values along this axis. The separation of these four genera on the basis the variation in codon usage pattern explained by first axis of COA on RSCU values is statistically significant ($F_{3,\ 612} = 161.35$, $p < 10^{-7}$). This observation indicates that so far as the synonymous codon usage are concerned, members of atadenovirus and siadenovirus are more close to one another, while those in aviadenoviruses are far away from these two genera. The members of mastadenovirus are distributed over a large region, suggesting that this group follows more heterogeneous patterns in synonymous codon usage than the other adenoviruses. These observations are in accordance with the cluster analysis on the extent of divergence in codon usage, which yields two major clusters - the adenoviruses having relatively higher GC-content, e. g., aviadenoviruses are branched together in the upper cluster and those with relatively AT-rich genomes, such as atadenoviruses and siadenoviruses, appear in the lower cluster (Fig. 2). Members of mastadenovirus, having widely varying GC-content, are dispersed through both the clusters. It is worth mentioning that the distribution of different adenoviral genomes along the axis1 (Fig. 1) and their segregation in two clusters (Fig. 2) are consistent with their genomic G+C-content and also with their phylogenetic distribution, as determined by Davison et al (2003).[2]

Table 2. Non-parametric tests of association between the first two axes of COA on RSCU and multiple synonymous base usage parameters and hydropathy of encoded proteins

Parameters	Atadenovirus		Aviadenovirus		Mastadenovirus		Siadenovirus	
	Axis 1	Axis2	Axis 1	Axis2	Axis 1	Axis2	Axis 1	Axis2
A_{3S}	-0.49***	0.83***	-0.42**	0.46***	-0.77***	0.39***	-0.35*	0.66***
T_{3S}	-0.52***	-0.48***	-0.86***	-0.50***	-0.90***	-0.19**	-0.57***	-0.61***
G_{3S}	0.55***	-0.76***	0.04 NS	-0.81***	0.54***	-0.67***	0.51**	-0.60***
C_{3S}	0.81***	0.28*	0.89***	0.74***	0.89***	0.43***	0.82***	0.47**
GC_{3S}	0.95***	-0.35**	0.92***	0.17NS	0.97***	-0.06 NS	0.90***	-0.05 NS
Gravy	0.14 NS	-0.51***	-0.45**	-0.09 NS	-0.41***	-0.03 NS	0.02 NS	-0.25*

Notable significant relationships are marked by *** P < 0.0001;** P < 0.001;* P < 0.01;NS Nonsignificant.

The correlation coefficient between the second axis and GC_{3S} is relatively small, as compared to that between the axis1 and GC_{3S} (Table 2). But it is worth mentioning that the axis2 exhibits strong negative correlation with G_{3S} and positive correlation with C_{3S} for all four genera of adenoviruses (Table 2). These observations indicate that G_{3S} and C_{3S} interact synergistically in the first principal axis resulting in the increase of GC_{3S} content, but antagonistically in the second principal axis so that increase in the frequency of C_{3S} is accompanied by a decrease in G_{3S} and vice-versa.

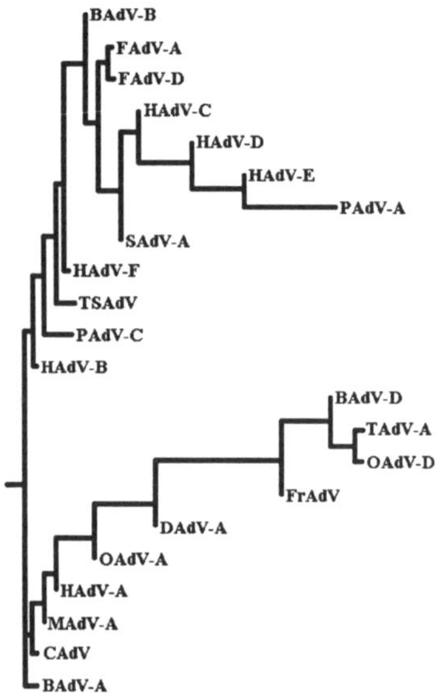

Figure 2: A dendrogram representing the extent of divergence in synonymous codon usage of 22 completely sequenced adenoviruses constructed by neighbor-joining method.

This antagonistic behavior might be due to the fact that after GC_{3S} reach some saturation value, frequency of one (G_{3S} or C_{3S}) of them can increase further only at the expense of the other one (C_{3S} or G_{3S}), i.e., occurrence of one base excludes the other in order to maintain the overall G+C-content at synonymous positions under some threshold value.

3.3 Differential codon usage in structural and nonstructural genes

With a view to examine whether the presence of any selection pressure(s) is responsible for intra-genome heterogeneity in codon usage patterns in adenoviruses, we have compared the average RSCU values of structural (i.e. major core protein, minor core protein, hexon and hexon-associated protein etc.) and nonstructural (i.e. DNA polymerase, transcription activators etc.) genes of four different adenoviral genera separately. There are several codons (mostly G- or T-ending), usages of which are significantly higher among the structural genes. On the other hand, several codons (C- or A-ending) are over represented in nonstructural genes as compared to the codon usage in structural genes (data not shown). This indicates that the selection force resulting in the differential codon usage patterns in structural and non-structural genes might not be the simple G+C-bias. As the structural genes in viruses are generally highly expressed than the nonstructural genes,[23] a natural selection, putatively operating at the level of translation might also be responsible for differential usage of synonymous codons in structural and non-structural genes of adenoviruses. However, in adenoviruses, no significant correlation is found between codon usage bias (as measured by Nc values of genes) and the gene length.

3.4 *Correlation between synonymous base compositions and hydropathy of encoded proteins*

An important finding derived from our study on adenoviral genomes is that for each group of adenoviruses, one of the principal axes generated by COA on RSCU values of the genes exhibits significant correlation with the hydropathy of the encoded proteins (as determined by the Gravy Score of the gene-products) (Table 2). With a view to find out the reason behind this apparently unexpected association between nucleotide usage in third codon position and protein hydropathy, we have calculated the correlations between hydrophobicity and synonymous base usage in four adenovirus genera separately. It is found that the hydrophobicity exhibits highest positive correlations with G_{3S} in atadenoviruses ($r = 0.43$, $p < 0.001$) and siadenoviruses ($r = 0.36$, $p < 0.001$), whereas with T_{3S} in aviadenoviruses ($r = 0.37$, $p < 0.001$) and mastadenoviruses ($r = 0.54$, $p < 0.001$) (Fig. 3).

Usage of individual nucleotides at synonymous sites of genes in members of different genera of adenoviruses reveals a mixed and indefinite nature of the correlation with hydrophobicity of respective gene-products (Table 3). In general (with a few exceptions), the hydropathy levels of encoded proteins shows positive correlation with both G_{3S} and T_{3S} in atadenoviruses, with only T_{3S} in aviadenoviruses and mastadenoviruses, and with only G_{3S} in siadenoviruses.Earlier it is found that as compared to hydrophilic proteins, there is an increase in usage of the G-ending codons and decrease in that of the C-ending codons in hydrophobic proteins of *Mycobacterium tuberculosis* and *Mycobacterium leprae*.[24] Existence of significant correlation between hydrophobicity of encoded proteins and the base composition of third codon positions have also been reported in some other prokaryotes and several eukaryotes.[25] However, there is no report of such correlation in any of the viral genomes studied so far. To our knowledge, this is for the first time; a correlation has been demonstrated between the synonymous codon usage in genes and this physico-chemical property of corresponding gene-products in a group of viral genomes.

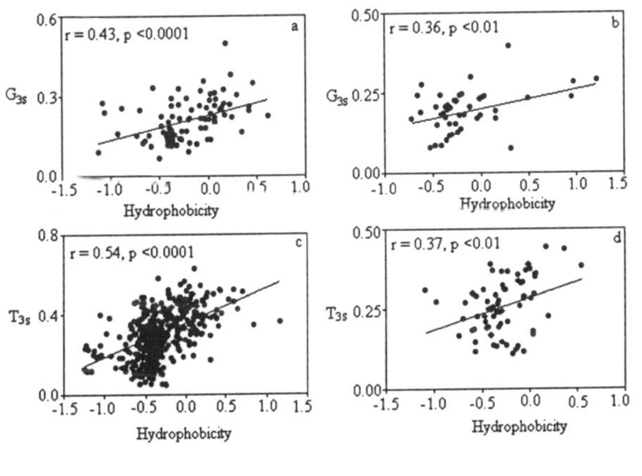

Figure 3: Hydropathy values (Gravy scores) of the encoded proteins plotted against (a) G_{3S} in atadenoviruses, (b) G_{3S} in siadenoviruses, (c) T_{3S} in mastadenoviruses and (d) T_{3S} in aviadenoviruses.

Table 3. Correlations between the protein hydropathy and synonymous base compositions in adenoviruses

Genera	Virus	Correlation coefficients (r)			
		A3S[a]	T3S[b]	G3S[c]	C3S[d]
Atadenovirus	BAdV-D	-0.73**	0.51**	0.61**	-0.43*
	DAdV-A	-0.44*	0.09	0.25	0.15
	OAdV-D	-0.54**	0.51**	0.41*	-0.33
Aviadenovirus	FAdV-A	-0.07	0.28	-0.02	-0.27
	FAdV-D	0.22	0.43*	-0.01	-0.38*
Mastadenovirus	BAdV-A	-0.06	0.58**	-0.41*	-0.32
	BAdV-B	0.19	0.72**	-0.24	-0.43*
	CAdV	0.01	0.60**	0.02	-0.59**
	HAdV-A	-0.04	0.54**	-0.05	-0.57**
	HAdV-B	0.34*	0.36*	-0.39*	-0.22
	HAdV-C	-0.01	0.63**	-0.20	-0.49**
	HAdV-D	0.43*	0.44*	-0.42*	-0.31
	HAdV-E	0.32	0.58**	-0.48**	-0.41*
	HAdV-F	0.29	0.72**	-0.39*	-0.53**
	MAdV-A	-0.44*	0.41*	0.08	-0.11
	OAdV-A	0.47*	0.60**	-0.43*	-0.58**
	PAdV-A	0.13	0.41*	0.09	-0.31
	PAdV-C	0.25	0.58**	-0.30	-0.47*
	SAdV-A	0.39*	0.65**	-0.42*	-0.56**
	TSAdV	0.37*	0.76**	-0.31	-0.66*
Siadenovirus	FrAdV	0.09	-0.06	0.12	-0.48*
	TAdV-A	-0.33	-0.24	0.58**	0.05

Significant relationships are marked by ** $P < 0.01$; * $P < 0.05$

Therefore, the present analysis indicates that the selection of nucleotide at synonymous sites in adenoviral genes might affect or be affected by the hydropathy levels of the encoded products. However, the cause and effect relation of this correlation is not clear. It is known that the hydrophobicity of amino acid residues plays an important role in protein folding.[25] It has also been reported that the codons overrepresented in alpha-helix are underrepresented in beta-sheet and vice versa and this discrepancy may be related to the particular translation kinetics necessary to ensure the proper folding of nascent peptide.[24] With a view to predict the plausible biological origin of the correlation between synonymous codon usage and protein hydropathy, we have, therefore, predicted the secondary structure of five most hydrophobic proteins (i.e., gene-products with highest Gravy scores) and five most hydrophilic proteins (i.e., gene-products with lowest Gravy scores) from each adenovirus species. It is found that the regions of the proteins with high propensity of formation of alpha-helices are significantly over represented (t-test, p-value$<10^{-5}$) in proteins with high hydrophilicity (i.e., low Gravy Score), whereas the regions predicted to be beta-sheet were found in significantly higher frequencies (t-test, p-value$<10^{-5}$) in proteins with high hydrophobicity (i.e., high Gravy Score). Therefore, the differential synonymous base usage by hydrophobic and hydrophilic proteins in adenoviruses may have its origin in their propensities for secondary structure formation. Further studies are required to get the deeper insights on the biological factors underlying this relationship in viruses.

In summary, the trends in synonymous codon usage in adenoviruses is found to be governed by several factors - the virus-specific directional mutational bias, natural selection putatively operating at the level of translation and more interestingly, hydrophobicity of the gene-products. Apparently, the trends in synonymous codon selection do not exhibit any host-specificity. No correlation is found between codon usage bias and gene-length. Furthermore, the antagonistic behavior of G_{3S} and C_{3S} along the second major axis of COA on RSCU values of adenoviral genes suggests the existence of a constraint on the extent of GC-bias in third codon position of any specific adenoviral genome. Such findings on trends in synonymous codon usage in adenoviruses might provide not only the valuable information for better understanding of the evolution of adenoviral genomes, but also provide clues to development of an efficient gene delivery/expression systems based on adenoviral vectors.

References

1. R. N. de Jong, P. C. van der Vliet and A. B. Brenkman. Adenovirus DNA replication: protein priming, jumping back and the role of the DNA binding protein DBP. *Curr. Top. Microbiol. Immunol.*, 272;187–211, 2003.
2. A. J. Davison, M. Benko and B. Harrach. Genetic content and evolution of adenoviruses. *J. Gen. Virol.*, 84;2895-2908, 2003.
3. J. J. Rux and R. M. Burnett. Adenovirus structure. *Hum. Gene. Ther.*, 15;1167-1176, 2004.
4. C. M. Lai, Y. K. Lai, and P. E. Rakoczy. Adenovirus and adeno-associated virus vectors. *DNA Cell Biol.*, 21;895-913, 2002.
5. P. M. Sharp and W. H. Li. An evolutionary perspective on synonymous codon usage in unicellular organisms. *J. Mol. Evol.*, 24;28-38, 1986.
6. A. Pan, C. Dutta, and J. DAS. Codon usage in highly expressed genes of *Haemophilus influenzae* and *Mycobacterium tuberculosis*: translational selection versus mutational bias. *Gene,* 215;405-413, 1998.
7. S. Das, A. Pan, S. Paul, and C. Dutta. Comparative Analyses of Codon and Amino Acid Usage in Symbiotic Island and Core Genome in Nitrogen-Fixing Symbiotic Bacterium Bradyrhizobium japonicum. *J. Biomol. Struct. Dyn.*, In Press., 2005b.
8. H. Grosjean and W. Fiers. Preferential codon usage in prokaryotic genes: the optimal codon-anticodon interaction energy and the selective codon usage in efficiently expressed genes. *Gene*, 18;199-209, 1982.
9. A. R. Kerr, J. F. Peden, and P. M. Sharp. Systematic base composition variation around the genome of Mycoplasma genitalium, but not *Mycoplasma pneumoniae*. *Mol. Microbiol.*, 125;1177-1179, 1997.
10. J.O. McInerney. Replicational and transcriptional selection on codon usage in *Borrelia burgdorferi*. *Proc. Natl. Acad. Sci. USA,* 95;10698-10703, 1998.

11. H. Romero, A. Zavala, H. Musto. Codon usage in *Chlamydia trachomatis* is the result of strand-specific mutational biases and a complex pattern of selective forces. *Nucleic Acids Res.,* 28;2084-2090, 2000.
12. S. Das, S. Paul, S. Chatterjee and C. Dutta. Codon and Amino Acid Usage in Two Major Human Pathogens of Genus *Bartonella* – Optimization Between Replication-Transcriptional Selection, Translational Control and Cost Minimization. *DNA Res.,* 12;91–102, 2005a.
13. G. A. Singer and D. A. Hickey. Thermophilic prokaryotes have characteristic patterns of codon usage, amino acid composition and nucleotide content. *Gene,* 317;39-47, 2003.
14. F. J. van Hemert, and B. Berkhout. The tendency of lentiviral open reading frames to become A-rich: constraints imposed by viral genome organization and cellular tRNA availability. *J. Mol. Evol.,* 41;132-140, 1995.
15. D. B. Levin and B. Whittome. Codon usage in nucleopolyhedroviruses. *J. Gen. Virol.,* 81;2313-2325, 2000.
16. C. R. Pringle and A. J. Easton. Monopartite negative strand RNA genomes. *Seminars in Virology,* 8;49-57, 1997.
17. K. N. Zhao, W. J. Liu and I. H. Frazer. Codon usage bias and A+T content variation in human papillomavirus genomes. *Virus Res.,* 98;95-104, 2003.
18. G. M. Jenkins and E. C. Holmes. The extent of codon usage bias in human RNA viruses and its evolutionary origin. *Virus Res.,* 92;1-7, 2003.
19. F. Wright. The 'effective number of codons' used in a gene. *Gene,* 87;23-29, 1990.
20. J. Kyte and R. F. Doolittle. A simple method for displaying the hydropathic character of a protein, *J. Mol. Biol.,* 157;105-132, 1982.
21. J. Garnier, J. F. Gibrat and B. Robson. GOR method for predicting protein secondary structure from amino acid sequence. *Methods Enzymol.,* 266;540-553, 1996.
22. N. Saitou and M. Nei. The neighbor-joining method: a new method for reconstructing phylogenetic trees. *Mol. Biol. Evol.,* 4;406-425, 1987.
23. W. Gu, T. Zhou, J. Ma, X. Sun and Z. Lu. Analysis of synonymous codon usage in SARS Coronavirus and other viruses in the Nidovirales. *Virus Res.,* 101;155-161, 2004.
24. A. B. de Miranda, F. Alvarez-Valin, K. Jabbari, W. M. Degrave and G. Bernardi. Gene expression, amino acid conservation, and hydrophobicity are the main factors shaping codon preferences in *Mycobacterium tuberculosis* and *Mycobacterium leprae. J. Mol. Evol.,* 50;45-55, 2000.
25. G. D'Onofrio, K. Jabbari, H. Musto and G. Bernardi. The correlation of protein hydropathy with the base composition of coding sequences. *Gene,* 238;3-14, 1999.

MICROARRAY MISSING VALUE IMPUTATION BY ITERATED LOCAL LEAST SQUARES *

ZHIPENG CAI[†], MAYSAM HEYDARI[†] AND GUOHUI LIN [‡]

Bioinformatics Research Group, Department of Computing Science, University of Alberta. Edmonton, Alberta T6G 2E8, Canada.

Emails: zhipeng,maysam,ghlin@cs.ualberta.ca

Microarray gene expression data often contains missing values resulted from various reasons. However, most of the gene expression data analysis algorithms, such as clustering, classification and network design, require complete information, that is, without any missing values. It is therefore very important to accurately impute the missing values before applying the data analysis algorithms. In this paper, an *Iterated Local Least Squares Imputation method* (ILLsimpute) is proposed to estimate the missing values. In ILLsimpute, a similarity threshold is learned using known expression values and at every iteration it is used to obtain a set of coherent genes for every target gene containing missing values. The target gene is then represented as a linear combination of the coherent genes, using the least squares. The algorithm terminates after certain iterations or when the imputation converges. The experimental results on real microarray datasets show that ILLsimpute outperforms three most recent methods on several commonly tested datasets.

Keywords: Microarray gene expression data, Missing value imputation, Local least squares

1. Introduction

DNA microarray experiments are extensively used to monitor the expression of a large amount of genes under various conditions. Associated with mathematical analysis methods, DNA microarray has important applications in biological and clinical studies. The data generated from a set of microarray experiments is usually expressed as a large matrix, with the expression levels of genes in rows, and the experimental conditions ordered in columns.[7] In other words, let $G_{m\times n} = (g_{ij})_{m\times n}$ be the expression matrix of m genes in n experiments. Then g_{ij} records the expression level of the i-th gene in the j-th experiment. One frequent issue that affects microarray data analysis is the existence of missing values, that is, matrix $G_{m\times n}$ could contain many entries with unknown expression levels. A number of reasons could lead to the missing data, including insufficient resolution, image corruption, or even dust and scratches on the slides.[7] On the other hand, most of the microarray data analysis algorithms, such as gene clustering, disease (experiment) classification, and gene network design, require the complete information. In other words, they

*This work is supported by NSERC and CFI.

[†]Z.C. and M.H. contributed equally to this work.

[‡]To whom correspondence should be addressed. Fax: (780) 492-1071. Email: ghlin@cs.ualberta.ca.

require matrix $G_{m\times n}$ contains no missing values. It is therefore very important to accurately estimate the missing values in matrix $G_{m\times n}$, if any, before we can apply the data analysis algorithms. One straightforward solution is to repeat the experiments a sufficient amount of times[1,7] and use a combination of them to obtain a complete expression matrix. It is easily seen that such an approach is very costly and inefficient. Moreover, a portion of expression information would be wasted.

There are several proposals on effectively imputing the missing values, without extra experiments, through taking advantage of modern mathematical and computational techniques. To name a few, Troyanskaya *et al.*[7] proposed a *weighted K-nearest neighbor* method (KNNimpute) and a *singular value decomposition* method (SVDimpute) to impute the missing values. In more details, KNNimpute method selects for every target gene K nearest neighboring genes from the entire set of genes (one measure of distance will be detailed in Section 2). It then uses weighted linear combinations of these neighboring genes to predict the missing values in the target gene. In SVDimpute method, from the expression matrix $G_{m\times n}$, a set of mutually orthogonal expression patterns (called *eigengenes*) is obtained that can be linearly combined to approximate the expression levels of all the genes. Subsequently, K most significant eigengenes are selected to estimate the missing values. It has been shown[7] that KNNimpute works well on static microarray data and noisy time series microarray data; SVDimpute performs better on time series microarray data with low noise levels.[7] In year 2003, Oba *et al.*[4] proposed a novel missing value estimation method based on *Bayesian Principal Component Analysis* (BPCA), which estimates a probabilistic model and latent variables within the framework of Bayesian inference. More recently, Kim *et al.*[3] successfully applied *Local Least Squares Imputation* (LLSimpute) method to estimate the missing values. In LLSimpute, a target gene with missing values is modeled as a linear combination of K coherent genes that were found using nearest neighbor method, where K is learned using known expression values.

In this paper, we propose a novel iterated way of using local least squares to more accurately impute the missing values — *Iterated Local Least Square Imputation* (ILLsimpute). Note that in most existing imputation methods, a constant K coherent genes are selected for a target gene. This constant is usually pre-specified, though in LLSimpute it is learned out of the microarray dataset. But they select for every target gene K coherent genes. In our ILLsimpute, we do not put a hard constraint on the numbers of coherent genes picked for target genes, i.e., they may vary. In fact, the known expression values in the microarray dataset are used to learn a *threshold* of similarity to obtain a set of coherent genes for every target gene. Subsequently, the target gene is represented by a linear combination of those coherent genes using the local least squares, and the missing values are estimated as done in LLSimpute method. The process is repeated for a number of iterations or terminates when the imputed values converge. The detailed steps of operations in ILLsimpute will be presented in Section 2. We have compared the performance of ILLsimpute with three other most recent methods, namely, BPCA and LLSimpute, and KNNimpute method on six microarray datasets that we were able to obtain using a number of different percentages of missing values. The detailed experimental results are summarized in Section 3. Finally, we conclude the paper in Section 4 with some future works and discussions.

2. Iterated Local Least Squares Imputation — ILLsimpute

Again, let $G_{m\times n}$ denote the gene expression matrix for m genes in n experiments. Note that in general we have $m \gg n$. g_{ij} denotes the expression level of the i-th gene in the j-th experiment. The *Iterated Local Least Squares Imputation* (ILLsimpute) method is made up of two parts: In the first part, a similarity *threshold* is estimated using the known expression values in $G_{m\times n}$. In the second part, the *threshold* is used in LLSimpute method for several iterations to obtain the final estimated values for the missing entries in $G_{m\times n}$. In what follows, we introduce how LLSimpute method[3] works and then how we estimate the similarity *threshold* for finding a set of coherent genes for each target gene.

First of all, we introduce a distance measure between two genes, which is adopted throughout the paper in finding coherent genes for a target gene. A *target* gene is one that contains missing values to be estimated. To determine its coherent genes, for every other gene, the missing values are filled with the average of the known expression levels of the gene, called *row average*. Then, ignoring those entries in the gene that correspond to missing value columns in the target gene, as well as those missing value columns in the target gene, we have two complete vectors of (known or row average) expression levels, whose Euclidean distance is taken as the distance between the candidate gene and the target gene. For example, if the target gene is $(U, 1.5, U, 2.0, -0.5, U, 3.3)$, where U stands for unknown, then for gene $(1.5, 1.4, U, U, -0.5, -3.9, 3.5)$ in which the unknowns are estimated to be $\frac{1}{5}(1.5 + 1.4 - 0.5 - 3.9 + 3.5) = 0.4$, the distance to the target gene is the Euclidean distance between $(1.5, 2.0, -0.5, 3.3)$ and $(1.4, 0.4, -0.5, 3.5)$, which is $\sqrt{2.61} \approx 1.61$.

2.1. Local Least Squares Imputation — LLSimpute

Using LLSimpute method to estimate missing values in a target gene, one first chooses K nearest neighboring genes using the distance measure defined in the above (K to be determined, Section 2.3). These genes are regarded as coherent genes to the target gene. The missing values in these coherent genes are filled with their respective row averages.

To explain how local least squares imputation works, we assume without loss of generality that gene 1 is the target gene and it has missing values at the first $n - n'$ positions. That is, $g_{11}, g_{12}, \ldots, g_{1,n-n'}$ are unknown. Suppose the K nearest neighboring genes of gene 1 are genes $s_1, s_2, \ldots, s_K$. Denote the submatrix of $G_{m\times n}$ containing rows $1, s_1, s_2, \ldots, s_K$ as $G'_{(K+1)\times n}$. We rewrite $G'_{(K+1)\times n}$ as:

$$\begin{pmatrix} g_1 \\ g_{s_1} \\ \vdots \\ g_{s_K} \end{pmatrix} = \begin{pmatrix} g_{1\times(n-n')} & w^T \\ B_{K\times(n-n')} & A_{K\times n'} \end{pmatrix} = \begin{pmatrix} g_{11} & \cdots & g_{1,n-n'} & w_1 & \cdots & w_{n'} \\ B_{11} & \ldots & B_{1,n-n'} & A_{1,1} & \cdots & A_{1,n'} \\ \vdots & \vdots & \vdots & \vdots & \vdots & \vdots \\ B_{K1} & \ldots & B_{K,n-n'} & A_{K,1} & \cdots & A_{K,n'} \end{pmatrix}.$$

We then proceed to compute a K-dimensional coefficient vector x such that the square

$|A^T x - w|^2 = (A^T x - w)^T (A^T x - w)$ is minimized, that is,

$$\min_x |A^T x - w|^2.$$

Let x^* denote the vector such that the square is minimized, that is,

$$w \simeq A^T x^* = x_1^* a_1 + x_2^* a_2 + \ldots + x_K^* a_K.$$

Therefore, we may take

$$g_{1\times(n-n')}^* = B^T x^* = x_1^* B_1 + x_2^* B_2 + \ldots + x_K^* B_K$$

as an estimation for the missing values $g_{1\times(n-n')}$.

2.2. Measures of Performance

The performance of an imputation method, or the *imputation accuracy*, is generally measured by the *normalized root mean squared error* (NRMSE). Let S be the set of expression matrix entries containing missing values. Since these missing values were simulated (see Section 3.1 for more details), every entry $x_i \in S$ has its true expression value a_i. The imputed value for this entry is a_i^*. The difference $|a_i^* - a_i|$ is the imputation error associated with this entry x_i. Let μ denote the mean of all squares of errors, i.e. $\mu = \frac{1}{|S|}\sum_{x_i \in S}(a_i^* - a_i)^2$, and σ denote the standard deviation of all the true expression values for these entries, i.e. $\sigma = \sqrt{\frac{1}{|S|}\sum_{x_i \in S}(a_i - \bar{a})^2}$, where $\bar{a} = \frac{1}{|S|}\sum_{x_i \in S} a_i$ is the mean of these true expression values. Then, NRMSE is defined to be

$$\text{NRMSE} = \frac{\sqrt{\mu}}{\sigma}.$$

Clearly, the lower NRMSE, the better performance the method has.

2.3. Nearest Neighboring Gene Determination

In LLSimpute method, there is a stage to determine the value for parameter K, before actually doing the imputation. For every target gene, it first replaces every missing value in the other genes by its row average, as is done in the distance calculation. Then, for every gene, a certain number of known expression levels are erased to create the so-called *known missing values*. For every value of K ranging from 1 to the total number of genes in the dataset, it runs LLSimpute once to estimate these *known missing values* and calculates the imputation accuracy, measured by NRMSE. The value that achieves the best imputation accuracy is chosen for K. It's worth mentioning that when there are at least 400 complete genes, only the complete genes are considered as candidate neighboring genes.

Note that once K is determined, LLSimpute method finds exactly K coherent genes for every target gene, regardless however quality difference is for different target genes. We have observed that some target genes seem to have closer coherent genes while for the others the coherent genes are not necessarily similar. Therefore, it is more reasonable to assume that different target genes have different numbers of coherent genes. We have decided not to constrain the number of coherent genes, but to set up a distance *threshold* to cut off dissimilar genes. That is, only those genes within distance *threshold* to the target gene are selected as coherent genes. We set *threshold* to $mean \times ratio$, where $mean$ is the average distance of all other genes to the target gene and $ratio$ is a constant to be determined.

We have tried several ways to determine $ratio$. In one way, we followed the determination of K in LLSimpute method. That is, we filled in the original missing values by their respective row averages, and then erased a certain number of known expression levels to create the *known missing values*. We then tried different values for $ratio$, ranging from 0.5 to 1.5 with an increment of 0.1, and picked the one that achieves the best imputation accuracy through running LLSimpute method once to be the value for $ratio$. In another way, we tried a greedy fashion in which $ratio$ was set initially to 1.0, and then it was increased or decreased by 0.1 depending on the direction leading to better imputation accuracy. The process was terminated when there was no better imputation accuracy could be achieved and the final $ratio$ could be considered as a local optimum. Note that in the second way, we allowed $ratio$ to go beyond 0.5 and 1.5.

In the third and the fourth ways of $ratio$ determination, we first calculated the percentage of missing values in the dataset, i.e. the *missing rate*, then removed genes containing missing values from the dataset to obtain a complete dataset, and created a same percentage of *known missing values*. This new dataset, though smaller, was used to determine $ratio$, again by two approaches as in the last paragraph.

Among these four ways of $ratio$ determination, we found out that the first way leads to the best performance. The reported results in Section 3 were obtained using $ratio$ determined in the first way.

2.4. Iterated Local Least Squares Imputation — ILLsimpute

Using the determined $ratio$ (or equivalently, similarity *threshold*), in the first iteration, our method ILLsimpute selects the coherent genes for every target gene and then runs LLSimpute to estimate the missing values. Afterwards, at each iteration, ILLsimpute uses the imputed results from last iteration to *re-select* the coherent genes for every target gene, using the same $ratio$. Note that the difference in the first iteration is that row averages were used to select the coherent genes. ILLsimpute then applies again LLSimpute method once to *re-estimate* each of the missing values. ILLsimpute terminates after a pre-specified number of iterations or when the re-imputed values in the current iteration have no differences to the imputed values in the preceding iteration, that is, the imputed values converge. In our implementation, we found out that convergence usually took hundreds of iterations and according to literature discussions we have decided to set the number of iterations to 5.

3. Experimental Results

We compared the performance of our method ILLsimpute to three other most recent methods, namely, BPCA, LLSimpute, and KNNimpute method.

3.1. Datasets

We have obtained six microarray datasets for our comparison purpose. The first four datasets are from Spellman *et al.*,[5] which were used for identification of cell-cycle regulated genes in yeast *Saccharomyces cerevisiae*. These datasets were obtained from the file 'CDCDATA.txt' following link `http://genome-www.stanford.edu/cellcycle/data/rawdata/`. There are three parts in the file: Alpha-part, Cdc part and Elu part. There are 6178 genes in the original file. The first dataset *alpha-dataset* and the second *elu-dataset* are the Alpha-part and the Elu-part in the file, respectively, obtained by removing genes with missing values in any part. Both datasets contain 4304 genes, with alpha-dataset in 18 experiments and elu-dataset in 14 experiments, respectively.

Again from the original file, consider only those C-genes (i.e., YAC, YBC, ..., YPC genes) in the last 14 columns. Removing genes with missing values gives us *cyc-a-dataset* that contains 2865 genes in 14 experiments. Another way is to remove genes as long as they contain a missing value in any column in the original file. This gives a much smaller dataset *cyc-b-dataset* that contains 242 genes in 14 experiments.

The fifth dataset was from a study of response to environmental changes in yeast[2] and can be retrieved through link `http://www-genome.stanford.edu/Mec1/data/DNAcomplete_dataset/`. It contains 6167 genes in 52 experiments. We first removed experiments/columns that have more than 2% missing values, and then removed genes still containing missing values, to obtain *env-dataset* that contains 5431 genes in 13 experiments.

The sixth dataset is the cDNA microarray data relevant to human colorectal cancer (CRC) studied in Takemasa *et al.*,[6] called *ta.crc-dataset* and containing 758 genes in 50 samples.

We note that alpha-dataset, elu-dataset, and ta.crc-dataset have been used in the studies of BPCA[4] and LLsimpute.[3]

3.2. Threshold Determination in ILLsimpute

As explained in Section 2.3, we have tried 4 different ways to determine the best value for $ratio$ and we have decided to go with the first way. Figure 1 plots the NRMSE values achieved by ILLsimpute using different $ratio$ values on elu-dataset and cyc-b-dataset both with 10% missing rate. We have tested all $ratio$ values from 0.5 to 1.5 with an increment of 0.1. We remark that the best value for $ratio$ is dataset dependent, for example, it is 0.6 and 0.9 for elu-dataset and cyc-b-dataset, respectively. (The greedy way of $ratio$ determination gave both 1.0 for elu-dataset and cyc-b-dataset, and the subsequent NRMSE was 0.246 and 0.283, respectively.)

3.3. Number of Iterations Determination in ILLsimpute

Though we expect that the imputed missing values converge, we found out that ILLsimpute method took a large number (in hundreds) of iterations when the maximum number of iterations was not specified. We have also observed that sometimes even at convergence point ILLsimpute method didn't necessarily achieve the best NRMSE values. Figure 2 plots the NRMSE values achieved by ILLsimpute method using different maximum numbers of iterations, ranging from 1 to 10, on elu-dataset with 10% missing rate and $ratio$ set to 0.6 and cyc-b-dataset with 10% missing rate and ratio set to 0.9 — the best values for $ratio$. On both cases, the best NRMSE values were achieved in 5 iterations. Therefore, according

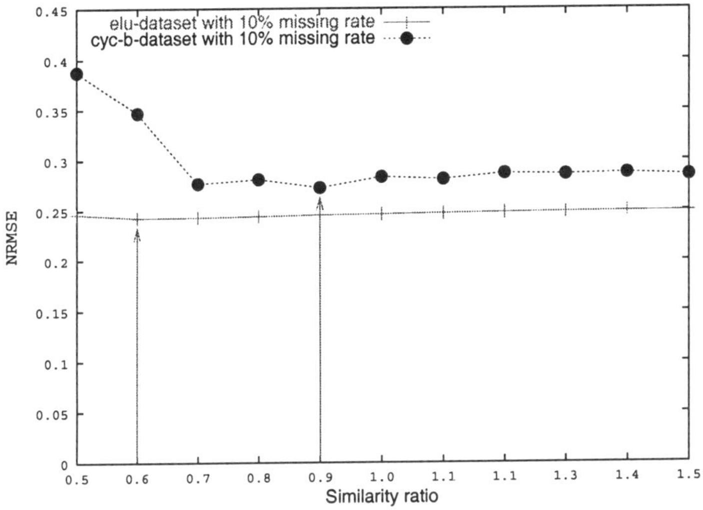

Figure 1. NRMSE values for ILLsimpute on elu-dataset and cyc-b-dataset both with 10% missing rate, w.r.t. different $ratio$ values.

to these results and some discussions in the literature we have chosen 5 iterations to be used in the other experiments.

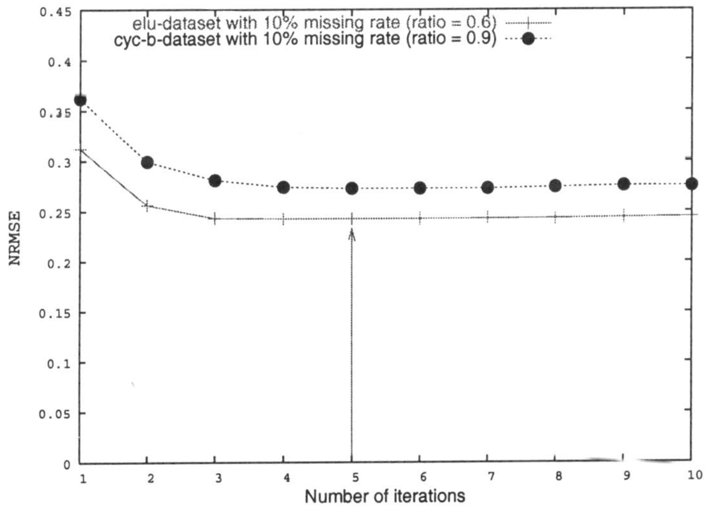

Figure 2. NRMSE values for ILLsimpute on elu-dataset and cyc-b-dataset with 10% missing rate, after different numbers of iterations.

3.4. Imputation Accuracy Comparison

The six datasets we have at hand were obtained through throwing away genes containing missing values from the original files. That is, they no longer contain any missing values. In the experiments, we randomly removed some percentage, i.e. missing rate, of expression levels to create missing values, then applied four imputation methods, KNNimpute, ILLsimpute, BPCA, and LLSimpute, to estimate them. The performances of the methods were measured by NRMSE values. Recall that for each dataset we needed to estimate the value for $ratio$ in ILLsimpute (estimate the value for K in LLSimpute as well), but the number of iterations was fixed at 5. Table 1 summarizes the best values for $ratio$ (and the average number of coherent genes for all target genes in all 5 iterations) in ILLsimpute and the best values for K in LLSimpute on cyc-b-dataset with different missing rates. We have tried several different values for K in KNNimpute, and found out that $K = 10$ gave the best accuracy. The reported results for KNNimpute method were obtained using $K = 10$. It is interesting to note that in ILLsimpute method the average numbers of coherent genes for all target genes in each iteration only differ at most 1 (results not reported), though for individual targets their numbers of coherent genes vary a lot. Also is interesting is that these average numbers differ quite a lot from those K values in LLSimpute method, which could contribute to the improved imputation accuracies of ILLsimpute.

Table 1. The best values for $ratio$ in ILLsimpute and the resultant average numbers $\overline{K}$ of coherent genes for all target genes, the best values for K in LLSimpute, and NRMSE values for all four methods, on cyc-b-dataset with different missing rates. Accuracies in bold are the best ones among all four (cf. Figure 3(d)).

	1%	2%	3%	4%	5%	10%	15%	20%
$ratio$	0.9	0.9	1.2	0.9	1.3	0.8	1.4	1.3
$\overline{K}$	139	139	181	139	190	119	197	190
ILLsimpute NRMSE	0.118	**0.098**	**0.123**	**0.135**	**0.139**	**0.281**	**0.318**	**0.356**
BPCA NRMSE	**0.086**	0.111	0.225	0.339	0.368	0.409	0.398	0.400
KNNimpute NRMSE	0.552	0.534	0.516	0.511	0.519	0.531	0.539	0.541
LLSimpute NRMSE	0.165	0.184	0.225	0.238	0.247	0.353	0.373	0.393
K	140	140	140	140	210	210	210	210

From the plots of NRMSE values (Figure 3) achieved by all four methods on six datasets, we can see that KNNimpute method always performs the worst. For all the other three methods, they perform equally well on env-dataset and ta.crc-dataset. In fact, from Figures 3(e) and 3(f), it is hard to tell which one of them performs better than the other two except KNNimpute. All three methods again perform equally well on alpha-, elu-, and cyc-a-datasets when the missing rate is small, i.e. less than 5% (cf. Figures 3(a), 3(b), and 3(c)). However, their performances differ when the missing rate is large. Typically, our method ILLsimpute performs very close to BPCA, though still a little better, and both of them outperform LLSimpute. On cyc-b-dataset, except 1% missing rate, our method

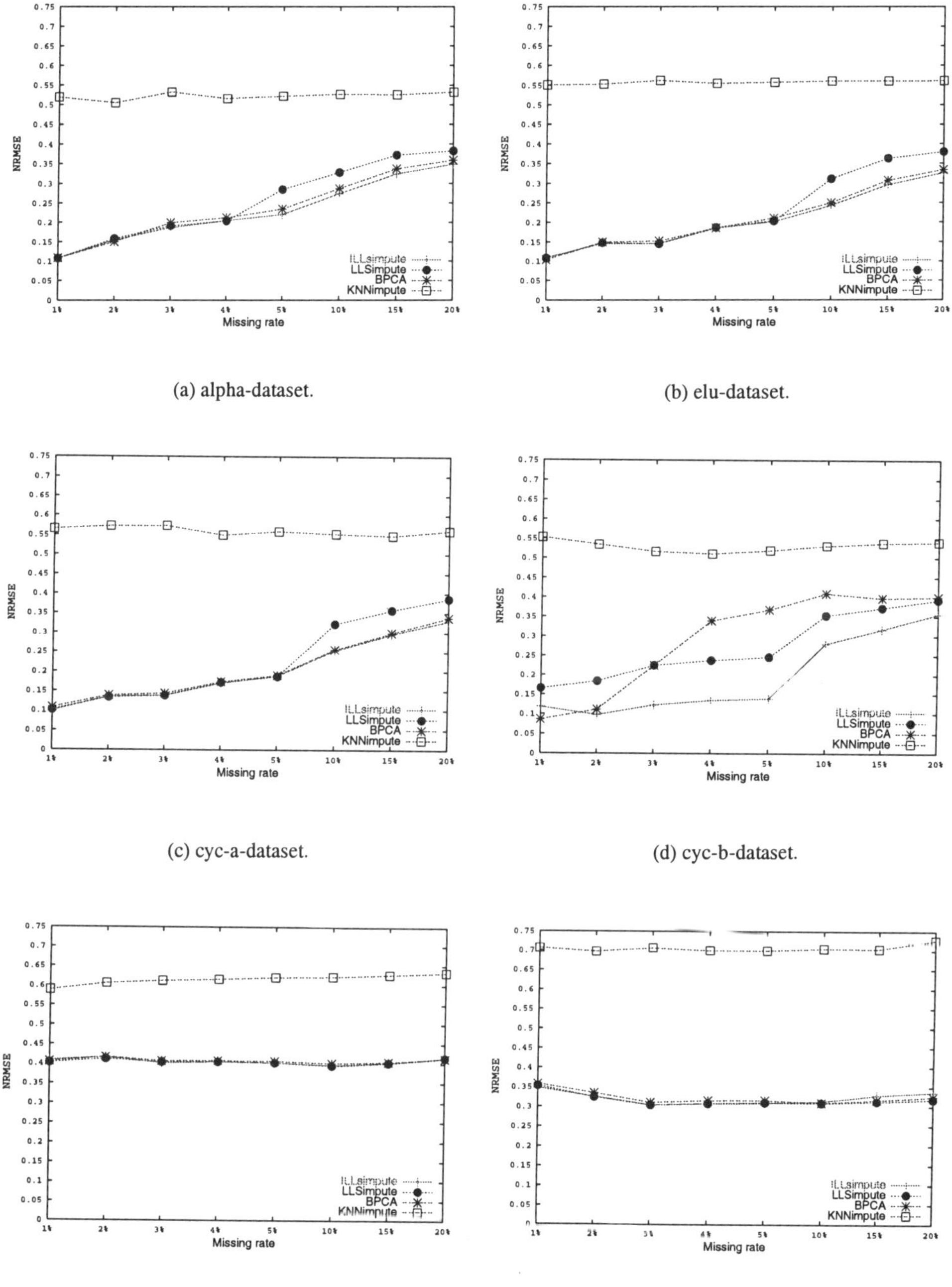

(a) alpha-dataset.

(b) elu-dataset.

(c) cyc-a-dataset.

(d) cyc-b-dataset.

(e) env-dataset.

(f) ta.crc-dataset.

Figure 3. NRMSE comparisons for ILLsimpute, BPCA, LLSimpute, and KNNimpute on six datasets with various percent of missing values.

IlLsimpute outperforms BPCA, LLSimpute, and KNNimpute, and the difference is larger at the practical cases where the missing rate is 5–10%.

From all these results, we might be able to claim that our method ILLsimpute performs better than both BPCA and LLSimpute, the two most recent imputation methods, or at least as well as they perform.

4. Conclusions

We have proposed a novel iterated version of Local Least Squares Imputation (ILLsimpute) method to estimate the missing values in microarray data. In ILLsimpute, the number of nearest neighbors for every target gene is automatically determined, rather than prespecified in most existing imputation methods. The experimental results on six real microarray datasets show that ILLsimpute outperforms three most recent imputation methods BPCA, LLSimpute, and KNNimpute, or at least equally well, on all datasets with simulated missing values.

References

1. A. J. Butte, J. Ye, G. Niederfellner, K. Rett, H. Hring, M. F. White, and K. P. White. Determining significant fold differences in gene expression analysis. *Pacific Symposium on Biocomputing*, 6:6–17, 2001.
2. A. P. Gasch, M. Huang, S. Metzner, D. Botstein, S. J. Elledge, and P. O. Brown. Genomic expression responses to DNA-damaging agents and the regulatory role of the yeast ATR homolog Meclp. *Molecular Biology of the Cell*, 12:2987–3003, 2001.
3. H. Kim, G. H. Golub, and H. Park. Missing value estimation for DNA microarray gene expression data: Local least squares imputation. *Bioinformatics*, 20:1–12, 2004.
4. S. Oba, M. Sato, I. Takemasa, M. Monden, K. Matsubara, and S. Ishii. A Bayesian missing value estimation method for gene expression profile data. *Bioinformatics*, 19:2088–2096, 2003.
5. P. T. Spellman, G. Sherlock, M. Q. Zhang, V. R. Iyer, K. Anders, M. B. Eisen, P. O. Brown, D. Botstein, and B. Futcher. Comprehensive identification of cell cycle-regulated genes of the yeast *saccharomyces cerevisiae* by microarray hybridization. *Molecular Biology of the Cell*, 9:3273–3297, 1998.
6. I. Takemasa, H. Higuchi, H. Yamamoto, M. Sekimoto, N. Tomita, S. Nakamori, R. Matoba, M. Monden, and K. Matsubara. Construction of preferential cDNA microarray specialized for human colorectal carcinoma: molecular sketch of colorectal cancer. *Biochemical and Biophysical Research Communications*, 285:1244–1249, 2001.
7. O. Troyanskaya, M. Cantor, G. Sherlock, P. Brown, T. Hastie, R. Tibshirani, D. Botstein, and R. B. Altman. Missing value estimation methods for DNA microarrays. *Bioinformatics*, 17:520–525, 2001.

PROPERTY-DEPENDENT ANALYSIS OF ALIGNED PROTEINS FROM TWO OR MORE POPULATIONS

STEINAR THORVALDSEN, ELINOR YTTERSTAD, TOR FLÅ

Dept of Mathematics and Statistics, Faculty of Science, University of Tromsø, 9037 Tromsø - Norway.

Multiple sequence alignments can provide information for comparative analyses of proteins and protein populations. We present some statistical trend-tests that can be used when an aligned data set can be divided into two or more populations based on phenotypic traits such as preference of temperature, pH, salt concentration or pressure. The approach is based on estimation and analysis of the variation between the values of physicochemical parameters at positions of the sequence alignment. Monotonic trends are detected by applying a cumulative Mann-Kendall test. The method is found to be useful to identify significant physicochemical mechanisms behind adaptation to extreme environments and uncover molecular differences between mesophile and extremophile organisms. A filtering technique is also presented to visualize the underlying structure in the data. All the comparative statistical methods are available in the toolbox *DeltaProt*.

1 Introduction

Comparative analysis of proteins and proteomes derived from their genome data has already proven powerful in gene identification and in prediction of structure and function. Multiple sequence alignment can also provide information for comparative physicochemical analysis of proteins. Analysis of variations in amino acids at the different alignment positions allows inferences to be made about the pair- and multiple wise relationships between sequences or populations of sequences. The main approach in this paper is based on estimation and analysis of the variation between the values of a physicochemical parameter at positions of a sequence alignment.

The description and definition of chemical similarity and dissimilarity of molecules has long been an active area of study in experimental, theoretical and computational chemistry [1]. The descriptors may take the form of measured or computed physical properties such as topological or constitutional indices. There are several approaches based on counting shared features. Such features include atom or element types, bonds, topological torsions, etc. In formulating other descriptions of quantitative physicochemical distance, one is obliged to make approximations and to use heuristically derived solutions. There have also been several attempts to consider the measured biological properties of the amino acids as a basis for the diversity.

Microorganisms are found almost everywhere on earth. Some are able to tolerate extreme conditions such as low and high temperatures, low and high pH, high salt concentrations, high pressure and high radiation levels. These organisms are commonly referred to as *extremophiles*. Habitats with high temperature harbour the *thermophilic* organisms, which favour temperatures between 45-100° Celsius, and even higher [2]. At the other end of the temperature scale are the *psychrophiles*, or cold-loving organisms, that are able to live even below the freezing point of water [3].

Clearly, to survive at low temperatures, the organisms have to face different challenges. One of the most important tasks is that the metabolism and the enzymes must be able to maintain adequate activity, which otherwise would slow down dramatically at low temperatures. Higher viscosity of water reduces the diffusion rates of substrates and products. Many enzymes from cold sources are shown to be more thermolabile and have a higher catalytic efficiency when compared to orthologous enzymes from warmer sources [4]. A commonly stated hypothesis is that this increased activity is caused by a more flexible structure at low temperature. Several articles concerning sequence comparisons, which aim to find common denominators for cold adaptation, have been published in recent years [4]. But the mechanism still remains unclear. Cold-adapted enzymes have been reported to have some general collective characteristics such as fewer salt bridges, reduction in the R/(R +K) ratio and the number of R (considered stabilizing because it has a long side chain which is both hydrophobic and can form salt bridges and hydrogen bonds), and a lower fraction of larger aliphatic residues expressed by the (I+L)/(I+L+V) ratio indicating a reduced core packing. But at the compositional level the differences in cold-adapted populations seem to be marginal, with overlapping standard deviation intervals.

To study more closely the mechanisms involved in protein cold adaptation on a molecular level, the enzyme *Uracil-DNA N-glycosylase* (UNG; E.C.3.2.2.3) has been chosen from related mesophile and psychrophile bacteria. UNG is an important intracellular, monomeric enzyme which recognizes and removes uracil occurring in DNA [5,6]. A disturbance of this reparation system results in occurrence of a number of diseases, including various kinds of a cancer. The enzyme consists of a classic single domain alpha/beta fold with a central four stranded beta-sheet surrounded by ten alpha-helices. By using a bioinformatics approach, we attempt to reveal the trends in temperature adaptations of this enzyme.

2 Materials and Methods

2.1 Sequence Data

Orthologues of UNG protein sequences from gamma-proteobacteria were collected from various genome sequence projects around the world, and a total of 32 amino acid sequences were found. The optimum growth temperature, T_{opt}, was determined by studying the literature, and by searching The Prokaryotic Growth Temperature Database, PGTdb [7]. Sequence identity is in the range from 38 to 98%. They were divided in *three populations* defined by temperature adaptation: *mesophilic* (T_{opt}=31-40°C, 20 sequences), *psychrotolerant* (T_{opt}=21-30°C, 9 seq.), and *psychrophilic* (T_{opt}=5-20°C, 3 seq.). Temperature is the main environmental trait that separates these populations, not factors such as highly concentrated salts or toxics.

Structural alignments of the data sets were created using the crystal structure from *E.coli* UNG (1FLZ.pdb) as guide. A structural alignment is a set of matched pairs or blocks where there is a meaningful correspondence between the data points in one population and those in the other. This gives us the possibility of investigating the physicochemical measurements in the sequences by statistical methods.

Furthermore, for the sake of a more specific analysis, the amino acid data was decomposed into structural elements in two different ways, and sectioned along the sequence according to these criteria:

- 2D structure region (alpha, beta, loop)
- 3D structure location (surface, twilight zone, core)

For this purpose the secondary (2D) structure was downloaded from the DSSP server, and solvent Accessible Surface Area (ASA) was calculated using the program GETAREA [8] with the crystal structure of *E.coli* as template and default settings. The spatial (3D) location was attached according to solvent accessible surface area of the side chain, where *surface*: 100-50% exposed side chain, *twilight zone*: 50-10%, and *core*: 10-0%. This method makes it possible to analyze the data relative to both its 2D and some of its 3D structural constraints.

2.2 Comparing Sequence Properties by Statistical Methods

Protein sequences are, with rare exceptions (e.g. long fibrous proteins like collagen or silk), quasi-random strings of amino acids with scant evidence of order or periodicity. As a first approximation we will consider these strings as random and independent sampling from a pool of amino acids with a specified probability distribution. The dataset in general will be unbalanced, in the sense that there are different numbers of sequences in each population.

In all the statistical analyses, it is important that the significant differences found (or not found) are due to the different conditions of the populations and not due to the organisation and conservation of the particular enzyme in the study. Therefore, the statistical tests automatically discard sites with no differences (the conserved sites in the alignment) from the analysis.

Each amino acid has many different indices, ranging from molecular mass to helix formation parameter. Sixty physicochemical, steric and other numerical properties of the amino acids were downloaded from the database AAindex release 6 [9] or collected from the literature [10]. The statistical tests will be performed for each of these properties, and the properties are assumed to be additive in the protein structure. For some properties (like molecular mass) this is self-evident, for others (like water-accessible surface area), it may only be an approximation for mean values, and for some (like heat capacities) it may be decided by experiments [11]. We applied *three* univariate statistical tests in the analysis. The main question of interest is whether there are any significant differences between the populations defined by temperature adaptation.

One-way ANOVA

In a previous study [10] properties of amino acids were averaged in unaligned sequences from different temperature populations, but no other statistical analyses than regression studies were performed. When the number of sequences in each temperature population is different, a statistical comparison may also be made by unbalanced one-way analysis of variance (parametric ANOVA).

Let the amino acids in a sequence s be x_j, $j = 1,2,\ldots.n$; where n is the number of non-conserved sites, and let the measurement of a particular property of the amino acids of the same sequence be $q(x_j)$, $j = 1,2,\ldots.n$. This yields real values when we assume that we have a table of quantitative chemical values, q, for each amino acid. There are a total of M sequences from three temperature populations. The test is based on mean values from each sequence i:

$$\bar{q}_i = \tfrac{1}{\mathrm{n}} \sum_{j=1}^{\mathrm{n}} q(x_j), \quad i = 1,2,\ldots,\mathrm{M}$$

Since the sample size n is moderately large, and by assuming independence between the sites, the normality assumption of the ANOVA is fulfilled. An ANOVA will test whether one or more of the population means are unequal, see Figure 2.

Matched-pair tests for step change

We want to improve this strategy above by focusing on *positions* of aligned sequences, rather than averaging over unaligned sequences. Our data (amino acids) can be grouped according to both population and position in the aligned sequence they belong to. The second grouping factor, position, is included in the model to take into account possible differences between positions along the aligned sequence.

For amino acids x and x', we may define a chemical *difference measure* for the two amino acids:

$$d(x,x') = q(x') - q(x)$$

This measure is an expression of the diversity between the amino acids, and the choice of measures to be used depends on the property we want to test.

Let s_m, $m = 1, 2, \ldots$M, be the M aligned amino acid sequences, and let the amino acid at site j in s_m be denoted by $x_{m,j}$. We define the difference between the sequences from population $p_{(1)}$ and $p_{(2)}$ at site j by averaging the measurements at site j within each population:

$$d_j(p_{(1)}, p_{(2)}) = \bar{q}(x_{Pop\,2,j}) - \bar{q}(x_{Pop\,1,j}), \quad j = 1,2,\cdots,\mathrm{n}$$

By this random variable d we measure n assumed independent and identically distributed differential effects between population 1 and 2, where n is the length of the non-conserved alignment. The underlying distributions of the variables d are very seldom known, and for many physicochemical properties there are distinct indications that the distributions are non-normal. In such situations the use of the standard parametric methods assuming normality may be criticized regarding validity and optimality. However, nonparametric methods based on ranks are valid for a broad family of underlying distributions. This gives rise to the possibility of using a data-adaptive test. A standard goodness of fit test was used for to screen the chemical data for deviations from normality (Kolmogorov-Smirnov test) with a significance level of 0.25. By this some physicochemical properties are found to be normal, and some are not (like Kyte-Doolittle hydrophobicity with more than one peak). In general, non-parametric tests will need larger sample size than the corresponding tests based on normality.

In the case of *normality* we approximate d with the normal distribution, and apply the *paired t-test*. In the non-normality case we may apply the paired *Wilcoxon signed-rank* test [12].

These two tests, based on paired data, define a useful and reliable statistical method when we are investigating a variable along the sequence in two population groups. In practice we select the most extreme temperature populations for this test (psychrophiles and mesophiles). The tests can be used on all continuous type of paired data with symmetric distributions. The variances of each of the mean values may be unequal, which will be the case for unbalanced data sets. Results of the Wilcoxon test will still be valid if the distributions have different symmetrical shapes and a common mean.

Cumulative Mann-Kendall test for monotonic trend

A statistical test for trend in the population levels was also performed. Our data consists of temperature-ordered populations, and this ordering may be utilised in the statistical approach by applying a general trend test. The step change presented above is a special case of a more general type of trend often called *monotonic* trends. A monotonic trend indicates that the properties shift monotonically with temperature, but does not specify if this occurs continuously, linearly, in one or more discrete steps, or in any other explicit pattern.

Mann-Kendall tests are a group of nonparametric tests for detection of monotonic relationship between two variables [13]. The collected data are separated into n different sites. Within each site j in the alignment we have M observations of property value q and population group p obtained in pairs:

$$(q_i, p_i), \quad i = 1,2,\ldots,\mathrm{M}$$

The test uses only the relative magnitude of the data rather than their measured values. By the Mann-Kendall test we rank both the property data and the grouping data at each site, and base the test statistic on these ranks. In case of tied data (equal-values), we use the average ranking. The Mann-Kendall statistic K_j at site j is computed by comparing each of the M(M-1)/2 possible pairs of observations, and examine if the two variables are ranked in the same order or in reverse order:

$$K_j = \sum_{m=1}^{M-1} \sum_{i=m+1}^{M} sign\left[(q_i - q_m)(p_i - p_m)\right]$$

where

$$sign(x) = \begin{cases} +1, & if\ x > 0 \\ 0, & if\ x = 0 \\ -1, & if\ x < 0 \end{cases}$$

This is the number of positive differences minus the number of negative differences. The distribution of K_j has been studied by Kendall, and its variance is known [12,13].

Next we need a way to consider all sites in the alignment simultaneously. In a cumulative test the Mann-Kendall test is applied to each site separately, and then the results are combined for an overall test:

$$K = \sum_{j=1}^{n} K_j$$

Under the null hypothesis (no trend), K is asymptotically normally distributed with mean zero and variance:

$$Var(K) = \sum_{j=1}^{n} Var(K_j)$$

Each site by itself may show a positive trend, none of which is significant, but the overall cumulative Mann-Kendall statistic may still be significant.

Visualization

The differences between the populations can be compared graphically as well as statistically, and we used a smoothing technique to be able to recover and visualize underlying structure in the data set [14]. We use a 2D rectangular box-filter where the vertical filter size is all the amino acids in the aligned sequence position of the population, and the horizontal window size can be varied. This 2D filter can be used to plot smoothed line of amino acid properties, such as comparative plots shown in Fig. 1.

3 Applications and Results

3.1 Difference in Physiochemical Properties

We have compared amino acid sequences from three populations by statistical tests and found that it provides interesting results. A summary of the results are shown in Table 1. Results are only reported in the table if the P-values are found to be significant ($P<0.05$) in at least two of the three tests described above. We observe interesting differences especially in the surface and twilight zone, and in the loop regions of the molecule.

We observe decreasing trends with cold adaptation for hydrophobicity, isoelectric point, and long range non-bounded energy. The exterior of the molecules is more negatively charged and will increase their solubility in water. The loop and exterior of the psychrophilic enzymes has a more negative potential (low isoelectric point)

Table 1. List of main differences from mesophile to psychrophile population. P-values are obtained by the ANOVA test (P_{ANOVA}), the paired tests ($P_{t/W}$), the cumulative Mann-Kendall test (P_{cumMK}), and are reported in the given order. Details in the references to amino acid properties can be found in [9, 10].

Property	Trend	2D region P-values	3D region P-values	Entire sequence P-values	Ref.
Hydrophobicity	↓		Surface: 0.05/0.31/0.003	0.04/0.04/0.003	Ponnus-wamy 1993
Isoelectric point	↓	Loop: 0.001/0.03/0.002	Twilight: 0.03/0.04/0.20	0.0001/0.01/0.03	Zimmerman 1968
Molecular weight	↓↑		Twilight: ↓ 0.01/0.04/0.01 Surface: ↑ 0.10/0.04/0.01	0.93/0.82/0.40	Fasman 1976
Volume	↓↑		Twilight: ↓ 0.02/0.03/0.005 Surface: ↑ 0.10/0.003/0.007	0.91/0.40/0.58	Gunsteren-Mark, 1992
Energy long range	↓	Loop: 0.007/0.30/0.0001	Surface: 0.007/0.23/0.0004	0.02/0.12/0.005	Oobatake-Ooi, 1977
Heat capacity	↑		Surface: 0.14/0.04/0.003	0.28/0.30/0.01	Hutchens, 1970
Compressibility	↓	Loop: 0.09/0.02/0.03	Surface: 0.03/0.0007/0.0005	0.18/0.30/0.01	IqbalVerrall 1988
ΔH (unfolding enthalpy change	↑		Twilight: 0.02/0.05/0.10	0.14/0.39/0.20	Oobatake-Ooi, 1993
-TΔS (unfolding entropy change)	↓		Twilight: 0.01/0.06/0.05	0.07/0.36/0.19	Oobatake-Ooi, 1993
ΔG (Gibbs free energy change)	↑		Twilight: 0.004/0.09/0.004	0.57/0.53/0.49	Oobatake-Ooi, 1993
Shape (pos. of branch point)	↓	Loop: 0.01/0.08/0.02	Twilight: 0.0002/0.002/0.0007	0.02/0.06/0.02	Gunsteren-Mark, 1992

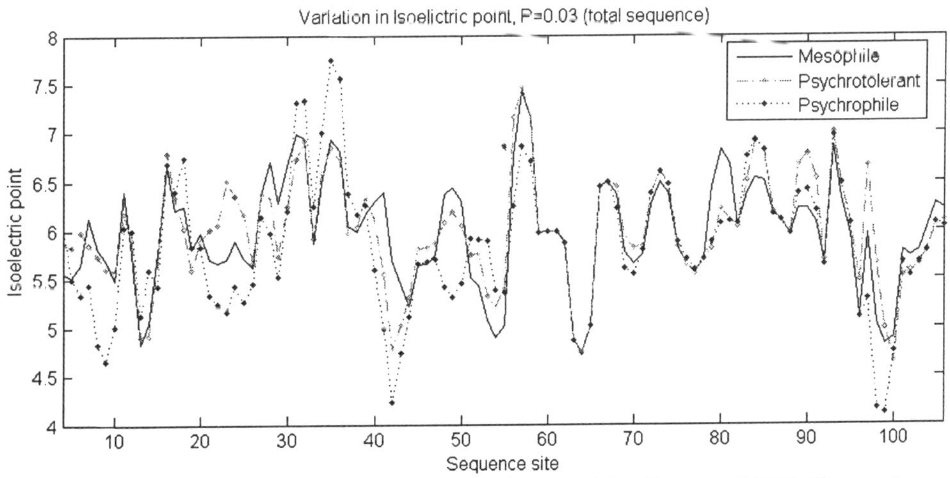

Figure 1. Variation of the isoelectric point property in the first part of the alignment of UNG. We use a box filter of size m x 3 to recover the underlying structure in the data, where m is the number of sequences in the population. The psychrophilic population has 3 sequences, psychrotolerant 9, and mesophilic 20. In average the psychrophilic population appears to have lower values than the mesophilic counterparts.

surrounding the active sites where DNA binds, than the other enzymes. The substrate for UNG is the negatively charged DNA. However, the active site of the UNG enzymes are dominated by conserved positively charged residues where DNA binds. A highly negative charged surface may lead to slow binding of the enzyme to the substrate. But on the other hand, this may contribute to a faster catalysis, as the negative potential forces the product quickly away from the enzyme. Hence, a lower potential will promote weaker electrostatic interactions between the DNA and the enzyme, but in addition it may show a tendency to optimize the electrostatics around the active site.

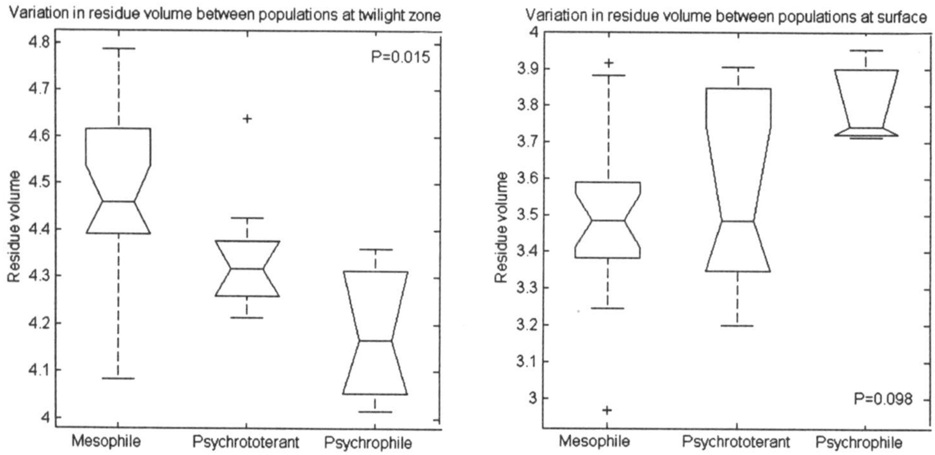

Figure 2. Boxplots showing variations in the volume within and between the populations at the twilight zone (left) and at the surface (right). The analyses show an interchange between the two regions. The one-way ANOVA test is used to compute each P-value.

The heat capacity is one of the fundamental parameters describing thermodynamic properties of a system, and for the amino acids the heat capacity is known to decrease with temperature. In our data the heat capacity is increased from the mesophilic to the psychrophilic population at the surface, but also slightly in general. We interpret this modification to be compensatory to the inherent effects of temperature change. The compressibility parameter also follows an inverse change compared to heat capacity. Most studies of the stability of proteins are concentrated on evaluation of the Gibbs free energy of unfolding, a parameter that provides a measure of thermodynamic stability of the protein molecule. We found no significant overall difference for this parameter between the populations, only a local increase in the twilight zone. However, the Gibbs energy, ΔG, consists of two terms describing the enthalpic, ΔH, and entropic, $-T\Delta S$, contribution, i.e. $\Delta G=\Delta H-T\Delta S$. The enthalpic and entropic contributions for a given system appear to have a close relationship, the so-called enthalpy/entropy compensation. In some cases the enthalpy/entropy compensation is significantly close to obscure the occurrence of the changes in a system, if the analysis is done only in terms of Gibbs energy. The differences between the separate changes in the enthalpy and entropy may be quite significant, as we found it to be in our data analysis, where ΔH goes up and –TΔS down (Table 1). Both ΔH and ΔS are dependent on the heat capacity of the involved amino acids and will decrease with temperature, so the changes found in enthalpy and entropy are fully consistent with the observed increase in heat capacity of residues found in the psychrophilic population.

In Table 1, it is also of importance to note that the shape property, defined as position of branch point in the side chain (e.g. ranging from 0 in Alanin to 5 in Argenin), decreases with adaptation to cold. This observation is a direct extension in agreement with earlier results found for thermophilic proteins [10], and may indicate a more flexible exterior because of early or no branching. Surprisingly, we observe no significant changes in the alpha or beta structures.

4 Some Conclusions

We have applied and expanded the methods of comparative analysis of proteins. The improved strategy is partly extensions of traditionally used statistics [10], e.g., residue frequencies, residue properties, but applied to positions of aligned sequences rather than averaged over unaligned sequences. In this paper a unified framework for context-sensitive and property-dependent analysis of alignments is developed, including data representation and efficient computations using statistical methods. We have demonstrated how alignment data can be incorporated into a Mann-Kendall trend test. In this *cumulative* Mann-Kendall test the alignment sites are tested individually, and then combined into one overall test result. We extracted significant differences into several distinct physicochemical factors.

In the present study of UNG, we found that the properties shape (defining length to the first branch point of side chain), and isoelectric point are generally the most important properties for adaptation to cold. The exterior of the molecules is more negatively charged and will increase their solubility in water and provide weaker electrostatic interactions with the negative substrate (DNA). But small areas around the active site have a positive potential, which possibly acts to improve the interactions.

Furthermore, the shape parameter is decreasing mainly in loop regions and at the twilight zone, indicating weaker medium range interactions, and an increased flexibility at the surface and between the secondary structure elements. This may indicate that the cold adapted protein are characterised by an improved flexibility of the structural components involved in the catalytic cycle. In addition the heat capacity, unfolding enthalpy, and unfolding entropy are found to be different in a direction that compensates the inherent chemical effects of temperature change.

Hence, the ability to be active at temperatures that are close to the freezing point of water requires an array of minor adaptations to compensate for the temperature loss and maintain the enzymatic function. However, these results are based on bioinformatics results only and need to be verified by more detailed analyses in the laboratory.

Some of the features observed may be specific to groups of proteins, and different enzymes may have different strategies for cold adaptation. But also the same enzyme may have different strategies depending on its working environment, and more sequence families should be analyzed to detect both general and special molecular determinants of cold adaptation. A multivariate extension of the present analysis may also be of interest.

All analyses reported in this work were implemented in Matlab. Our toolbox *DeltaProt* can be downloaded from our web-site at: http://www.math.uit.no/bi/deltaprot/

Acknowledgments

The sequence alignment of UNG was kindly provided by Nils P. Willassen.

References

1. N. Nikolova, J. Jaworska. Approaches to measure chemical similarity - A review. *QSAR & Combinatorial Science,* (9-10): 1006-1026, 2004.
2. K. Kashefi and D. R. Lovley. Extending the upper temperature limit for life. *Science*, 301(5635): 934, 2003.
3. K. Junge, H. Eicken, et al. Bacterial Activity at -2 to -20 degrees C in Arctic wintertime sea ice. *Appl Environ Microbiol.*, 70(1): 550-7, 2004.
4. G. Feller and C. Gerday. Psychrophilic enzymes: hot topics in cold adaptation. *Nat Rev Microbiol* 1: 200-208, 2003.
5. S.S. Parikh, C.D Mol, D.J Hosfield and J.A Tainer. Envisioning the molecular choreography of DNA base exicions repair. *Curr Opin Struct Biol* 9(1):37-47, 1999.
6. I. Leiros, E. Moe, A.O. Smalås and S. McSweeney. Structure of the uracil-DNA N-glycosylase from Deinococcus radiodurans. *Acta Cryst.*, D61:1049-1056, 2005.
7. L.C.W Huang, K.H. Laing, K.T. Pan and J.T. Horng. PGTdb: a database providing growth temperatures of prokaryotes. *Bioinformatics* 20(2): 276-278, 2004.
8. R. Fraczkiewicz and W. Braun. Exact and efficient analytical calculation of the accessible surface areas and their gradients for macromolecules. *J. Comp. Chem.*, 19: 319–333, 1998.
9. S. Kawashima, H. Ogata and M. Kanehisa. AAindex: amino acid index database. *Nucleic Acids Res.*, 27, 368-369, 1999.
10. M.M. Gromiha, M. Oobatake and A. Sarai. Important amino acid properties for enhanced thermostability from mesophilic to thermophilic proteins. *Biophysical Chemsitry*, 82, 51-67, 1999.
11. M. Hackel, H. J. Hinz and G. R. Hedwig. Additivity of the partial molar heat capacities of the amino acid side-chains of small peptides: Implications for unfolded proteins. *Physical Chemistry Chemical Physics*, 2 (23): 5463-5468, 2000.
12. M. Hollander and D. A. Wolfe. *Nonparametric Statistical Methods*, 2.ed. Wiley, 1999.
13. M.G. Kendall and J.D Gibbons. *Rank Correlation Methods,* 5th ed. Edward Arnold, 1990.
14. A. V. Oppenheim and R.W. Schafer. *Discrete-Time Signal Processing,* Englewood Cliffs, NJ: Prentice-Hall, 1989.

A GENERALIZED OUTPUT-CODING SCHEME WITH SVM FOR MULTICLASS MICROARRAY CLASSIFICATION

LI SHEN ENG CHONG TAN

School of Computer Engineering, Nanyang Technological University, Nanyang Avenue, Singapore, 639798, Singapore

Multiclass cancer classification based on microarray data is described. A generalized output-coding scheme combined with binary classifiers is used. Different coding strategies, decoding functions and feature selection methods are combined and validated on two cancer datasets: GCM and ALL. The effects of these different methods and their combinations are then discussed. The highest testing accuracies achieved are 78% and 100% for the two datasets respectively. The results are considered to be very good when compared with the other researchers' work.

1 Introduction

DNA microarrays can contain thousands of gene expression levels in one single experiment. Obtaining gene expression levels from tumor tissues can help us to understand the activities of genes underlying different cancers. Therefore, these expression data may also be used to identify types or subtypes of cancers.

Applying machine learning techniques to microarray data for cancer classification has been intensively researched in recent years. Most of the work is in the field of binary classification and very high accuracy can be obtained [1]. However, it is suggested by some authors that multiclass classification tasks are more difficult than binary ones [2].

In this paper, a generalized output-coding scheme has been applied to multiclass microarray classification. With this, different coding strategies and decoding functions can be put into one single framework. The validity of various combinations has been verified. Support Vector Machine (SVM) was chosen as the binary classifier, which has been successfully applied to microarray classification. It is one of the state-of-the art machine learning techniques and has strong theoretical foundation.

Because microarray data has the characteristics that the number of genes is much larger than the number of samples, feature selection is also important before classification. Three major categories of feature selection methods have been tested: gene ranking, recursive feature elimination (RFE) and dimension reduction.

2 Methods

2.1 Output-Coding for Multiclass Classification

Assume we have a set of m microarray samples: $(\mathbf{x}_i, y_i)$, $i = 1,2,\ldots m$, where $\mathbf{x}_i \in \Re^n$ is a vector of length n representing gene expression levels and $y_i \in \{1,2,\ldots,k\}$ is the class label of the i th sample. The multiclass classification algorithm aims to find the mapping $\mathrm{M}: \Re^n \to \{1,2,\ldots,k\}$ using the m training samples. Output-coding methods solve the multiclass problem by decomposing the k -class problem into a set of l binary subproblems, training the resulting l base classifiers and then combining the l outputs to predict the class label. We have adopted the generalized scheme proposed by [3]. It begins with a given coding matrix

$$\mathbf{M} \in \{-1,0,+1\}^{k\times l}$$

for which each row $\mathbf{r}_i$ ($i=1,2,\ldots k$) represents the codeword of the i th class and each column $\mathbf{s}_j$ ($j=1,2,\ldots l$) represents the j th base classifier. Each row $\mathbf{r}_i$ must be unique for its corresponding class. $\mathbf{M}(i,j)=1$ or -1 means that the i th class should be considered as positive or negative for the j th base classifier, respectively. If $\mathbf{M}(i,j)=0$, the i th class is simply ignored by the j th base classifier. Any binary classifier can be used to solve the induced two-class problem.

Now let f_s ($s=1,2,\ldots l$) denote the l base classification functions. Given a microarray sample $\mathbf{x}$, let $\mathbf{f}(\mathbf{x})=(f_1(\mathbf{x}),f_2(\mathbf{x}),\ldots f_l(\mathbf{x}))$; then its class label y is predicted as

$$y = \arg\min_i d(\mathbf{r}_i, \mathbf{f}(\mathbf{x})) \qquad (1)$$

where d is called the decoding function. By adopting this generalization scheme, we can combine some of the researchers' work into one single system.

There are various methods to generate coding matrices. Different coding matrices may have substantial effect on classification accuracy. Probably the simplest coding approach is to set $\mathbf{M}$ as a square matrix of size $k\times k$. Let all diagonal elements of $\mathbf{M}$ be 1 and all the other elements be -1 . Therefore, it equals the method that creates one binary problem for each of the k classes. This is called one-versus-all (OVA) approach.

Another approach, proposed by [4], is to use the binary classifier to distinguish one pair of classes at a time. Meanwhile, the other classes are simply ignored. So there are totally C_k^2 base classifiers to induce. This is called the all-pairs (AP) approach.

Error correcting output codes (ECOC) was proposed by [5]. They argue that if the minimum hamming distance between a pair of rows of the coding matrix is c , the output codes can have the ability to correct $\lfloor (c-1)/2 \rfloor$ errors of the base classifiers. Two major coding strategies called random coding and exhaustive coding are given:

- *Random coding*. Let $l=\lceil 10\log_2(k)\rceil$ as suggested by [5]. Each element of the coding matrix is then assigned a value from $\{-1,1\}$ uniformly at random. After the coding matrix is generated, a hill-climbing procedure given by [6] is followed. The hill-climbing method can usually improve the averaged and minimum hamming distances between pairs of rows of the coding matrix so that the classification accuracy may be improved.

- *Exhaustive coding*. Firstly let $l=2^{k-1}$. The codeword for the first class is all $+1$. For i , $2\le i\le k$, the codeword for the i -th class is constructed by repeating a pattern 2^{i-2} times, which is a length- 2^{k-i} block of $+1$'s followed by a length- 2^{k-i} block of -1 's. Because the first column is thus assigned $+1$ for all of its elements, it is deleted from the coding matrix. This makes $l = 2^{k-1} - 1$. It is easy to see that the minimum hamming distance is $\left\lfloor \frac{2^{k-1} - 1}{2} \right\rfloor$. The disadvantage of exhaustive coding is that l increases exponentially with k. If k is a large number, that would make the computation intractable.

Decoding functions determine how the distance between the outputs of base classifiers and codeword is calculated. One way of doing this is to count the number of positions s in which the codeword entry differs from the sign of the prediction $f_s(\mathbf{x})$. Formally the distance measure is given as

$$d_H(\mathbf{r}, \mathbf{f}(\mathbf{x})) = \sum_{s=1}^{l} \left(\frac{1 - sign(r_s f_s(\mathbf{x}))}{2} \right) \tag{2}$$

where $sign(z) = +1$ if $z > 0$, -1 if $z < 0$ and 0 if $z = 0$. r_s is the entry of codeword $\mathbf{r}$ at position s. This is called the *hamming distance* decoding. A disadvantage of this decoding function is that it totally ignores the output values of base classifiers.

A second decoding function takes into account the confidence of the predictions. It utilizes a loss function L which is algorithm-specific. The loss function calculates the "loss" of the prediction given the output values and the codeword. The loss function for SVM is defined as

$$L(y, f) = (1 - yf)_+ \tag{3}$$

where y is an entry of the codeword and f is the output of SVM. z_+ is defined as $\max(z, 0)$. The distance measure can now be written as

$$d_L(\mathbf{r}, \mathbf{f}(\mathbf{x})) = \sum_{s=1}^{l} \left(L(r_s, f_s(\mathbf{x})) \right) \tag{4}$$

This is called the *loss based* decoding.

There is another decoding function that takes account of the prediction confidence by simply calculating the inner product of the codeword and the vector of classifier outputs. This is defined as

$$d_I(\mathbf{r}, \mathbf{f}(\mathbf{x})) = -\sum_{s=1}^{l} \left(r_s f_s(\mathbf{x}) \right) \tag{5}$$

thus the name *inner product* decoding.

Finally we introduce a decoding function that is based on the probability of the prediction. Given the assumption that the base classifiers are independent, the class of which the codeword gives the maximum joint probability is the one predicted. Negative log-likelihood can be used to define the decoding function as

$$d_P(\mathbf{r}, \mathbf{p}(\mathbf{x})) = -\sum_{s=1}^{l} \frac{1 + r_s}{2} \log(p_s(\mathbf{x})) - \sum_{s=1}^{l} \frac{1 - r_s}{2} \log(1 - p_s(\mathbf{x})) \tag{6}$$

where $\mathbf{p}(\mathbf{x}) = (p_1(\mathbf{x}), p_2(\mathbf{x}), \dots p_l(\mathbf{x}))$. p_s is the probabilistic output of the base classifier s. A parametric model can be used to estimate the probability of SVM outputs as suggested by [7]

$$p_s(\mathbf{x}) = \frac{1}{1+\exp(A_s f_s(\mathbf{x}) + B_s)} \tag{7}$$

where $f_s(\mathbf{x})$ is the output of SVM which is trained as base classifier s. Three-fold cross-validation (CV) is used in this paper to fit A_s and B_s.

2.2 Feature Selection

There are three major categories of feature selection methods:

- *Gene Ranking*. Intuitively one would select those genes that are correlated with a class but are uncorrelated with the other classes. We choose a gene ranking method that is based on the ratio of their between-group to within-group sums of squares. For a gene j, this ratio is

$$BW(j) = \frac{\sum_i \sum_k I(y_i = k)(\bar{x}_{kj} - \bar{x}_{.j})^2}{\sum_i \sum_k I(y_i = k)(x_{ij} - \bar{x}_{kj})^2} \tag{8}$$

  where $\bar{x}_{.j}$ and $\bar{x}_{kj}$ denote the average expression level of gene j across all classes and across samples belonging to class k only. $I(\cdot)$ is the indicator function. The base classifiers are built using the genes with the largest BW values.
- *Recursive Feature Elimination*. RFE was first proposed by [8] to do feature selection in binary classification. The genes with the smallest corresponding weights are dropped and the process can be executed recursively. In multiclass context, the RFE is executed on each base classifier independently so that the best performance and the smallest gene subset can be obtained concurrently. Three fold CV is used to evaluate the goodness of gene subset.
- *Partial Least Squares (PLS) and Principal Components Analysis (PCA)*. Dimension reduction methods have been proposed to tackle the "curse of dimensionality" problem. It is prohibitive to use some of the statistical methods when $m < n$ because of excessive computation time. Dimension reduction is also used as the preprocessing step to make these methods feasible. PLS [9] and PCA [10] have been proven to be effective for microarray classification [11-12] and have been used in this paper.

3 Results

3.1 Datasets and Experimental Setup

We chose two multiclass microarray datasets for our experiments. The first is the *GCM* dataset published by [13]. It consists of 144 training samples and 54 testing samples of 15 common cancer classes. Each sample has 16063 gene expression levels. For simplicity, we dropped the 8 metastatic samples from the testing dataset because they are not present in the training dataset. Therefore, 46 testing samples and 14 cancer classes are considered in this paper. The distribution of training and testing samples among the 14 classes is listed in Table 1. The second is the *ALL* dataset published by [14]. It consists of 163 training samples and 85 testing samples of 6 subtypes of acute lymphoblastic leukemia. Each sample has 12558 gene expression levels. The distribution of training and testing samples among the 6 subtypes is listed in Table 2. All data are log-transformed. All genes are normalized to have zero mean and unit standard deviation. No other preprocessing steps are applied. For GCM data, three coding strategies are used: AP, OVA, and random. We did not use exhaustive coding because l would be equal to $2^{13}-1=8191$ and this will make the computation intractable. For ALL data, AP, OVA and exhaustive coding strategies are used.

Table 1. GCM: number of samples per cancer class. BR=Breast, PR=Prostate, LU=Lung, CO=Colorectal, LY=Lymphoma, BL=Bladder, Melanoma=ME, UT=Uterus, LE=Leukemia, RE=Renal, PA=Pancreas, OV=Ovary, ML=Mesothelioma, CNS=Brain.

Cancer Class	*BR*	*PR*	*LU*	*CO*	*LY*	*BL*	*ME*	*UT*	*LE*	*RE*	*PA*	*OV*	*ML*	*CNS*
Training	8	8	8	8	16	8	8	8	24	8	8	8	8	16
Testing	3	2	3	3	6	3	2	2	6	3	3	3	3	4

Table 2. ALL: number of samples per subtype.

Subtype	*BCR-ABL*	*E2A-PBX1*	*Hyperdiploid>50*	*MLL*	*T-ALL*	*TEL-AML1*
Training	9	18	42	14	28	52
Testing	6	9	22	6	15	27

According to the suggestion of [2], 250 top genes are selected from BW ratio ranking. We also tested the data without feature selection, which is denoted as NO in the following. For RFE, the gene subset for each base classifier is determined by three fold CV. For PLS and PCA, all components that can be extracted are used. All programs are written in MATLAB codes. The software package written by Steve Gunn is used for the SVM algorithm. It is available at: http://www.kernel-machines.org/. The regularization parameter for SVM is set to 1 for all the experiments.

3.2 Experimental Results

Figs. 1-3 show the results of the GCM dataset and Figs. 4-6 show the results of the ALL dataset. We have the following observations:

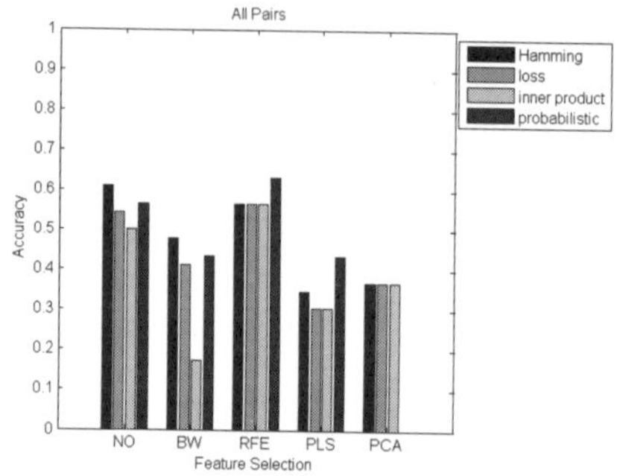

Fig. 1 GCM data, all-pairs coding.

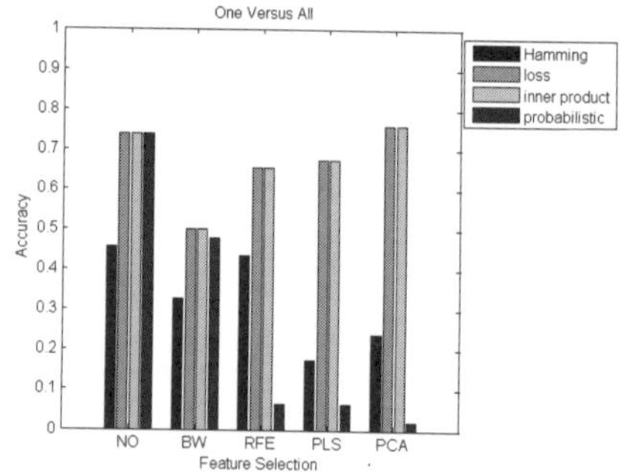

Fig. 2 GCM data, one-versus-all coding.

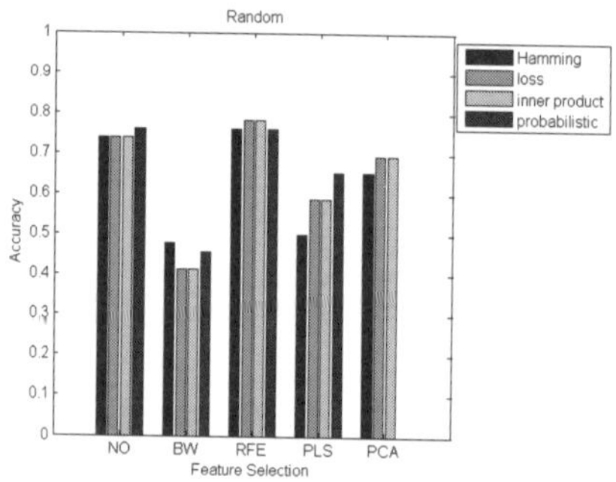

Fig. 3 GCM data, random coding.

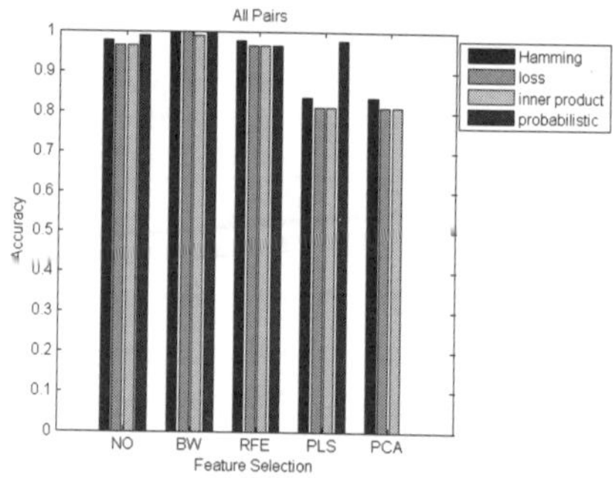

Fig. 4 ALL data, all-pairs coding.

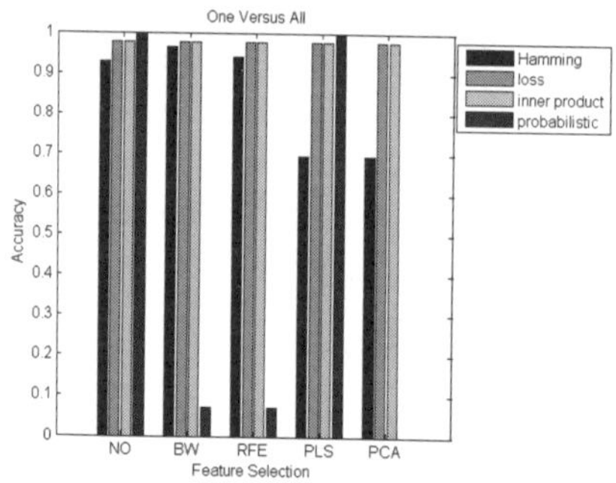

Fig. 5 ALL data, one-versus-all coding.

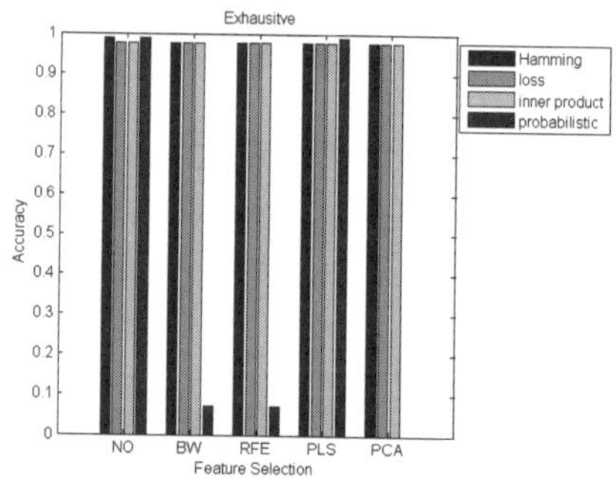

Fig. 6 ALL data, exhaustive coding.

- The ECOC coding strategies generally outperform the other coding strategies. The highest accuracy on the GCM data is achieved by random coding. Combining with loss based and inner product decoding functions and RFE, a 78% testing accuracy has been obtained. On the ALL data, exhaustive coding has achieved almost perfect accuracy for most decoding functions and feature selection methods. There are some exceptions on probabilistic decoding function. This can be attributed to the ability of ECOC to correct errors for weak base classifiers.
- The AP coding strategy works quite well with the ALL data, but it is the worst coding strategy with the GCM data. All testing accuracies on GCM data are below

70%. Learning from Table 1 we know that some classes of the GCM data are very small. It is hard for base classifiers to perform well on these pairs of small cancer classes. A lot of errors may occur for base classifiers, thus the multiclass classification accuracy is degenerated.

- The probabilistic decoding function is very sensitive to coding strategies and feature selections. It fails to work with PCA. It also fails when OVA, exhaustive coding strategies and BW, and RFE feature selections are used on the ALL data. Meanwhile, it achieves 100% accuracy on the ALL data when AP and BW are used, etc. It is known that fitting sigmoid parameters by solving (7) is sensitive to the distribution of samples of two classes. So the unequal distribution of classes of microarray data may lead to the failure of probabilistic decoding function.
- The hamming distance decoding function is not suitable with OVA. It is because many ties will happen when the base classifiers do not give enough high prediction confidence and they are just solved by random assignment. It is better to integrate prediction confidence when OVA is used. However, hamming distance decoding works well with AP and ECOC. This is because the base classifiers of AP usually have high prediction confidence and ECOC has the ability to correct errors if base classifiers are weak. It is noticed that loss based and inner product decoding give very similar results.
- Feature selection by BW ratios performs poorly with GCM data but rather well with ALL data. This is consistent with the results by [2]. BW does not tell information about the class labels so it may select genes that only contain information of several classes without regards to the rest. It is also noticed that results are usually good when no feature selection is used.
- PCA outperforms PLS on the GCM data except using the probabilistic decoding function. However, it has been validated that PLS is usually a better dimension reduction method than PCA [11-12]. It is known from the experiments that the PLS components extracted from GCM data is only 19 while the number of PCA components is 143. We then deduce that the small sizes of some classes in GCM data prevent the PLS to extract enough components so that some information is lost. On ALL data, PLS performs better than PCA.

4 Conclusions

The output coding scheme from machine learning has been successfully applied to multiclass microarray classification in this paper. Usages of different coding matrices, decoding functions and feature selection methods have been discussed. It has been shown that a good coding matrix can lead to high accuracy of multiclass microarray classification. Better coding strategies are required to further improve the performance of the output coding scheme.

Though gene ranking and dimension reduction methods have been shown to be effective for multiclass classification, it is seen that sometimes without feature selection, the results are even better. RFE is good for binary classification but for output coding based multiclass classification, it can only be used to enhance base classifiers. Data over-fitting can easily happen and the variances of outputs would be large especially when class sizes are small. This can degrade the multiclass accuracy in the end. It is better to use the CV errors of multiclass classification as feedback to select genes. Some algorithms like genetic algorithm could be considered.

References

1. S.B. Cho and H.H. Won. Machine Learning in DNA Microarray Analysis for Cancer Classification. *First Asia-Pacific Bioinformatics Conference*, Adelaide, Australia. Yi-Ping and Phoebe Chen Ed, 2003.
2. T. Li, C. Zhang and M. Ogihara. A comparative study of feature selection and multiclass classification methods for tissue classification based on gene expression. *Bioinformatics*, 20, 15, 2429-2437, 2004.
3. E.L. Allwein, R.E. Schapire, Y. Singer. Reducing Multiclass to Binary: A Unifying Approach for Margin Classifiers. *J. Machine Learning Research*, 1, 113-141, 2000.
4. T. Hastie and R. Tibshirani. Classification by pairwise coupling. *The Annuals of Statistics*, 26(2), 451-471, 1998
5. T.G. Dietterich and G. Bakiri. Solving Multiclass Learning Problems via Error-Correcting Output Codes. *J. Artificial Intelligence Research*, 2, 263-286, 1995.
6. F. Ricci and D.W. Aha. Error-Correcting Output Codes for Local Learners. *In Proceedings of the 10th European Conference on Machine Learning*, 1998.
7. J. Platt. Probabilistic outputs for support vector machines and comparison to regularized likelihood methods. *Advances in Large Margin Classifiers*, MIT Press, 1999.
8. I. Guyon, J. Weston, S. Barnhill, V. Vapnik. Gene selection for cancer classification using support vector machines. *Maching learning*, 46, 389-422, 2002.
9. J.A. Wegelin. *A survey of Partial Least Squares (PLS) methods, with emphasis on the two-block case* (Technical Report). Department of Statistics, University of Washin, 2000.
10. G.H. Golub and C.F. Van Loan. *Matrix Computations*. The Johns Hopkins University Press, 1996.
11. D.V. Nguyen and D.M. Rocke. Multi-class cancer classification via partial least squares with gene expression profiles. *Bioinformatics*, 18, 9, 1216-1226, 2002.
12. L. Shen and E.C. Tan. Dimension Reduction Based Penalized Logistic Regression for Cancer Classification Using Microarray Data. *IEEE/ACM Trans. Computational Biology and Bioinformatics*, 2, 166-175, 2005.
13. S. Ramaswamy, P. Tamayo, R. Rifkin, S. Mukherjee, C.H. Yeang, M. Angelo, C. Ladd, M. Reich, E. Latulippe, J.P. Mesirov *et al*. Multiclass cancer diagnosis using tumor gene expression signatures. *Proc. Natl Acad. Sci., USA*, 98, 15149–15154, 2001.
14. E.J. Yeoh, M.E. Ross, S.A. Shurtleff, W.K. Williams, D.P. Rami Mahfouz, F.G. Behm, S.C. Raimondi, M.V. Relling, A. Patel, C. Cheng *et al*. Classification, subtype discovery, and prediction of outcome in pediatric acute lymphoblastic leukemia by gene expression profiling. *Cancer Cell*, 1, 133-143, 2002.

TECHNIQUES FOR ASSESSING PHYLOGENETIC BRANCH SUPPORT: A PERFORMANCE STUDY

DEREK RUTHS LUAY NAKHLEH

Department of Computer Science, Rice University, Houston, Texas 77005, USA.
{druths,nakhleh}@cs.rice.edu

The inference of evolutionary relationships is usually aided by a reconstruction method which is expected to produce a reasonably accurate estimation of the true evolutionary history. However, various factors are known to impede the reconstruction process and result in inaccurate estimates of the true evolutionary relationships. Detecting and removing errors (wrong branches) from tree estimates bear great significance on the results of phylogenetic analyses. Methods have been devised for assessing the support of (or confidence in) phylogenetic tree branches, which is one way of quantifying inaccuracies in trees. In this paper, we study, via simulations, the performance of the most commonly used methods for assessing branch support: bootstrap of maximum likelihood and maximum parsimony trees, consensus of maximum parsimony trees, and consensus of Bayesian inference trees. Under the conditions of our experiments, our findings indicate that the actual amount of change along a branch does not have strong impact on the support of that branch. Further, we find that bootstrap and Bayesian estimates are generally comparable to each other, and superior to a consensus of maximum parsimony trees. In our opinion, the most significant finding of all is that there is no threshold value for any of the methods that would allow for the elimination of wrong branches while maintaining all correct ones—there are always weakly supported true positive branches.

1. Introduction

The accuracy and validity of most comparative genomic studies rely on the quality of an underlying "guiding" phylogenetic tree. Such a tree is often inferred using a phylogeny reconstruction method. However, such methods are bound to make errors in the inferred tree (by inferring wrong branches and missing correct ones) due to a host of reasons such as biological processes that may not be modeled by a single tree (e.g., recombination and horizontal gene transfer) or "data issues" (e.g., incomplete taxon sampling, insufficient data, wrong assumptions). Various methods have been introduced for estimating the support of (or confidence in) tree branches; two of the most commonly used methods are the *bootstrap* method[19] and Bayesian inference techniques. The bootstrap method is usually coupled with the maximum parsimony (MP) or maximum likelihood (ML) heuristic searches, and amounts to estimating many trees over subsamples of the dataset and using the the percent of trees containing a branch to be its support. Bayesian inference uses statistical inference techniques whose final outcome is a set of trees, each coupled with probabilities associated with its branches to reflect their support. Further, MP heuristics often compute a large set of optimal trees. The number of trees in which a given branch appears can be taken to be its support (these are referred to as the "consensus" methods). After support values are computed, a threshold is chosen and branches with support lower than that threshold are contracted. The hope is that a threshold exists such that erroneous branches will be removed while correct ones will be retained.

Existing simulation-based performance studies of branch support measures have considered Maximum Likelihood with bootstrap and Bayesian Inference,[5,6] as well as the statistical properties of the bootstrap.[2,3,4] Other studies considered the performance of the various branch-support estimation methods on biological datasets, in which case the true phylogeny is usually unknown.[12,1,18] One exception is the work of Taylor *et al.* in which they studied the accuracy of the bootstrap and Bayesian approaches in reconstructing the phylogeny of several strains of yeast.[23] The results focused on studying the effect of evolutionary rate consistency and tree shape on accuracy.

In this paper, we evaluate both the absolute and relative correctness of each method under the conditions of the study by evaluating the performance of branch support assessment methods via simulations. We generate random phylogenetic trees, and simulate the evolution of DNA sequences down these trees. We study the accuracy of the trees and the support of their branches, as calculated by the methods, by comparing their estimates to the true (known) phylogenetic trees. In this study, we focused on the performance of the most prevalent branch support-labeling algorithms. Thus, we have omitted the less common distance-based branch support assessment methods, such as bootstrap with *neighbor joining*. We also do not explicitly consider the error resulting from the fact that the heuristics we consider do not actually converge to the true tree under all conditions.[20,14,22]

We focus on three main questions. (1) Is the support (as calculated by each of the methods) of a correct branch significantly higher than that of a wrong branch? (2) Is there a clear threshold for each of the methods that would allow for contracting wrong branches while retaining all correct ones? (3) Is there any correlation between the support of a branch and the actual amount of evolution along that branch?

Under the conditions of our experiments, bootstrap and Bayesian techniques outperform a consensus of MP trees, with respect to the first question. Further, we find that the support of a correct branch as computed by each of the techniques remains largely unaffected by amount of evolution along that branch. However, with respect to the second question, the answer is not very promising. Under the conditions of our experiments, any choice of threshold for any of the methods involves a significant tradeoff between the number of wrong branches contracted and the number of correct branches retained.

2. Methods

In this study we considered three different phylogenetic estimation methods—Maximum Parsimony[9] (MP), Maximum Likelihood[7] (ML), and Bayesian Inference[10] (BI). Since MP and ML estimation methods do not produce trees that have support-labeled branches, these methods are used in conjunction with a bootstrap algorithm in order to generate support values for tree branches. Another prevalent method for generating support-labeled trees is to take the majority consensus of the top scoring trees returned by MP, which we also considered in our study.

2.1. *Phylogeny Estimation Methods*

Two of the most commonly used and most accurate criteria for phylogeny reconstruction are *maximum parsimony* (MP) and *maximum likelihood* (ML). They are both hard optimization criteria for which various accurate heuristics have been devised.

The MP criterion is based on the assumption that "evolution is parsimonious", i.e., the best evolutionary trees are the ones that minimize the number of changes along the branches of the tree. In our study, we used the PAUP*[21] MP heuristic (starts with a random tree, and traverses the tree space using TBR moves). The ML problem seeks the tree T and its associated parameters (such as branch-lengths, rates of evolution for each site, etc.) that maximize the probability of generating the given set of sequences. In our study, we used the PAUP*[21] ML heuristic (starts with a random tree, and traverses the tree space using TBR moves). *Bayesian Inference* seeks the tree that maximizes the estimated posterior probability of the tree τ_i given the sequences **X**. The MrBayes tool is a heuristic that uses the Markov chain Monte Carlo method to approximate the posterior probability.[11] We used this application for inferring trees (we used 100,000 generations with a burn-in period of 10,000 generations).

2.2. *Bootstrap*

The bootstrap technique is commonly used to add support-labelings to the output of MP and ML estimation methods. This technique subsamples the original sequence data to produce "new" input data of the same length in which some of the original sites appear duplicated and some do not appear at all. The technique constructs the new datasets in such a way that they remain statistically similar to the original input data. The bootstrap technique constructs these datasets and the associated best MP or ML tree a specified number of times. Following this, a support-labeled tree is constructed by taking the majority consensus of the set of trees created during the iterations. For our study, we used the bootstrap techniques with MP and ML, as implemented in PAUP*,[21] where the number of repetitions we considered was 100.

2.3. *Consensus Trees & Branch Contraction*

The *p-consensus tree*, τ_c, for a set of trees, T, is the tree containing only those branches that occur in at least p percent of the trees in T. Associated with each branch in the consensus tree is the percent of trees that contain that branch—this is considered the support for that branch. A *strict consensus tree* is a consensus tree for which $p = 100$. Therefore, it contains only those branches that occur in all of the trees in T. On the other end of the spectrum, the *majority consensus trees* is the consensus tree for which $p = 50$, containing only those branches that occur in at least half of the trees in T.

In a strict consensus tree, the minimum support of any branch in the tree is 100. Also the maximum support any branch can possibly have in any tree is 100. As a result, all branches in a consensus tree have the same, maximum support value.

In a majority consensus tree, the support for any branch can range between 50 and 100. After constructing such a majority consensus tree, we may want to remove all branches that have a support value below a certain threshold. This threshold-based removal procedure is called *branch contraction*. Assuming that some branches are removed by such a process, the result of branch contraction is an unresolved (non-binary) tree in which all remaining branches have a support value greater than or equal to the threshold.

2.4. *Tree Comparison*

Given two trees, the model τ_M and the estimate τ_e, the distance is reported in terms of *false positives*, the number of branches in τ_e that are not in the model τ_M, and *false negatives*, the number of branches in τ_M that are missing from τ_e. The false negative and false positive values are divided by the number of branches in τ_e, so that both error rates fall between 0 and 1. In this study, we used the false positives (FP), false negatives (FN), and their average (also known as the Robinson-Foulds measure[16]) to quantify the error between the model and inferred trees.

3. Experimental Design

Sequence Dataset Generation. We generated five different fully resolved, 20-taxon trees using the r8s tree generation tool.[17] We then deviated each tree from ultrametricity by scaling each branch length by a random value $r = e^x$ where $-2 \leq x \leq 2$. These deviated trees were designated the model trees. For each model tree, and each sequence length of 250, 500, 1000, and 1500 nucleotides, we generated 40 DNA sequence datasets using Seq-gen with the scaling time-reversible model and a substitution rate of 0.6.[15]

True Tree Calculations. For the purpose of our study, we differentiate between the *model tree* and the *true tree*. The branch lengths of the former are the *expected* numbers of changes along the branches, whereas the branch lengths of the latter are the *actual* numbers of changes along the branches. Since we want to study the performance of support assessment techniques as a function of the actual branch lengths, we generated true trees by relabeling the branch lengths of the model trees t_M with the actual substitution rates (which are known in simulations).

Generating ML Bootstrap, MP Boostrap and BI Results. We used PAUP*[21] to generate ML and MP bootstrap results (100 repetitions), and MrBayes[11] for Bayesian inference. We ran each of the methods on each sequence dataset individually.

Generating MP Consensus Trees. Majority consensus MP trees were generated by a series of steps. First, we ran the PAUP* implementation of MP (described above) and reported all trees. We separated the trees into levels, where the top level corresponded to trees with the lowest parsimony score (the best trees), and each subsequent level contained trees of increasing parsimony score. We calculated the majority consensus trees for each sequence dataset using trees from just level 1; levels 1 and 2; levels 1,2, and 3; and levels 1,2,3, and 4.

4. Results

In order to characterize the performance of the different estimation methods, we chose to study the relationship between the substitution rate along a branch and each method's support for that branch as well as the interplay between the contraction threshold and three different measurements of tree errors (false positives, false negative, and average error). In Section 5, we compare the results of each method. The standard deviations for the results of each method were small (MP ≤ 0.084, ML ≤ 0.083, and MB ≤ 0.047) and will not be shown in figures to enhance readability.

4.1. *Selection of Optimal MP Consensus Method*

Recall from Section 3 that we generated results for MP majority consensus for four combinations of top tree levels (1; 1 and 2; 1, 2 and 3; 1, 2, 3, and 4). Therefore, for any given

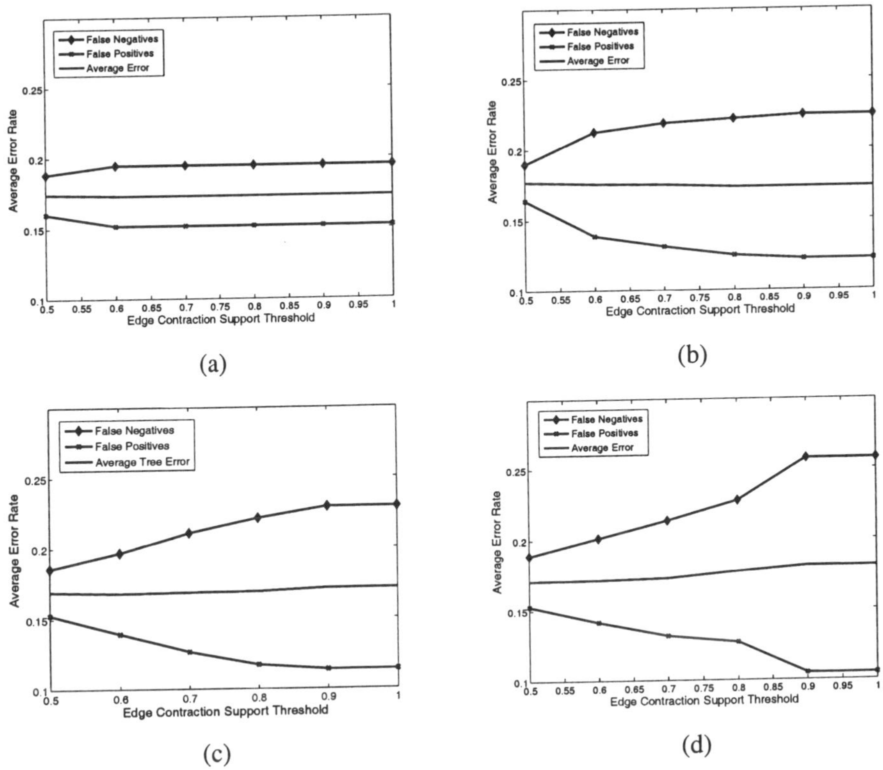

Figure 1. The average error of the estimated trees constructed using MP with Majority Consensus on the datasets with a sequence length of 1500. The x-axis is the range of possible contraction support threshold values. Errors are calculated with respect to the true tree for the dataset. The errors reported are the average of the error for each tree constructed for each dataset. The graphs show Majority Consensus using the (a) top one level, (b) top two levels, (c) top three levels, and (d) top four levels.

sequence dataset there are four MP consensus trees, corresponding to each of these level combinations. In the remainder of the analysis, we compare only the best of the four MP consensus level sets with MP bootstrap, ML bootstrap, and BI. Figure 1 shows the average performance of this method over all datsets with sequence length of 1500 for the four choices of levels. While average total error is nearly identical for all choices of trees, consensus trees built from all four levels contain the fewest false positives, yielding a tree with fewer wrong relationships than other trees. As a result, we chose the 4-level MP consensus trees to be representative of the MP consensus method in the remainder of this study. A significant observation is that regardless of the threshold value chosen, the average error rate of the majority consensus tree does not drop below 16%.

4.2. *Branch Support vs. Substitution Rate*

Within a "reasonable" range of substitution values (well below the point of saturation), it is usually the case that a larger number of substitutions along a branch is correlated to a higher probability of inferring that branch. Therefore, branches in the true tree (whose branch lengths are well below the point of saturation in our experiments) with high substitution rates should have a stronger phylogenetic signal and hence higher probability of being inferred. We tested this hypothesis by grouping branches in the true trees by their actual substitution rate, creating five bins for branches with substitution rates in the ranges 0—0.1, 0.1—0.2, 0.2—0.3, 0.3—0.4, and 0.4—0.5. For each method, we collected the support values generated for the branches in each dataset. The resulting distributions of support values in each bin for datasets with a sequence length of 1500 are shown in Figure 2.

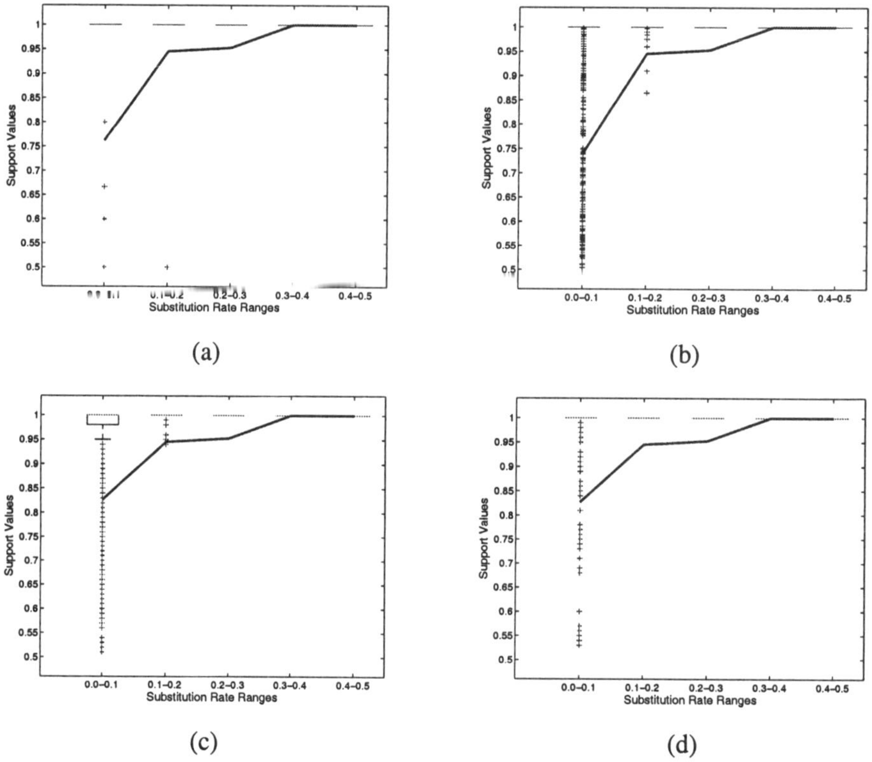

Figure 2. Whisker-box plots indicating the distribution of support values for each substitution rate range: (a) MP consensus, (b) MP bootstrap, (c) ML bootstrap, and (d) BI. The + marks indicate outliers. The trend lines indicate the percentage of correct branches (true positives) in each bin that were predicted in the estimated trees on average.

The trend lines indicate the percentage of correct branches (true positives) in the bin that were predicted by the each method. As expected, the percent of true branches predicted is higher for branches with greater substitution rates.

In Figure 2, the whisker-box plots[13] and average support values for all substitution ranges for all methods are compressed into a very small region around 1.0, indicating that the majority of support values for branches, regardless of true substitution rate or method, were close to 1. In the lower substitution rate range, these were much higher support values than we expected to see. Surprisingly, repeating the same test on the 250 bp datasets (not shown) yields similarly high average support values – greater than 90% in the lowest substitution rate bin for all methods. Thus, MP bootstrap, ML bootstrap, and BI methods all characteristically assign high support values even to branches with low substitution rates – implying that when one of these methods detects a true branch, it obtains a strong signal, regardless of the true substitution rate along that branch. However, observe that for branches with low substitution rates, consensus and bootstrap MP trees have a false negatives rate of about 25%, and for bootstrap ML and BI trees have a false negatives rate of about 15%. In other words, while these methods are computing high support of very short branches, they are missing a sizable portion of the true branches. Further, the MP consensus trees have the least number of outliers for very short branches; yet, this comes at the expense of higher false negatives rate. The other methods have higher numbers of outliers and lower false negative rates.

4.3. *Effects of Branch Contraction on Accuracy*

The benefit of having support-weighted branches is that branches with low support (defined as appropriate for the intended use of the tree) can be removed through branch contraction. In order to characterize how branch contraction can be used to derive more accurate phylogenetic trees, we calculated the error of each estimated support-labeled tree for various choices of a branch contraction threshold. These results are shown in Figure 3. The figure shows the error measured in false positives, false negatives, and average error (as defined in Section 2) for all four methods. There are several trends evident in the graphs:

False positives monotonically fall with higher branch contraction thresholds. This trend can be seen in all four plots shown in Figure 3 and is expected since we assume that the noise in the data giving rise to the prediction of incorrect branches is minimal, leading to those wrong branches having small support values. This is precisely what is observed.

Low-supported branches are evenly split between true and false positives. Despite the fact that the number of false positives falls with higher branch contraction thresholds, the number of false negatives rises. On all four plots, the slope of the false positives line is mirrored by the slope of the false negatives line. This indicates that approximately as many true branches receive low support values as do false branches. An ideal method would have a falling false positive score and a constant false negative score for increasing contraction thresholds.

Overall average error modestly increases with higher branch contraction thresholds. Due to the fact that as the branch contraction threshold increases, the false positives decrease and the false negatives increase at similar rates, we expect that the overall error will not change significantly. In fact, for all methods, as the contraction threshold is increased, the average error increases slightly, seen most prominently in the MP bootstrap (Figure 3(b)) and ML boostrap (Figure 3(c)) methods. This should not be interpreted as implying that the trees are of equal correctness. On the contrary, as will be discussed in Section 5, the overall average error is not the best error metric to use when evaluating the correctness of a tree.

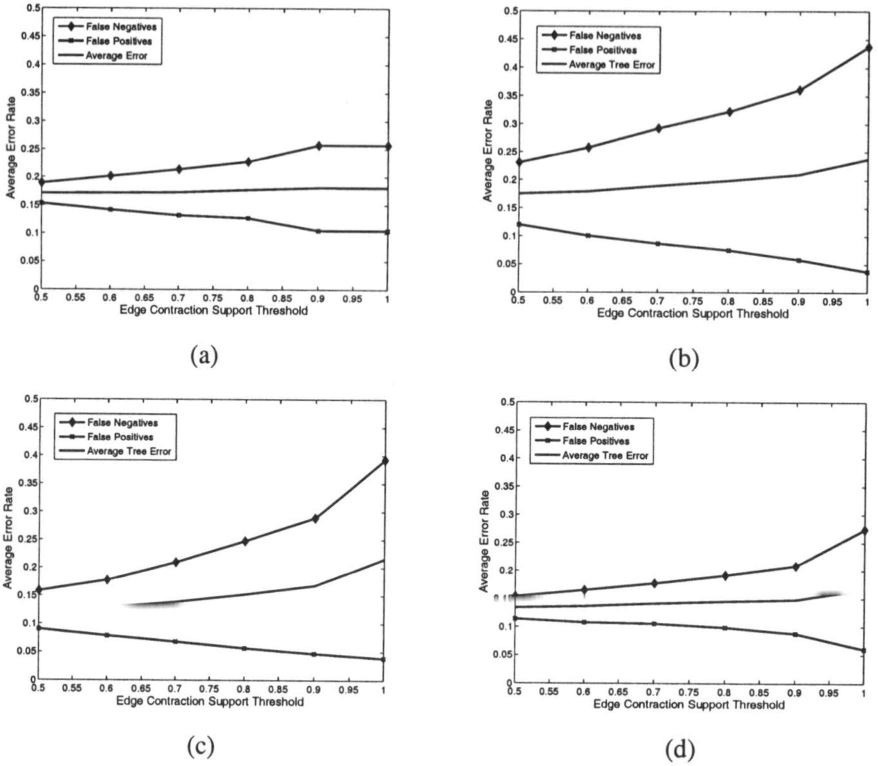

Figure 3. The average error of estimated trees, branch contracted according to the threshold shown along the x-axis, as compared to the true tree for the dataset: (a) MP consensus, (b) MP bootstrap, (c) ML bootstrap, and (d) BI. All figures shown were constructed using only the results from the 1500 sequence length datasets.

The MP consensus method has few moderately supported branches. Unlike the MP bootstrap, ML boostrap, and BI methods (Figures 3(b—d)), the numbers of false positives and negatives for the MP consensus method hardly change for different values of the contraction threshold. This is an indicator that branches in MP concensus trees characteristically have extreme support values - either close to 0.5 or 1 - resulting from the population of trees is generally too small in size to provide sufficient diversity to generate good support values. In spite of this limitation, the MP consensus method still generates trees with comparable overall average error levels, albeit with undesirably high false positive rates for high contraction thresholds.

5. Discussion

The overarching goal of this project was to find out how support values generated by various phylogenetic estimation methods can be used to estimate more accurate trees. Based on the results presented in Section 4, specifically those discussed in Section 4.3, we have several observations (true under the conditions of our experiments) to offer:

The MP consensus method does not produce informative support-labeled trees. Though the method does produce good trees in terms of all three forms of error measured in this project, Figure 3(a) reveals that support-values cannot be used to significantly improve the majority consensus tree.

MP bootstrap, ML bootstrap, and BI perform very similarly. BI produces the most resolved trees of the three methods (evident from its significantly lower false negative rates), whereas MP and ML both have slightly lower false positive rates (which are less significant than the false negative difference in BI).

Strict consensus gives the most correct tree. We make this observation from the perspective of minimizing the false positives. As discussed earlier in Section 2, false negatives lead to conservative trees, missing some resolution in the relationships between taxa whereas false positives are relationships that do not actually exist. While Figure 3 shows that strict consensus trees will contain more errors than majority consensus trees, the strict consensus trees will be conservative estimates as opposed to majority consensus which contain wrong relationships.

It is impossible to construct a fully resolved (binary) tree with 100% certainty. As the figures show, attempting to maintain a more resolved tree requires the admission of more false positives into the tree. In order to eliminate these errors, the tree must become less resolved. Because of this trade-off, phylogenetic analysis methods must be designed to operate on non-binary trees. The alternative is to accept greater accumulated error in the results.

6. Conclusions

In this project, we studied four different phylogenetic estimation methods for constructing support-labeled trees. The contribution of this paper bears a significant impact on the understanding of the relative merits of the different algorithms we studied and of the trade-off involved in choosing a branch contraction threshold. In addition, our results support the observation that strict consensus trees will always yield more correct trees, if the goal is to minimize the number of wrong branches in the estimated tree. Further, our results show that even with sophisticated methods such as Bayesian inference, obtaining a fully resolved accurate tree is very hard. Therefore, phylogenetic analysis tools that assume the trees are always binary (fully resolved) may have a serious shortcoming in their applicability.

This study has also identified the trend of methods ascribing low support values to equal numbers of true and false branches in the estimated tree. What remains unclear is why true branches receive low support values and whether there are ways to improve this true branch confidence. Such improvements would directly impact the accuracy of estimated trees.

References

1. M. Alfaro, S. Zoller, and F. Lutzoni. Bayes or Bootstrap? A Simulation Study Comparing the Performance of Bayesian Markov Chain Monte Carlo Sampling and Bootstrapping in Assessing Phylogenetic Confidence. *Mol. Biol. Evol.*, 20(2):255–266, 2003.
2. V. Berry and O. Gascuel. On the Interpretation of Bootstrap Trees: Appropriate Threshold of Clade Selection and Induced Gain. *Mol. Biol. Evol.*, 13(7):999-1011, 1996.
3. V. Berry, O. Gascuel, and G. Caraux. Choosing the tree which actually best explains the data: another look at the bootstrap in phylogenetic reconstruction. *Computational Statistics & Data Analysis*, 38:273–283, 2000.

4. V. Berry, D. Bryant, T. Jiang, P. Kearney, M. Li, T. Wareham, and H. Zhang. A Practical Algorithm for Recovering the Best Supported Edges of an Evolutionary Tree. In *Proc. 11th Annual ACM-SIAM Symposium on Discrete Algorithms*, 287-296, 2000.
5. M. P. Cummings, S. A. Handley, D. S Myers, D. L. Reed, A. Rokas, and K. Winka. Comparing Bootstrap and Posterior Probability Values in the Four-Taxon Case. *Syst. Biol.*, 52(4):477–487, 2003.
6. P. Erixon, B. Svennblad, T. Britton, and B. Oxelman. Reliability of Bayesian Posterior Probabilities and Bootstrap Frequencies in Phylogenetics. *Syst. Biol.*, 52(5):665–673, 2003.
7. J. Felsenstein. Evolutionary trees from DNA sequences: a maximum likelihood approach. *Journal of Molecular Biology*, 17:368–376, 1981.
8. J. Felsenstein. Confidence limits on phyogenies: an approach using the bootstrap. *Evolution*, 39:783-791, 1985.
9. J. Felsenstein. *Inferring Phylogenies*. Sinauer Associates, Inc., Sunderland, MA, 2003.
10. J. Huelsenbeck, B. Rannala, J. Masly. Accomodating phylogenetic uncertainty in evolutionary studies. *Science*, 288:2349–2350, 2000.
11. J. Huelsenbeck and F. Ronquist. MrBayes: Bayesian inference of phylogenetic trees. *Bioinformacs*, 17(8):754–755, 2001.
12. A. Leache and T. Reeder. Molecular Systematics of the Eastern Fence Lizard (Sceloporus undulatus): A comparison of Parsimony, Likelihood, and Bayesian Approaches. *Syst. Biol.*, 46:523–536, 2002.
13. R. McGill, J. W. Tukey, and W. A. Larsen. Variations of Boxplots. *The American Statistician*, 32:12-16, 1978.
14. M. Nei, S. Kumar, and K. Takahashi. The optimization principle in phylogenetic analysis tends to give incorrect topologies when the number of nucleotides or amino acids is small. *Proc. Natl. Acad. Sci. USA*, 95: 12390–12397, 1998.
15. A. Rambaut and N. C. Grassly. Seq-Gen: an application for the Monte Carlo simulation of DNA sequence evolution along phylogenetic trees. *Comput. Appl. Biosci.*, 13:235–238, 1997.
16. D.R. Robinson and L.R. Foulds. Comparison of phylogenetic trees. *Math. Biosciences*, 53:131–147, 1981.
17. M. Sanderson. r8s software package, Available from http://loco.ucdavis.edu/r8s/r8s.html, 2001.
18. M. Simmons, K. Pickett, and M. Miya. How Meaningful AER Bayesian Support Values? *Mol. Biol. Evol.*, 21(1): 188–199, 2004.
19. P. Soltis and D. Soltis. Applying the Bootstrap in Phylogeny Reconstruction. *Statistical Science*, 18(3):256–267, 2003.
20. Y. Suzuki, G. Glazko, and M. Nei. Overcredibility of molecular phylogenies obtained by Bayesian phylogenetics. *PNAS*, 99(25): 16138–16143, 2002.
21. D. Swofford. PAUP*: Phylogenetic analysis using parsimony (*and other methods). Version 4.0b10. Sinauer Associates, Sunderland, Mass, 2002.
22. K. Takahashi and M. Nei. Efficiencies of Fast Algorithms of Phylogenetic Inference Under the Criteria of Maximum Parsimony, Minimum Evolution, and Maximum Likelihood When a Large Number of Sequences Are Used. *Mol. Biol. Evol.*, 17(8):1251–1258, 2000.
23. D. Taylor and W. Piel. An Assessment of Accuracy, Error, and Conflict with Support Values from Genome-Scale Phylogenetic Data. *Mol. Biol. Evol.*, 21(8):1534–1537, 2004.

ANALYZING INCONSISTENCY TOWARD ENHANCING INTEGRATION OF BIOLOGICAL MOLECULAR DATABASES*

YI-PING PHOEBE CHEN[1,2] AND QINGFENG CHEN[1]

[1]Faculty of Science and Technology, Deakin University, Melbourne, VIC 3125, Australia,
[2]Australia Research Council Centre in Bioinformatics, Australia

Abstract: The rapid growth of biological databases not only provides biologists with abundant data but also presents a big challenge in relation to the analysis of data. Many data analysis approaches such as data mining, information retrieval and machine learning have been used to extract frequent patterns from diverse biological databases. However, the discrepancies, due to the differences in the structure of databases and their terminologies, result in a significant lack of interoperability. Although ontology-based approaches have been used to integrate biological databases, the inconsistent analysis of biological databases has been greatly disregarded. This paper presents a method by which to measure the degree of inconsistency between biological databases. It not only presents a guideline for correct and efficient database integration, but also exposes high quality data for data mining and knowledge discovery.

1 Introduction

In recent years, advanced experiment methods have resulted in the rapid growth of life science databases. Many biological databases have been developed for different purposes, such as GenBank and NCBI [1-2]. The enormous data in databases are meaningful for the exploration of their life origin and evolution, and to predict the function and structure of life systems. They have been commonly usedby biologists during data analysis.

Due to the increasingly complex and specific nature of biological databases, a complicated biological question has to be answered by consulting multiple biological databases. However, the knowledge of life systems is too detailed and complex to be completely comprehended. Such complexity presents a big challenge to merge knowledge from diverse databases. The heterogeneity of databases blocks the accessibility to them [3-4]. In other words, the inconsistent structures and terminologies of biological databases result in a significant lack of interoperability. Thus, it creates a demand for data preprocessing.

As an important cleaning action, the integration of biological databases is significant when dealing with the heterogeneity of biological databases. However, the twisted and deformed biological data often demand additional knowledge so that the values held in databases can be specified and constrained. This causes considerable difficulties for data integration.

---

* This work is partially supported by a Discovery Grant DP0559251 from the Australian Research Council.

Technical and semantic problems are two key issues which present themselves when integrating biological databases. The former can be solved because most current biological databases are implemented on relational database management systems (RDBMS) that provide standard interfaces like JDBC and ODBC for data and metadata exchange [5-6]. Nevertheless, the solution of semantic problems remains unsolved..

Modern bioinformatics demand knowledge extracted from databases for communication purposes. For example, a user's query of a protein kinase may refer to hundreds of databases. There are two options to integrate knowledge from databases: (1) standardising the nomenclature of diverse databases; and (2) creating bridges between databases even if they differ radically in structure and nomenclature. The former have encountered resistance from database maintainers and specialists who hesitate to change preferred terminology [4]. As a tradeoff scheme, the latter has been commonly applied in the integration phase of biological databases. Among them, ontology-based biological database integration is one of the representative methods designed to capture knowledge from databases.

There have been many attempts to develop standards that can be applied tobioinformatics ontologies and which subsequently exploit biological information. For example, EcoCyc ontology [7] covers E.coli gene, metabolism, regulation and signal transduction, and Gene ontology (GO) [8] describes drosophila, moused and yeast gene function, process and cellular location and structure. Recently, ontology-based semantic integration of biological databases was presented in [2, 9]. Philippi [6] proposed a method for the ontology-based semantic integration of life science databases using XML technology. To enhance semantic interoperability, there have been considerable efforts to solve nomenclature-mapping problems and standardise the naming of functional relations and processes and their arguments such as ontology-mapping in GO community [10]. It provides a comprehensive list of synonyms that can be used immediately to improve indexing and search over the literature. However, no effort has been made to analyse the inconsistency of biological databases which would effectively lead to the enhancement of database integration.

This paper presents a method by which to analyse inconsistencies between biological databases using ontology. The method is able to find out the databases that are inappropriate for integration or need to be further improved. This not only reduces the search space but also generates high quality data for accurate and efficient data mining and knowledge discovery. Algorithms and experiments are presented to further demonstrate our approaches.

The remainder of this paper is organised as follows. Section 2 presents basic concepts. The approaches by which to analyse the inconsistency between biological databases are presented in Section 3. In Section 4, experiments are presented. Section 5 concludes this paper.

2 Basic Concepts

2.1 Problem Description

The increasing biological databases relating to genome sequences and protein structures and functions are challenging the traditional approaches for knowledge acquisition. To answer a complex biological question, hundreds of biological databases can be consulted. It is critical to guarantee accessibility to the databases. However, the discrepant structures and nomenclature of databases have an effect on their communication capabilities.

Although some biological data publishing and collection use HTML (Hypertext Markup Language) format, this method cannot describe complex structure documents. Besides, the varied organisation, storage and publication of biological data leads to different information types. For example, the representative database NCBI (National Center for Biotechnology Information) adopts mostly the binary ASN.1 [1], while flatfiles are used in GenBank [2]. The differences in the information types result in heterogeneities between biological databases and prevent us from obtaining high quality data for data analysis. Additionally, the information derived only from a single database does not enable us to obtain a comprehensive understanding, and the knowledge acquisition is inconvincible.

There have been some efforts to establish a link between disparate databases, such as data warehousing and database federation. Nevertheless, the increasing new data and databases have led efforts to reach a terminological impasse whereby databases have to agree on nomenclature and compatible formats before a link is able to be built. Nevertheless, database maintainers and specialists in certain research fields find it difficult to accept such a link. Ontology has been recently used to create bridges between biological databases. However, there are still some problems in the ontology-based integration of biological information. Problems include:

- Ontologies with independent terminologies and structures are often incompatible. This causes difficulties when acquiring knowledge from databases.
- Heterogeneities such as synonym result in a significant lack of interoperability among biological databases. This blocks the generation of high quality data.
- Semantic inconsistency has been widely ignored. The integration of biological databases with high discrepancies cannot guarantee efficient data mining.

Such inconsistencies surrounding biological databases and when the databases are appropriate for further processing, present major ambiguities. The analysis of the inconsistent nature of biological databases assists us in sorting out the appropriate databases from which high quality data can be derived. Hence, it is imperative to develop approaches by which to measure the inconsistency of biological databases and ensure reliable integration of biological databases.

2.2 Symbols and Formal Semantics

Suppose A and L denote atom symbols and proposition formulae respectively. In particular, A can contain α and $\neg\alpha$ for some atoms α. Let $\wedge$, $\neg$ and $\rightarrow$ be logical connectives. Let C, $c \in A$ be concepts such as *Gene* and *Protein*, *CV* for control vocabulary, r for relationship, and α, β and γ for attributes in general. Let $\equiv$ be logical equivalence. A model of a formula ϕ is a possible set of atoms where ϕ is true in the usual sense.

Controlled vocabularies are a set of named concepts that may have an identifier. The concepts or their identifiers are often used as database entries. Its definition is as follows.

DEFINITION 2.1. Suppose *t*, *def*, *id* and *sn* present term, definition, identifier and synonyms respectively. Let *C* be the set of concepts of *databases. Hence, we have*

$$\textit{Controlled Vocabulary } CV := \{c \mid c = (t, def, id, sn) \in C\}$$

An example of *Gene Ontology* (*GO*) is as follows. Each concept (*biological process*) has a term (*recommended name*), an identifier (*id: GO: number*), definition (*explanation and references*) and synonyms (*other names*). The *definition* of each *biological process* is provided by brief description and references to relevant literature or web links.

Ontology includes relationships as well as concepts. The relationships consist of '*is-a*' (*Specification relationships*) and '*part-of*' (*Partitive relationships*), by which concepts can correlate with each other. Although '*part-of*' relationship can be defined, only the transitive '*is-a*' hierarchy is required for querying databases. For example, '*Enzyme* is one kind of *Protein*', '*Protein* is one kind of *Macromolecule*' and '*Membrane* is part of *Cells*'. Therefore, ontology can be viewed as a tree, where the nodes and directed edges present concepts and relationships respectively.

DEFINITION 2.2. *Let O be ontology, and r be relationships that link concepts. Ontology can be defined as a set of tuples.*

$$\textit{Ontology } O := \{(c_1, c_2, r) \mid c_1, c_2 \in CV, \text{ and } r : c_1 \rightarrow c_2\}$$

where $c_1 \rightarrow c_2$ presents a relationship r from c_1 to c_2, such as 'c_1 *is-a* c_2'.

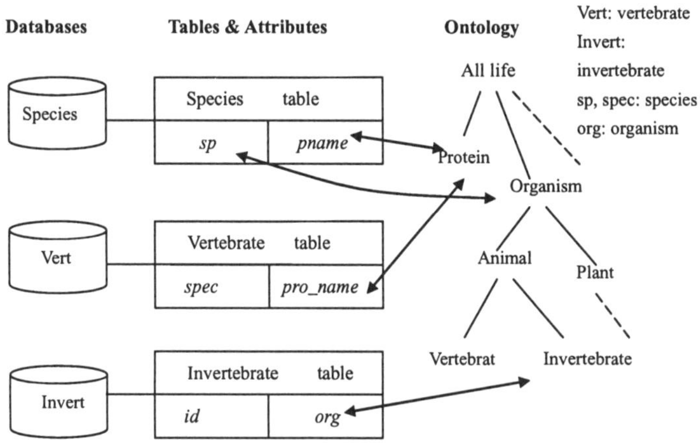

Figure 1. Biological database attributes are linked to ontology concepts. Attributes *pname* and *pro_name* from databases *Species* and *Vert* have different attribute names, but they are correlated by a common concept *protein* of the ontology.

EXAMPLE 2.1. In Figure 1, Vertebrate, Animal, Plant and Organism are connected by transitive '*is-a*' relation. (*Animal, Organism, Animal* $\rightarrow$ *Organism*), (*Plant, Organism, Plant* $\rightarrow$ *Organism*), (*Vertebrate, Animal, Vertebrate* $\rightarrow$ *Animal*) *and* (*Invertebrate, Animal, Invertebrate* $\rightarrow$ *Animal*) represent tuples of ontology.

To analyse the inconsistency of biological databases, the above need to be defined semantically using ontology. One of the key processes is to link tables and attributes to a specified ontology. Subsequently, users can execute queries via hierarchies, such as '*is-a*', to derive information from databases. Four operators to describe the interactions among attributes, tables and ontology are given below. Let $Att_1 \in DB_1$ and $Att_2 \in DB_2$ be database attributes. Let CV_1 and CV_2 be controlled vocabularies.

(1) **Mapping:** Let $Att \in DB$ be database attributes. Let O and c be ontology and concepts respectively. '*maps*(O, Att, c)' states the attribute Att in DB can be mapped to a corresponding concept c via *ontology* O.
(2) **Cross-reference:** Let CV be controlled vocabulary. '*cross-reference*$(CV, (Att_1, Att_2), c)$' states that if Att_1 and Att_2 can be linked to a common concept c by *cross-reference* of CV, they are semantically equivalent owing to c.
(3) **Translation:** '*translates*$((CV_1, CV_2), (Att_1, Att_2), c)$' states that database attribute Att_1 and Att_2 can be translated to a common concept c using the controlled vocabularies CV_1 and CV_2. Thus, it is feasible to relate database entries that use different terms for the same thing, such as the *English species name* and *Systematic species name* in Figure 2.
(4) **Taxonomy:** Let c_i *and* c_j be concepts. '*is-a*(c_i, c_j)' states that c_i is a sub-concept of c_j and c_j is a sup-concept of c_i. For simplicity, the operator '*is-a*' below implies both '*is-a*' and '*part-of*' relationships mentioned above. Actually, the '*is-a*' relationship holds transitivity. Hence, we have

$$\forall\, c_1, \ldots, c_n \in \mathrm{O},\ is\text{-}a(c_1, c_2) \wedge is\text{-}a(c_2, c_3) \wedge \ldots \wedge is\text{-}a(c_{n-1}, c_n) \rightarrow\ is\text{-}a(c_1, c_n)$$

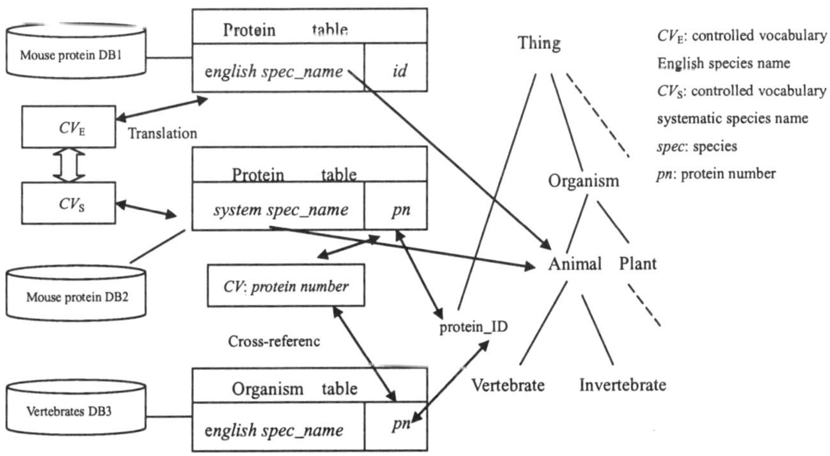

Figure 2. Translation by mapping synonymous concepts of controlled vocabularies is used to link databases with synonyms. Database attributes corresponding to the same concept and sharing the same controlled vocabulary can be viewed as cross-references of attributes.

The above axioms describe possible processes in response to a user's query on biological databases. Ontology plays a central role in mapping database attributes to common concepts or translating attributes between different controlled vocabularies, such as *English controlled vocabulary* and *Systematic controlled vocabulary* in Figure 2. Additionally, *queries* operator usually intends to search in *attribute* for specified *terms* as mentioned above. Hence, a user's query can be classified into two categories in terms of entries regarding *attribute*:

- if the queried attribute is found in databases, it will be mapped to a corresponding concept of ontology, and will enable other database attributes to be linked together;
- if no database attribute is defined as the queried attribute, a corresponding concept of ontology is selected. Its sub-concepts and super-concepts will be searched to find the attribute.

Although the latter is complex, it can eventually get back to the former pathway via ontology. In either case, the queries bring about a collection of results, which can be used to measure the inconsistency found in biological databases. Usually, users specify a term *T* along with queries. *T* is able to reduce the searched concepts that are irrelevant to the queries. Suppose *Att* is the queried attribute by users, and its mapping concept of ontology *O* is *C*. Hence, we have

1. $sub(C, T) = \{c \mid \forall c, is\text{-}a(c, C), c \sqsupseteq T\}$
2. $sup(C, T) = \{ c \mid \forall c, is\text{-}a(C, c), c \sqsupseteq T \}$

where $\sqsupseteq$ denotes a inclusion relationship in view of semantics.

EXAMPLE 2.2. Suppose a queried database attribute is *Animal* with a specified term *parrot*. Hence, in Figure 1, we have *sub(Animal, parrot)* = {*Vertebrate*}, *sup(Animal, parrot)* = {*Organism*}. Without the term *parrot*, *sub(Animal, parrot)* = {*Vertebrate, Invertebrate*}, *sup(Animal, parrot)* = {*Organism*}.

From the observation, the database attributes should be semantically defined as specific as possible, which can avoid searching unrelated databases.

DEFINITION 2.3. *Let* $ATT_{DB} = \{a_1, a_2, \ldots, a_n\}$ *be a set of attributes of biological database DB. The set of attributes derived from reference database and compared databases are denoted by* ATT_R and ATT_C *respectively.*

The reference database consists of multiple databases containing the queried attribute or the attribute that can be mapped to concepts of $sub(C, T) \cup sup(C, T)$. It is used to decide whether or not the attributes found in compared databases are consistent with the specified attribute. An example regarding ATT_R and ATT_C is given below.

EXAMPLE 2.3. In Figure 1, $ATT_{Species}$ = {*sp, pname*}, ATT_{Vert} = {*pr_ name, spec*} and ATT_{Invert} = {*id, org*}. If users query attribute *pname,* then $ATT_R = ATT_{Species}$ = {*sp, pname*} and $ATT_C = ATT_{Vert} \cup ATT_{Invert}$ = {*pro name, spec, id, org*}..

DEFINITION 2.4. *Let* $\models$ *be a supporting relationship. For a set of database attributes* ATT_{DB}, $ATT_{DB} \models$ *is defined as follows.*

(1) if the queried database attribute α is found in current databases, we have
  - $ATT_R \models \alpha$ iff ATT_R contains α
  - $ATT_c \models \neg\alpha$ iff ATT_c contains β that is a database attribute of compared

databases, which has a common concept with α.

(2) if the queried database attribute α is not found in databases but can be mapped to a concept *C* in ontology, we have

- $ATT_R \models \alpha_1$ iff ATT_R *contains* α_1 and $maps(O, \alpha_1, c)$
- $ATT_C \models \neg\alpha_1$ iff ATT_c *contains* α_2 that is the corresponding database attribute of *c* in compared databases

Here *O* and *c* present the ontology and concepts in $sub(C, T) \cup sup(C, T)$ respectively. α_1 denotes a mapping attribute in reference database from *c*.

EXAMPLE 2.4. Suppose *'pname : mouse'* and *'animal : mouse'* are two queries on Figure 1, in which *pname* and *animal* are queried attributes, and *mouse* is a term that locates databases. For the query *'pname : mouse'*, database *Species* can be viewed as the reference database. The term *mouse* reduces the search space to database *Species* and *Vert*. Hence, we have $ATT_{Species} \models pname$ and $ATT_{Vert} \models \neg pname$. For the query *'animal : mouse'*, no database attribute is defined as *animal*. Nevertheless, the sub-concepts and super-concepts of *animal* in ontology can be mapped to this attribute. The search will be limited to *Vertebrate* and *Species* due to the term *mouse* that is mapped to attribute *spec* of *Vert* and *sp* of *Species*. *Vert* is selected as the reference database so $ATT_{Vert} \models spec$. The attribute *sp* of *Species* is viewed as a negative attribute of *spec*, namely $ATT_{Species} \models \neg spec$.

3 Analyzing Inconsistency of Biological Databases

3.1 Models of Queried Biological Databases Attributes

DEFINITION 3.1. *Suppose* $ATT \in \wp(L)$, $X \in \wp(A)$. *Let* ATT_{DB} *be attributes derived from* $DB \in \{R, C\}$. *Let* $X \models ATT$ *denote that* $X \models \alpha$ *holds for every* α *in ATT.*

$$model(ATT) = \{X \in \wp(A) \,/\, X \models ATT\}$$

where *ATT* denotes a set of database attributes. The model of *ATT* actually presents a set of atoms that support *ATT*.

For measuring inconsistency, we use compatibility of biological databases. The *consistentset* of a model is the set of database attributes that have identical names with corresponding reference attributes. The *conflictset* of a model consists of (1) the set of database attributes that are semantically equivalent; and (2) the *null* attribute that presents no attribute is semantically equivalent to the reference attribute. Actually, some databases may not contain the queried attribute at all.

DEFINITION 3.2. *Let* δ *be a selected reference attribute from reference database R. Let* $Y \in \wp(A)$ *be a model of* δ. *The consistentset and conflictset are defined below.*

- $Consistentset(\alpha) = \{\alpha \mid \alpha \in Y, \alpha \equiv \delta\}$
- $Conflictset(\alpha) = \{\alpha \mid \alpha \in Y, \alpha \equiv \neg\delta, \text{ or } \alpha = null\}$

Based on *consistentset* and *conflictset* from minimal models, a measurement can be used to compute the inconsistency of minimal models in respect to specified database attributes.

DEFINITION 3.3. *The compatibility function from A into* [0, 1], *is defined below when* α *is not empty, and* $Compatibility(\varnothing) = 0$.

$$Compatibility(\alpha) = \frac{|Consistenset(\alpha)|}{|Consistentset(\alpha)| + |Conflictset(\alpha)|} \times 100\%$$

where $|Consistentset(\alpha)|$ and $|Conflictset(\alpha)|$ are the cardinality of $Consistentset(\alpha)$ and $Conflictset(\alpha)$ respectively. If $Compatibility(\alpha) = 0$, then we can say that the model Y has no opinion upon α and vice versa; if $Compatibility(\alpha) = 1$, it indicates that there are no negative attributes $\neg\alpha$ in the model Y; if $0 < Compatibility(\alpha) < 1$, it presents the model Y as partially inconsistent/consistent with respect to α.

The compatibility function quantifies the inconsistency of biological databases in relation to the queried database attributes. A queried attribute in biological databases is regarded as compatible or consistent in the case that the compatibility regarding this attribute is equal to, or greater than, the threshold minimal compatibility (*mincomp*) given by users or experts. Here, let $mincomp = 0.5$. Then

- *consistent* if $Compatibility(\alpha) \geq mincomp$
- *inconsistent* if $Compatibility(\alpha) < mincomp$

Ideally, we would like to achieve $Compatibility(\alpha) = 1$. Nevertheless, the inconsistency is the objective existence of discrepant terminologies and ontology used in diverse biological databases. If biological databases are found to be inconsistent, it is possible that they contain too many incompatible database attributes or most of current databases do not contain the queried database attribute. Therefore, the method presented in this paper is a prerequisite for the integration of biological databases.

3.2 Experiments

Table 1 represents the definition of attribute fields for five biological databases and refers to Gene ontology and NCBI databases. Among them, DB_2 and DB_5 are databases which regard *Species*, DB_1 and DB_4 are databases with respect to *Vertebrate*, and DB_3 is about *Invertebrate*. All database attributes under *DNA sequence* are linked to *DNA sequence* concept of the ontology. For the attributes under *Organism*, *org* is related to *Vertebrate* concept but *spec* is linked to *Organism* concept. The *null* value in this table means no such attribute is defined in corresponding biological databases, such as attributes of DB_1 under *Description* column. In particular, the attribute $spec_s$ of DB_5 is a systematic species name. A translation is, therefore, needed to search for this attribute.

Table 1. Attributes of biological databases.

DB	Author	DNA Sequence	Description	Identifier	Organism	Enzyme
DB_1	*au*	*dns*	*null*	*id*	*org*	*null*
DB_2	*author*	*seq_dna*	*desc*	*id*	*spec*	*ename*
DB_3	*au*	*seq_dna*	*desc*	*mid*	*org*	*ename*
DB_4	*au*	*seq_dna*	*null*	*gid*	$spec_s$	*enzyme*
DB_5	*au*	*seq_dna*	*null*	*id*	*org*	*ec_nr*

Measuring the inconsistency of biological databases mainly comprises of three steps: (1) input queried database attributes; (2) compute the compatibility of databases in relation to queried attributes; and (3) determine the consistency of databases. Two experiments are presented below. One is to query attribute '*enzyme* : *mouse*' via cross-reference, and the other is to query attribute '*animal* : *mouse*' using translation.

In the former, DB_3 is ignored for it does not meet the constraint *mouse*. DB_5 is selected because the reference database for *enzyme* is found in DB_5, which is mapped to concept *protein* of the ontology in Figure 1. According to the ontology, the attributes under *Enzyme* of DB_2, DB_4 and DB_5 use different terminology to represent the same concepts. The common concept *Enzyme* can be used for cross-reference among them. According to Definition 3.2, we can obtain *Consistentset*(*enzyme*) = {*enzyme*} from DB_5, and *Conflictset*(*enzyme*) = {*null*, *enmae*, *ename*} from DB_1, DB_2 and DB_4. Both *null* and *ename* are regarded as $\neg$*enzyme* when computing the compatibility of biological databases. Finally, we obtain *Compatibility*(*enzyme*) = 1 / 4 = 0.25 < *mincomp*. Therefore, the biological databases are inconsistent in relation to the database attribute *enzyme*.

As for the latter case, DB_3 is ignored in the same way. The database attribute *org* of DB_1 and DB_4 is linked to *Vertebrate* concept in the ontology, and *spec* of DB_2 and DB_5 is linked to *Organism* concept in Figure 2. Among them, the *spec* attribute in DB_2 and DB_3 needs to be translated to the corresponding attribute *specs* in DB_5 for it is a systematic species name. Hence, the *model*(*ATT*) = {*org*, *spec*, *org*, *spec*}. There are two possibilities by which to select the reference attribute here: (1) *org*; and (2) *spec*. If we use *org* as the reference attribute, we have *Consistentset*(*org*) = {*org*, *org*} with respect to DB_1 and DB_4, and *Conflictset*(*org*) = {*spec*, $spec_s$} in relation to DB_2 and DB_5. Therefore *Compatibility*(*org*) = 2 / 4 = 0.5 $\geq$ *mincomp*. Therefore, the biological databases are consistent in respect to the queried database attribute *animal* : *mouse* using *org*. On the other hand, if we use *spec* as the reference attribute, we have *Consistentset*(*spec*) = {*spec*}, *Conflictset*(*spec*) = {*org*, *org*, $spec_s$} and *Compatibility*(*spec*) = 1 / 4 = 0.25 < *mincomp*. Thus, they are inconsistent in respect to the database attribute (*animal*, *mouse*) using *spec*.

4 Conclusions

Knowledge acquisition from biological databases plays a nontrivial role in biological studies. However, the heterogeneity of biological databases has resulted in a significant lack of interoperability between them. The integration of biological databases is critical when dealing with heterogeneity but suffers from the twisted and deformed nature of biological data. Ontology-based integration of biological databases is an efficient way to capture knowledge from multiple sources. Nevertheless, no effort has been made to analyse the inconsistency in biological databases. This paper proposes a method to measure the inconsistency of biological databases via ontology. It assists in obtaining high quality data for data mining and knowledge discovery. We demonstrate our method by conducting experiments.

References

1. http://www.ncbi.nlm.nih.gov/.

2. Benson DA., Karsch-Mizrachi I., Lipman DJ., Ostell J. and Wheeler DL., GenBank Update, *Nucleic Acids Research*, vol 32 (Database issue), pp 23-26, 2004.
3. Stevens R., Goble C., Horrocks I. and Bechhofer S., OILing the Way to Machine Understandable Bioinformatics Resources, *IEEE Trans Inf Technol Biomed*, 6(2), pp 129-134, 2002.
4. Williams N., Bioinformatics: how to get databases talking the same language, *Science*, 275(5298), pp 301-302, 1997.
5. Kohler J., Philippi S. and Lange M., SEMEDA: ontology based semantic integration of biological databases, *Bioinformatics*, 19(18), pp 2420-2427, 2003.
6. Philippi S. and Kohler J., Using XML Technology for the Ontology-based Semantic Integration of Life Science Databases, *IEEE Trans Inf Technol Biomed*, 8(2), pp 154-160, 2004.
7. Karp P.D., Riley M., Saier M., Paulsen I.T., Paley S.M. and Pellegrini-Toole A., The EcoCyc and MetaCyc Databases, *Nucleic Acids Research*, 30(1), pp 59-61, 2000.
8. Yeh I., Karp PD., Noy NF. and Altman RB., Knowledge acquisition, consistency checking and concurrency control for Gene Ontology (GO), *Bioinformatics*, 19(2), pp 241-248, 2003.
9. Kohler J., Philippi S. and Lange M., SEMEDA: ontology based semantic integration of biological databases, *Bioinformatics*, 19(18), pp 2420-2427, 2003.
10. http://www.geneontology.org/.

A NOVEL APPROACH FOR STRUCTURED CONSENSUS MOTIF INFERENCE UNDER SPECIFICITY AND QUORUM CONSTRAINTS

CHRISTINE SINOQUET

LINA, Université de Nantes,
2 rue de la Houssinière, BP 92208, 44322 Nantes Cedex, France,
E-mail: christine.sinoquet@univ-nantes.fr

We address the issue of structured motif *inference*. This problem is stated as follows: given a set of n DNA sequences and a quorum q (%), find the optimal structured consensus motif described as gaps alternating with specific regions and shared by at least $q \times n$ sequences. Our proposal is in the domain of metaheuristics: it runs solutions to convergence through a cooperation between a sampling strategy of the search space and a quick detection of local similarities in small sequence samples. The contributions of this paper are: (1) the design of a stochastic method whose genuine novelty rests on driving the search with a threshold frequency f discrimining between specific regions and gaps; (2) the original way for justifying the operations especially designed; (3) the implementation of a mining tool well adapted to biologists'exigencies: few input parameters are required (quorum q, minimal threshold frequency f, maximal gap length g). Our approach proves efficient on simulated data, promoter sites in Dicot plants and transcription factor binding sites in *E. coli* genome. Our algorithm, Kaos, compares favorably with MEME and STARS in terms of accuracy.

1. Introduction

In the last fifteen years a lot of work has appeared in the literature which addresses the general topic of *consensus motif inference* (CMI) in a set of biological sequences (Gibbs sampling,[10,16] Pratt,[9] CONSENSUS,[8] Teiresias,[14] MEME,[2] PROJECTIONS,[4] Smile,[12,6] Winnower,[13] Splash,[5] STARS,[11] MODEL,[7] ...). The reader is directed to some recent surveys[3,11,7]. Though the problem is incontestably not new, it remains an important and difficult one in sequence analysis. Exact algorithms are not convenient for large datasets or long sequences. Therefore, approximate algorithms have to be designed. There was room for an approach dedicated to the specific problem of *structured* motif inference. Let us first remind of a peculiar instance of the CMI problem, called the local multiple alignment problem (LMA). In its general form, this problem may be stated as follows: identify a set of sub-words $\{o_{i_1}, o_{i_2}, \cdots, o_{i_{r_i}}\}$, called occurrences, under the four following constraints: (1) any o_{i_j} verifies a similarity constraint with any other o_{i_p}; (2) the occurrences have the same length; (3) each sequence s_i contains 0 or 1 such occurrence o_{i_k}; (4) there are at least $q \times n$ sequences s_i contributing to the set of occurrences (quorum constraint). Generally, LMA algorithms output the set of occurrences, a position-specific scoring matrix ($pssm$) M and the consensus motif. For any character c of the DNA sequence alphabet $\mathscr{A}$, $M[c, j]$ yields the frequency of character c at position j over the set of occurrences. The consensus motif is computed as the word with the most frequent characters in the $pssm$ M. But a central issue in post-genomics is the automatic discovery of functional biological motifs with the aim of identifying a structure common to a set of sequences. The structure may be described as specific regions alternating with gaps. Gaps are regions which contain no intrinsic information. Thus we are interested in structured local multiple alignments (SLMA$_s$). A structured motif is a word built on alphabet $\mathscr{A} \cup \{\mathrm{x}\}$, where x denotes the wild-card character. A wild-card character in position j of the motif indicates that the frequencies of all characters in $\mathscr{A}$ are below a given threshold frequency, say, f. A gap is made up of contiguous wild-card characters. Here our concern is the retrieval of structured motifs such as ACTGxxxxCTTxxGGxxxAAGA, for example.

Naively deriving a structured consensus motif from the $pssm$ built by a classical LMA method does not yield the optimal solution for large datasets or long sequences. Nevertheless, it is wise to benefit from an existing LMA algorithm for local similarity search. On the other hand, for systematic data mining purpose, one can not waste time with successive guesses at the putative structure. The number of input parameters for a search must be reduced: quorum (q); threshold frequency (f); minimal motif length ($minMotifLength$), maximal gap length (g). Thirdly, we are aware of the robustness of stochastic methods in the domain of CMI (Gibbs sampler,[10] MEME,[2] PROJECTIONS,[4] STARS,[11]). Finally, we wish to design an algorithm with a low memory cost. These motivations lead to our investigating a stochastic sampling strategy cooperating with an LMA algorithm.

The remainder of the paper is organized as follows. Section 2 introduces specific terminology and notions we use subsequently. Section 3 is dedicated to the presentation of our algorithm, Kaos. The discussion of the results obtained on simulated and biological data may be found in Section 4.

2. Definitions

For DNA sequences, the alphabet $\mathscr{A}$ is $\{A,\ C,\ T,\ G\}$.

Definition 2.1. *(*LMA*)* Given n sequences $s_1,\ s_2 \cdots s_n$ built on alphabet $\mathscr{A}$, an integer k and a similarity criterion, an LMA is a set of substrings $o_1,\ o_2 \cdots o_n$ of $s_1,\ s_2 \cdots s_n$ such that $o_1,\ o_2 \cdots o_n$ have common length k and maximize the similarity criterion. Note that an LMA is easily represented with a matrix. In the following, we will also call LMA any process yielding such a set of substrings.

Definition 2.2. **(pssm, support, hits)** Let $\mathcal{S} = s_1,\ s_2 \cdots s_n$ be n sequences built on alphabet $\mathscr{A}$, let $\mathcal{O} = \{o_1,\ o_2 \cdots o_n\}$ be an LMA of length k built from these sequences considering a given similarity criterion and let $minSeq$ and $minMotifLength$ be two integers. A *pssm* M associated with this LMA is any matrix of reals of size $|\mathscr{A}| \times l$ ($l \geq minMotifLength$) induced from a sub-matrix L of the LMA matrix as follows: L has at least $minSeq$ lines and $minMotifLength$ columns; M compiles the frequencies of the characters of $\mathscr{A}$, for each column of L. The **hits** of the $pssm$ M are the locations in the sequences of the sub-words aligned in L. These sub-words are the **support** of the *pssm*.

Definition 2.3. **(specific character)** Let f be a given threshold frequency, $0 \leq f \leq 100$. If the highest frequency at column j, $M^+[j]$, is over f and is obtained for character c, then

c is a *specific* character and is refered to as $char(M^+[j])$.

Definition 2.4. (wild-card character, mask) Let us choose x ($\notin \mathscr{A}$) as the *wild-card character*. We denote $mask(M)$ the string $\in \mathscr{A}^l \cup \{x\}$ verifying:

forall $1 \leq j \leq l,\ mask(M)[j] = \begin{cases} x\ if\ M^+[j] < f \\ char(M^+[j])\ otherwise. \end{cases}$

Definition 2.5. (gap) A gap is any word built on alphabet $\{x\}$ which is one of the longest substrings of contiguous wild-card characters in $mask(M)$.

The only constraint on the motif structure is g, the maximal gap length allowed.

Definition 2.6. (property $\mathcal{G}(f,g)$) Property $\mathcal{G}(f,g,M)$ holds if and only if the length of the longest gap in M's mask is less than or equal to g for the threshold frequency f and this mask has no gap at either extremity.

3. Cooperation between a sampling strategy and an LMA

3.1. *Moving in the $pssm$ search space*

To solve a combinatorial optimization problem, all metaheuristics find a balance between search intensification (identifying solutions of better quality from known solutions) and search diversification (escaping from local optima). Our approach successively infers solutions through iterations considering small samples of m sequences chosen at random among the n initial sequences. If at least $q \times n$ sequences share a similarity, it is likely that $q \times m$ sequences in the samples share this similarity on average. The searched space $\mathcal{E}$ consists of $pssm_s$. But we would emphasize that the definite originality of our method lies in the following points: (1) where Gibbs sampling, PROJECTIONS, MEME and CONSENSUS (at each of its iterations) consider $pssm_s$ for a *given* motif length and $pssm_s$ supported by the *same* number of sequences, on the contrary, our method considers the space of $pssm_s$ with minimal second dimension $minMotifLength$ and minimal support size $minSeq$; (2) moreover, for structured motif inference purpose, we impose straightaway that we only move in the $pssm$ sub-space where all elements verify property $\mathcal{G}(f,g)$. Each $pssm$ represents a similarity shared by, say, r sequences belonging to the initial set. The higher r is, the more likely is the $pssm$ the optimal solution. The objectives for the search are increasing the support size (quorum constraint) and optimizing a criterion $\mathscr{C}$ designed to evaluate the specificity of solutions.

3.2. *Sketch of the algorithm*

3.2.1. *Sequence sampling and first intensification level*

The sketch of the search is given in algorithm 1. Any iteration begins with the generation at random of two sequence samples S_1 and S_2 (line 3). The plug-in LMA software tuned with length w yields two LMA matrices of size $m \times w$ (line 4). An exact procedure is implemented to scan each of these matrices and identify the largest sub-matrices (with respect to the numbers of lines and columns) whose $pssm_s$ verify the constraint $\mathcal{G}(f,g)$. This procedure will be described in an extended version of the paper.

At line 5, operator $\otimes$ processes each pair $(s_{1_1},\ s_{1_2})$ of $\mathcal{S}_{1_1} \times \mathcal{S}_{1_2}$ shifting one matrix with regard to the other one to identify a local similarity. As a result, the elementary operation $s_{1_1} \times s_{1_2}$ outputs the **unique** $pssm$ s_2 (if it exists) which satisfies property $\mathcal{G}(f,g)$ and optimizes criterion $\mathscr{C}$. s_2 is computed according to the formula: $s_2[c,j] = \frac{n_{1_1} s_{1_1}[c,j_1] + n_{1_2} s_{1_2}[c,j_2]}{n_{1_1}+n_{1_2}}$. j_1 and j_2 are the columns of matrices s_{1_1} and s_{1_2} which contribute to the column j of matrix s_2. n_{1_1} and n_{1_2} are the support sizes of s_{1_1} and s_{1_2}. Indeed, such computations need only be done for some shifts. The other shifts are efficiently rejected through straightforward focuses on the gaps in s_{1_1} (or s_{1_2}) (starting with the longest ones), and the corresponding regions in s_{1_2} (or s_{1_1}). Thus it is likely that the longest gaps in the potential solution corresponding to the current shift will be pointed out. For a given shift, optimization is performed through scanning $\mathscr{L}$, the list of pairs ($begin$, end) for all gaps in s_{1_1} and s_{1_2}. This list is sorted in decreasing order with respect to the gap lengths. The principle of the optimization will be detailed in the extended version of the paper.

3.2.2. *Starting points and second intensification level*

If they are of sufficient quality with respect to criterion $\mathscr{C}$, some elements of $\mathcal{S}_2$ will be chosen as "starting points" for moving in the search space $\mathcal{E}$ (line 6). A starting point from $\mathcal{S}_2$ will replace a "bad" current solution in $\mathcal{S}_3$. Moving in $\mathcal{E}$ is performed with the objective of maximizing the support sizes (see definition 2.2) of the current solutions in $\mathcal{S}_3$. Here, intensification is obtained with a second operator $\oplus$ (line 6) which compares the elements of all pairs in $\mathcal{S}_3 \times \mathcal{S}_2$: if a solution s_2 in $\mathcal{S}_2$ has the same *mask* as a solution s_3 belonging to $\mathcal{S}_3$, then the support and frequencies of s_3 are updated. $\oplus$ is a specialized version of $\otimes$. Furthermore, if updating the frequencies for a given solution s^* with support size above $q \times n$ only entails variations within a given percentage, say, 2%, then the convergence criterion is satisfied (line 7), the search stops and it yields the pair $\{s^*,\ \mathcal{S}_3\}$ which will be processed afterwards. Now the reader can understand that a "bad" solution s_3 in $\mathcal{S}_3$ replaced with a starting point s_2 is characterized as follows: (1) it was not much reinforced by operator $\oplus$ through successive iterations and therefore it has a low support size; (2) s_2 scores over s_3 with regard to criterion $\mathscr{C}$. To sump up, diversification by generation of starting points and intensification in $\mathcal{E}$ are simultaneous processes iterated until a convergence criterion is satisfied for a solution s^*.

3.3. *Justification of the operators implemented and the criterion optimized*

To evaluate the pertinency of operator $\otimes$, we must check that it is unlikely that false positive solutions corresponding to local similarities might be retained in $\mathcal{S}_2$. If we call f_1 and f_2 the two frequencies involved in the formula $s_2[c,j] = \frac{n_{1_1} s_{1_1}[c,j_1] + n_{1_2} s_{1_2}[c,j_2]}{n_{1_1}+n_{1_2}}$, Figure 1 shows variations of f_2 *versus* f_1 for 4 values of f and different values of ratio $\frac{n_{1_1}}{n_{1_2}}$ under the constraint $\frac{n_{1_1} f_1 + n_{1_2} f_2}{n_{1_1}+n_{1_2}} \geq f$. The ratio values considered here are the 49 our

Algorithm 1 $Search(g, f, q)$

Input: a set S of n sequences; maximal gap length g; minimal frequency f; quorum q; sample size m.
Output: answer 'Yes'/'No'; if 'Yes', a mask $\mathcal{M}ask$, the corresponding *pssm* M, hits $hits$ for at least $q \times n$ sequences.
Search space: $\mathcal{E}$, the set of $pssm_s$ verifying constraint $\mathcal{G}(f, g)$
Solution sets: $\mathscr{S}_{1_1}$, $\mathscr{S}_{1_2}$, $\mathscr{S}_2$ *and* $\mathscr{S}_3$
Operators: $\otimes$ and $\oplus$: $\mathscr{P}(\mathcal{E}) \times \mathscr{P}(\mathcal{E}) \rightarrow \mathscr{P}(\mathcal{E})$ ($\mathscr{P}(E)$: set of all subsets of E)

$\mathscr{S}_3 \leftarrow \emptyset$
repeat
  Step 1: generate at random from S two samples of m sequences respectively, S_1 and S_2.
  identify at most n_1 best solutions $\mathscr{S}_{1_1}$ and at most n_1 best solutions $\mathscr{S}_{1_2}$ running plug-in algorithm LMA respectively on samples S_1 and S_2, under constraint $\mathcal{G}(f, g)$ and using criterion $\mathscr{C}$.
  Step 2: identify at most n_2 best solutions $\mathscr{S}_2$ from $\mathscr{S}_{1_1} \otimes \mathscr{S}_{1_2}$, under constraint $\mathcal{G}(f, g)$ and with respect to criterion $\mathscr{C}$.
  Step 3: update $\mathscr{S}_3$ with $\mathscr{S}_3 \oplus \mathscr{S}_2$ and possibly new starting points from $\mathscr{S}_2$, under constraint $\mathcal{G}(f, g)$ and using criterion $\mathscr{C}$.
until (convergence criterion is satisfied for a solution s^* in $\mathscr{S}_3$ verifying quorum constraint) or timeout is reached
if such a solution s^* exists **then** motifAssembling(s^*, $\mathscr{S}_3$); **return** 'Yes', $\mathcal{M}ask$, M, $hits$
return 'No'

implementation works with $(n_{1_1}, n_{1_2} \in [4, 10]$; $m = 10; minSeq = 4$, see definition 2.2). As intuitively expected, the probability that the frequency $f_1 = 25\%$ of an aleatory character might be compensated by a sufficiently high value of $f_{2min} = h(f_1) = (f - f_1)\,\frac{n_{1_1}}{n_{1_2}} + f$, to yield a false specific character, drastically decreases as f increases. We draw vertical lines $x = 25 \pm 5\%$ and horizontal lines $y = 25 \pm 5\%$ to delimit areas for frequencies near to 25%. Through $f = 60\%$ to $f = 90\%$, the two areas delimited intersect a decreasing number of lines $y = h(x)$.

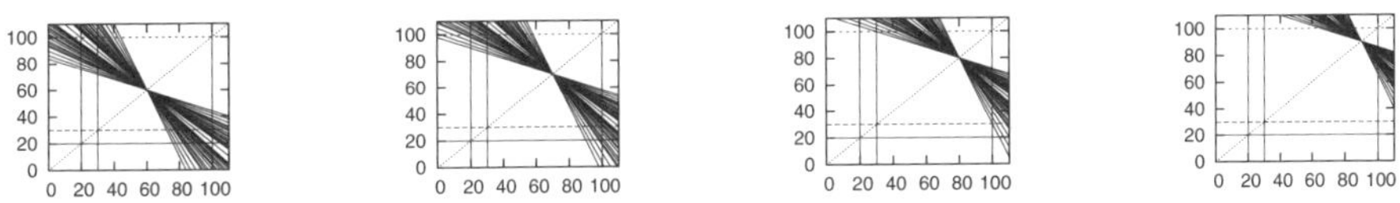

f = 60% f = 70% f = 80% f = 90%

Figure 1: Probability for reinforcement of a specific character c through operation $\otimes$, under various values for the threshold frequency f ($f \geq 60\%$). We draw the lines corresponding to $f_{2min} = h(f_1) = (f - f_1)\,\frac{n_{1_1}}{n_{1_2}} + f$ for 49 values of the ratio $\frac{n_{1_1}}{n_{1_2}}$; $n_{1_1}, n_{1_2} \in [4, 10]$.

Conclusion 3.1. For DNA sequences, if the threshold frequency f is greater than or equal to 70% and if all 49 possibilities for $\frac{n_{1_1}}{n_{1_2}}$ with n_{1_1}, $n_{1_2} \in [4, 10]$ are equiprobable, then the probability that an aleatory character in s_{1_1} compensates the same character in s_{1_2}, to yield a specific character in s_2, decreases from 0.45 to 0 (0.29 for $f = 72\%$; 0.12 for $f = 75\%$).

With this knowledge, we study the behaviour of operator $\otimes$ in the three cases: TP $\times$ TP, FP $\times$ TP, FP $\times$ FP, where FP denotes a false positive solution (a local similarity) and TP corresponds to a sub-optimal solution. The optimal solution is a similarity shared by $n \times q$ sequences at least and a TP only differs from the optimal solution by some false wild-card or specific characters. Table 3 in Appendix details our reasoning, which is based on conclusion 3.1. We draw the following conclusion from the comparison of columns 3, 3', 5 and 5' of Table 3 (see lines 3, 6 and 9):

Conclusion 3.2. Table 3 shows that crossing operands one of which at least is a false positive solution is likely to entail the generation of wild-card characters whereas specific characters are generated instead if both operands are true positive solutions.

Operator $\otimes$ implements shifts of a $pssm$ with respect to another $pssm$. If one of the operands at least is a FP, it is unlikely that many specific characters of both $pssm_s$ correspond. Were it the case for a pair of specific characters of both operands for a given shift, it is unlikely that these specific characters would be identical. So the cases mentioned at (6,2) and (9,2) in Table 3 are highly improbable. As a consequence of this and conclusion 3.2, conclusion 3.3 is stated as follows:

Conclusion 3.3. Crossing operands $(s_{1_1} \times s_{1_2})$ one of which at least is a false positive solution is likely to yield a $pssm$ whose mask has wild-card characters in the majority. The resulting $pssm$ will be rejected because of the presence of gaps at extremities or because they do not verify the maximal gap length constraint.

The criterion $\mathscr{C}$ can not be the usual log-likelihood ratio[1] because contrary to other LMA methods, our $pssm_s$ are **not** supported by the same numbers of sequences. Conclusion 3.3 gives a strong motivation to reward solutions with the following criterion: $\sum_{j=1,\ M^+[j]\geq f}^{l} M^+[j]$.

3.4. *Final processing*

3.4.1. *Motif assembling*

At line 8, among the solutions in $\mathscr{S}_3$, some may be false positive solutions while others may correspond to sub-regions of the final consensus motif, either strictly included in the sub-region corresponding to s^*, overlapping it or totally disjointed from it. Procedure *motifAssembling* chooses s^* as a first "reference" and refines it. Then it finds the solution in $\mathscr{S}_3$ having the greatest number of co-occurring hits with the reference and satisfying quorum constraint. This solution is refined in its turn and is used as the new reference. This process is repeated until no such reference can be identified. The procedure will be detailed in the extended version of this paper.

3.4.2. *Estimation of the motif statistical significance*

It is not a rule that all functional biological motifs should necessarily be less represented than other words in a dataset. Anyway, we must estimate the statistical significance of the motif predicted. We consider a motif of length l with ns specific characters. The probability that such a motif occurs by chance at least one time in each of n sequences of maximal size t, with at most d mismatches relative to the specific characters, is approximated by $\left(1-\left(1-\sum_{m=0}^{d}\binom{ns}{m}\left(\frac{|\mathscr{A}|-1}{|\mathscr{A}|}\right)^m\left(\frac{1}{|\mathscr{A}|}\right)^{ns-m}\right)^{t-l+1}\right)^n$.

3.5. *Complexity*

We chose MoDEL[7] as the plug-in LMA software for its rapidity. The complexity of an LMA performed with MoDEL is $O(m\,t\,w\,b)$ where w is the **chosen length** for the local alignment,

b is the intrinsic number of iterations for this method (default value is 45), m is the size of the sequence samples and t is the maximal length for the n sequences. We suppose that the convergence for procedure $Search(g,\ f,\ q)$ is obtained at u^{th} iteration. The complexity is then $O(u\ m\ t\ w\ b + \lambda^2\ {n_3}^2\ n)$ where λ is the maximal number of hits per sequence for any of the n_3 solutions in $\mathscr{S}_3$. The memory cost is $O(2\ n_1 + n_2 + n_3)\ (n\ \lambda + w)$ where n_1 and n_2 are the sizes of $\mathscr{S}_{1_1}$ and $\mathscr{S}_2$. Due to space limitation the corresponding proofs will be published in the extended version of the paper.

4. Experimental results

4.1. *Simulations*

Generating a consensus motif at random, we also generated $n \times q$ occurrences, blurred them with mismatches (in the specific regions only) and inserted them in n aleatory generated sequences. We compared the retrieved consensus motif with the real one and computed statistics: $cover$ (Is the whole motif retrieved?) and $exactness$ (Are there errors?) (not detailed here). Table 1 includes more details on the false wild-card or the false specific characters predicted. For some tests (b, c, d and e), we adapted to our concerns the so-called challenge motif problem stated by Pevzner and Sze[13]. The first conclusion to draw from Table 1 is that our method is accurate. The lowest average exactness is 0.92, obtained for motif a and c under rather hard conditions: $q = 70\%$; $f = 80\%$. The cover is quite satisfying too: it never drops below 0.90. On the other hand, we never encounter more than one false wild-card character or one false specific character on average, except for b_2 and c_2 (respectively 1.03 and 1.02 false wild-card characters).

A second benchmark was designed to study whether tuning the parameter g (maximal gap length) with different values alters the results. The benchmark consists of 50 sequences generated at random (length in [50, 300]) and containing CTACTxxxATCCTTGGGxxxxxxxxTCGTxxxAA ACTTGCTAGATTCAGGGAxxxGAC-GGAGGGTA whose longest gap has length 8. Successively tuning g to 8, 15 and 25 does not alter the quality of the ouputs obtained for 20 runs.

4.2. *Biological benchmarks*

Finally we ran Kaos on the biological datasets collected for STARS evaluation[17,18]. The motifs considered consist of (1) the Tata box TATAxATA in 131 sequences from various Dicot plants (32880 nucleotides, sequence length 251), (2) the binding site TGTAAxxxxxxT-TxAC for TyrR protein in *E. coli* genome (5 sequences, 3585 nucleotides, sequence length in [251-2021]) and (3) the binding site CTGTAxAxxxAxxCAG for LexA protein in *E. coli* genome (16 sequences, 28941 nucleotides, sequence length in [100-3842]). We compared MEME, STARS and 20 runs of Kaos under the constraints $q = 100\%$ and $f = 80\%$. To test our method with quorum $q = 70\%$ (and $f = 80\%$), we replaced 30% of the sequences in the initial datasets (1) and (3) with as many sequences of the same lengths chosen at random in the adequate genomes. Then we compared 20 runs of Kaos with MEME, which allows the retrieval of zero or one occurrence of the motif per sequence. We had to convert

MEME's outputs into structured motifs for comparison purpose. We compared the lengths of the motifs and the locations of false wild-card/specific characters for the motifs predicted by Kaos and any other algorithm. We conclude that in both cases ($q = 100\%$ and $q = 70\%$) Kaos is as efficient as MEME and STARS (see extended version).

Then we tested the behaviour of Kaos when the quorum decreases. We chose a difficult case: the binding sites for PurR protein (18 sequences, 44592 nucleotides) (see Table 2). From 90% to 70% all solutions found by Kaos are consistent with the real motif, with each other and with MEME's results. 60% is the quorum value below which MEME and Kaos retrieve a consensus depending on the sequences generated at random.

Table 1: Performances of Kaos with 2 quorum values and 2 frequency values for various motifs.

	(a1)	(a2)	(a3)	(a4)	(b1)	(b2)	(c1)	(c2)	(d1)	(d2)	(e1)	(e2)
q	100	70	100	70	100	70	100	70	100	70	100	70
f	100	100	80	80	80	80	80	80	80	80	80	80
co	0.97	0.97	0.91	0.90	1.0	1.0	1.0	0.97	0.92	0.93	0.91	0.95
ex	1.0	1.0	0.98	0.92	0.98	0.96	0.95	0.92	0.96	0.95	1.0	1.0
fw	0.49	0.5	0.58	0.57	0.5	1.03	0.2	1.02	0.37	0.47	0.11	0.3
fs	0.38	0.37	0.13	0.9	0	0.44	0.9	0.3	0.22	0.21	0.2	0.95

	n	t	l	ns	$q = 100\%$			$q = 70\%$		
					$d = 2$	$d = 4$	$d = 7$	$d = 2$	$d = 4$	$d = 7$
(a)	100	50	14	10	2.4 E-182	5.4 E-29	0.999	1.4 E-96	2.2 E-4	0.999
(a)	100	300	14	10	1.3 E-95	0.720	1.0	3.3 E-43	1.0	1.0
(b)	20	600	16	12		$d = 3$: 1.7 E-14			$d = 3$: 2.5 E-6	
(c)	20	600	18	14		$d = 4$: 1.4 E-15			$d = 4$: 5.1 E-7	
(d)	20	600	20	16		$d = 5$: 4.8 E-17			$d = 5$: 5.8 E-8	
(e)	20	600	22	18		$d = 6$: 9.1 E-19			$d = 6$: 4.3 E-9	

(a) GCGxxAAGCAxxCC (b) TGATxxTGAxxACGCC (c) TTTxCTCxCGxCCGxgag
(d) AATTTxxTCCTAGxTxTACG (e) CTTGGACxCGAxCCTCxxCGCC

Note: The maximal gap length g is set to 5. (a), (b), (c), (d) and (e) refer to various motifs. In each subcase (a_i), 100 sequences of lengths ranging from 50 to 300 have been generated under quorum $(q\%)$ and minimal frequency $(f\%)$ constraints. In cases (b), (c), (d) and (e), 20 sequences of length 600 have been generated. Average values for cover (co), exactness (ex) and number of false wild-card/specific characters (fw/fs) predicted have been computed for 100 runs. Bottom section: statistical significance. n is the number of sequences of size t, l is the motif length, ns is the number of specific characters, d is the maximal number of mismatches observed per occurrence. For motif (a) different values of d were encountered in the worst cases ($f = 70\%$).

Table 2: Robustness of Kaos with respect to quorum q; comparison with other methods.

q	100%	90%	80%	70%
MEME	ACGCAAACGTTTGCGTT	see 100%	ACGCAAACGTTTGCGT	ACGCAAACGTTTACGTT
MEME(∗)	AxGCAAACGxTTxCxT	see 100%	AxGCAAACGTTTxCxT	AxGCAAACGTTTACTT (8) AxGCAAACGTTTxCGT (3) AxGCAAACGTTTCxT (9)
STARS	GCxAxCGTTTTC			
Kaos	GxAAxCGxTTxC (7) AxGxAAACGTTTxCxT (13)†	AxGxAAxCGxTTxC (10) AxGxAAACGxTTxC (1) CGCAAxCGxTTxC (9)	GxAAxCGxTTxCxT (12) CGxAAACGTTTxCxT (8)	AxGxAAACGxxTxC (6) ‡ CAAAxGTxTCxGT (7) CxAACGTxTTxGT (7)

Note: $f = 80\%$. The maximal gap length is 5. The dataset contains sequences with PurR protein binding sites and random sequences. The known motif is AxGxAAxCGxTTxCxT. Sequences have lengths in range [299-5864], with average 2477. (*) For an easier comparison, the ouputs of MEME have been converted into structured motifs. The numbers in brackets indicate how many runs over 20 output the corresponding result. Statistical significance: † 1.2 E-69; $d = 0$ ‡ 0.999; $d = 4$. d is the maximal number of mismatches observed per occurrence.

5. Conclusion

We presented a novel method for SLMA under minimal specific character frequency, maximal gap length and quorum constraints. *pssm* convergence is obtained in an original way: strengthening (literally *merging*) the best candidates satisfying frequency and gap constraints, which definitely distinguishes our algorithm from all known methods. Kaos is robust with regard to the following criteria: maximal gap length specified with too high a value, parameters q and f. Besides, our method has a low memory cost. Finally, our software is easy to use since only three input parameters are required and the maximal gap length may be overestimated. Our results show that it is worth doing future work. The next step will consist in examining which parts of the algorithm may be speeded up. A more thorough examination of the choice for scoring functions is also one of our future tasks.

Acknowledgements

We wish to thank A. Mancheron for kindly putting at our disposal the datasets he collected. Thanks are also due to D. Hernandez for providing the beta version of the MoDEL software.

Appendix

Table 3: Study of the behaviour of operator $\otimes$ *w.r.t.* the type of operands (FP or TP, see text, subsection 3.3).

			1	2	3	3'	4	4'	5	5'	6	6'
		s_{opt}	x	c	c_1	c_1	x	x	c	c	x	x
		s_{1_1}	x	c	c_1	c_2	c_1	c_2	c	x	c	x
		s_{1_2}	x	c	c_2	c_1	c_2	c_1	x	c	x	c
1	TP×TP	s_{1_1}			$1c_1$	$1c_2$	c_1	c_2	c	$c+$	c	c eq
					$2c_2$	$2c_1$	$c_2?$	$c_1?$				
2		s_{1_2}			$1c_2$	$1c_1$	c_2	c_1	$c+$	c	c eq	c
3		s_2	x	c	c_1	c_1	x	x	c	c	x	x
4	TP×FP	s_{1_1}			$1c_1$	$1c_2$	$1c_1$	see 4	c	$c+$	c	c eq
					$2c_2?$	$2c_1?$	$2c_2?$					
5		s_{1_2}			$1c_2$	$1c_1$	$1c_2$	see 4	$c?$	c	$c?$	c
6		s_2	x	c	x	x	x	x	x	c	x	x
7	FP×FP	s_{1_1}			c_1	see 3	see 3	see 3	c	x	see 5	see 5
					$c_2?$							
8		s_{1_2}			c_2				x	c		
9		s_2	x	c	x				x	x		

Note: This table relies on conclusion 3.1. The operands are s_{1_1} and s_{1_2}. The elementary operator for $\otimes$ is denoted $\times$. We give the most probable result for $s_2 = s_{1_1} \times s_{1_2}$, considering the 2 contributing columns of s_{1_1} and s_{1_2} and the resulting column for s_2. x denotes a wild-card character in the masks of the $pssm_s$ considered. c, c_1 and c_2 denote specific characters. s_{opt} is the optimal solution. Indicating the character of s_{opt} is obviously only of concern for operations involving TP$_s$. Notations: $'1\ c_1'$ recalls that c_1 **has** the highest frequency rank in the column of the $pssm$ considered; $'2\ c_2'$ means that c_2 is **likely** to have the second frequency rank; $'c_1+'$ denotes the character c_1 is likely to have a high frequency rank; $'c_1?'$ means that there is no possible guess as to what the rank for c_1 can be; 'c_1 eq' means that c_1 has a frequency averaging $\frac{1}{|\mathcal{A}|}$ as all equiprobable characters. We comment line TP×TP and columns 3 and 3' to show how to read this table: if one of the TP$_s$ has c_1 as a specific character, corresponding to the specific character of s_{opt}, and the other TP has c_2 as a specific character, since both are TP$_s$, it is likely that c_1 appears with the second frequency rank for the second TP. The cases at lines and columns (4-6,5) and (4-6,5') are not symmetrical. c may be encountered in a FP with either a low or high probability (though $< f$ since the corresponding character of the mask is a wild-card character). Thus the notation $'c?'$ is adequate. On the contrary, there is a high probability that the character c is high-ranked if s_{1_1} has a false wild-card character and s_{opt} has character c in its mask. In this case, the adequate notation is $c+$.

References

1. T.L. Bailey, Likelihood vs. information in aligning biopolymer sequences, USCD technical report cs93-318. University of California, San Diego, 1993.
2. T.L. Bailey, C. Elkan, Unsupervised learning of multiple motifs in biopolymers using expectation maximization. *Machine learning*, 21, 51–80, 1995.
3. B. Brejová, C. DiMarco, T. Vinar, S.R. Hidalgo, G. Hoguin, C. Patten, Finding patterns in biological sequences. Tech. Rep. CS798g, University of Waterloo, 2000.
4. J. Buhler, M. Tompa, Finding motifs using random projections. In *Proceedings of the Fifth International Conference on Computational Molecular Biology (RECOMB)*, 69–76, Montréal, Canada, ACM Press, apr, 2001.
5. A. Califano, SPLASH: Structural pattern localization analysis by sequential histograms. *Bioinformatics*, 16(4), 341–357, 2000.
6. A.M. Carvalho, A.T. Freitas, A.L. Oliveira, M.-F. Sagot, A highly scalable algorithm for the extraction of cis-regulatory regions. In Yi-Ping Phoebe Chen and Limsoon Wong, ed., *Proceedings of the 3rd Asia Pacifi c Bioinformatics Conference (APBC)*, volume 1 of *Advances in Bioinformatics and Computational Biology*, Imperial College Press, 273–282, 2005.
7. D. Hernandez, R. Gras, R. Appel, MoDEL: an efficient strategy for ungapped local multiple alignment. *Computational Biology and Chemistry*, 28, 2, 119-128, apr, 2004.
8. G. Hertz, G. Stormo, Identifying DNA and protein patterns with statistically significant alignments of multiple sequences. *Bioinformatics*, 15, 563–577, 1999.
9. I. Jonassen, Efficient discovery of conserved patterns using a pattern graph. *Computer Applications in the Biosciences*, 13, 509-522, 1997.
10. C.E. Lawrence, S.F. Altschul, M.S. Boguski, J.S. Liu, A.F. Neuwald, J.C. Wootton, Detecting subtle sequence signals: a Gibbs sampling strategy for multiple alignment. *Science*, 262, 208–214, oct, 1993.
11. A. Mancheron, I. Rusu, Pattern discovery allowing gaps, substitution matrices and multiple score functions. In *Proceedings of the Third Workshop of Algorithms in Bioinformatics (WABI)*, 2812, 129–145, Budapest, Hungary, Springer-Verlag, LNBI, sep, 2003.
12. L. Marsan, M.-F. Sagot, Algorithms for extracting structured motifs using a suffix tree with application to promoter and regulatory site consensus identification. *Journal of Computational Biology*, 7, 345–360, 2000.
13. P.A. Pevzner, S.-H. Sze, Combinatorial algorithm for finding subtle signals in DNA sequences. In *Proceedings of the Eighth International Conference on Intelligent Systems for Molecular Biology (ISMB)*, 269–278, San Diego, California, aug, 2000.
14. I. Rigoutsos, A. Floratos, Combinatorial pattern discovery in biological sequences: The TEIRESIAS algorithm. *Bioinformatics*, 14(1), 55–67, 1998.
15. T.D. Schneider, G.D. Stormo, L. Gold, and A. Ehrenfeucht, Information content of binding sites on nucleotide sequences. *J. Mol. Biol.*, 188, 415–431, 1986.
16. W. Thompson, E.C. Rouchka, C.E. Lawrence, Gibbs Recursive Sampler: finding transciption factor binding sites. *Nucleic Acids Research*, 31, 13, 3580–3585, 2003.
17. http://www.softberry.com/berry.phtml
18. http://arep.med.harvard.edu/ecoli_matrices/

A RANDOMIZED ALGORITHM FOR LEARNING MAHALANOBIS METRICS: APPLICATION TO CLASSIFICATION AND REGRESSION OF BIOLOGICAL DATA

C. J. LANGMEAD*

Department of Computer Science, Department of Biological Sciences,
Carnegie Mellon University, Wean Hall 4212,
5000 Forbes Ave.,
Pittsburgh, PA 15213, USA
E-mail:cjl@cs.cmu.edu

We present a randomized algorithm for semi-supervised learning of Mahalanobis metrics over $\mathbb{R}^n$. The inputs to the algorithm are a set, U, of unlabeled points in $\mathbb{R}^n$, a set of pairs of points, $S = \{(x, y)_i\}; x, y \in U$, that are known to be similar, and a set of pairs of points, $D = \{(x, y)_i\}; x, y \in U$, that are known to be dissimilar. The algorithm randomly samples S, D, and m-dimensional subspaces of $\mathbb{R}^n$ and learns a metric for each subspace. The metric over $\mathbb{R}^n$ is a linear combination of the subspace metrics. The randomization addresses issues of efficiency and overfitting. Extensions of the algorithm to learning non-linear metrics via kernels, and as a pre-processing step for dimensionality reduction are discussed. The new method is demonstrated on a regression problem (structure-based chemical shift prediction) and a classification problem (predicting clinical outcomes for immunomodulatory strategies for treating severe sepsis).

1. Introduction

Many classification, clustering, and regression algorithms depend on a metric over the input space. For example, k-means clustering, k-nearest-neighbor classification/regression, radial basis function networks, and kernel methods, such as SVMs, need to be given good metrics that accurately reflect the important relationships between points. Many common distance metrics, such as the Euclidean metric, assume that each feature is not only independent of the others, but also equally important. Both assumptions are routinely violated in real-world applications. To address this problem, a number of (semi)supervised distance metric learning algorithms have been proposed (e.g., [15, 16]). Here, the user supplies a set of pairs of instances that are known to be similar and/or dissimilar to each other; the distance metric learning algorithm finds a metric that respects these relationships. In contrast to techniques like Multidimensional Scaling [6] and Principal Components Analysis (PCA) [9], which simply find an embedding of the test points, a distance metric learning algorithm learns a true metric, that can be incorporated into *any* learning algorithm that uses metrics, pseudo-metrics, or (dis)similarity measures.

This paper introduces a randomized approach to metric learning. In particular, given any algorithm, λ, for learning a Mahalanobis metric, M, the new method can construct a new metric $\hat{M}$, by calling λ as a subroutine on samples of the training data and subspaces of $\mathbb{R}^n$. The randomization serves two purposes: (1) The computational complexity of metric learning algorithms generally depend on the number of training samples, z, and the dimensionality of the input space, n. The new algorithm builds a metric by combining a constant number of metrics over random subspaces of $\mathbb{R}^n$, each trained on a small sample

*This work is supported the Young Pioneer Award from the Pittsburgh Life Sciences Greenhouse

of the training data. While our algorithm has the same computational complexity as λ, in practice, the decrease in training times is dramatic. Moreover, the subspace metrics can be trained in parallel, yielding another constant factor speed-up. (2) In the language of statistical learning theory [e.g., 8], random sampling of training instances and features reduces variance and thus guards against overfitting. Our approach to metric learning is influenced by the Random Forest algorithm [2] which uses similar strategies for classification and regression, but is not a technique for learning metrics.

We demonstrate the accuracy and efficiency of the new algorithm for representative problems in biology and medicine. In particular we construct distance metrics over (i) nuclear electronic environments and (ii) clinical data from patients with severe sepsis. We then use these metrics in k-nearest neighbor regression and classification, respectively. Our results demonstrate that the randomized algorithm is not only much faster than the deterministic algorithm (λ), but often produces metrics that improve the accuracy of k-nearest neighbor regression and classification. This indicates that the random algorithm resists overfitting and produces metrics that better generalize to novel instances.

2. Distance Metric Learning

Consider a set of points, $U = \{x_i\}_{i=1}^{l} \subseteq \mathbb{R}^n$, and a distance metric of the form

$$d_M(x_i, x_j) = \|x_i - x_j\| = \sqrt{(x_i - x_j)^T \mathbf{M}(x_i - x_j)}, \tag{1}$$

where $\mathbf{M}$ is a symmetric positive semi-definite $n \times n$ matrix. If $\mathbf{M}$ is an identity matrix, then Eq. (1) is simply the un-weighted Euclidean distance. If $\mathbf{M}$ is merely diagonal, then Eq. (1) is a diagonally weighted Euclidean distance. More generally, $\mathbf{M}$ parameterizes a family of Mahalanobis distances over $\mathbb{R}^n$ and encodes the weights and correlations among variables.

The distance metric learning problem is to learn $\mathbf{M}$ given U and a set, $S = \{(x, y)_i\}; x, y \in U$. The set S represents pairs of points in U that a domain expert has asserted are "similar", but has not defined precisely how or why they are similar. Optionally, the domain expert may also provide a set of dissimilar pairs of points, $D = \{(x, y)_i\}; x, y \in U; S \cap D = \{\}$.

There are a number of approaches to learning $\mathbf{M}$, given U, S, and, D. Xing *et al* [16] pose the following optimization problem:

$$\hat{M} = \underset{\mathbf{M}}{\text{argmin}} \sum_{(x_i, x_j) \in S} \| x_i - x_j \|_{\mathbf{M}}^2 \tag{2}$$

$$s.t. \quad \sum_{(x_i, x_j) \in D} \| x_i - x_j \|_{\mathbf{M}} \geq \alpha \tag{3}$$

$$\mathbf{M} \succeq 0, \tag{4}$$

where α is an arbitrary constant ≥ 1. The constraints are convex and the objective is linear in $\mathbf{M}$, thus Eq. 2 can be solved using convex-optimization techniques. Xing *et al* introduce two iterative algorithms solving for $\hat{M}$. The first algorithm deals with the case when $\mathbf{M}$ is diagonal, the second algorithm solves the problem when $\mathbf{M}$ is an arbitrary positive semi-definite matrix. Tsang and Kwok [15] pose a different optimization problem (not shown here due to space considerations) and derive a quadratic programming problem using the technique of Lagrangian multipliers. The convex programming formulation of the problem has no user parameters (other than the sizes of S and D), but requires an iterative solution. The quadratic programming formulation of the problem can be solved more efficiently but has several user parameters.

Both approaches learn a metric over the entire input space. This raises two possible concerns: computational complexity and overfitting. The complexity of [16] is $O(I^2n^3)$, where I^2 reflects the nested iterations required to converge, and the n^3 term is the time needed to diagonalize an $n \times n$ matrix; the complexity of [15] is $O(z^3+n^2)$, where $z = |D|$ and the z^3 term is the time to solve a quadratic programming problem with z variables. The second concern is overfitting; if the sizes of S and D are small, it becomes difficult to estimate the parameters of the metric. Similarly, if the input features are particularly noisy, spurious correlations among features may unduly influence the parameter estimates. Both issues can be addressed using randomization.

3. Algorithm

Our algorithm is presented in Algorithm 1. Briefly, the sets S and D are randomly sampled (with replacement) to construct sets S^b and D^b, where b is the number of samples in the subsets. Next, S^b and D^b are modified using the function `Random-Features`(S^b, D^b, m), where m is the dimensionality of the subspace ($m < n$). The resulting sets, S^b_m and D^b_m, contain points in the same random m-dimensional subspace of $\mathbb{R}^n$. The function `Random-Features` also returns a vector, **f**, whose components encode which dimensions were sampled. S^b_m and D^b_m comprise a subspace training set. This training set is passed to the given metric-learning algorithm, λ, which returns a metric, $\mathbf{M}^f$, over the random subspace. $\mathbf{M}^f$ is, by definition, an $m \times m$ symmetric positive semi-definite matrix. The choice of λ dictates the properties of the subspace metric. For example, the algorithms presented in [16] and [15] are guaranteed to converge to an optimal solution for a given training set.

The next step is to update the matrix encoding the metric over $\mathbb{R}^n$. The matrix $\hat{\mathbf{M}}$ is initialized as a $n \times n$ identity matrix. Thus, $\hat{\mathbf{M}}$ is also a symmetric positive semi-definite matrix. Recall that the components of vector **f** indicate which dimensions of $\mathbb{R}^n$ were sampled. That is, the components of **f** correspond to rows and columns of $\hat{\mathbf{M}}$. Consider the auxiliary $n \times n$ matrix, $A : A(\mathbf{f}(i), \mathbf{f}(j)) = \mathbf{M}^f(i,j); \forall\, i, j \leq m$. The function `update`$(\hat{\mathbf{M}}, \mathbf{M}^f, \mathbf{f})$ returns the matrix $\hat{\mathbf{M}} + A$. This whole procedure is repeated c times, with different random samples of the training data and features. It can be shown that $\hat{\mathbf{M}} + A$ is a symmetric positive semi-definite matrix. That is, the symmetric, positive semi-definiteness of $\hat{\mathbf{M}}$ is an invariant property of the algorithm. Thus, $\hat{\mathbf{M}}$ satisfies all the properties of a metric.

The computational complexity of the algorithm is a function of the size of U, the parameters c, b, and m, and the complexity of the underlying metric learning algorithm, λ. Let $z = |U|$. The parameters c and b are generally set so that the product $cb = O(z)$. The parameter m is generally set so that the product $cm = O(n)$. Let $g_\lambda(b, m)$ be the computational complexity of λ on b instances of m-dimensional data. Note that $g_\lambda(b, m)$ varies

Algorithm 1 Random Metric Learning Algorithm pseudocode. See text for details.

Input:

U: The unlabeled set of points in $\mathbb{R}^n$
$S = \{(x, y)_i\}; x, y \in U$: a set of pairs of similar points
$D = \{(x, y)_i\}; x, y \in U$: a set of pairs of dissimilar points (*optional*)
λ: an algorithm for learning metrics
c: the number of subspace metrics to learn
b: the number of subspace training samples
m: the dimensionality of the subspace

/* *Initialize the metric* */
$\hat{\mathbf{M}}_{n\times n} \leftarrow \mathbf{I}_{n\times n}$: an $n \times n$ identity matrix

for $i = 1$ to c **do**
  /* *Select random training instances* */
  $S^b \leftarrow$ `Random-Sample` (S, b)
  $D^b \leftarrow$ `Random-Sample` (D, b)

  /* *Select a random subspace* */
  $S^b_m, D^b_m, f \leftarrow$ `Random-Feature` (S^b, D^b, m)

  /* *Learn the subspace metric* */
  $\mathbf{M}^f_{m\times m} \leftarrow \lambda(S^b_m, D^b_m)$

  /* *Update the complete metric* */
  $\hat{\mathbf{M}}_{n\times n} \leftarrow$ `update` $(\hat{\mathbf{M}}_{n\times n}, \mathbf{M}^f_{m\times m}, f)$
end for
return $\hat{\mathbf{M}}$

by the choice of learning algorithm but is trivially lower-bounded by $\Omega(max(b^2, m^2))$ since the $m \times m$ matrix $\mathbf{M}$ must be explicitly realized and b^2 pairwise similarities must be computed. The complexity of our algorithm is $O(cn^2 + cg_\lambda(b, m))$. In practice, we set $b \approx z/10$, $m \approx n/2$, and $c \geq 10$. Thus, c is a constant and since $b = O(z)$ and $m = O(n)$, our algorithm has the same complexity as λ which, effectively, has $c = 1$, $b = z$ and $m = n$. However, in practice, the randomized algorithm is much faster because b and m are significantly smaller than z and n, respectively. Finally, note that each subspace metric can be trained in parallel giving a constant factor speedup for p processors (i.e., $O(cn^2 + \lceil \frac{c}{p} \rceil g_\lambda(b, m))$).

An important feature of the algorithm is that the c subspace metrics are trained on different samples and different features. Each subspace metric is optimal for that subspace and the data used to construct it. The global metric is *not* optimal for U, despite being built from optimal subspace metrics. However, the random algorithm can be thought of as an *ensemble* method for metric learning. Ensemble methods in machine learning are well studied and have a number of desirable properties (see, e.g., [7]). In particular, ensemble methods often generalize better to novel instances than non-ensemble methods. Briefly, if the members of the ensemble are diverse, that is, not highly correlated in their erroneous predictions, then it becomes very unlikely that two "bad" predictors can conspire to affect the outcome. Conversely, two or more "good" predictors have a better chance of affecting the outcome. In our algorithm, diversity is enforced by sampling both the training instances and the features. We present empirical evidence in Sec. 5 that demonstrates that the randomized algorithm always produces metrics that are as good as, and often better than, the metrics produced using deterministic algorithms.

4. Experiments

We tested our metric learning algorithm in the contexts of regression and classification problems. The data for each experiment were split into randomly selected, non-overlapping training and test sets. The training set was used to learn the metric over the feature space. That metric was then used to compute the distance of each test instance to the training instances. The predicted value (or class) for each test instance was then computed using k-nearest neighbor regression (or classification). A variety of parameter settings (i.e., b, c, m, z) were tested. The entire procedure was repeated 20 times to obtain error estimates for a given configuration of parameters.

The base metric learning algorithm (i.e., λ) in our experiments was the technique of Xing *et al* [16]. We will refer to our algorithm as λ_r for the purpose of comparison with λ. Both algorithms were implemented in Matlab and the experiments were conducted on a 3 GHz Pentium 4 workstation running Linux. We did not use the parallel training option for λ_r. Finally, λ_r and λ were trained and tested on the same training and test sets, respectively, to ensure a fair comparison.

4.1. *Chemical Shift Prediction*

The first study comprised 5 separate experiments for predicting the real-valued chemical shifts for backbone nuclei (H^N, H^α, $^{13}C^\alpha$, $^{13}C'$, ^{15}N) given a structural model of the protein.

4.1.1 Background

Chemical shifts are the most fundamental of spectral parameters in Nuclear Magnetic Resonance (NMR) spectroscopy. Briefly, a spin $I = 1/2$ nucleus in an external magnetic field will have two spin states whose energy difference is linearly proportional to the strength of the applied magnetic field. Each nucleus, however, is influenced by the electrons in its vicinity. Relevant factors include covalent bonding, the orientations of nearby aromatic rings, hydrogen bonding, solvent interactions, and ionization constants. Thus, each nucleus feels a slightly different field because it has a different local electronic environment. The experimentally measured chemical shift for a given nucleus is proportional to the field felt by that nucleus.

Chemical shifts can most accurately be predicted from structure models using quantum mechanics. A purely quantum mechanical approach to chemical shift prediction involves determining the wave function of the molecule; this is computationally intractable for large molecules, like proteins. Consequently, hybrid methods for chemical shift prediction are used in practice; these methods combine quantum calculations for local interactions and either classical or empirical calculations for long-range interactions. There are a number of practical and important applications of structure-based chemical shift prediction including: resonance assignment (e.g. [10]), structure determination and model refinement (e.g., [3, 5]), high-throughput fold recognition (e.g. [11, 13]), as well as a variety of assays for probing mutations and structure-activity relationships.

4.1.2 Features and Training/Test Data

The regression model for these experiments included variables (i.e., features) that represent quantum and empirical factors. These contributions were computed in the manner of [17] and include the effects of torsion angles, residue type, ring-currents, electrostatics, and hydrogen bonding. The experimental data for these experiments were obtained from the REFDB [18] database. REFDB is a carefully re-referenced subset of the BioMagResBank (BMRB) [14] that corrects some systematic errors within the BMRB. The REFDB also includes a mapping to specific Protein Data Bank [1] (PDB) IDs; these structures were downloaded from the PDB and used to compute the quantum and empirical features.

The resulting data set contained between 20,000 and 47,000 instances, depending on nucleus type, across 454 different proteins[a]. Each instance comprised an experimentally measured chemical shift (re-referenced, as needed), and a feature vector containing the quantum and empirical calculations. The various nuclei types are sensitive to different affects, thus the specific feature-vector size varied per nucleus type; the nuclei H^N, H^α, $^{13}C^\alpha$, ^{13}C', and ^{15}N had feature vectors of size 5, 7, 6, 7, and 9, respectively. The goal is to learn a metric over these features.

4.2. *Patient Outcome*

In a different experiment, we examined the performance of our algorithm in a classification task. In particular, we used a data set created by Clermont *et al* [4] for *in silico* modeling of immunomodulatory interventions for severe sepsis. The goal here is to predict an outcome for each of 1000 simulated patients. The four possible outcomes are (i) helped by intervention; (ii) harmed by intervention; (iii) lives regardless of intervention; (iv) dies regardless of intervention. The input space consists of 26 features that include the levels, ratios, and products of a number of biomarkers at the time of the disease detection and one hour following detection. The Clermont data set is divided into separate training and test sets. The partitions between these two sets were maintained during our experiments.

5. Results

We first compared the rates of increase in training times for the random and deterministic algorithms. Figure 1(a) plots the accuracy and the number of seconds required to train λ and λ_r for $z = 100, 200, ..., 1000$ samples on the patient outcome data set. The test set consisted of 200 randomly selected instances and the classifications were made with $k = 1$ nearest neighbor classification. Similar results are obtained for the other data sets. The

[a]RefDB actually contains data from 601 proteins. We used a subset of proteins that did not include protein complexes.

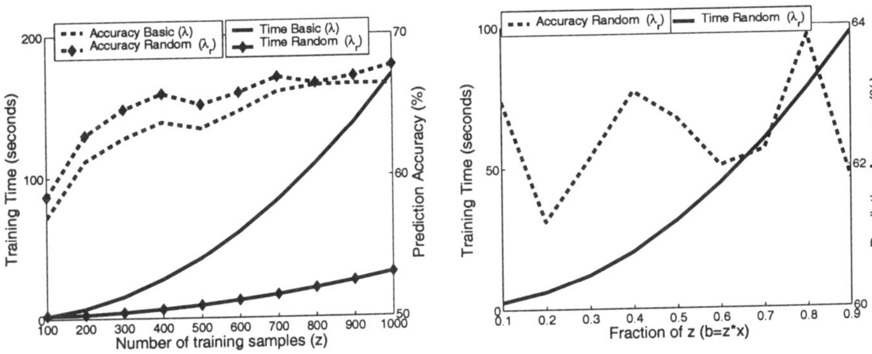

(a) (b)

Figure 1. Left: Training times (solid lines) and accuracies (dashed lines) for λ and λ_r for varying z. Right: Training times (solid line) and accuracies (dashed line) for λ_r for varying b. Note that left- and right hand axis scales are not the same in the two figures.

parameters for λ_r were $c = 20, b = z/10$, and $m = n/2$. Recall that the parameters for λ are, effectively, $c = 1, b = z$, and $m = n$. The rates of increase in training time are dramatically different, as expected. The accuracies under both metric increase with the size of z. We note that the accuracies under the λ_r metric are either statistically identical or *higher* than those under the λ metric. That is, λ_r provides the same level of accuracy, or better, than λ while drastically reducing training times. The accuracies under the λ_r metric were statistically higher than those under the λ metric for $z = 300, 400, 500, 1000$. These results are consistent with the notion that randomization may be an effective means for resisting overfitting a given training set.

Figures 1(a), 2(a), and 2(b) plot the accuracy and training times for λ_r various values of b, c, and m, respectively, while fixing the remaining variables ($z = 200, c = 20, b = z/10$, and $m = n/2$). Training times increase dramatically with increasing b, but there is no statistical difference in prediction accuracy between $b = z \times 0.1$ and $b = z \times 0.9$, suggesting that small values of b are sufficient. Similarly, training times increase linearly with increasing c, but there is no statistical difference in accuracy between $c = 10$ and $c = 100$. Training times rise with increasing m, but not as quickly as for increasing z, c, or b. Interestingly, there is a statistically significant *decrease* in accuracy ($p < 0.001$) as m increases. This result further supports the hypothesis that randomization can lead to better metrics.

Finally, we evaluated the accuracy of the random algorithm on larger test sets. In these experiments the parameters of the random algorithm were $z = 1000, c = 20, b = z/10$, and $m = n/2$. Using the complete training and test sets for the patient outcome data (1,000 instances each), the classification accuracy for $k = 1$ nearest neighbor was 73% using the λ_r metric versus 68% for using the λ metric. These results are in agreement with the smaller test cases, described above. However, Clermont *et al* were able to obtain an 84% percent accuracy on the same training/test data using logistic regression [4]. We then decided to test the performance of our method in the context of logistic regression. We first constructed a classifier from the training data using simple logistic regression. The classification accuracy on the test data was 83% — very similar to the results obtained by Clermont *et al.* Next we considered projecting the training data into the learned metric space. Note

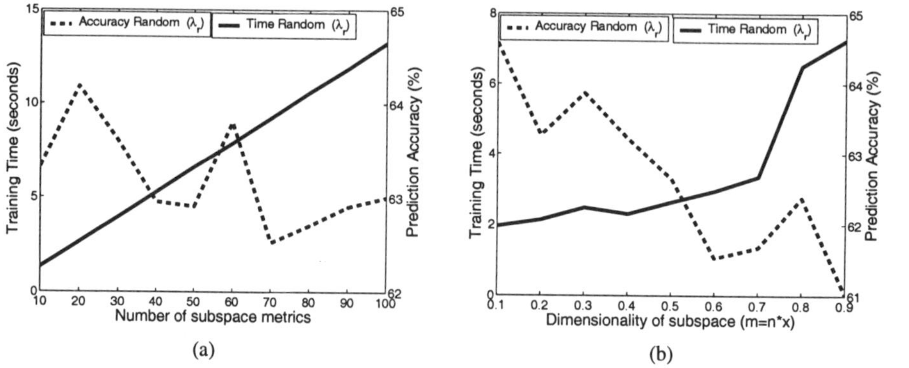

Figure 2. Left: Training times (solid line) and accuracies (dashed line) for λ_r for varying c. Right: Training times (solid line) and accuracies (dashed line) for λ_r for varying m. Note that left- and right hand axis scales are not the same in the two figures.

Table 1. **Chemical Shift Prediction:** Accuracy and summary of training and test data for 5 different nuclei. Column 2: the total number of instances for a given nucleus; 1,000 randomly selected instances were selected for training, the remaining were used to test the learned metric using $k = 5$ nearest neighbor regression. Columns 3-4, the mean shift value, and range (in ppm). Columns 5-7, the root mean squared error (in ppm) for the program SHIFTS and k-NN under the λ metric and under the λ_r metric, respectively. Errors are estimated based 20 random partitions into training and test sets.

				SHIFTS	λ Metric	λ_r Metric
Nucleus	Instances	Mean Shift	Range	RMSE	RMSE	RMSE
H^N	47,401	8.29	5.44	0.647	0.611	**0.607**
H^α	39,009	4.41	4.19	1.13	0.340	**0.336**
^{15}N	34,196	119.62	35.18	5.16	3.87	**3.87**
$^{13}C^\alpha$	30,174	58.10	26.46	1.74	1.67	**1.67**
$^{13}C'$	19,877	176.15	15.40	1.88	1.54	**1.53**

that because the matrix $\mathbf{M}$ is positive semi-definite, we can write $\mathbf{M}$ as $\mathbf{AA}^T$. The matrix $\mathbf{A}$ can be thought of as a projection into a different feature space. Using the metric learned over the training data, we projected the training data into the corresponding feature space. We then constructed a different classifier on the transformed training data using simple logistic regression. We then projected the test data into the same space and performed the classification. The resulting accuracy was 87%. Thus, the projection into the new space does confer advantages in the context of logistic regression. The accuracy increases further to 90% when we use a logistic regression algorithm with a ridge-estimator[12]. Here we also noticed that the convergence of that algorithm is much faster on the transformed data.

The test sets for the chemical shift predictions contained between 19,000 to 46,000 instances, depending on nucleus type. Accuracies are reported in terms of root mean squared error (RMSE) in *parts per million* (ppm), the standard units of chemical shifts. The RMSEs for the chemical shifts vary widely by nucleus type. This is due to the fact that the average magnitude and range of different nuclei varies widely. For example the H^α nucleus has an average chemical shift of ≈ 4.4 ppm and a range of ≈ 4.2 ppm; the ^{15}N nucleus, on the other hand, has an average chemical shift of ≈ 119.6 ppm and a range of ≈ 35.2 ppm. Table

1 summarizes the training sets and reports the RMSE for the program SHIFTS [17], and for $k = 5$ nearest neighbor regression under the λ and λ_r metrics. The predictions errors under the λ_r metric are statistically significantly lower than those of the SHIFTS program for all nuclei ($p \ll 0.01$). The predictions errors under the λ_r metric are statistically lower than the λ metric for ^{13}C', H$^{\alpha}$, and H^{N} ($p < 0.01$); the errors for ^{13}C$^{\alpha}$ and H$^{\alpha}$ are statistically identical under both metrics.

6. Discussion

The primary finding of these experiments is that the randomized algorithm produces metrics that are equivalent, and sometimes better, than those produced by a deterministic metric learning algorithm. The randomized algorithm, however, is much faster because the subspace metrics can be trained on smaller sets of data. Indeed, the size of the training set (U) has the greatest effect on both training time and accuracy. An interesting finding is that accuracy drops when m, the dimensionality of the subspace, increases. This suggests that the sampling of random subspaces may guard against overfitting.

There are a number of extensions to this work that are worthy of investigation. Xing *et al* have noted that it is also possible to consider non-linear metrics by introducing a non-linear feature map, ϕ,

$$d_M(x, y) = \sqrt{(\phi(x) - \phi(y))^T \mathbf{M}(\phi(x) - \phi(y))}. \tag{5}$$

This can also be thought of as learning a metric in a feature space associated with a kernel function $k(x_i, x_j) = \phi(x_i)\mathbf{M}\phi(x_j)$ [15]. The exploration of non-linear metrics is an interesting area for future research.

Another interesting area of investigation is the use of dimensionality reduction schemes. Standard techniques for dimensionality reduction (e.g., via PCA) can be applied either before or after the metric is learned. Dimensionality reduction prior to learning the metric will reduce training times, as demonstrated above. However, applying dimensionality reduction after learning the metric is also worthy of consideration. As previously mentioned, matrix $\mathbf{M}$ can be written as $\mathbf{M}$ as $\mathbf{A}\mathbf{A}^T$, and matrix $\mathbf{A}$ can be thought of as a projection into a different feature space. The potential advantage to doing the dimensionality reduction in the feature space is that the user implicitly defines the kinds of relationships that are important by constructing S and D. For some problems, there may be several kinds of relationships that are of interest. For example, the patient data defined "similar" in terms of mortality and the response to the intervention. But there may also be other kinds of similarity that may be of interest, such as comorbidity. Here, the expert can define a different S and D that reflect the similarity of interest; the same training data can be used, but a different metric will be learned. This additional flexibility is an attractive feature of any metric learning algorithm.

7. Conclusion

We have introduced a randomized algorithm for learning metrics over an input space. The new algorithm compares favorably with deterministic metric learning algorithms in terms of accuracy, generalization, and efficiency for both classification and regression tasks.

Acknowledgments

We would like to thank Dr. Gilles Clermont for use of his data and Dr. Eric Xing for use of his code for his metric learning algorithm.

References

1. Berman, H., Westbrook, J., Feng, Z., Gilliland, G., Bhat, T., Weissig, H., Shindyalov, I., and Bourne, P. The Protein Data Bank. *Nucl. Acids Res. 28* (2000), 235–242.
2. Breiman, L. Random forests. *Machine Learning 45*, 1 (2001), 5–32.
3. Case, D., Darden, T., Cheatham III, T., Simmerling, C., Wang, J., Duke, R., Luo, R., Merz, K., Wang, B., Pearlman, D., Crowley, M., Brozell, S., Tsui, V., Gohlke, H., Mongan, J., Hornak, V., Cui, G., Beroza, P., Schafmeister, C., Caldwell, J., Ross, W., and Kollman, P. AMBER 8. University of California, San Francisco, 2004.
4. Clermont, G., Bartels, J., Kumar, R., Constantine, G., Vodovotz, Y., and Chow, C. In silico design of clinical trials: A method coming of age. *Crit. Care Med. 32*, 10 (2004), 2061–2070.
5. Cornilescu, G., Delaglio, F., and Bax, A. Protein backbone angle restraints from searching a database for chemical shift and sequence homology. *J Biomol NMR 13*, 3 (1999), 289–302.
6. Cox, T., and Cox, M. *Multidimensional Scaling*. Chapman and Hall, 1994.
7. Dietterich, T. Ensemble methods in machine learning. *Proc. of the First International conference on Multiple Classifier Systems, Lecture Notes in Computer Science* (2000), 1–15.
8. Hastie, T., Tibshirani, R., and Friedman, J. *The Elements of Statistical Learning: Data Mining, Inference, and Prediction*. Springer-Verlag, 2001.
9. Jolliffe, I. *Principal Component Analysis*. Springer-Verlag, New York, 1989.
10. Langmead, C. J., and Donald, B. R. An Expectation/Maximization Nuclear Vector Replacement Algorithm for Automated NMR Resonance Assignments. *J. Biomol. NMR. 29*, 2 (2004), 111–138.
11. Langmead, C. J., and Donald, B. R. High-Throughput 3D Homology Detection via NMR Resonance Assignment. *Proc. IEEE Computer Society Bioinformatics Conference (CSB), Stanford University, Palo Alto, CA* (2004), 278–289.
12. le Cessie, S., and van Houwelingen, J. Ridge estimators in logistic regression. *Applied Statistics 41* (1992), 191 – 201.
13. Mielke, S., and Krishnan, V. Protein structural class identification directly from NMR spectra using averaged chemical shifts. *Bioinformatics 19*, 16 (2003), 2054–64.
14. Seavey, B., Farr, E., Westler, W., and Markley, J. A Relational Database for Sequence-Specific Protein NMR Data. *J. Biom. NMR 1* (1991), 217–236.
15. Tsang, I. W., and Kwok, J. Distance Metric Learning with Kernels. In *Proceedings of the International Conference on Artificial Neural Networks (ICANN)* (Cambridge, MA, June 2003), pp. 126–129.
16. Xing, E., Ng, A., Jordan, M., and Russell, S. Distance Metric Learning, with application to clustering with side-information. In *Advances in Neural Information Processing Systems 15* (Istanbul, Turkey, 2002), MIT Press.
17. Xu, X., and Case, D. Automated prediction of 15N, 13C‘alpha‘, 13C‘beta‘ and 13C’ chemical shifts in proteins using a density functional database. *J. Biomol. NMR 21* (2001), 321–333.
18. Zhang, H., Neal, S., and Wishart, D. A Database of Uniformly Referenced Protein Chemical Shifts. *J. Biomol. NMR 25*, 3 (2003), 173–195.

DISENTANGLING THE ROLE OF TETRANUCLEOTIDES IN THE SEQUENCE-DEPENDENCE OF DNA CONFORMATION: A MOLECULAR DYNAMICS APPROACH

MARCOS J. ARAÚZO-BRAVO

Department of Biosciences and Bioinformatics, Kyushu Institute of Technology, Iizuka, Fukuoka, 820-8502, Japan, E-mail: marara@bse.kyutech.ac.jp

SATOSHI FUJII

Department of Chemistry and Biochemistry, Kyushu University, Fukuoka, Japan, E-mail: fujii@takenaka.cstm.kyushu-u.ac.jp

HIDETOSHI KONO

Neutron Research Center and Center for Promotion of Computational Science and Engineering, Japan Atomic Energy Research Institute, 8-1, Umemidai, Kizu-cho, Soraku-gun, Kyoto, 619-0215 PRESTO, Japan Science and Technology Agency, 4-1-8, Honcho, Kawaguchi City, Saitama, 332-0012, Japan, E-mail: kono@apr.jaeri.go.jp

AKINORI SARAI

Department of Biosciences and Bioinformatics, Kyushu Institute of Technology, Iizuka, Fukuoka, 820-8502, Japan, E-mail: sarai@bse.kyutech.ac.jp

Sequence-dependence of DNA conformation plays an essential role in the protein-DNA recognition process during the regulation of gene expression. Proteins recognize specific DNA sequences not only directly through contact between bases and amino acids, but also indirectly through sequence-dependent conformation of DNA. To test to what extent the DNA sequence defines the DNA structure we analyzed the conformational space of all unique tetranucleotides The large quantity of data needed for this study was obtained by carrying out molecular dynamics simulations of dodecamer B-DNA structures. Separate simulations were performed for each of the possible 136 unique tetranucleotides at the dodecamer centers and the simulated trajectories were transformed into the DNA conformational space. This allowed us to explain the multimodal conformational state of some dinucleotides as aggregations of tetranucleotide conformational states that have such a dinucleotide inside their center. We proposed simple models to express in a linear way how the different bases that embrace a central dinucleotide perturb its conformational state, emphasizing how the conformational role of each base depends on its relative position (left, central, right) in the final tetranucleotide, and how the same peripherical base plays a different role depending on which is the central dinucleotide. These models allow us to establish an index to quantify the degree of context-dependence, observing an increasing context-dependence from the average base-pair step conformations AA/TT, CG, AC/GT (context-independent), AG/CT, AT, GC, GG/CC (weakly context-dependent), and GA/TC, CA/TG, TA (context-dependent).

1. Introduction

The idea that sequence defines DNA structure has gained acceptance, and thus the root of sequence dependent conformational variations has become an important problem. Results from crystallographic screens to address this problem indicate that variations from mean structural features may provide proteins with the information required for indirect read-out, and for specifying altered structures.[16] Coarse preditions of the DNA structure from nucleic sequence using knowledge-based techniques[2] are possible, but such an approach requires data of enough quantity and quality. To test to what extent the DNA structure is determined by its sequence we made a systematic analysis of an interaction range of 3 base-pair steps long —tetranucleotide— level. We analyzed the conformational space of the all the 136 unique tetranucleotides. Since in the current structure databases there are not enough data to perform a reliable statistical analysis over all the possible tetranucleotides, we generated the large quantity of data necessary for this study by Molecular Dynamics (MD) simulations. We tried to envisage the perturbations induced in every central dinucleotide conformational state by all the possible bases that embrace a central dinucleotide and to analyze the reasons for the multimodal conformational states underlined by several authors through the study of crystal structures and computational techniques.[13]

2. Methods

We have generated dodecamer B-DNA sequences 5'-CGCGW_lXYZ_rCGCG-3', where $\{W_l, X, Y, Z_r\} \in \mathcal{N} = \{\text{A,C,G,T}\}$. Each sequence has one of the 136 unique tetranucleotides at its center, and the terminals are always the CGCG tetranucleotide that gives higher stability to the ensemble. Initial DNA structures were built based on the Arnott B-DNA model[3] with the nucgen module in the AMBER packages 6 and 7.[14,8] Using the Leap module of the package, the initial DNA structures were solvated with the TIP3P water molecules[9] so that the DNA molecule could be covered with at least a 9 Å water-layer in each direction in a truncated octahedral unit cell. For the neutralization of the system, 22 K^+ ions were added at favorable positions and then 17 K^+ and 17 Cl^- ions were added so that the salt concentration of the system would be 0.15 M. First a 1000-steps minimization for water molecules and ions with fixed DNA structure was taken, followed by a further 2500-steps minimization for the entire system to remove the large strains in the system. The cutoff used for the van der Waals interactions was 9.0 Å. The particle mesh Ewald method (PME)[7] was used for calculating the full electrostatic energy of a unit cell. After the minimization, the entire system was linearly heated up from zero to 300 K with a weak harmonic restraint to the initial coordinates on DNA (10 kcal/mol) during 20 ps of MD simulation under NVT condition. Further, a 100 ps of molecular simulation was carried out, keeping the weak DNA restraint for the equilibration of the system under NPT condition at 300 K. MD simulation for each of the 136 unique sequences was then carried out to sample the DNA conformations for 2 ns with NPT condition. The temperature was controlled to be 300 K by Berendsen's algorithm[4] with a coupling time of 1 fs, which was set to be the same as the time step of the MD simulation to produce a canonical ensemble of DNA conformations.[11] The SHAKE algorithm[15] was used on bonds involving hydrogen.

The force field parameters used for the MD was from Wang *et al.* (parm99).[17]

A sampling period of 2 ns is not always enough time to reach the stationary state. For the case of the AATT and ACGA, 10 ns simulations were performed instead of 2 ns. Thus, we confirmed that 2 ns were enough to stabilize the AAT structure, but for the ACGA at least 5 ns were necessary. More MD are being carried out to optimize the sampling period for each one of the 136 different tetranucleotide structures. In all cases, to obtain the final ensemble, we used the last 1 ns trajectories, where the system was sampled at every 1 ps (1000 conformations).

To perform the conformational analysis, the DNA molecule was approximated as an elastic object, with 6 degrees of freedom θ_i within a fixed geometry of bases. The local conformation of the DNA was identified at each location of a base-pair (from complementary strands) in terms of known deformations such as base-pair step translations Shift, Slide, Rise, and base-pair step rotations Tilt, Rolls and Twist.[12,6] In the current analysis we use the conformational parameters of the central dinucleotide calculated with the program 3DNA.[10] Since symmetric properties exist, from all the possible 256 tetranucleotides a subset of 136 are unique. Similarly, from all the possible 16 dinucleotides only 10 are unique. Since the conformational coordinates are calculated using one of the DNA strands,[10] the Shift and Tilt coordinates of the other DNA strand are inverted for the symmetric steps. Then, special care should be taken in the case of Shift and Tilt conformational coordinates when dealing with symmetries.

In order to reproduce the dinucleotide conformational states from the tetranucleotide ones, the dinucleotide XY MD data are calculated as the union of all the tetranucleotides $W_l XY Z_r$ that have the dinucleotide XY at their center, $\{W_l, X, Y, Z_r\} \in \mathcal{N} =$ {A,C,G,T}

$$XY = \bigcup_{l}^{\mathcal{N}} \bigcup_{r}^{\mathcal{N}} W_l XY Z_r \tag{1}$$

3. Results

3.1. *Statistical Analysis of the Aggregation of Tetranucleotide Conformational States*

In order to study how the tetranucleotide conformational states aggregate to produce the dinucleotide ones, for each set of the 1000 states in which each one of the 136 unique tetranucleotides evolves in its MD simulated trajectory, we calculated the gravity center of each 6 base-pair conformational coordinates. Then we aggregated the tetranucleotide data that have the same central dinucleotide using Eq. (1). For the 6 conformational coordinates of the 10 dinucleotide aggregates we calculated the gravity center μ, the standard deviation σ (of the gravity center of the tetranucleotide set that forms the aggregate), the tetranucleotide Tet_{max} that induces the maximum perturbation Δ_{max}, where the perturbation is $\Delta = |\mu - \mu_{Tet}|$ (μ_{Tet} is the gravity center of the tetranucleotide *Tet*). All these values are summarized in Table 1.

At first glance, from an observation of the average values μ of the conformational state of each dinucleotide in Table 1, it is clear that each DNA sequence induces a different structural conformational state, e.g. the Shift ranges from -0.45 Å for GA to 0.18 Å for AC, or the Twist ranges from 25.87° for CA to 36.64 ° for GC.

In the longer tetranucleotide range, we observe how the bases that embrace the central dinucleotide, to form a tetranucleotide, perturb the conformational state of their central dinucleotide in a non-uniform way. This phenomenon is quantified through the standard deviation σ, e.g. the CG Twist has a high dispersion of 4.8°, where the most disturbing tetranucleotide is GCGG, whereas for the AG Twist the dispersion is only 1.7°.

3.2. *Multimodal Conformation State of the Central Base-Pairs*

The breaking down of the dinucleotide conformational space within the tetranucleotide space allows us to explain the multimodal behavior of several dinucleotide steps already pointed out in the literature.[13] To disentangle the dinucleotide conformational space we used scatterplots and analyzed the conformational distribution pattern of all the tetranucleotides that aggregate at the same central dinucleotide.

The bidimensional scatterplots of the coordinates pairs with more salient features were chosen from all 15 possible pairs of combinations of the 6 conformational coordinates θ_i, shown in Figure 1. The left side panels of the figure present examples with unimodal conformational distributions, whereas the examples in the right side show multimodal distributions. The histograms and the equipotential ellipses were also calculated in the scatterplots. The ellipses are projections of the six-dimensional equi-potential surfaces on the respective base-pair plane obtained from the 2×2 covariance matrices; these contours correspond to energies of 4.5 k_BT ("$3\Delta\theta$ ellipses").[10] We emphasize the role of the different tetranucleotides that have the same central dinucleotide, coloring their dot distribution with the same color. The color code grades in the scale from blue to red for ordered couples of peripherical bases (AXYA, AXYC, AXYG, AXYT, CXYA, CXYC, CXYG, CXYT, GXYA, GXYC, GXYG, GXYT, TXYA, TXYC, TXYG, TXYT). We use the same color scheme for the corresponding "$3\Delta\theta$ ellipses". We observe in the right side panels of Figure 1 how the ellipses that lie in a dissimilar way to the global distribution surround peripherical dots with a uniform color. Thus, the peripherical conformational states belong to the same tetranucleotides. Then, the trajectory of each DNA structure evolves generally around the same conformational energy local minimum, and the same structure does not oscillate between different local minima. The aggregation of the trajectories around different gravity centers produced by structures with the same dinucleotide center but with different neighbors is the cause for emerging multimodal distributions in the MD dinucleotides conformational states. The bimodal (GA, GG, CG) and three-modal (TA) distributions are due to the superposition of tetranucleotide modes with different gravity centers. This means that the modes of some dinucleotides are split by their tetranucleotide modes.

The bistable behavior of the steps involving G|C nucleotides (CG, GC and GG/CC) has been already reported based both on computational models[13] and on MD simulations.[5] Packer *et al.*[13] proposed the electrostatic interactions as the reason for this behavior. Our

Table 1. Average μ, standard deviation σ, maximum perturbation Δ_{max}, and tetranucleotide of maximum perturbation Tet_{max} for the 6 conformational coordinates of each central unique dinucleotide. The symbol / separates symmetric dinucleotides.

		Shift, Å	Slide, Å	Rise, Å	Tilt °	Roll °	Twist °
		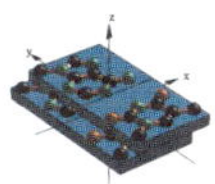	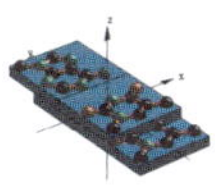		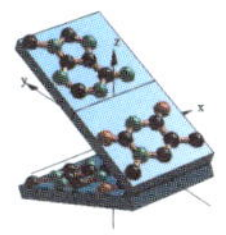	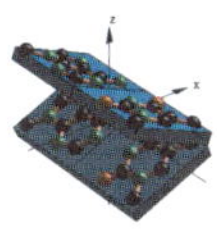	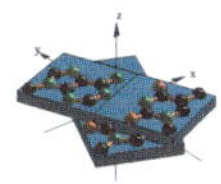
CG	μ	-0.012	-0.336	3.268	-0.044	9.029	26.989
	σ	0.435	0.290	0.235	1.494	1.512	4.838
	Δ_{max}	0.731	0.693	0.454	3.478	2.188	8.248
	Tet_{max}	GCGC	ACGT	ACGT	GCGC	ACGA	GCGG
CA	μ	-0.287	-0.443	3.260	0.578	9.606	25.865
	σ	0.336	0.286	0.208	1.094	1.417	4.690
	Δ_{max}	0.692	0.578	0.482	2.629	2.761	10.661
	Tet_{max}	TCAC	ACAT	ACAT	TCAC	TCAC	TCAA
TA	μ	-0.119	-0.268	3.249	-0.209	8.455	29.026
	σ	0.441	0.347	0.154	1.153	1.834	2.713
	Δ_{max}	0.834	0.705	0.392	2.535	3.278	4.969
	Tet_{max}	ATAA	ATAT	ATAT	TTAC	TTAC	ATAT
AG	μ	-0.193	-0.907	3.451	-2.294	3.030	32.214
	σ	0.189	0.274	0.085	0.809	1.472	1.741
	Δ_{max}	0.373	0.558	0.180	1.647	3.224	4.033
	Tet_{max}	CAGG	TAGC	TAGC	CAGG	CAGG	CAGG
GG	μ	-0.179	-0.961	3.547	0.160	4.995	33.024
	σ	0.388	0.538	0.112	1.388	1.175	2.840
	Δ_{max}	0.728	0.903	0.219	3.310	1.838	6.660
	Tet_{max}	GGGC	CGGA	CGGA	GGGC	AGGA	CGGT
AA	μ	-0.228	-0.495	3.345	-2.394	1.196	34.465
	σ	0.254	0.231	0.071	0.593	1.474	2.484
	Δ_{max}	0.506	0.384	0.132	1.207	4.344	6.356
	Tet_{max}	GAAG	TAAG	TAAG	TAAT	TAAG	TAAG
GA	μ	-0.447	-0.253	3.392	-1.095	2.401	36.280
	σ	0.373	0.489	0.081	1.785	1.549	2.685
	Δ_{max}	0.723	0.901	0.238	3.057	3.451	6.442
	Tet_{max}	CGAT	CGAG	AGAT	CGAT	GGAT	AGAA
AT	μ	0.126	-0.970	3.268	0.404	0.045	31.148
	σ	0.218	0.171	0.069	0.895	1.733	2.434
	Δ_{max}	0.351	0.272	0.139	2.082	4.072	4.838
	Tet_{max}	TATA	TATG	TATG	TATA	AATG	TATG
AC	μ	0.176	-0.858	3.360	-0.835	1.072	32.022
	σ	0.266	0.190	0.077	1.061	1.368	1.923
	Δ_{max}	0.764	0.396	0.158	2.725	2.559	3.952
	Tet_{max}	AACC	CACG	AACT	AACC	AACG	AACC
GC	μ	-0.061	-0.610	3.422	0.165	0.311	36.635
	σ	0.312	0.353	0.098	0.989	1.608	2.456
	Δ_{max}	0.541	0.658	0.211	1.563	3.701	4.429
	Tet_{max}	TGCA	AGCC	GGCC	AGCG	GGCC	AGCG

MD simulations results agree in general with the results of Packer *et al.*[13] as is shown in Araúzo *et al.*[1] The explanation of how the dinucleotide multimodal conformational states

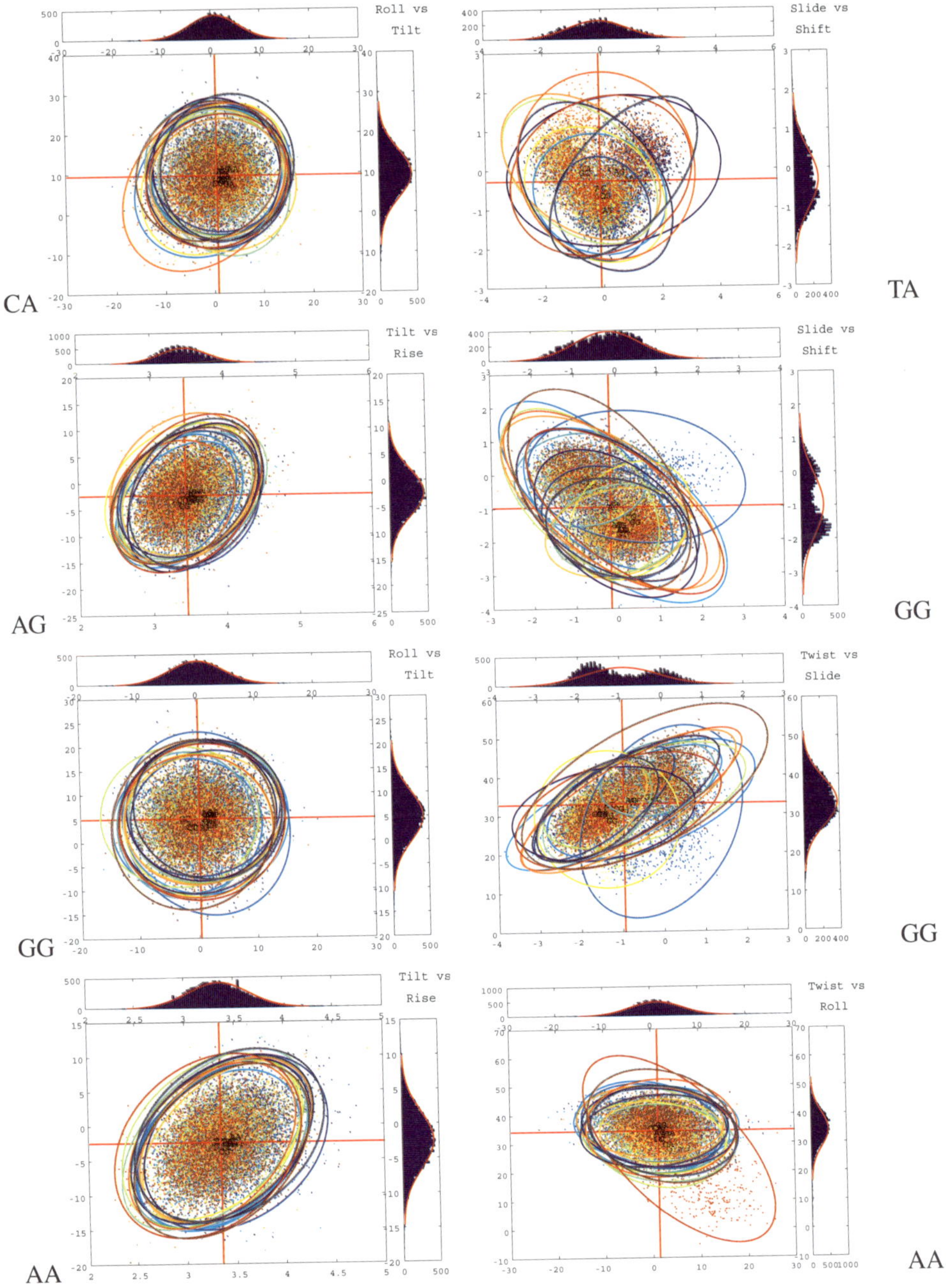

Figure 1. Scatterplots of some unimodal (left) and multimodal (right) dinucleotide conformational distributions as aggregations of tetranucleotides.

arised from the perturbations induced by their neighbors is complementary to the molecular mechanism of the sequence-dependence based on electrostatic interactions during the stacking process, proposed by Packer *et al.*,[13] for the dinucleotide steps such as GG/CC with an intrinsic bimodal feature due to electrostatic interaction. Our results suggest that the final conformational energy local minimum of the central dinucleotide could be induced by the interactions with its neighbors.

3.3. *Quantification of the Influence of the Neighbor Bases over the Central Base-Pairs*

To measure the degree to which every set of 3 dinucleotide steps interacts to form the conformational state of each tetranucleotide, we propose simple linear models. These models inverse the dinucleotide aggregation Eq. (1) under the hypothesis that each tetranucleotide conformational state can be explained as a function of 3 dinucleotides

$$W_l XY Z_r = f_{XY}\left(W_l X, XY, Y Z_r\right) \tag{2}$$

As an initial approach, we model such a function as a linear one and use the minimal square method to estimate the linear combination coefficients. We are interested to measure the degree to which each of all the possible dinucleotides that can embrace a central dinucleotide interacts to perturb the conformational state of the central one. This allows reinterpreting of the dinucleotide aggregation Eq. (1) as a function of the dinucleotides that perturb a central one instead of the original function of aggregation of tetranucleotides. This is done substituting in Eq. (1) the tetranucleotide expression given by Eq. (2)

$$XY = \bigcup_{l}^{\mathcal{N}} \bigcup_{r}^{\mathcal{N}} f_{XY}\left(W_l X, XY, Y Z_r\right) \tag{3}$$

where to shorten the notation, the 6-dimensional conformational states of the peripherical dinucleotides $W_l X$ and $Y Z_r$ will be denoted from now on as W_l and Z_r, respectively. With this notation we try to emphasize how the left and right neighbors perturb the conformational state of the central dinucleotide. Approximating the functions f_{XY} with linear models, finally we obtain

$$XY \approx \sum_{l}^{\mathcal{N}} w_l \cdot W_l + xy \cdot XY + \sum_{r}^{\mathcal{N}} z_r \cdot Z_r \tag{4}$$

where each uppercase symbol, W_l, XY, Z_r, represents the 6-dimensional conformational vector of the corresponding left, central and right dinucleotides, whereas the lowercase symbols, w_l, xy, z_r, stand for the regression coefficients estimated with the minimal square method. With the symbol $\approx$ we want to emphasize that this method is only an approximation, since we are interested in obtaining a rough idea of the contribution of each dinucleotide in the perturbation of the central one, and not to do prediction of DNA conformational states. For such a task, non-linear techniques such as neural networks can be more accurate. We perform 10 linear regressions, one for each unique dinucleotide XY. Each model has 9 parameters, 4 (a_l, c_l, g_l, t_l) accounting for the perturbations that the 4 different bases in the left side can induce in the central dinucleotide, 1 (xy) accounting for the way in which the central dinucleotide counteracts the perturbation, and other 4 (a_r, c_r, g_r, t_r), accounting for the perturbations induced from the right side. Thus, we estimate 90 parameters in total. In order to obtain these parameters, we group all the tetranucleotides with the same central dinucleotide in the same model. Thus, groups of 16 or 10 members arise depending on the symmetries. In each model we use simultaneously the 6 conformational coordinates. To estimate the model parameters, the dependent term is the average conformational state μ_{Tet} of the tetranucleotide (data shown in Araúzo *et al.*,[1]) the independent terms are the average conformational states μ shown in the first row of each dinucleotide in Table 1. For example, a model without symmetric components, such as AA, has 16 members, thus providing 96 data to estimate its 9 parameters. A model with symmetric components, such as AT, provides 60 data. With this procedure we obtain finally the following 10 linear models

$$
\begin{aligned}
AA &= -0.03A_l - 0.03C_l - 0.11G_l - 0.05T_l + 0.99AA_c + 0.07A_r + 0.07C_r + 0.03G_r + 0.10T_r \\
AC &= -0.12A_l - 0.12C_l - 0.11G_l - 0.07T_l + 1.06AC_c + 0.07A_r + 0.05C_r + 0.05G_r + 0.05T_r \\
AG &= -0.07A_l - 0.10C_l - 0.10G_l - 0.10T_l + 1.01AG_c + 0.03A_r + 0.08C_r + 0.10G_r + 0.07T_r \\
AT &= +0.16A_l + 0.18C_l + 0.08G_l + 0.11T_l + 0.92AT_c + 0.02A_r - 0.05C_r - 0.10G_r - 0.08T_r \\
CA &= +0.32A_l + 0.14C_l + 0.10G_l + 0.08T_l + 0.99CA_c - 0.26A_r - 0.14C_r - 0.15G_r - 0.07T_r \\
CG &= +0.08A_l - 0.12C_l - 0.08G_l - 0.10T_l + 1.00CG_c - 0.00A_r + 0.04C_r + 0.06G_r - 0.04T_r \\
GA &= -0.16A_l - 0.07C_l - 0.11G_l - 0.04T_l + 1.28GA_c - 0.30A_r - 0.20C_r - 0.19G_r - 0.19T_r \\
GC &= -0.20A_l - 0.10C_l - 0.15G_l - 0.14T_l + 1.08GC_c + 0.09A_r + 0.03C_r + 0.08G_r + 0.14T_r \\
GG &= -0.20A_l - 0.19C_l - 0.22G_l - 0.18T_l + 1.26GG_c - 0.12A_r - 0.03C_r - 0.15G_r - 0.01T_r \\
TA &= +0.26A_l + 0.10C_l + 0.17G_l + 0.19T_l + 1.02TA_c - 0.26A_r - 0.21C_r - 0.23G_r - 0.12T_r
\end{aligned}
\tag{5}
$$

These equations summarize the disentangling of the perturbation of each of the 10 unique dinucleotides by all their possible neighbors in our MD simulation data. They show how the conformational role of each base depends on its relative position (left, central, right) in the final tetranucleotide, e.g. an A to the left side of AC (a_l=-0.12) causes a global decrease of the native conformational coordinates of AC, whereas an A to the right side of AC (a_r=+0.07) increases the coordinates. Also, Eqs. 5 show how the same peripherical base plays a different role depending on which is the central dinucleotide, e.g. a C to the left side of CA (c_l=+0.14) increases the coordinates, whereas a C to the left side of GG (c_l=-0.19) decreases the coordinates. The mean absolute errors (MAE) of the models range from 0.58 for AA to 1.08 for CG.

The 10 linear models, Eqs. 5, allow us to establish a simple index δ that quantifies the degree of context-dependence of each central dinucleotide. This is done subtracting from each central linear regression parameter of each model xy the absolute value of the sum of the peripherical parameters w_l, z_r, and normalizing dividing by the central parameter

$$\delta_{xy} = \left(xy - \sum_{l}^{\mathcal{N}} |w_l| - \sum_{r}^{\mathcal{N}} |z_r| \right) / xy \qquad (6)$$

The higher the δ_{xy} is, the more independent is the central dinucleotide conformational state of its neighbors. Thus, this δ_{xy} allows us to classify on a quantitative basis the dinucleotides in the following way, according to the increasing context-dependence: AA/TT, CG, AC/GT (context-independent), AG/CT, AT, GC, GG/CC (weakly context-dependent), and GA/TC, CA/TG, TA (context-dependent). Currently, we are in the process of validating Eqs. (5) with crystal structure data. When more crystal structures become available in structural databases, Eqs. (5) can also be derived from real data (at the actual growth speed of such databases this can happen quite soon). In theory, it is also possible to perform the above analysis for each independent conformational state, by modeling each conformational state with a different model. In this way 60 models will arise. Here such a problem is not tackled since we are interested in the analysis of the global conformational state, but such an approach can be interesting in order to build conformational prediction models.

4. Conclusions

This work described an analysis of the deformability along 6 general base-pair step conformational coordinates of all 136 distinct DNA tetranucleotide duplex sequences based on MD simulations. It complements previous statistical efforts for experimental dinucleotide duplexes by Olson *et al.*[12] The MD results show that the multimodality in the conformational state of several dinucleotide steps observed in crystal data can be explained as the aggregration of the conformational states of the tetranucleotides that had at their center the same dinucleotide. Even for the cases in which the bistability of GG/CC seemed to be an intrinsic dinucleotide property derived from the bimodal distribution of the electrostatic interaction, [13] the different neighbors pushed the conformational state to one of the two local minima. These results suggest that sequence defines structure, but does in a complex way, since the same neighbor perturbs the conformational state of each central dinucleotide in a different manner. The conformational multimodality plays an important role in the DNA recognition since the different conformational modes induced by the neighbors of a central base-pair step can work as a signal for the binding of protein or other ligand. Currently, we are carrying out an analysis to classify the different types of perturbations that emanate in 3 dinucleotide interactions assembling each of the 136 unique tetranucleotides.

Acknowledgments

M.J. Araúzo-Bravo would like to acknowledge the Japanese Society for the Promotion of Science (JSPS) for supporting him for this research. This work is supported in part by Grants-in-Aid for Scientific Research 16014219 and 16041235 (A. Sarai) and 16014226 (H. Kono) from Ministry of Education, Culture, Sports, Science and Technology in Japan. We thank Prof. N. Go for encouraging this work and providing useful comments. Part of the MD calculations were carried out using ITBL computer facilities at JAERI.

References

1. M. J. Araúzo-Bravo, S. Fujii, H. Kono, S. Ahmad, and A. Sarai. Sequence-dependent conformational energy of DNA derived from molecular dynamics simulations: Toward understanding the indirect readout mechanism in protein-DNA recognition. *Journal of the American Chemical Society*, 2005. In press.
2. M. J. Araúzo-Bravo and A. Sarai. Knowledge-based prediction of DNA atomic structure from nucleic sequence. *Genome Informatics*, 16(2), December 2005. In press.
3. A. Arnott and D. W. Hukins. Refinement of the structure of B-DNA and implications for the analysis of X-ray diffraction data from fibers of biopolymers. *Journal of Molecular Biology*, 81(2):93–105, December 1973.
4. H. J. C. Berendsen, J. P. M. Postma, W. F. van Gunsteren, and A. DiNola. Molecular dynamics with coupling to an external bath. *Journal of Chemical Physics*, 81:3684–3690, 1984.
5. D. L. Beveridge, G. Barreiro, K. S. Byun, D. A. Case, S. B. Dixit T. E. Cheatham III and, E. Giudice, F. Lankaš, R. Lavery, J. H. Maddocks, R. Osman, E. Seibert, H. Sklenar, G. Stoll, K. M. Thayer, P. Varnai, and M. A. Young. Molecular dynamics simulations of the 136 unique tetranucleotide sequences of DNA oligonucleotides. I. Research design and results on d(C_pG) steps. *Biophysical Journal*, 87:3799–3813, December 2004.
6. R. E. Dickerson, M. Bansal, C.R. Calladine, S. Diekmann S., W. N. Hunter, O. Kennard, E. Kitzing, R. Lavery, H. C. M. Nelson, W.K. Olson, and W. Saenger. Definitions and nomenclature of nucleic acid structure parameters. *Nucleic Acids Research*, 17(5):1797–1803, 1989.
7. U. Essmann, L. Perera, M. L. Berkowitz, T. Darden, H. Lee, and L. G. Pedersen. A smooth particle mesh Ewald method. *Journal of Chemical Physics*, 103:8577–8593, 1995.
8. T. E. Cheatham III and M. A. Young. Molecular dynamics simulation of nucleic acids: Successes, limitations and promise. *Biopolymers*, 56:232–256, 2001.
9. W. L. Jorgensen. Transferable intermolecular potential functions for water, alcohols and ethers. Application to liquid water. *Journal of the American Chemical Society*, 103:335–340, 1981.
10. X. J. Lu and W. K. Olson. 3DNA: A software package for the analysis, rebuilding and visualization of three-dimensional nucleic acid structures. *Nucleic Acids Research*, 31(17):5108–5121, 2003.
11. T. Morishita. Fluctuation formulas in molecular-dynamics simulations with the weak coupling heat bath. *Journal of Chemical Physics*, 113(8):2976–2982, 2000.
12. W. K. Olson, M. Bansal, S. K. Burley, R. E. Dickerson, M. Gerstein, E. C. Harvey, U. Heinemann, X. J. Lu, S. Neidle, Z. Shakked, H. Sklenar, M. Suzuki, C. S. Tung, E. Westhof, C. Wolberger, and H. M. Berman. A standard reference frame for the description of nucleic acid base pair geometry. *Journal of Molecular Biology*, 313(1):229–237, 2001.
13. M. J. Packer, M. P. Dauncey, and C. A. Hunter. Sequence-dependent DNA structure: Dinucleotide conformational maps. *Journal of Molecular Biology*, 295:71–83, 2000.
14. D. A. Pearlman, D. A. Case, J. W. Caldwell, W. R. Ross, T. E. Cheatham III, S. DeBolt, D. Ferguson, G. Seibel, and P. Kollman. AMBER, a computer program for applying molecular mechanics, normal mode analysis, molecular dynamics and free energy calculations to elucidate the structures and energies of molecules. *Computer Physics Communications*, 91:1–41, 1995.
15. J. P. Ryckaert, G. Ciccotti, and H. J. C. Berendsen. Numerical integration of the cartesian equations of motion of a system with constraints: molecular dynamics of n-alcanes. *Journal of Computational Physics*, 23:372–336, 1977.
16. A. Sarai and H. Kono. Protein-DNA recognition patterns and predictions. *Annual Review of Biophysics and Biomolecular Structure*, 34:379–398, June 2005.
17. J. M. Wang, P. Cieplak, and P. A. Kollman. How well does a restrained electrostaic potential (RESP) model perform in calculating conformational energies of organic and biological molecules? *Journal of Computational Chemistry*, 21:1049–1074, 2000.

A NEW NEURAL NETWORK FOR B-TURN PREDICTION: THE EFFECT OF SITE-SPECIFIC AMINO ACID PREFERENCE

ZHONG-RU XIE and MING-JING HWANG

Institute of Bioinformatics, National Yang-Ming University,
Institute of Biomedical Sciences, Academia Sinica,
Taipei, Taiwan

Abstract

The prediction of β-turn, despite the observation that one out of four residues in protein belongs to this structure element, has attracted considerably less attention comparing to secondary structure predictions. Neural network machine learning is a popular approach to address such a problem of structural bioinformatics. In this paper, we describe a new neural network model for β-turn prediction that accounts for site-specific amino acid preference, a property ignored in previous training models. We showed that the statistics of amino acid preference at specific sites within and around a β-turn is rather significant, and incorporation of this property helps improve the network performance. Furthermore, by contrasting with a previous model, we revealed a deficiency of not incorporating this site-specific property in previous models.

Introduction

β-turn

Prediction of protein secondary structure is an intermediate step in the prediction of its tertiary structure. Most secondary structure prediction methods predict only three states — α-helix, β-sheet and coil [1]. However, in addition to these three repetitive structural states, tight turn is a significant element frequently occurring in protein structures. Based on the number of their constituent amino acid residues, tight turns are categorized as δ-, γ-, β-, α- and π- turns [1]. Of these five tight turns, the occurrence of β-turn is the most frequent, constituting approximately 25% to 30% of the residues in globular proteins [2]; in contrast, the second most frequently occurring tight turn, γ-turn, takes up only 3.4% of the total residues [3]. β-turn formation is also an important stage in protein folding [4], and because β-turns usually occur on solvent-exposed surfaces, they often participate in molecular recognition processes in the interactions between peptide substrates and receptors [5].

Despite that β-turn is a common and critical structure element, and that a great number of secondary structure prediction methods have been developed, β-turn prediction algorithms are surprisingly few. Most of the β-turn prediction methods are early statistical approaches, which achieve limited accuracy [1]. As accurate β-turn prediction would increase the accuracy and reliability of secondary structure prediction, which in turn would contribute to improve the prediction of tertiary structure and the identification of structural motifs such as β-hairpin, there is a need to explore more sophisticated β-turn prediction algorithms.

β-turn Prediction

The widely accepted definition for β-turn is: A β-turn comprises four consecutive residues where the distance between Cα(i) and Cα(i+3) is less than 7 Å, and the tetrapeptide is not in a helical conformation [1]. Based on these criteria, a number of β-turn prediction algorithms have been developed. They can be categorized as: 1) Site-Independent Model, 2) 1–4 and 2–3 Residue-Correlation Model, 3) Sequence-Coupled Model, and 4) Others [2].

Because a β-turn is consisted of four consecutive amino acid residues, the prediction for β-turn can be performed based on the probabilities of the 20 amino acid residues occurring at each of the 4 oligopeptide subsites.

The Site-Independent Model is a simple prediction method that multiplies the probability of each kind of the 20 amino acids occurring at each of the four subsites. Different from the Site-Independent Model, both the 1–4 and 2–3 Residue-Correlation Model and the Sequence-Coupled Model do not consider the occurrences of the 4 residues as completely independent incidents. The 1–4 and 2–3 Residue-Correlation Model is based on the observation that when a tetrapeptide folds into a β-turn, the interaction between 1st and 4th as well as between 2nd and 3rd residues becomes remarkable. Particularly, a hydrogen bond may form between the backbone carbonyl oxygen of the 1st residue and the backbone amino hydrogen of the 4th residue. The Sequence-Coupled Model also incorporates conditional probabilities. However, it is a residue-coupled model that calculates the conditional probabilities of 1–2, 2–3 and 3–4 residues.

As β-turn prediction has only two outcomes — β-turn and non-β-turn, the former should take up ~25% of the occurrences according to what is observed in protein structures — it is not sufficient to evaluate the performance of a prediction algorithm based only on prediction accuracy, which could be misleading when, for example, a method is biased to give more non-β-turn prediction outcomes.

Therefore, the four parameters commonly used to measure the performance of β-turn prediction algorithms are: 1) Qtotal (Qt): total prediction accuracy, 2) Qpredicted (Qp): percentage of correct positive prediction, 3) Qobserved (Qo): sensitivity, and 4) MCC: Matthews Correlation Coefficient, which accounts for both over- and under-predictions. They are defined in the equations given below, where "p" denotes the number of correctly predicted β-turn residues, "n" the number of correctly predicted non-β-turn residues, "o" the number of incorrectly predicted β-turn residues (false positives), "u" the number of incorrectly predicted non-β-turn residues (false negatives), and "t" the total number of residues predicted. "Qpredicted" and "Qobserved" are the proportion of false positive prediction results and that of false negative results, respectively. The MCC value is an overall evaluation parameter, which is dimensionless. MCC has a theoretical value between 0 (for random prediction) and 1 (for perfect prediction).

$$Q_{total} = \left(\frac{p+n}{t}\right) \times 100$$

$$MCC = \frac{pn - ou}{\sqrt{(p+o)(p+u)(n+o)(n+u)}}$$

$$Q_{predicted} = \left(\frac{p}{p+o}\right) \times 100$$

$$Q_{observed} = \left(\frac{p}{p+u}\right) \times 100 \qquad (1)$$

Machine Learning Approaches

Most of the recent algorithms that generally outperform earlier statistical approaches in the prediction of protein structure states have been developed via machine learning, neural networks and support vector machines (SVM) being most notable. Neural network algorithms usually use a segment of peptide sequence as the basis for prediction, where it automatically looks for subtle correlations between the input amino acids and their structural preference via a back-propagation training process. In these approaches, each of the segment residues is transformed into 20 (or 21) nodes of numeric data, which are then used as 20 (or 21) numerical values for the input nodes (or neurons) of the neural network. During the training process, the correlations between each set of the input nodes and output data are automatically adjusted to be in line with the relationship between the structure and the preference of amino acids.

In 2003, Kaur and Raghava proposed a neural network method for the prediction of β-turns utilizing multiple sequence alignments [6]. They constructed two serial feed-forward back-propagation networks, both of which have an input window of 9 residues wide (21 nodes in each residue) and a single hidden layer of 10 units (nodes). The first layer, a sequence-to-structure network, is trained with the multiple sequence alignment in the form of PSI-BLAST [7]-generated position-specific scoring matrices. The preliminary predictions from the first network along with PSIPRED [8]-predicted secondary structure states are then used as input to the second, structure-to-structure network to refine the predictions. They achieved a MCC value of 0.37 using multiple sequence alignment on the first layer and 0.43 overall using the first-layer results plus secondary structure prediction on the second layer. Their results are among the best reported in the literature for β-turn predictions.

However, in Kaur and Raghava's network, the group of 20 nodes, representing the 20 kinds of amino acids, for the central residue of the peptide segment is adjusted to merely fit the general correlations between the structure and the amino acid preference; site-specific amino acid preference is not taken into account. Here we show that a statistical analysis on the occurrence of the 20 amino acids at each of the four sites of the β-turn, and of its adjacent sites also, revealed marked site-specific preference, and incorporation of this preference improved network performance.

Materials and Methods

The Data Set

The data set in this study is consisted of 426 non-redundant protein structures as originally established by Guruprasad and Rajkumar (2000) [3]. Selected from Protein Data Bank [9], the data set was obtained using the program PDB_SELECT [10] such that no two chains of the selected representative proteins have > 25% sequence identity. All the structures selected are determined by X-ray crystallography at 2.0 Å resolution or better. Each chain contains at least one β-turn, and the β-turn assignment is based on the annotation of PDBsum [11].

Previous Neural Network Training Methods vs. Site-specific Amino Acid Preference Based Training Method

A back-propagation training procedure is used to optimize the weights of the neural network. During training, the network response at the output layer is compared to a supplied set of known answers (training targets). The errors are computed and back-propagated through the network in an attempt to improve the network response. The nodal weight factors are then adjusted by the amounts determined by the training algorithm. The iterative procedure of processing the inputs through the network, computing the errors and back-propagating the errors to adjust the weights constitutes the learning process.

Previous neural network methods for structure-state prediction of proteins (e.g. secondary structure prediction and turn prediction) stipulate that the structure of a residue is dependent upon its adjacent amino acid sequences. According to most of these methods, patterns are presented as windows of a certain number (n) of residues, in which a prediction is made for the central residue (ith residue) [6, 8] or a residue in a specific position of the window [12], as shown in Figure 1A. In this way, the group of 20 nodes for the central residue is adjusted to merely fit the general correlations between the structure state of this residue and the amino acid preference deduced for each site on this structure fragment. As the central residue is the point of focus, these methods generally do not care if the adjacent groups of nodes do not fit a certain structure state. In other words, a residue may be predicted as a β-turn residue even if its neighboring residues are not. In addition, site-specific amino acid preference is not considered.

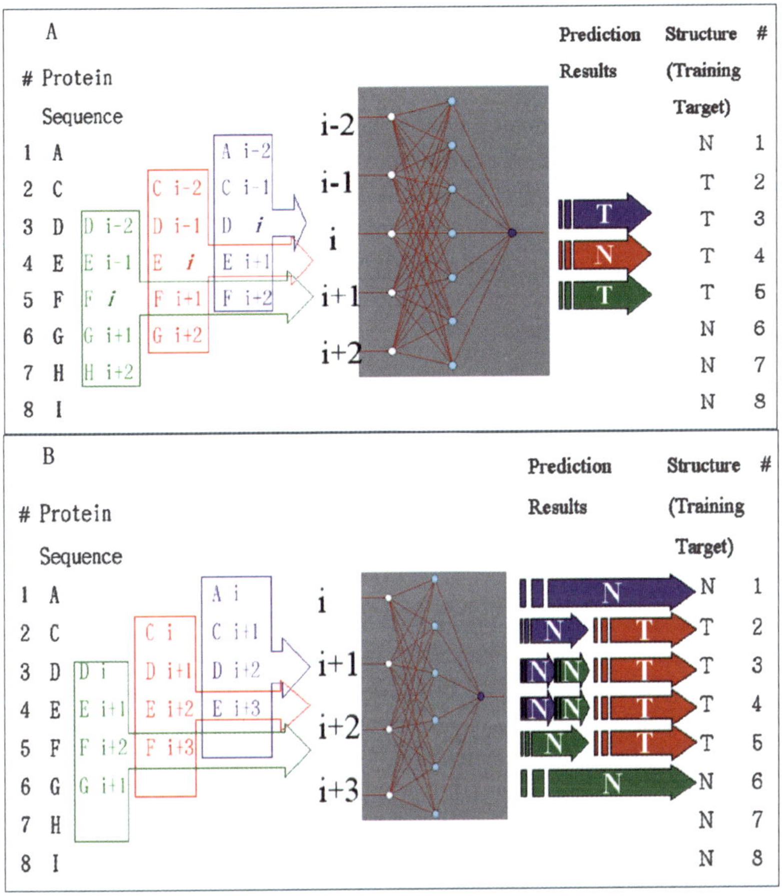

Figure 1. Illustrations of the differences between two neural network training methods: (A) A demonstration of the previous methods: The sequence ACDEFGHI on the left indicates a fragment of consecutive 8 amino acid residues in an input protein sequence. The sequence NTTTTNNN on the right indicates the known structure state of the corresponding residues: N for non-β-turn residue and T for β-turn residue. Note that T always occurs 4 or more times consecutively, as a β-turn contains exactly 4 consecutive residues. In this illustration, the neural network contains 5 groups (i-2, i-1, i, i+1, i+2; actual implementation has more) of input nodes (white dots in the gray rectangle), with each group comprising 20+1 units (such as probabilities of 20 amino acids derived from multiple sequence alignment plus an additional piece of information) to code for the residues, and 1 output node (the dark blue dot in the gray rectangle) for β-turn prediction of site i. The blue, red and green boxed and arrowed rectangles indicate 3 input data, and the "T" or "N" labeled blue, red and green arrows are the corresponding prediction results of the 3 input data. (B) A demonstration of the network model used in the present study: The network contains 4 groups of input nodes (9 residues actually used), i, i+1, i+2, i+3, and 1 output node, the result of which is either a "T" or a "N" assigned to all of the 4 predicted sites. The outputs are indicated by arrows using the same color of their corresponding inputs. Note that if a residue receives a "T" prediction from any of the input, it will be assigned as T (β-turn residue).

In this study, we proposed a new model to produce a training process in which the weights of each group of the nodes are adjusted to fit the preference patterns on each site of the β-turn and of the neighboring residues as well. As shown in Figure 1B, if the (i)th amino acid residue of the input window occurs, as in the case of the target (i.e. true answer), exactly on the 1st site of the β-turn, while the (i+1)th residue occurs on the 2nd site, and so on, the neural network will perform a positive training. When the input window shifts, e.g. the (i)th residue occurs on the 2nd site of the β-turn, and the (i+1)th residue on the 3rd site, and so on, the neural network will perform a negative training. As a result, each group of the nodes will be trained to fit the preference patterns on specific sites within and around the β-turn.

Neural Network Architecture

Besides the implementation to account for site-specific preference, our network architecture follows that of Kaur and Raghava [6]. Briefly, two serial feed-forward back-propagation networks with a single hidden layer were used. The number of hidden nodes was optimized and the two networks used were a sequence-to-structure network in the first layer and a structure-to-structure network in the second layer. The first network had the input window containing information of 9 residues and 24 nodes in the single hidden layer (these numbers of residues and nodes produced best performance among several combinations tested). The input to the first network was a multiple alignment profile. The target output was a single continuous number, which was converted to a binary number — one for β-turn and zero for non-β-turn. The window was shifted residue by residue through the protein chain, yielding N patterns for a chain with N residues. The prediction results obtained from the first layer network along with the secondary structure prediction results from PSIPRED were used as input to the second layer. Specifically, besides the first layer output, each of the 9 residues of the 2nd network input window was given reliability indices of the three secondary structure states (helix, strand and coil).

Results

Statistics of Amino Acid Preference at Specific Sites of β-turn

In this study, the occurrence probability of the 20 kinds of amino acids contained in the non-redundant dataset of 426 proteins on sites within and in the vicinity of a β-turn (sites i to i+3 corresponding to the 1st to 4th residue of the β-turn, and sites i-3 to i-1 and i+4 to i+6 corresponding to the three residues preceding and following the β-turn) and their occurrence probability in the whole dataset were calculated. The one-sample test for binomial proportion [13] was performed on the occurrence probability of the 20 kinds of amino acids on these sites. Table 1 shows the z-value results. In this table, a z value > 2 or < -2 indicates the occurrence frequency of a certain amino acid at a certain site is significantly higher or lower than its occurrence frequency in the dataset. The larger the absolute z-value, the more significant the difference is. As may be seen from Table 1, different sites, particularly the four sites of β-turn, have very different preference patterns for different kinds of amino acids. For example, both the 1st (i) and 2nd (i+1) site have a strong preference for proline, whereas the 3rd (i+2) site does not and in fact selects against it. In contrast, glycine appears to be significantly preferred at the3rd (i+2) and 4th (i+3) site, but not at the 2nd (i+1) site. There are many other notable preference patterns. Thus, the amino acid preference patterns on different specific sites indeed differ significantly. This provides a basis for the new neural network training strategy, which allows neural network to more precisely adjust the weights of each group of the input nodes to fit the preference patterns on the specific sites of β-turn in the training process.

Table 1. z values of amino acid preference on the sites within (site i to i+3) and around a β-turn produced by one-sample test for binomial proportion. Those discussed in the text are highlighted.

Residue\Site	i-3	i-2	i-1	i	i+1	i+2	i+3	i+4	i+5	i+6
A	-1.72	-4.05	-6.24	-5.31	-2.45	-12.78	-3.34	-7.63	-6.09	-2.10
C	1.52	1.95	4.32	5.13	-4.20	-2.86	2.15	-1.05	2.49	3.21
D	-4.41	-3.09	4.50	14.30	5.20	21.81	-0.81	-0.67	-2.55	-4.03
E	-3.81	-2.94	-5.68	-5.40	5.43	-3.35	-4.70	-1.91	-1.75	-2.28
F	3.26	3.30	2.50	-0.36	-5.68	-3.68	-1.69	-1.33	1.01	4.67
G	-0.93	0.70	-0.91	1.91	-0.56	35.21	18.80	-0.61	-2.12	-1.77
H	0.29	3.25	2.24	2.11	-0.11	2.98	0.52	-0.73	0.69	0.50
I	3.16	2.08	0.99	-7.39	-8.59	-12.67	-6.40	-4.13	2.38	2.51
K	-2.28	-1.51	-2.19	-3.98	6.81	-3.43	2.64	6.58	-0.20	-3.42
L	2.29	-2.38	-1.06	-5.57	-10.80	-13.54	-6.34	-6.36	0.71	-0.89
M	-0.88	-0.09	-0.14	-2.69	-6.79	-6.45	-2.19	-3.12	-0.96	-2.52
N	-0.26	-1.80	1.49	9.12	2.36	24.72	1.13	2.28	-1.44	-2.11
P	-2.03	2.13	-0.57	12.78	33.19	-7.18	2.96	15.41	6.96	1.17
Q	-1.28	-2.16	-3.64	-4.87	-2.84	-3.33	-0.49	-1.43	3.08	-2.84
R	-1.08	-1.21	1.15	-5.82	-0.57	-3.50	-0.45	1.89	-1.59	-2.72
S	-0.23	-1.02	-1.00	6.20	4.66	2.98	0.50	3.29	-1.32	0.10
T	0.73	0.64	1.41	2.02	-2.40	-1.20	3.63	6.37	0.78	3.80
V	4.18	4.73	1.92	-7.14	-8.63	-13.23	-4.91	-3.51	4.56	5.31
W	1.32	1.57	3.26	-1.12	-1.80	-1.89	-0.64	-0.86	2.97	1.66
Y	3.80	4.48	4.09	0.01	-4.25	-2.56	-1.31	-1.13	2.55	4.51
No. of Res.	7042	7072	7101	7129	7129	7129	7129	7079	7040	7015

Prediction Using Multiple Sequence Alignment in the First Layer

Our first-layer network was trained using input of multiple sequence alignment profiles generated from PSI-BLAST [12], as was done in the study of Kaur and Raghava [6]. The main difference is the new neural network model we used to fit site-specific amino acid preference, as described above. We performed a seven-fold cross validation, and the results, in comparison with those of BetaTPred2 (the current version of Kaur and Raghava's program for predicting β-turn [6]), were presented in Table 2. As may be seen, our results were significantly better. Specifically, our network achieved an MCC value of 0.402, which is significantly higher ($p < 10e-8$) than that (0.37) of the first layer network of BetaTPred2. The values of Qtotal and Qpredicted were also improved, though at the cost of slightly degraded Qobserved. These data indicate that the proportion of false positive prediction results has been significantly decreased with our model. In other words, the probability of correct prediction is significantly increased.

Table 2. Comparisons of results from the first layer between this study and that of Kaur and Raghava (BetaPred2) [6]. SD: standard deviation.

	BetaTPred2 [6]		This study	
	Average	SD	Average	SD
MCC	0.37	0.01	0.402	0.01
Qt	73.5	1.5	74.9	1.9
Qp	47.2	1.9	53.2	2.4
Qo	64.3	2.2	62.6	6.3

Prediction Using First Layer Output Plus Secondary Structure Information in the Second Layer

Again, following the procedures of Kaur and Raghava [6], our second layer was trained with the first layer output and the secondary structure prediction results from PSIPRED [10]. Cross-validation results shown in Table 3 yielded an MCC value of 0.443, which is just a bit higher than that (0.43) of BetaTPred2. Similar to the results of the first layer (Table 2), we improved on Qtotal and Qpredicted, but not Qobserved.

Table 3. Comparisons of results from the second layer between this study and that of Kaur and Raghava (BetaPred2) [6]. SD: standard deviation.

	BetaTPred2 [6]		This study	
	Average	SD	Average	SD
MCC	0.43	0.01	0.443	0.01
Qt	75.5	1.7	76.4	2.3
Qp	49.8	2.0	55.6	3.5
Qo	72.3	2.6	66.6	7.5

Discussion

In this study, we have developed a new neural network model to account for site-specific amino acid preference for β-turn predictions. We showed that site-specific preference is statistically significant and when incorporated in the neural network training can improve the network performance. In fact, ignoring site-specific preference may be a source of errors for previous models such as that of Kaur and Raghava [6]. For example, as shown in Table 1, Cysteine frequently occurs but Lysine rarely occurs on the 1st site of β-turn (z values 5.13 and -3.98), whereas on the 2nd site, the occurrence preference for the two amino acids is reversed (z values -4.20 and 6.81). In the training process of previous models, the (i)th group of neurons must fit all of the amino acids preferred on four sites simultaneously. If the residue of the 1st site of β-turn is the input to the (i)th group of neurons, the neuron weight of Cysteine will be increased and that of Lysine will be decreased. However, if the residue of the 2nd site of β-turn is the input to the (i)th group of neurons, the neuron weights of Cysteine and Lysine will be adjusted in the opposite way. This extreme example indicates possible interference of training data subsets using previous models. As the weights of a particular group of neurons are not adjusted to fit the amino acid preference on specific sites, but are merely updated as a general pattern to fit most of the preference, the prediction power would be compromised. This is corroborated by the observation that our main improvement (for the first layer) was achieved by increasing the value of Qp (Table 2), or reducing the false positive rate. Additionally, because only one residue is predicted in each prediction process using the previous models, the prediction results of consecutive residues in a sequence taken together are likely to conflict with each other; with the site-specific model (Figure 1B), contradictory adjacent predictions are eliminated.

The less-than-expected improvement by the second layer (MCC from 0.402 to 0.443), as opposed to that (MCC from 0.37 to 0.43) of Kaur and Raghava's model (Table 3 vs. Table 2), revealed a possible role of the second layer in previous network models. Many secondary structure prediction methods use two serial neural networks for prediction, where even if the second layer network does not involve other data except for the initial prediction results from the first layer, significantly greater improvement from the first layer is still achieved [8, 14]. Our study suggests that the function of the second layer network in these models is likely to reconcile or filter the initially disaccord results, whereas in our site-specific model, this is already achieved to a large extent in the first layer.

Tight turns are usually classified as coil in secondary structure assignment. However, its structural and functional significance is no less than that of α-helix or β-sheet, and could play a prominent role in the prediction of tertiary structures. Indeed, despite that the accuracy of secondary structure prediction methods has exceeded 75% [14], that for terminals of α-helix and β-strand has not yet reached a satisfactory level. Accurate tight turn predictions could remedy this problem as they could complement nicely with existing secondary structure predictions. This study demonstrated the merit of incorporating site-specific amino acid preference for β-turn prediction and provided insight into a deficiency of previous models. The same idea should be applicable to other structure-state predictions with beneficial results.

References

1. K. C. Chou. REVIEW: Prediction of tight turns and their types in proteins. *Analytical Biochemistry*, 286:1–16, 2000.
2. H. Kaur and G. P. S. Raghava. An evaluation of β -turn prediction methods. *Bioinformatics*, 18:1508–1514, 2002.
3. K. Guruprasad and S. Rajkumar. β - and γ -turns in preteins revisited: A new set of amino acid turn-type dependent positional preferences and potentials. *J. Biosci.*, 25:143–156, 2000.
4. K. Takano, Y. Yamagata and K. Yutani. Role of amino acid residues at turns in the conformational stability and folding of human lysozyme. *Biochemistry*, 39:8655–8665, 2000.
5. G. D. Rose, L. M. Gierasch and J. A. Smith. Turns in peptides and proteins. *Adv. Protein Chem.*, 37:100–109, 1985.
6. H. Kaur and G. P. S. Raghava. Prediction of β -turns in proteins from multiple alignment using neural network. *Protein Science.*, 12:627–634, 2003.
7. S. F. Altschul, T. L. Madden, A. A. Schaffer, J. H. Zhang, Z. Zhang, W. Miller and D. J. Lipman. Gapped BLAST and PSI-BLAST: a new generation of protein database search programs. *Nucleic Acids Research,* 25:3389–3402, 1997.
8. D. T. Jones. Protein secondary structure prediction based on position-specific scoring matrices. J. Mol. Biol., 292:195–202, 1999.
9. H. M. Berman, J. Westbrook, Z. Feng, G. Gilliland, T. N. Bhat, H. Weissig, I. N. Shindyalov and P. E. Bourne. The Protein Data Bank. *Nucleic Acids Research.*, 28:235–242, 2000.
10. U. Hobohm and C. Sander. Enlarged representative set of protein structures. *Protein Sci.,* 3:522–524, 1994.
11. R. A. Laskowski. PDBsum: summaries and analyses of PDB structures. *Nucleic Acids Research,* 29:221–222, 2001.
12. M. Kuhn, J. Meiler and D. Baker. Strand-loop-strand motifs: prediction of hairpins and diverging turns in proteins. *PROTEINS: Structure, Function, and Bioinformatics*, 54:282–288, 2004.
13. B. Rosner. *Fundamentals of Biostatistics.* (5th ed.). Boston: Harvard University Press.
14. B. Rost. Review: Protein Secondary Structure Prediction Continues to Rise. *Journal of Structural Biology.* 134:204–218, 2001

IDENTIFICATION OF OVER-REPRESENTED COMBINATIONS OF TRANSCRIPTION FACTOR BINDING SITES IN SETS OF CO-EXPRESSED GENES

SHAO-SHAN HUANG,[1,2,*] DEBRA L. FULTON,[1,2,*] DAVID J. ARENILLAS,[1,2,3] PAUL PERCO,[4] SHANNAN J. HO SUI,[1,2] JAMES R. MORTIMER[5] AND WYETH W. WASSERMAN[1,2,3,#]

[1]*Centre for Molecular Medicine and Therapeutics,*
[2]*Child and Family Research Institute,*
[3]*Department of Medical Genetics,*
University of British Columbia, Vancouver, Canada
[4]*Department of Nephrology, Medical University of Vienna, Vienna, Austria*
[5]*Merck Frosst Centre for Therapeutic Research, Kirkland QC, Canada*
[*]*These authors contributed equally to this work.*
[#]*Corresponding author. E-mail: wyeth@cmmt.ubc.ca*

Transcription regulation is mediated by combinatorial interactions between diverse trans-acting proteins and arrays of cis-regulatory sequences. Revealing this complex interplay between transcription factors and binding sites remains a fundamental problem for understanding the flow of genetic information. The oPOSSUM analysis system facilitates the interpretation of gene expression data through the analysis of transcription factor binding sites shared by sets of co-expressed genes. The system is based on cross-species sequence comparisons for phylogenetic footprinting and motif models for binding site prediction. We introduce a new set of analysis algorithms for the study of the combinatorial properties of transcription factor binding sites shared by sets of co-expressed genes. The new methods circumvent computational challenges through an applied focus on families of transcription factors with similar binding properties. The algorithm accurately identifies combinations of binding sites over-represented in reference collections and clarifies the results obtained by existing methods for the study of isolated binding sites.

1. Introduction

The interaction between transcription factor (TF) proteins and transcription factor binding sites (TFBS) is an important mechanism in regulating gene expression. Each cell in the human body expresses genes in response to its developmental state (e.g. tissue type), external signals from neighboring cells and environmental stimuli (e.g. stress, nutrients). Diverse regulatory mechanisms have evolved to facilitate the programming of gene expression, with a primary mechanism being TF-mediated modulation of the rate of transcript initiation. Given a finite collection of protein structures capable of binding to specific DNA sequences and the diversity of conditions to which cells must respond, it is logical and well-documented that combinatorial interplay between TFs drives much of the observed specificity of gene expression. The arrays of TFBS at which the interactions occur are often termed cis-regulatory modules (CRM).[1]

The sequence specificity of TFs has stimulated development of computational methods for discovery of TFBS on DNA sequences. Well established methods represent aligned collections of TFBS as position weight matrices (PWM). Sequence specificity of individual PWM profiles can be quantified by information content, and scoring a sequence against the PWM of a TF gives a quantitative measure of the sequence's similarity to the binding profile (for review see Wasserman and Sandelin[16]). Searching for high scoring motifs in putative regulatory sequences with a collection of profiles (for instance, JASPAR[10]) can suggest the binding sites in the sequence and the associated TF. However, this methodology is plagued by poor specificity due to the short and variable nature of the TFBS. Phylogenetic footprinting filters have been demonstrated repeatedly to improve specificity.[6] Such filters are justified by the hypothesis that sequences of biological importance are under higher selective pressure and will thus accumulate DNA sequence changes at a slower rate than other sequences. Based on this expectation, the search for potential TFBS can be limited to the most similar non-coding regions of aligned orthologous gene sequences from species of suitable evolutionary distance. Further, one might expect that genes which are coordinately expressed are under the control of the same TFs, suggesting that over-represented TFBS in the co-expressed genes are likely to be functional. These concepts are implemented by Ho Sui *et al.* in the web service tool oPOSSUM,[3] which, when given a set of co-expressed genes, can identify the TFBS motifs that are over-represented with respect to a background set of genes. This approach has achieved success in finding binding sites known to contribute to the regulation of reference gene sets.

Prior methods that attempt to address the known interplay between TFs at CRMs can be difficult to interpret.[2,5,12] We introduce a new approach rooted in the biochemical properties of TFs, which allows greater computational efficiency and improved interpretation of results. The resulting method is assessed against diverse reference data to demonstrate its utility for the applied analysis of gene expression data. Supplementary information is available at http://www.cisreg.ca/oPOSSUM2/supplement/.

2. Methods

2.1. *Background: the oPOSSUM database*

Ho Sui *et al.*[3] describe the creation of the oPOSSUM database which stores predicted, evolutionarily conserved TFBS to support over-representation analysis of TFBS for single TFs. Briefly, human-mouse orthologs are retrieved from Ensembl. TFBS profiles from the JASPAR database are used to identify putative TFBS within the conserved non-coding regions from 5000 base pairs (bp) upstream to 5000 bp downstream of the annotated transcription start site (TSS) on both strands. The oPOSSUM database stores the start and end positions and the matrix match score (> 70 %) of each site. This data is used by the oPOSSUM II algorithm in searching for over-represented TFBS combinations (described below).

2.2. *Overview and rationale of oPOSSUM II algorithm*

Finding over-represented combinations of TFBS presents several new issues that are not encountered in single site analysis. We address two of the main challenges: computational complexity and TFBS class redundancy. Firstly, the number of possible combinations of size n from m TFBS ($n \leq m$) increases combinatorially with respect to both m and n, which greatly impacts computing time. Secondly, several TFs have similar binding properties, thus subsets of profiles may be effectively redundant. Consequently, an exhaustive search is not an efficient method to find over-represented combinations of patterns.

To address both problems we introduced two approaches. First, we used a novel method to group the profiles into classes. Rather than using protein sequence similarity, a hierarchical clustering procedure was applied to group the profiles into classes according to their quantitative similarity. One representative member was selected from each class for further analysis. We then searched for the occurrences of class combinations in both co-regulated genes (foreground) and a set of background genes. We considered unordered combinations and applied an inter-binding site distance (IBSD) constraint to avoid exhaustive enumeration of all combinations, since many co-operative TFBS are found to occur in clusters without strict ordering constraints.[1] Thus, we only need consider each set of TFBS where all IBSDs satisfy the distance parameter. This approach can dramatically reduce the search space when evaluating any combination size. A scoring scheme was adopted from the Fisher exact test to compare the degree of over-representation of the class combinations. The highly over-represented class combinations were re-assessed using all possible profile combinations within the indicated classes.

The overall scheme of oPOSSUM II analysis is shown in Figure 1. The sections below describe the details of each step.

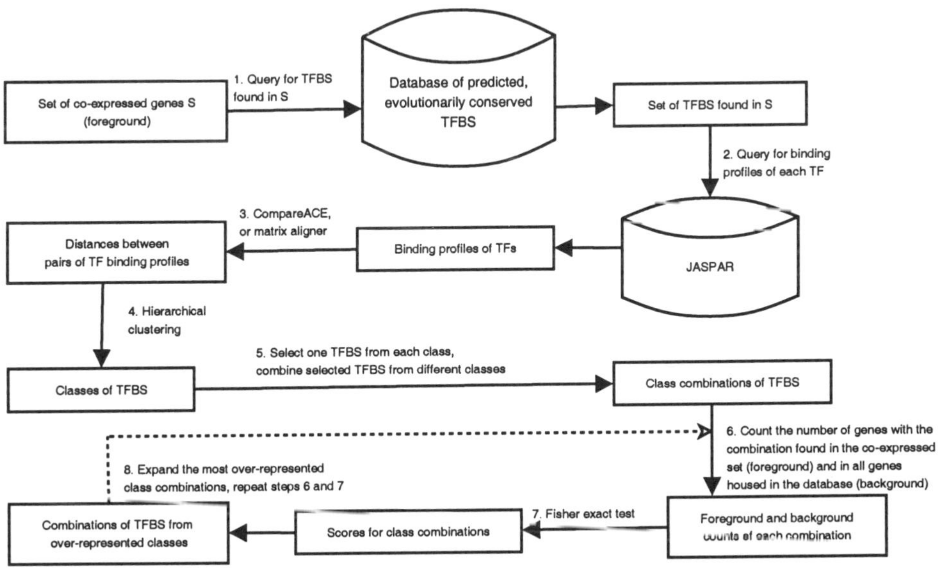

Figure 1. Overview of the oPOSSUM II analysis algorithm. Steps are numbered in the order executed. The database of predicted TFBS is identical to that of the oPOSSUM analysis system (Ho Sui *et al.*[3]).

2.3. *TFBS in foreground gene set*

When presented with a set of co-expressed genes S, oPOSSUM II queries the oPOSSUM database for all putative TFBS T present in S within a maximum of 5000 bp upstream and 5000 bp downstream from the TSS on each gene. The analysis may be restricted to those TFs found in selected taxonomic subgroups (plant, vertebrate and insect are currently available), or TFs whose profiles exceed a minimum information content.

2.4. *Classification of TFBS profiles*

Binding profiles for T were retrieved from the JASPAR database. A profile comparison algorithm, either CompareACE[4] (default) or matrix aligner,[11] calculated the pairwise similarity scores of all the profiles using profile alignment methods. The similarity score $s(t_i, t_j)$ between profiles t_i and t_j was converted to distance $d(t_i, t_j)$ by $d(t_i, t_j) = 1 - s(t_i, t_j)$. A distance matrix M was formed from these pairwise distances. From M, an agglomerative clustering procedure produced a hierarchy of clusters (subsets) of T. The complete linkage method was used since it tends to find cohesive classes. Cutting the cluster tree at a specified height thr_H partitioned T into classes.

2.5. *Selection of TFBS and enumeration of combinations*

For each class C, we selected the profile that is the most similar to other profiles in C as the class representative. We chose this approach as we could not identify an adequate procedure that would generate a consensus profile with comparable specificity to the matrices within the class. To identify the class representative, we first calculated the sum of pairwise similarity score σ_i between a profile t_i and other profiles in C, i.e., $\sigma_i = \sum_{t_i, t_j \in C} s(t_i, t_j)$. The profile with the maximum sum of similarity score was chosen. From the selected TFBS, unordered combinations of specified size (*cardinality*) were created. oPOSSUM II then searched the foreground gene set (the co-expressed genes) and the background gene set (default is all the genes in the database) for occurrences of these combinations. Let max_d be the maximum inter-binding site distance. For each gene, the occurrences of the combinations were found using a sliding window of width equal to max_d within the required search region. We counted the number of genes with a combination in both the foreground and background gene sets.

2.6. *Scoring of combinations*

The Fisher exact test detects non-random association between two categorical variables. We adopted the Fisher P-values to rank the significance of non-random association between the occurrence of a combination and the foreground gene set, i.e., over-representation of the combination in the foreground compared to background. For each combination, a two-dimensional contingency table was constructed from the foreground and background count distributions:

	Number of genes with a given combination	Number of genes without a given combination
Foreground	a_{11}	a_{12}
Background	a_{21}	a_{22}

For $i, j = 1, 2$, row sum $R_i = a_{i1} + a_{i2}$ and column sum $C_j = a_{1j} + a_{2j}$, and the total count $N = \sum_i R_i = \sum_j C_j$. From the hypergeometric probability function, the conditional probability P_{cutoff} given the row and column sums is

$$P_{\text{cutoff}} = \frac{(C_1!C_2!)(R_1!R_2!)}{N! \prod_{i,j=1,2} a_{ij}}.$$

We calculated the P-values for all other possible contingency tables with row sums equal to R_i and column sums equal to C_j. The Fisher P-value is the sum of all the P-values less than or equal to P_{cutoff}, which represent equal or greater deviation from independence than the observed table.

Caution must be taken when interpreting these Fisher P-values. First, the foreground and background genes are allowed to overlap, which is a violation of an assumption for the statistical test. Secondly, the Fisher exact test model may not precisely characterize the data sets being analyzed. As a result, the Fisher P-values were used purely as a measure to compare the degree of over-representation between different combinations. We will hereafter refer to the P-values as "scores". Although the scores do not describe the probabilistic nature of the over-representation, the ranking they provide is shown to be useful.[3]

2.7. *Finding significant TFs from over-represented class combinations*

Let thr_C be the maximum score for which a TFBS combination may be considered significant. Our empirical studies of reference collections suggested that a default maximum score value of 0.01 detects relevant TF combinations. Let x_i be any TFBS class combination with a score less than or equal to thr_C and X is the set of distinct class combinations that satisfy the score threshold: $X = \{x_i | score(x_i) \leq thr_C\}$. For each combination x_i, let each of $C_1, C_2, \ldots, C_h$ be a set of TFBS profiles that are represented by each of the h class profiles in that combination. Compute the Cartesian product $\mathbb{C}_p$ of $C_1, \ldots, C_h$. We call this "expanding the TFBS classes" from the class representatives. The enumeration and ranking procedures were repeated for the h-tuples in $\mathbb{C}_p$.

2.8. *Random sampling simulations of foreground genes*

oPOSSUM II needs to accommodate input gene sets of different cardinalities, so we wished to investigate the relationship between gene set size and the false positive rate. 100 random samples of r genes were selected from the background and given to oPOSSUM II as foreground genes. For each sample, oPOSSUM II reported the scores for all the class combinations. As these random samples of genes were not expected to be co-regulated, any combination was a false positive. Let $(0, max_s]$ be the interval over which false positives are accumulated. We recorded the number of false positive class combinations for a range of max_s when $r = 20, 40, 60, 80, 100$.

2.9. *Validation*

Three reference sets of human genes were used as input to oPOSSUM II to assess the performance of the algorithm. Two independent sets of skeletal muscle genes were tested. The first set (muscle set 1) was compiled from the reference collection identified by Wasserman and Fickett[15] and updated by a review of recent literature. A second set (muscle set 2) combines the results of microarray studies of Moran *et al.*[8] and Tomczak *et al.*[14] The third set contains smooth muscle-specific genes experimentally verified by Nelander *et al.*[9] All sets were validated with max_d=100, matrix score threshold=75%, and conservation level=1.

As a further comparison to the methods in Kreiman,[5] which were validated in part against the yeast CLB2 gene cluster,[13] the yeast CLB2 cluster was analyzed using the yeast oPOSSUM database (Ho Sui, unpublished).

3. Results

3.1. *TFBS classification*

Since the three reference gene sets were restricted to vertebrates, the first step in oPOSSUM II analysis was to cluster the available vertebrate TFBS. We cut the hierarchical cluster tree at a height of 0.45 ($thr_H = 0.45$) because the majority of resulting clusters correlated well with the structural families defined in JASPAR (cluster tree available in web supplement). Most notably, binding profiles from FORKHEAD, HMG and ETS families were grouped according to classifications. However, as we expected, the zinc finger profiles were dispersed into new groupings due to their divergent binding profile composition. Using this approach, the 68 vertebrate TFBS in JASPAR were partitioned into 32 classes. This step produced a considerable reduction in the search space. For example, in the analysis of pair combinations, the search space was reduced by a factor of four.

3.2. *Validation with reference data sets*

3.2.1. *Yeast CLB2 cluster*

The yeast CLB2 gene cluster contains genes whose transcription peaks at late G2/early M phase of the cell cycle. Transcription of these genes is regulated by the TF FKH, which is a component of the TF SFF, and which interacts with the TF MCM1. Each of the top ten scoring class combinations found by oPOSSUM II contained the binding sites of the ECB class, of which MCM1 is a member. The highest ranked combination was {ECB, FKH1}, which is consistent with the literature and the results of Kreiman.[5] The complete results are available on the supplementary web site.

3.2.2. *Three human reference gene sets*

Figure 2 lists the top five over-represented class combinations for each of the three human reference gene sets. The score values for these combinations were less than 2.0E-3. Also listed are the five most over-represented TFBS classes in the total 32 classes created, as reported by oPOSSUM single site analysis.

Prior studies involving muscle set 1[15] have identified the occurrence of clusters of muscle regulatory sites including MEF2, SRF, Myf/MyoD, SP1 and TEF. The classes that contain MEF2 and SP1 dominated the top combinations in both skeletal muscle sets (Figure 2a and 2b). Yin-Yang modulates SRF-dependent, skeletal muscle expression. Thing1-E47 is a bHLH TF localized to gut smooth muscle in adult mice, therefore, the presence of class

Combination TFBS Pairs	Single TFBS
8 (Bsap); 20 (MEF2*)	1 (Myf*)
8 (Bsap); 29 (SRF*)	8 (Bsap)
1 (Myf*); 31 (Yin-Yang*)	29 (SRF*)
20 (MEF2*); 28 (SP1*)	26 (RREB-1)
20 (MEF2*); 29 (SRF*)	28 (SP1*)

a. Skeletal Muscle Set 1

Combination TFBS Pairs	Single TFBS
20 (MEF2*); 28 (SP1*)	20 (MEF2*)
20 (MEF2*); 32 (Thing1-E47*)	25 (Androgen)
20 (MEF2*); 21 (MZF_5-13)	29 (SRF*)
8 (Bsap); 20 (MEF2*)	1 (Myf*)
1 (Myf*); 20 (MEF2*)	7 (Spz1)

b. Skeletal Muscle Set 2

Combination TFBS Pairs	Single TFBS
28 (SP1*); 29 (SRF*)	29 (SRF*)
21 (MZF_5-13); 29 (SRF*)	26 (RREB-1)
29 (SRF*); 31 (Yin-Yang*)	20 (MEF2*)
29 (SRF*); 7 (Spz1)	7 (Spz1)
29 (SRF*); 32 (Thing1-E47*)	1 (Myf*)

c. Smooth Muscle Set

Figure 2. The top five over-represented pair combinations of TFBS classes reported by oPOSSUM II and over-represented single TFBS sites reported by oPOSSUM for the skeletal and smooth muscle sets. The numbers are the class identifiers and enclosed in parentheses is the name of a TF within that class, which is either known to mediate transcription in the assessed tissue (*) or is a class representative.

32 in the list may be linked to other myogenic factors in the bHLH superfamily (such as Myf). Bsap and MZF are not muscle specific. The Bsap motif is long (20 bp) and exhibits an unusual pattern of low information content distributed across the entire motif, suggesting that it may behave differently than other binding profiles. The inclusion of this profile in the JASPAR database is under review (B. Lenhard, personal communication).

For the smooth muscle genes, the SRF class appeared in each of the top five combinations, consistent with established knowledge.[7] The top combination, {SP1, SRF}, is required for the expression of smooth muscle myosin heavy chain in rat. Yin-Yang can stimulate smooth muscle growth. Spz1 acts in spermatogenesis, and has no known role in muscle expression.

For all three reference sets, the top scoring combinations suggested new classes not found in the single site analysis. In all cases, there were relevant TFBS identified only in the combination analysis.

3.3. *Effect of set size on false positive rate*

The result of random sampling simulation of foreground genes is shown in Figure 3, which plots the rate of false positive predictions for a range of gene set sizes as a function of max_s. The data suggested no dependency of the false prediction rate on set size. We also noted that at low score values, the proportion of false positives is low.

3.4. *Web interface*

oPOSSUM II web service is available at http://www.cisreg.ca/oPOSSUM2/opossum2.php. A user enters a set of putatively co-expressed genes and specifies the parameter values to be used in the analysis. Certain parameter values may produce lengthy runtimes. To accommodate this possibility, our web service will queue the analysis request and will notify the user via e-mail once the analysis is complete.

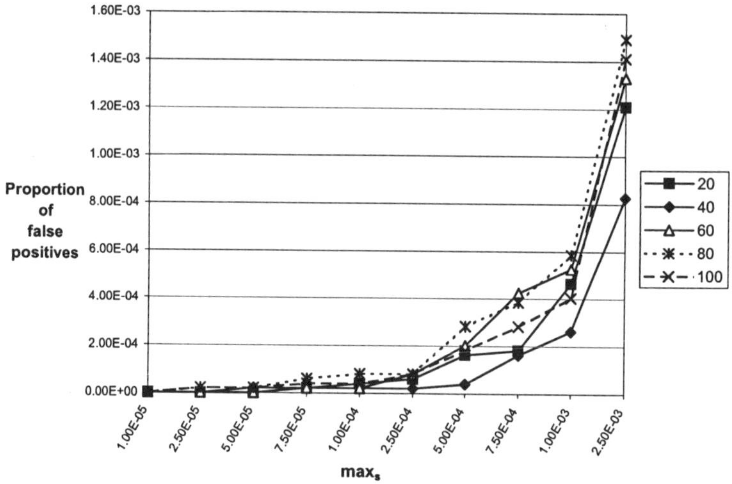

Figure 3. Effect of gene set size on false positive rate observed from pairwise TFBS combinations in randomly generated foreground gene sets.

4. Discussion

The analysis of over-represented combinations of TFBS in the promoters of co-expressed genes is motivated by biochemical and genetic studies which reveal the functional importance of cis-regulatory modules. In contrast to previously described methods which identify single over-represented motifs, the analysis of combinations must solve or circumvent the consequence of a combinatoric explosion, which can precipitate prohibitive runtimes. To reduce the search space, oPOSSUM II restricts its analysis to binding site combinations using biologically justifiable criteria, namely, TF profile similarity.

Our results suggest two important contributions over the existing single-site TFBS over-representation methods. Firstly, in each reference gene set, there is at least one relevant TF class that appears in multiple combinations, an observation that is not immediately obvious in single site analysis. Secondly, the algorithm finds functional TFBS that are not indicated in single site analysis. For instance with the yeast CLB2 gene cluster, members of the top scoring combination, ECB and FKH1, are ranked the first and eleventh in single site analysis. In the smooth muscle reference set, the SRF and SP1 combination is the most significant, but they are ranked the first and fourteenth in single site analysis. These results clearly demonstrate the power of combination site analysis.

Analysis of the microarray-based skeletal muscle reference set correctly implicates the combination of MEF2 and SP1 TFs in myogenesis. This result confirms the utility of high-quality microarray data for regulatory sequence analysis.

While our result for the yeast CLB2 cluster is comparable to that reported by Kreiman,[5] there are significant differences between the methods. Kreiman initially uses a motif discovery algorithm to identify new motif patterns from a gene set and then subsequently looks for over-represented combinations of motifs using both the new motif patterns and a TFBS profile database. In our interpretation, there is circular logic in looking for relevant motifs in a reference gene set and then identifying their over-represented combinations. For the CLB2 cluster, the profiles were taken from an existing database and our results are comparable. For the first skeletal muscle collection, Kreiman reports the top scoring combination as SP1, SRF, TEF and a motif drawn from the promoters of the positive gene set.

Although this paper presents the results for pairs of TFBS, the oPOSSUM II implementation is also able to evaluate combinations of higher cardinality. However, validation of larger combinations is seriously limited by the lack of robust reference data sets that include genes known to be regulated by multiple binding sites.

A few issues remain to be addressed by future research. First, the interpretation of analysis results is confounded by intra-class binding similarity. While this property facilitates the oPOSSUM II algorithm, users must be prepared to consider which proteins in a family are most likely to act within the tissue or under the condition studied. For instance, the fact that an E-box motif is over-represented in the skeletal muscle data does not directly lead the researcher to the MyoD protein; instead the user must consider the entire range of bHLH-domain TFs. Second, inter-class similarity can influence the results. Although oPOSSUM II does not allow overlap between TFBS in the analysis of a given combination, TFBS from different combinations can overlap. Thus two G-rich motifs may be reported as over-represented in different combinations (for instance, the SP1 and MZF motifs in Figure 2c) but highlight the same candidate TFBS within the sequences analyzed. A related issue is the compositional sequence bias in tissue specific genes,[17] which would motivate selection of a more refined background gene set. Finally, the required computing time is prohibitively long for a synchronous web service. Parallelization of the enumeration algorithm is a natural way to improve the running time.

5. Conclusion

oPOSSUM II utilizes putative TFBS identified from comparative genomic analysis, in conjunction with knowledge of co-regulated expression, to search for functional combinations of TFBS that may confer a given gene expression pattern. It uses a novel scheme to classify similar binding site profiles. Using this clustering approach, the oPOSSUM II method is able to circumvent the combinatorial challenge associated with the identification of significant TFBS combinations. Furthermore, the application of an IBSD constraint limits the number of possible combinations to analyze. Validation results suggest that TFBS combination site analysis can provide valuable information that is not available through a single-site analysis.

Acknowledgments

We thank Andrew Kwon for annotation of the muscle reference collections. We acknowledge operating support from the Canadian Institutes for Health Research (CIHR) and Merck Frosst; DF was supported by the CIHR/MSFHR Bioinformatics training program and the Merck Frosst Co-op program; WWW is supported as a Michael Smith Foundation for Health Research Scientist and a New Investigator of the CIHR.

References

1. M. I. Arnone and E. H. Davidson. The hardwiring of development: organization and function of genomic regulatory systems. *Development*, 124(10):1851–64, 1997.
2. N. Bluthgen, S. M. Kielbasa, and H. Herzel. Inferring combinatorial regulation of transcription in silico. *Nucleic Acids Res*, 33(1):272–9, 2005.
3. S. J. Ho Sui, J. R. Mortimer, D. J. Arenillas, J. Brumm, C. J. Walsh, B. P. Kennedy, and W. W. Wasserman. oPOSSUM: identification of over-represented transcription factor binding sites in co-expressed genes. *Nucleic Acids Res*, 33(10):3154–64, 2005.
4. J. D. Hughes, P. W. Estep, S. Tavazoie, and G. M. Church. Computational identification of cis-regulatory elements associated with groups of functionally related genes in *Saccharomyces cerevisiae*. *J Mol Biol*, 296(5):1205–14, 2000.
5. G. Kreiman. Identification of sparsely distributed clusters of cis-regulatory elements in sets of co-expressed genes. *Nucleic Acids Res*, 32(9):2889–900, 2004.
6. B. Lenhard, A. Sandelin, L. Mendoza, P. Engstrom, N. Jareborg, and W. W. Wasserman. Identification of conserved regulatory elements by comparative genome analysis. *J Biol*, 2(2):13, 2003.
7. C. S. Madsen, J. C. Hershey, M. B. Hautmann, S. L. White, and G. K. Owens. Expression of the smooth muscle myosin heavy chain gene is regulated by a negative-acting GC-rich element located between two positive-acting serum response factor-binding elements. *J Biol Chem*, 272(10):6332–40, 1997.
8. J. L. Moran, Y. Li, A. A. Hill, W. M. Mounts, and C. P. Miller. Gene expression changes during mouse skeletal myoblast differentiation revealed by transcriptional profiling. *Physiol Genomics*, 10(2):103–11, 2002.
9. S. Nelander, P. Mostad, and P. Lindahl. Prediction of cell type-specific gene modules: identification and initial characterization of a core set of smooth muscle-specific genes. *Genome Res*, 13(8):1838–54, 2003.
10. A. Sandelin, W. Alkema, P. Engstrom, W. W. Wasserman, and B. Lenhard. JASPAR: an open-access database for eukaryotic transcription factor binding profiles. *Nucleic Acids Res*, 32(Database issue):D91–4, 2004.
11. A. Sandelin, A. Hoglund, B. Lenhard, and W. W. Wasserman. Integrated analysis of yeast regulatory sequences for biologically linked clusters of genes. *Funct Integr Genomics*, 3(3):125–34, 2003.
12. R. Sharan, A. Ben-Hur, G. G. Loots, and I. Ovcharenko. CREME: Cis-Regulatory Module Explorer for the human genome. *Nucleic Acids Res*, 32(Web Server issue):W253–6, 2004.
13. P. T. Spellman, G. Sherlock, M. Q. Zhang, V. R. Iyer, K. Anders, M. B. Eisen, P. O. Brown, D. Botstein, and B. Futcher. Comprehensive identification of cell cycle-regulated genes of the yeast Saccharomyces cerevisiae by microarray hybridization. *Mol Biol Cell*, 9(12):3273–97, 1998.
14. K. K. Tomczak, V. D. Marinescu, M. F. Ramoni, D. Sanoudou, F. Montanaro, M. Han, L. M. Kunkel, I. S. Kohane, and A. H. Beggs. Expression profiling and identification of novel genes involved in myogenic differentiation. *FASEB J*, 18(2):403–5, 2004.
15. W. W. Wasserman and J. W. Fickett. Identification of regulatory regions which confer muscle-specific gene expression. *J Mol Biol*, 278(1):167–81, 1998.
16. W. W. Wasserman and A. Sandelin. Applied bioinformatics for the identification of regulatory elements. *Nat Rev Genet*, 5(4):276–87, 2004.
17. R. Yamashita, Y. Suzuki, S. Sugano, and K. Nakai. Genome-wide analysis reveals strong correlation between CpG islands with nearby transcription start sites of genes and their tissue specificity. *Gene*, 350(2):129–36, 2005.

A KNOWLEDGE-BASED APPROACH TO PROTEIN LOCAL STRUCTURE PREDICTION*

CHING-TAI CHEN, HSIN-NAN LIN, KUN-PIN WU, TING-YI SUNG† AND WEN-LIAN HSU

Institute of Information Science, Academia Sinica, Taipei, Taiwan
{caster, arith, kpw, tsung, hsu}@iis.sinica.edu.tw

Abstract

Local structure prediction can facilitate ab initio structure prediction, protein threading, and remote homology detection. However, previous approaches to local structure prediction suffer from poor accuracy. In this paper, we propose a knowledge-based prediction method that assigns a measure called the local match rate to each position of an amino acid sequence to estimate the confidence of our approach. To remedy prediction results with low local match rates, we use a neural network prediction method. Then, we have a hybrid prediction method, HYPLOSP (HYbrid method to Protein LOcal Structure Prediction) that combines our knowledge-based method with a neural network method. We test the method on two different structural alphabets and evaluate it by Q_N, which is similar to Q_3 in secondary structure prediction. The experimental results show that our method yields a significant improvement over previous studies.

1. Introduction

Protein local structure is a set of protein peptides that share common physiochemical and structural properties. Researchers usually cluster protein fragments by different local criteria, such as solvent accessibility, residue burial [8], and backbone geometry [9], and represent these fragment clusters by an *alphabet*, called a *local structure alphabet* (also known as a *structural alphabet* or *structural motifs*) [9]. Local structure prediction predicts the local structure of a protein fragment expressed by a *letter* of the structural alphabet from its amino acid sequence. Local structure prediction helps improve the performance of both profile and threading/fold-recognition methods for tertiary structure prediction [3, 6].

Various local structure libraries have been constructed, some of which focus on the reconstruction of protein tertiary structures. In such libraries, the number of letters in each structural alphabet is large, e.g., 100 in Unger et al. [16], 40 and 100 in Micheletti et al. [12], 100 in Schuchhardt et al. [15], and 25-300 with fragment lengths from 5 to 7 in Kolodny et al. [10]. Though large alphabet sets can better approximate protein tertiary

---

* This work is partially supported by the Thematic program of Academia Sinica under Grant 94B003 and by the National Science Council, Taiwan under Grant NSC94-2213-E-001-008.

† Correspondence to: Ting-Yi Sung, Institute of Information Science, 128 Sec. 2, Academia Rd, Nankang, Taipei, 115 Taiwan. E-mail: tsung@iis.sinica.edu.tw

structures, predicting protein local structures from amino acid sequences is much more challenging.

Thus, smaller structural alphabet sets have been proposed, and their associated local structure libraries have been constructed. Moreover, local structure prediction algorithms using these libraries have been developed. Bystroff et al. [2] generated a library called *I-site,* which contains 13 structural motifs of different length. Prediction is based on profile-profile alignment between each structural motif and the PSI-BLAST [1] result of the input sequence. They further proposed a new model, HMMSTR, to improve prediction accuracy. The *structural alphabet of HMMSTR*, denoted by *SAH*, is used in this paper to test our method. In [5], de Brevern et al. built their library, called *Protein Blocks* (*PB*), by clustering 5-mer protein fragments into a structural alphabet of 16 letters according to a torsion angle space. They then used a Bayesian probabilistic approach for prediction. Karchin et al. [9] constructed an *STR* library, in which the structural alphabet consists of 13 letters obtained from eight secondary structure states by dividing β-sheet into 6 types. They used a hidden Markov model (HMM) for local structure prediction.

The performance of local structure prediction depends on the definition of the underlying structural alphabet and the prediction algorithm. However, there is no unifying performance measure for evaluation. Bystroff et al. regard a local structure correctly predicted if the *MDA* (*Maximum Deviation of backbone torsion Angle*) of an eight-residue window is less than 120 degrees to their native structure [2, 4]. However, a straightforward evaluation measure, Q_N, is used in [5], which is similar to Q_3 used in secondary structure prediction. Q_N compares the predicted results with the encoded structural letter sequence, where N is the alphabet size, for example, N= 10 for SAH. Specifically, Q_N of a protein, *p*, is calculated as follows:

$$Q_N = \frac{\text{the number of residues of } p \text{ correctly predicted}}{\text{the number of all residues of } p} \times 100 .$$

In [5], the accuracy of Q_N is 40.7%. Q_N is also used by Karchin et al. in [8, 9]. Thus in this paper, we use Q_N to evaluate different prediction methods, as discussed in Section 3.

Previous studies indicate that accuracy is the main difficulty in local structure prediction. In this study, we propose a local structure prediction algorithm to improve the current accuracy. The proposed method is alphabet-independent, i.e., it is not designed for a specific structural alphabet. Furthermore, we use Q_N to evaluate the method and demonstrate its capability.

2. Methods

We propose a knowledge-based prediction method and use a measure called the *local match rate* to estimate the prediction confidence. The *local match rate* represents the amount of information at each position of an amino acid sequence acquired from the knowledge base. Empirically, by this method, a high match rate results in high prediction accuracy. To improve the low prediction accuracy of low-match-rate positions, we propose a neural network prediction method that also provides confidence from its output. We propose a hybrid method, called HYPLOSP (HYbrid method to Protein LOcal Structure Prediction), which combines the results of these two methods according to the local match rate and neural network confidence.

2.1 Knowledge-based approach

2.1.1 Construction of a sequence-structure knowledge base (SSKB)

Our knowledge base contains both local structure information and secondary structure information about peptides. The former is expressed by a structural alphabet (discussed in Section 3.1), and the latter is obtained from the DSSP database. For ease of exposition, we assume that we use a protein dataset with a known secondary structure and local structure based on a given structural alphabet.

The strength of a knowledge base depends on its size. Since the number of proteins with known secondary structures is relatively small, we amplify our knowledge base by finding homologous proteins to inherit the structural information of the given dataset. To this end, we utilize PSI-BLAST [1] to find proteins remotely homologous to a protein with a known structure, referred to as a *Query protein* in the PSI-BLAST output. While using PSI-BLAST, we set the parameter j to 3 (3 iterations), e to 10 (E-value < 10), and use the NCBI nr database as the sequence database. For each Query protein, PSI-BLAST generates a large number of homologous protein segments as well as their pairwise alignment called *high-scoring segment pairs, HSPs.* In each HSP, the counterpart sequence aligned with the Query protein is denoted by *Sbjct* in the PSI-BLAST output.

Performing PSI-BLAST on a Query protein, we obtain a large set of HSPs. Now we need to find the peptides in the Sbjct protein of each HSP that are similar to those of the Query protein so that similar peptides can inherit the structural information of the Query protein. We use a sliding window of length w to determine the peptides. In our experiments, we choose $w = 7$, which yields the best results among other lengths. Let p and q denote a pair of peptides in Query and Sbjct, respectively. We define the *similarity score*, S, between p and q as the number of positions that are identical or have a "+" sign in the sliding window. We call p and q *similar* if $S \geqq 5$. For the peptide q, which is similar to p, we define the *voting score* of q with respect to p as $(S \times A) / w$ to measure the confidence level for q to inherit the structural information of p, where A denotes the alignment score of the HSP reported in PSI-BLAST output. If p and q do not contain any gap, we add the record (q, the secondary structure of p, local structure of p, and voting score of q) to the knowledge base, in addition to the record (p, the secondary structure of p, local structure of p, and voting score of p). Otherwise, we discard this pair of similar peptides.

Figure 1 shows part of an HSP. The pair of peptides marked by a box have a similarity score of 5 and are thus considered similar. The voting score of the peptide in Sbjct with respect to that in Query is 180 (= 5×252 / 7). Suppose the structural alphabet is a set of {A, B, C, D, E, F}, and the secondary structure and local structure of peptide VLSPADK are CCHHHHC and BBEEECD, respectively. Since this peptide pair does not contain any gap, the record (MLTAEDK, CCHHHHC, BBEEECD, 180) is added to the knowledge base as shown in Table 1(a). Note that a peptide may inherit structural information from multiple peptides; if this is the case, we simply add new records to the existing record. For example, suppose the peptide MLTAEDK also inherits the structural information from another similar peptide with a voting score of 65. Then, the record of MLTAEDK in the knowledge base is updated, as shown in Table 1(b).

```
>sp|P08849|HBAD_ACCGE Hemoglobin alpha-D chain
 pir||A26544 hemoglobin alpha-D chain – goshawk   Length = 141
 Score =   252 bits (646), Expect = 1e-66

Query: 1    VLSPADKTNVKAAWGKVGAHAGEYGAEALERMFLSFPTTKTY-
FPHFDLSHGSAQVK
            +L+ DK  ++A W KV  H  ++GAEAL+RMF+++PTTKTYFPHFDLS
GS QV+
Sbjct: 1    MLTAEDKKLIQAIWDKVQGHQEDFGAEALQRMFITYPTTKTYF-
PHFDLSPGSDQVR
```

Figure 1. An example of HSPs found by PSI-BLAST

Table1. An example of knowledge base entries

Peptide fragment		M	L	T	A	E	D	K
Secondary Structure	H	0	0	180	180	180	180	0
	E	0	0	0	0	0	0	0
	C	180	180	0	0	0	0	180
Structural Alphabet	A	0	0	0	0	0	0	0
	B	180	180	0	0	0	0	0
	C	0	0	0	0	0	180	0
	D	0	0	0	0	0	0	180
	E	0	0	180	180	180	0	0
	F	0	0	0	0	0	0	0

(a)

Peptide fragment		M	L	T	A	E	D	K
Secondary Structure	H	0	0	180	180	245	245	65
	E	0	0	0	0	0	0	0
	C	245	245	65	65	0	0	180
Structural Alphabet	A	0	0	0	0	0	0	0
	B	245	245	0	0	0	0	0
	C	0	0	0	0	0	180	0
	D	0	0	0	0	0	0	180
	E	0	0	180	180	180	65	65
	F	0	0	65	65	65	0	0

(b)

2.1.2 Local structure prediction based on SSKB

Using the constructed knowledge base, SSKB, our knowledge-based local structure prediction method is comprised of the following steps:

Step 1: Use PSI-BLAST to find all HSPs with respect to a target protein (i.e., a protein whose secondary and local structures are unknown and to be predicted).

Step 2: Use similar peptides found in SSKB to vote for the local structure of each amino acid in the target protein.

In Step 1, the parameters and the sequence database used in PSI-BLAST are the same as those used in knowledge base construction. To define similar peptides stated in Step 2, we use the same sliding window length of 7, same voting score, and the same similarity score of 5 with no gap to define similar peptides as before. We match all peptides of the target protein and their similar peptides against SSKB. We then use the local structure information of the matched peptides in SSKB to vote for the local structure of the target protein. Let p be a peptide of the target protein. Throughout this section, we assume the structural alphabet is a set of $\{A_1, A_2,\ldots, A_n\}$. We associate each position, x, in p with n variables given by V^p_i, where $i = 1,\ldots, n$. Let q be p's counterpart peptide with similarity score S in an HSP with an alignment score A. If q is similar to p and can be found in SSKB, the voting score of q is added to that of p, which is updated as follows: For each position, x, compute

$$V^p_i(x) \leftarrow V^p_i(x) + V^q_i(x) \times (S \times A)/7, \; i=1,\ldots n,$$

and repeat the above calculation for all similar peptides. The local structure of x in p is given by the letter corresponding to $Max\,\{V^p_1(x), V^p_2(x),\ldots, V^p_n(x)\}$.

2.2 Neural network method

2.2.1 Neural network architecture

We use a standard feed-forward back-propagation neural network [14] with a single hidden layer. The number of hidden units in the hidden layer is 35, which has been found to be the most effective number in our training stage.

Taking each protein in the training set or testing set, we partition it into peptides by a sliding window of length 7. We also perform PSI-BLAST query to obtain the profile of the sequence, which is the Position-Specific Scoring Matrix (PSSM). Our neural network takes each peptide as input. Specifically, the input vector consists of the peptide's corresponding segment of PSSM as well as its secondary structure. So, the length of each input vector is 161, i.e., 7×20 for PSSM and 7×3 for the secondary structure. The output reports the results corresponding to the amino acid located at the center of the peptide (called the "peptide center" for short). Specifically, the output is a vector of size n, i.e., the size of the underlying structural alphabet, and each entry represents the confidence score of the peptide center to be assigned a specific alphabet letter.

2.2.2 Training procedure

An online back-propagation training procedure is used to optimize the weights of the network, whereby the weights are randomly initialized and updated with each input vector. The learning parameters of the hidden layer and the output layer are 0.075 and 0.05, respectively. In addition, the sum of square errors is used during back propagation.

In the training stage, the secondary structure information in the input vector is given by the *true* secondary structure from the DSSP database. The desired output is a vector with 1 at the entry corresponding to the *true* alphabet letter of the peptide center, and 0 elsewhere.

2.2.3 Local structure prediction based on a neural network

Our neural network prediction method consists of two steps:
Step 1: Perform secondary structure prediction on a target protein.
Step 2: Use the neural network method to predict the local structure of each amino acid in the target protein.

Unlike proteins in the training set, target proteins do not have secondary structure information. Thus, in Step 1 we use HYPROSP II [7] to predict the secondary structure. The predicted secondary structure and PSSM, extracted by a sliding window of length 7, constitute the input to the trained neural network. The letter with the highest confidence score in the output vector is then considered to be the local structure of the peptide center. Step 2 is repeated to predict all amino acids in the target protein.

2.3 Hybrid mechanism

Our knowledge-based method and the neural network method have different strengths. To better utilize their respective strengths, we propose a hybrid mechanism that uses the local match rate, to combine the two methods. At each position, *x*, of the target protein, we obtain from HSPs a set of similar peptides, *Q*(*x*), that contains the position *x*. The local match rate is defined as follows:

$$\text{Local Match Rate(x)} = \frac{|\mathrm{Q(x)} \cap \mathrm{SSKB}|}{|\mathrm{Q(x)}|} \times 100\% .$$

The local match rate represents the amount of information for each position *x* that can be extracted from the knowledge base. It is possible for the target protein to have a high local match rate in some positions and a low local match rate in others. Intuitively, a higher local match rate implies higher confidence in the result of the knowledge-based prediction method.

2.4 HYPLOSP: a hybrid method for protein local structure prediction

Our hybrid prediction method, HYPLOSP, combines the prediction results of the knowledge-based method and the neural network method at each position of the target protein. The neural network returns a confidence score for each output letter. In order to output these values to a text file, we normalize them into a range of 0 to 94, since there are only 95 printable ASCII codes. Then the neural network generates a set of normalized confidence scores $\{Conf\_NN_1, Conf\_NN_2, \ldots, Conf\_NN_n\}$ associated with each letter.

The knowledge-based method generates a set of voting scores, denoted by $\{V_1, V_2, \ldots, V_n\}$, associated with each position. We define the confidence score of letter A_i as

$$Conf\_KB_i = \mathrm{Min}\{ \frac{V_i}{\sum_j V_j} \times LocalMatchRate(x), 94 \}.$$

Using $Conf\_NN_i$ and $Conf\_KB_i$, we determine the final predicted structure at position x to be A_k if

$$Conf\_NN_k + Conf\_KB_k = \mathrm{Max} \bigcup_m (Conf\_NN_m + Conf\_KB_m) .$$

3. Experimental Results

3.1 Datasets

We downloaded 25,288 proteins from the DSSP database (dated 9/22/2004) as our first dataset. These proteins were divided into 46,745 protein chains. In our method, we use PSI-BLAST and pairwise sequence alignment to filter out protein chains with a pairwise sequence identity over 25%. Moreover, protein chains of length less than 80 are removed Finally, we have a non-redundant DSSP dataset, called *nrDSSP*, containing 3,925 unique protein chains along with their secondary structures. To evaluate our prediction methods, we transform nrDSSP into structural alphabets of our choice.

Furthermore, we use another dataset, containing new proteins for the period of Oct. 2004 to May 2005, to compare HYPLOSP with other prediction methods. Fifty-six protein chains remain after we filter out those with a sequence identity over 25% in this dataset and in nrDSSP.

We test our methods on two structural alphabets: SAH and PB. There are originally 11 alphabet letters in SAH, including 10 Φ–Ψ plane partitions for *trans* peptide and one for *cis* peptide. We follow Karchin's approach [9] and assign the *cis* residues among the other 10 partitions according to their Φ–Ψ values. We encode each amino acid with a SAH letter by assigning the letter of the Φ–Ψ plane that is the nearest to the amino acid. The PB alphabet is composed of 16 letters, each of which is 5-residue in length and represented by 8 dihedral angles. We use a sliding window of length 5 to extract peptides from amino acid sequences. The Root Mean Square Deviation on Angular values (RMSDA) between the peptide and each of the 16 PB letters is calculated, and the letter with the smallest RMSDA is assigned to the peptide center.

3.2 Cross-validation test of our methods

We perform 10-fold cross-validation experiments on each chosen structural alphabet to evaluate our knowledge-based (KB) method, neural network (NN) method, and the hybrid method, HYPLOSP. In each experiment, the dataset is randomly divided into ten sets. A set is selected as the testing set (containing *predicted* secondary structure information) and the other nine sets are integrated as the training set (containing *true* secondary structure information) for neural network training and the construction of SSKB. This process must be repeated for each set to be a testing set. In addition, we modify our methods that do not use secondary structure information as follows. For the knowledge-based method, we do not record secondary structure element (SSE) information while constructing SSKB, or while finding similar peptides in SSKB. For the neural network method, we do not take the SSE of a peptide as input for either training or testing (prediction); thus, the input of the network becomes a vector of size 140.

The performance results using SSE information are shown in Table 2. For the SAH alphabet, HYPLOSP reports a Q_N of 61.51% and outperforms our KB and NN methods (which report a Q_N of 56.7% and 59.53% on average) by approximately 5% and 2%, respectively. For the PB alphabet, our KB and NN methods achieve on average a Q_N of 57.79 % and 59.54%, respectively. Our hybrid method, HYPLOSP with an overall Q_N of 63.24% outperforms the KB and NN methods by 3.7% and 5.5%, respectively. In summary, HYPLOSP reports a Q_N over 60%, whether on the 10-letter SAH alphabet or the 16-letter PB alphabet.

The results not using SSE information are also shown in Table 2. Both the KB and NN methods suffer a considerable decrease in Q_N (between 3% and 5%). Therefore, the SSE information plays a role in these two methods. However, the Q_N of HYPLOSP is reduced by at most 1.37%, which is comparatively lower than the KB and NN methods. This implies that HYPLOSP is less sensitive to the absence of SSE and better utilizes both the neural network and knowledge-based methods.

Table 2. The performance of our methods on SAH and PB

	Using SSE		Not using SSE	
	Q_N on SAH	Q_N on PB	Q_N on SAH	Q_N on PB
NN	59.53%	59.54%	55.72%	54.65%
KB	56.70%	57.79%	53.14%	53.79%
HYPLOSP	61.51%	63.24%	60.14%	61.91%

3.3 Comparison with the previous studies

To compare HYPLOSP with the prediction methods used by the authors of SAH and PB, we use the second dataset (56 new proteins) for evaluation. The HYPLOSP model is trained on nrDSSP and tested on the testing dataset. We compare our methods with the HMMSTR server developed by Bystroff et al. [4] for the SAH alphabet, and with the LocPred server developed by de Brevern et al. [5] for the PB alphabet. Note that there are three models in LocPred server: Bayesian prediction, sequence families, and a new version of sequence families. We only compare HYPLOSP with the last one, since it is the best of the three.

The experimental results are shown in Table 3. HYPLOSP outperforms HMMSTR on the SAH alphabet by 4.4% and achieves a 13.24% improvement over LocPred on the PB alphabet. Furthermore, HYPLOSP demonstrates an alphabet-independent prediction capability and a relatively stable performance irrespective of the alphabet size. To be specific, HYPLOSP has a Q_N of 57.44% for the 10-letter SAH alphabet, and 55.17% for the 16-letter PB alphabet. Although the alphabet size grows by 60% ($(16-10) \div 10 \times 100\%$), Q_N only decreases by 2.27%.

Table 3. Comparison of HYPLOSP with other prediction methods

		Q_N
SAH	HMMSTR	53.04%
	HYPLOSP	57.44%
	Improvement	4.40%
PB	LocPred	41.93%
	HYPLOSP	55.17%
	Improvement	13.24%

5. Concluding Remarks

Existing local structure prediction methods show that prediction accuracy is a very challenging issue. We use two different prediction methods: one is knowledge-based and the other is neural network-based. To better utilize the advantage of these two methods, we propose a hybrid method called HYPLOSP, which is alphabet-independent. We select two current structural alphabets, SAH and PB, to evaluate HYPLOSP. We have performed a 10-fold cross-validation test on the nrDSSP dataset of nearly 4,000 protein chains to evaluate our KB, NN methods in comparison with HYPLOSP. In addition, we have also performed a test on 56 protein chains to compare HYPLOSP with the prediction methods used the authors of SAH and PB. The experimental results not only show better performance of HYPLOSP in terms of Q_N accuracy, but also demonstrate its capability to be alphabet-independent. We further analyze the relation between our prediction accuracy rate and the secondary structure. The analysis shows that improving current secondary structure prediction accuracy leads to a substantial improvement in local structure prediction.

References

1. Altschul SF, Madden TL, Schaffer AA, Zhang J, Zhang Z, Miller W, Lipman DJ. Gapped BLAST and PSI- BLAST: a new generation of protein database search programs. *Nucleic Acids Res.*, 25(17):3389-3402, 1997.
2. Bystroff C, Baker D. Prediction of local structure in proteins using a library of sequence-structure motifs. *J. Mol. Biol.*, 281: 565-577, 1998.
3. Bystroff C, Shao Y. Fully automated ab initio protein structure prediction using I-Sites, HMMSTR and Rosetta. *Bioinformatics*, 18: 54-61, 2002.
4. Bystroff C, Thorsson V, Baker D. HMMSTR: a hidden Markov model for local sequence-structure correlations in proteins. *J. Mol. Biol.*, 301: 173-190, 2000.
5. de Brevern AG, Etchebest C, Hazout S. Bayesian probabilistic approach for predicting backbone structures in terms of protein blocks. *Proteins*, 41:271-287, 2000.
6. Karplus K, Karchin R, Draper J, Casper J, Mandel-Gutfreund Y, Diekhans M, Hughey R. Combining local-structure, fold-recognition, and new fold methods for protein structure prediction. *Proteins*, 53:491-496, 2003.
7. Lin HN, Chang JM, Wu KP, Sung TY, Hsu WL. A knowledge-based hybrid method for protein secondary structure prediction based on local prediction confidence. *Bioinformatics*, 21:3227-3233, 2005.
8. Karchin R, Cline M, Karplus K. Evaluation of local structure alphabets based on residue burial. *Proteins*, 55:508-518, 2004.
9. Karchin R, Cline M, Mandel-Gutfreund Y, Karplus K. Hidden Markov models that use predicted local structure for fold recognition: alphabets of backbone geometry. *Proteins*, 51:504-514, 2003.
10. Kolodny R, Koehl P, Guibas L, Levitt M. Small libraries of protein fragments model native protein structures accurately. *J. Mol. Biol.*, 323:297-307, 2002.
11. Kuang R, Leslie CS, Yang AS. Protein backbone angle prediction with machine learning approaches. *Bioinformatics*, 20: 1612-1621, 2004.
12. Micheletti C, Seno F, Maritan A. Recurrent Oligomers in proteins: an optimal scheme reconciling accurate and concise backbone representations in automated folding and design studies. *Proteins*, 40: 662-674, 2000.
13. Rost B, Eyrich VA. EVA: large-scale analysis of secondary structure prediction. *Proteins*, 5:192-199, 2001.
14. Rumelhart, D., G. Hinton, and R. Williams. Learning internal representations by error propagation. In *Neurocomputing: Foundations of Research*, 675-695. Cambridge, MA: MIT Press, 1988.
15. Schuchhardt J, Schneider G, Reichelt J, Schomburg D, Wrede P. Local structural motifs of protein backbones are classified by self-organizing neural networks. *Protein Eng.*, 9: 833-842, 1996.
16. Unger R, Harel D, Wherland S, Sussman JL. A 3D building blocks approach to analyzing and predicting structure of proteins. *Proteins*, 5:355-373, 1989.

IDENTIFICATION OF MICRORNA PRECURSORS VIA SVM

LIANG HUAI YANG,[1,2] WYNNE HSU,[1] MONG LI LEE,[1] LIMSOON WONG[1]

[1]*School of Computing, National University of Singapore*
{yanglh, whsu, leeml, wongls}@comp.nus.edu.sg

[2]*School of Electronics Engineering and Computer Science,*
Peking University, P.R. China

MiRNAs are short non-coding RNAs that regulate gene expression. While the first miRNAs were discovered using experimental methods, experimental miRNA identification remains technically challenging and incomplete. This calls for the development of computational approaches to complement experimental approaches to miRNA gene identification. We propose in this paper a *de novo* miRNA precursor prediction method. This method follows the "feature generation, feature selection, and feature integration" paradigm of constructing recognition models for genomics sequences. We generate and identified features based on information in both primary sequence and secondary structure, and use these features to construct SVM-based models for the recognition of miRNA precursors. Experimental results show that our method is effective, and can achieve good sensitivity and specificity.

1. Introduction

Traditionally, the "Central Dogma" has decreed that genetic information flows linearly from DNA to RNA to protein, and never in reverse. The role of RNA in the cell has been limited to its function as mRNA, tRNA, and rRNA. The discovery of a diverse array of transcripts that are not translated to proteins but rather function as RNAs has changed this view profoundly. Now, it is increasingly hard to have a comprehensive understanding of cellular processes without considering functional RNAs. Efficient identification of functional RNAs—non-coding RNAs (ncRNAs) as well as cis-acting elements—in genomic sequences is, therefore, one of the major goals of current bioinformatics.

1.1. *Background*

MicroRNAs (miRNAs) are the smallest functional non-coding RNAs of animals and plants. They have been called "the biological equivalent of dark matter, all around us but almost escaping without detection." The mature miRNAs are synthesized from a longer precursor (pre-miRNA) forming a long hairpin structure that contains the mature miRNA in either of its arms. All reported mature miRNAs are between 17 and 29 nucleotides (nt) in length and the majority of them are about 21-25 nt long and have been found in a wide range of eukaryotes, from Arabidopsis thaliana and Caenorhabditis elegans to mouse and human.[3] MicroRNAs play an important regulatory functions in eukaryotic gene expression through mRNA degradation or translation inhibition. The regulatory functions of miRNAs range from cell proliferation, fat metabolism, neuronal patterning in nematodes, neurological diseases, modulation of hematopoietic lineage differentiation in mammals, development, cell death, cancer, and control of leaf and flower development in plants. An miRNA downregulates the translation of target mRNAs through base-pairing to these target mRNAs.[16,1] In animals, miRNAs tend to bind to the 3' untranslated region (3' UTR) of their target transcripts to repress translation. The pairing between miRNAs and their target mRNAs usually includes short bulges and/or mismatches. In contrast, in all known cases, plant miRNAs bind to the protein-coding region of their target mRNAs with three or fewer mismatches and induce target mRNA degradation[10] or repress mRNA translation.

1.2. *Related Works*

The experimental identification of miRNA is technically challenging and incomplete for two reasons. First, miRNAs tend to have highly constrained tissue- and time-specific expression patterns. Second, degradation products from mRNAs and other endogenous noncoding RNAs coexist with miRNAs and are sometimes dominant in small RNA molecule samples extracted from cells.

MicroRNAs and their associated proteins appear to be one of the more abundant ribonucleoprotein complexes in the cell. A single organism may have hundreds of distinct miRNAs, some of which are expressed in stage-, tissue- or cell type-specific patterns. Nonetheless, miRNAs whose expression is restricted to nonabundant cell types or specific environmental conditions could still be missed in cloning efforts. Thus, computational methods have been developed to complement experimental approaches to identify miRNA genes.

Many miRNAs have been predicted through various computational screens, such as comparative genomics, that can detect entirely new RNA families.[13,12] To date, over 1600 miRNAs have been identified in different organisms.[6]A variety of computational methods have been applied to several animal genomes, including Drosophila melanogaster, C. elegans and humans.[5,12,13] They use the following strategies:

(1) Homology searches for orthologs and paralogs of known miRNA genes. This strategy exploits the observation that some miRNAs are conserved across great evolutionary distances which indicates that their sequence is not arbitrary. Such sequence conservation in the mature miRNA and long hairpin structures in miRNA precursors facilitates genome-wide computational searches for miRNAs.
(2) Searching for a genomic cluster[15] in the vicinity of known miRNA genes. This strategy is important because some of the most rapidly evolving miRNA genes are present as tandem arrays within operon-like clusters, and the divergent sequences of these genes make them relatively difficult to spot if general approaches are used.
(3) Gene-finding approaches that do not depend on homology or proximity to known genes have also been developed and applied to entire genomes.[5,13,12,19] They typically start by identifying conserved genomic segments that both fall outside of predicted protein-coding regions and potentially could form stem loops and then scoring these candidate miRNA stem loops for the patterns of conservation and pairing that characterize known miRNAs genes.

MiRscan[12,15] and SRNALoop[5] have been systematically applied to nematode and vertebrate candidates, and miRseeker[13] has been systematically applied to insect candidates. Wang et al.[19] applied their method to plants. Dozens of new genes have been identified that were subsequently (or concurrently) experimentally verified. Other methods like profile-based detection of miRNA precursors[11] have also been proposed. In addition, several groups have developed computational methods to predict miRNA targets in Arabidopsis, Drosophila and humans.

1.3. *Paper Organization*

Notwithstanding its progress, *de novo* prediction is still a largely unsolved issue. Here, we follow the "feature generation, feature selection, feature integration" paradigm[14] of constructing recognition models for genomic sequences to develop a *de novo* method based on SVM for recognition of miRNA precursors. The paper is organized as follows: Section 2 details our methodology which includes the input data and feature generation. The data generation and experimental results are presented in Section 3 to demonstrate the effectiveness of our method and we conclude in Section 4.

2. Proposed Methodology

To predict new miRNAs by computational methods, we need to define sequence and structure properties that differentiate known miRNA sequences from random genomic sequence, and use these properties as constraints to screen intergenic regions/whole genome (introns excluding those protein encoding exons) in the target genome sequences for candidate miRNAs. Unlike protein coding genes, ncRNAs lack in their primary sequence common statistical signals that could be exploited for reliable detection algorithms. For miRNAs, different methods need to be contrived.

2.1. *Signals Used*

Computational gene-finding for protein-coding genes in both prokaryotic and eukaryotic genomes has been quite successful. These methods exploit genomic features such as long open-reading-frames and codon signatures. Many signal sensors have been designed to detect signals like splice sites, start and stop codons, branch points, promoters and terminators of transcription, polyadenylation sites, ribosomal binding sites, topoisomerase II binding sites, topoisomerase I cleavage sites, and various transcription factor binding sites and CpG islands.

However, it is not so easy for noncoding RNA (ncRNA) genes like miRNA. Usually only weakly-conserved promoter and terminator signals (and possibly other poorly known transcription binding sites) are present in ncRNA genes.[2] EST searches indicate that some human and mouse miRNAs are co-transcribed along with their upstream and downstream neighboring genes.[17] A recent study shows that microRNA genes are transcribed by RNA polymerase II.[9] This leads us to exploit some possible signals that might exist in the upstream and downstream of miRNA precursors. We distinguish the possible transcription of miRNA into two categories:

(1) Co-transcribed miRNAs: miRNAs located in the introns of annotated host genes. For this case, miRNAs share the same ±1000 up/downstream of the host genes.
(2) Independently transcribed miRNAs: These miRNAs are not far away from the annotated genes. We further divide them into two categories: (a) clustered miRNAs: we use the -1000 upstream of the first miRNA precursor in the cluster and the +1000 downstream of the last miRNA precursor in the cluster; (b) non-clustered miRNAs: we use the ±1000 up/downstream of the miRNAs precursor.

For the secondary category, it is observed that a prominent characteristic of animal miRNAs is that their genes are often organized in tandem, and are closely clustered on the genome.

Again the situation with miRNAs is more challenging. Far fewer miRNAs are available in the databases. MicroRNA sequences can be compared only at the nucleotide level—not as translated amino acids and miRNA sequences are quite short. As noted previously, the mature miRNA has only about 17-25nts and its precursor has about 100nts for animals. Consequently, distinguishing weakly conserved genes from random "hits" is more difficult when searching for miRNAs than for protein-coding genes. Moreover, even in cases where there are large RNA families, sequence conservation is often at the secondary-structure level, i.e., what is conserved are base pairing rather than the individual base sequence. Consequently, sequence alignment alone may fail to identify miRNAs that diverged too far apart in their primary sequence while retaining their base-paired structure.

To capture the information of secondary structure, we first fold the miRNA precursor using the Vienna RNA package RNAFold.[8] Next, to facilitate data processing, we encode the base-pairing by: A:U–"1", C:G–"2", G:C–"3", G:U–"4", U:A–"5", U:G–"6", Other-"0". An example cel-mir-1 miRNA precursor of C. elegans is shown in Figure 1. We ignore the loop part and mismatch starting part because of their large variations and low conservations.

```
>cel-mir-1
aaagugaccguaccgagcugcauacuuccuuacaugcccauacuauaucauaaaug
gauaUGGAAUGUAAAGAAGUAUGUAgaacggggugguagu

cut-off->:                                                      :<-cut-off     A:U--"1"
  aaagug:    ua    ag                c           gc        -    :au            C:G--"2"
         :accg  ccg   cu g c a u a c u u c  u u a c a u   c c a u a  cuau:   c G:C--"3"
         :||||  |||   | | | | | | | | | | | |  | | | | | |   | | | | |  ||||:   a G:U--"4"
         :uggu  ggc   gAUGUAUGAAG  AAUGUA   GGU a u  ggua:  u            U:A--"5"
    ---uga:    gg    aa                A           -A        a    :aa            U:G--"6"
cut-off->:                                                      :<-cut-off     Other-"0"
Encoding:12240022300 254 2 1 5 1 2 5 5 2 05 5 1 2 1 5 002 2 1 5 102615
```

Figure 1. Encoding the secondary structure.

Figure 2 shows the conceptual view of one input sequence. Each input consists of four components: upstream sequence, the primary sequence of miRNA precursor, the encoding sequence of the secondary structure of miRNA precursor, and the downstream sequence. Thus, the input contains the information of both primary sequence and secondary structure.

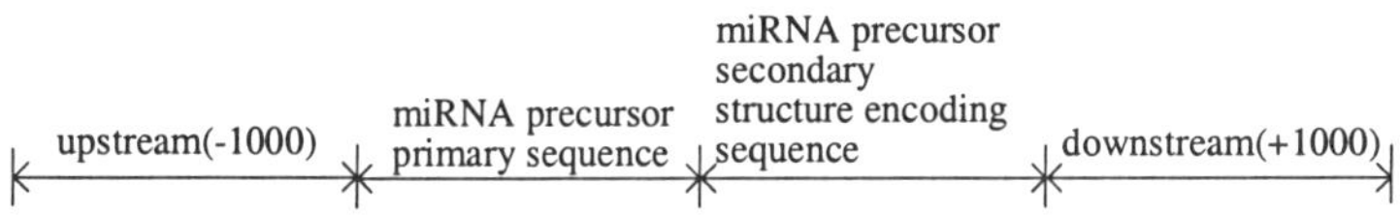

Figure 2. A conceptual view of input sequence.

2.2. *Feature Generation and Feature Selection*

To enable machine learning algorithms to learn from known miRNA sequences, we need to map the input sequence into a feature vector in the feature space. In this work, we follow the "feature generation, feature selection, feature integration".[14] In the "feature generation" process, we exploit the widely used so-called k-$gram$[14] frequency in our feature mapping.

Let Σ denote an alphabet, whose length is $|\Sigma| = L$. Let X be a sequence of letters from Σ. Given $1 \leq k < L$, a k-$gram$ is a k-length contiguous subsequence. We define our feature map as an indexed vector by all possible subsequences α of length-k from Σ^k. Formally, the feature map Φ_k: $X \rightarrow \mathbb{R}^{L^k}$ is defined as:

$$\Phi_k(X) = (\phi_\alpha(X))_{\alpha \in \Sigma^k}$$

where $\phi_\alpha(X)$ is the frequency count of α that occurs in X.

For our input data, the upstream, the primary sequence of the precursor and the downstream have the same alphabet $\sum$={A, C, G, U}. Given k=6, each sequence is coded into a vector with $\sum_{k=1}^{6} 4^k = 1364$ elements. The encoding sequence of the secondary structure of the precursor has an alphabet $\{1, 2, 3, 4, 5, 6\}$. We ignore the mismatch code "0". Let $k = 5$, the latter sequence is coded into a vector with $\sum_{k=1}^{5} 6^k = 1554$ elements. Hence, an input sequence will be mapped into a feature vector which will have a total of $3 * 1364 + 1554 = 5646$ elements. We use a suffix tree to accelerate the generation of features. Each depth-k node of the suffix tree stores a count of the number of leaf nodes it leads to.

The feature dimensionality is very large even for a small k. Most learning algorithms suffer from the "curse of dimensionality"— these methods typically require an exponential increase in the number of training samples with respect to an increase in the dimensionality of the samples in order to uncover and learn the relationship of the various dimensions to the nature of the samples. Hence, the selection of relevant informative features among the large collection of candidate features is necessary for machine learning tasks faced with high dimensional data. In the "feature selection" process, we use a correlation-based feature selection method based on the concept of entropy.[20]

2.3. *Support Vector Machines*

Support Vector Machines (SVMs) are a class of supervised learning algorithms first introduced by Vapnik.[18] Given a set of labelled training vectors (positive and negative input examples), an SVM learns a linear decision boundary to discriminate between the two classes. The result is a linear classification rule that can be used to classify new test examples. SVMs have exhibited excellent generalization performance (accuracy on test sets) in practice and have strong theoretical motivation in statistical learning theory.

In our application, we integrate the features selected previously into a model for classifying a candidate sequence as a miRNA precursor or as "other". This "feature integration" process is a typical application for SVMs.

3. Experiments: Classification of miRNA Precursors

In this section, we first describe how to generate the required data set for training and testing. Then we show the prediction result of the trained SVM.

3.1. *Data Generation*

All miRNA genes and precursors (Version 6; April2005) are downloaded from the microRNA Registry[6] which has 1650 precursors. Genome sequences for Caenorhabditis elegans and Caenorhabditis briggsae are available from WormBase at `ftp://ftp.wormbase.org`. Drosophila melanogaster and Drosophila pseudoobscura genome release 4.1 are obtained from FlyBase at `ftp://flybase.net/genomes`. Genomes and the corresponding annotation files of Homo sapiens, Mus musculus, Rattus norvegicus, and Gallus gallus are acquired from Ensembl at `http://www.ensembl.org/Download/`.

3.1.1. *Generating Positive Examples*

Animal miRNAs are often closely clustered together. We call two miRNAs on the same strand as "adjacent" if the number of nucleotides between the end of one miRNA and the start of the other is less than 1000 nts. If miRNAs $mr_1, mr_2, ..., mr_k$ satisfy $(mr_{i+1}.\text{start} - mr_i.\text{end}) < 1000\text{nts}$ for $i = 1, ..., k-1$, we say they form a miRNA cluster. The procedure of generating positive examples is as follows:

(1) For each species considered, we merge the adjacent miRNAs in the same strand to form clusters;
(2) According to the GFF annotations,
    - For miRNAs located in the introns of CDS, we obtain the –1000 upstream and +1000 downstream of the CDS, along with the miRNA precursor to form one input sequence;
    - For each independently transcribed miRNA, we extract the ± 1000 upstream/downstream of the miRNA or miRNA cluster, along with the miRNA precursor to form one input sequence;

3.1.2. *Generating Negative Examples*

It is an inherently difficult problem in bioinformatics to get negative examples. However, knowing that only a very very small fraction of non-annotated sequences correspond to "coding" sequences for miRNAs, we can generate negative examples of miRNA genes from inter-genic regions for learning. We make this assumption realizing that our negative examples might be somewhat contaminated with currently unknown miRNA genes. Hence, to alleviate the problem, we filter the negative examples in an iterative manner after making the initial predictions, i.e., we remove strongly predicted genes and re-train in order to purify our training examples.

Since all the miRNA precursors form a stem-loop secondary structure and each arm of the stem may contain the miRNA, we also require these negative examples to be as similar as possible to the true miRNA precursors. Otherwise, it will be trivial for the learning algorithm to detect these fake outliers. Specifically, when generating the negative examples, two conditions must be satisfied. First, they form a stem-loop. We use RNAfold[8] for folding the selected sequence using the C-libraries of the Vienna RNA package version 1.4. Second, the matching part of the stem is at least 15 nt long (currently the smallest miRNA is 17nt).

The procedure of generating negatives is as follows:

(1) With the help of GFF annotation file, we sort each CDS of the same strand according to its (start, end) position, and form the inter-genic regions;
(2) For each inter-genic region, we slide along the sequence and use a normal distribution $N(\mu, \sigma)$ to simulate the length of the precursor, where the μ, σ are estimated from the known miRNA precursors of the species in question. For instance, $\mu = 98, \sigma = 6.3$ for C. elegans.

During the generation of a sequence for stem-loops of a certain length, we may find two or more stem-loops on the same strand that has a large percentage of overlap. To avoid excessive overlap, when sliding along the intergenic region, we make a hop of about 50 nt by using a normal distribution $N(50, 20)$ with a large variation.

3.2. *Experimental Results*

We obtain a binary classification SVM on training sets by using the support vector machine library LIBSVM.[4] The input data for the SVM are scaled to $[-1, 1]$. We choose a radial basis function (RBF) kernel. All the experiments were performed in a PC with 1G RAM.

We present the results of three sets of experiments: training the SVM with one of three species D. melanogaster (dme), C. briggsae (cbr), and Mus musculus (mmu) separately and then use the resulting SVMs to predict other species. Due to memory restriction, we are not able to include a large number of negatives in the training set for feature selection. In the experiments, we only include 4000 negatives for feature selection.

Note that the choice of the negatives is an art since different combinations of negatives can lead to different selected feature sets. Hence, we test different combinations and keep those with good testing performance. For example, one data set may consists of 220 mmu

Table 1. Characteristics of training data for feature selection.

species	# of positives	# of negatives	# of features by CFS	# of features by CBFS
mmu	220	4000mmu	177	72
dme	78	4000dme	95	55
cbr	82	2000cbr+2000cel	134	55

Table 2. Experimental Results

Trained species	Test Species	Sensitivity(TP/(TP+FN)%)	Specificity(TN/(TN+FP)%)
dme(120dme150dps,39, 2, 2^{-1})	dps	62/(62+10)=86%	39666/(39666+2996)=93%
cbr(80cbr0cel, 44, 32, 2^{-9})	cel	88/(88+27)=76.52%	76661/(76661+3418)=95.73%
mmu(600mmu150hsa, 62, 8, 2^{-3})	rno	172/(172+13)=92.97%	77370/(77370+4842)=94.11%
mmu(0mmu350hsa, 62, 512, 2^{-7})	hsa	258/(258+63)=80.37%	69792/(69792+6518)=91.46%
mmu(600mmu450hsa, 62,32,2^{-3})	gga	110/(110+12)=90.16%	75069/(75069+4338)=94.54%
mmu(600mmu450hsa, 62, 32, 2^{-3})	ptr	57/(57+10)=85.08%	75203/(75203+3451)=95.61%

positives and 4000 mmu negatives; another data set may consists of 220 mmu positives, 2000 mmu negatives and 2000 hsa negatives. We also use the recursive feature selection method, i.e., we first obtain a feature set from a data set and then form a new data set by projecting the original data against this feature set. This method can put more instances into consideration. However, this method does not necessarily lead to better performance since the feature selection in the first step may be biased. In our experiments, we try two feature selection methods: CFS[7] and CBFS[20] for each combination. In general, the selected features are different for different data sets. The prominent property for all these feature sets is that they primarily consist of features from the encoded secondary structure. Some simple combinations of the negatives for feature selection are listed in Table 1.

Given one species, our purpose is to see if we can find a model to predict the miRNA precursors of another species. For this reason, during the training stage, we only use the positives of one species for training and hold out all the positives of the other species for testing. However, we use some negatives of the target species randomly chosen by assuming that most of the intergenic regions do not contain miRNA precursors. For the first experiment, we use all the known positives (78) of D. melanogaster (dme) and 4000 negatives to perform feature selection using CBFS. Among the 55 selected features, we choose the top 39 to train SVM models—we refer to it as dmeSVM. Among these models, we choose the one with larger area under the ROC curve (AUC). In general, we can get many models with equal AUC. Here, we report the model with 120 dme negatives and 150 dps negatives which has a sensitivity of 86% and a specificity of 93%. We optimize the parameters γ=0.5 and C=2. The prediction results of a species for its related species are given in Table 2. In the first column, the selected model is presented as *species(negative data combination, number of features used, C value, γ value).*

To see the relationship between these miRNA precursors $\mathcal{T}$ in the training set and the miRNA precursors $\mathcal{P}$ to be predicted in the testing set, we implemented a Needleman-Wunsch-based similarity computing algorithm with match score $=1$, mismatch $=-1$, and gap penalty $=1$, and the similarity is computed by the ratio of identities over the whole alignment length denoted as $sim(s,\mathcal{P})$, where $s \in \mathcal{T}$ and $sim(s,\mathcal{P}) =$

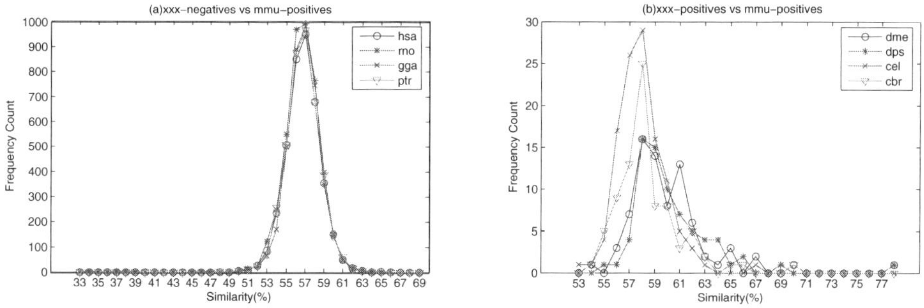

Figure 3. Similarity Histograms against the mmu Positives.

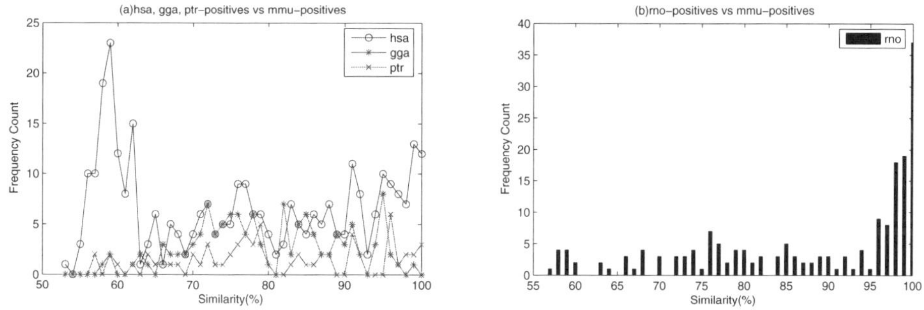

Figure 4. Similarity Histograms of hsa, gga and rno Positives against the mmu Positives.

$max\{sim(s,p) \mid p \in \mathcal{P}\}$. By sampling the negatives with a rate $\frac{1}{20}$ and taking the whole set of positives, we build the histograms of similarities to mmu of both negatives and positives of species other than mmu (Fig. 3). For the histogram of negatives of all species vs mmu positives (Fig. 3(a)), we know that the distribution is an approximate normal with their center around 56-58. This trend is also observed in the histogram of other positives of remote species against mmu-positives Fig. 3(b) which centers around 56-59. Only the later has a little bit longer tail. We show the similarity histogram of its related species in Fig. 4. The comparisons between other species are similar. For human(hsa), there are about 102 miRNA precursors with similarity around 53–62 %. Based on these observations, we can see that SVM's performance is not solely dependent on the primary sequence similarity in some sense. This point is reflected in the selected features.

We also check some false positives by looking at their conservations in their related species. We find that some false positives reach 88% identity in conservation. This indicates that the false positive may be a true positive.

4. Conclusion

In this work, we have described a SVM-based method to predict miRNA precursors. Based on the current number of candidates generated, the method performs well for related species. Future research directions include examining the selected features for biological explanations, investigate the performance for predicting unrelated species, and locating mature miRNA in its precursor.

References

1. J. E. Abrahante, A. L. Daul, M. Li, M. L. Volk, J. M. Tennessen, E. A. Miller, A .E. Rougvie. The Caenorhabditis elegans hunchback-like gene lin-57/hbl-1 controls developmental time and is regulated by microRNAs. Dev Cell 2003, 4:625-637.
2. J. Barciszewski, V. A. Erdmann(Ed.). Noncoding RNAs: Molecular Biology and Molecular Medicine. Kluwer Academic, 2004. Ch3 33-48: P. Schattner. Computational Gene-Finding for Noncoding RNAs.
3. D. P. Bartel. MicroRNAs: genomics, biogenesis, mechanism, and function. Cell, 116:281-297, 2004.
4. C. C. Chang and C.-J. Lin. LIBSVM: a library for support vector machines (2001).
5. Y. Grad, J. Aach, G. D. Hayes, B. J. Reinhart, G. M. Church, G. Ruvkun, J. Kim. Computational and experimental identification of C. elegans microRNAs. Mol Cell 2003, 11:1253-1263.
6. S. Griffiths-Jones. The microRNA Registry. Nucleic Acids Res, 32 Database issue:D109-11, 2004.
7. M. A. Hall. Correlation-based feature selection for machine learning. PhD thesis, The University of Waikato, 1999.
8. I. L. Hofacker, W. Fontana, P. F. Stadler, S. Bonhoeffer, M. Tacker, and P. Schuster. Fast folding and comparison of RNA secondary structures. Monatshefte fir Chemie, 125:167-188, 1994.
9. Y. Lee, M. Kim, J. Han, K. H. Yeom, S. Lee, S. H. Baek, V. N. Kim. MicroRNA genes are transcribed by RNA polymerase II. EMBO J. 23(20):4051-60, Epub 2004.
10. C. Llave, K. D. Kasschau, M. A. Rector, J. C. Carrington. Endogenous and silencing-associated small RNAs in plants. Plant Cell 2002, 14:1605-1619.
11. M. Legendre, A. Lambert, D. Gautheret. Profile-based detection of microRNA precursors in animal genomes. Bioinformatics 21:841-845, 2005.
12. L. P. Lim, N. C. Lau , E. G. Weinstein, A. Abdelhakim, S. Yekta, M. W. Rhoades, C. B. Burge, D. P. Bartel. The microRNAs of Caenorhabditis elegans. Genes Dev 2003, 17:991-1008.
13. E. C. Lai, P. Tomancak, R. W. Williams, G. M. Rubin. Computational identification of Drosophila microRNA genes. Genome Biol, 4:R42, 2003.
14. H. Liu and L. Wong. Data mining tools for biological sequences. J Bioinform Comput Biol., 1(1):139-67, 2003.
15. U. Ohler, S. Yekta, L. P. Lim, D. P. Bartel, C. B. Burge. Patterns of flanking sequence conservation and a characteristic upstream motif for microRNA gene identification. RNA, 10(9):1309-22,2004.
16. A. E. Pasquinelli, G. Ruvkun. Control of developmental timing by microRNAs and their targets. Annu Rev Cell Dev Biol 2002, 18:495-513.
17. N. R. Smalheiser. EST analyses predict the existence of a population of chimeric microRNA precursor-mRNA transcripts expressed in normal human and mouse tissues. Genome Biol 2003, 4:403.
18. V.N. Vapnik. Statistical Learning Theory. Springer, 1998.
19. X. J. Wang, J. L. Reyes, N. H. Chua, T. Gaasterland. Prediction and identification of Arabidopsis thaliana microRNAs and their mRNA targets. Genome Biol. 5(9):R65. Epub 2004.
20. L. Yu, H. Liu. Feature Selection for High-Dimensional Data: A Fast Correlation-Based Filter Solution. ICML, pp:856-863,2003.

GENOME-WIDE COMPUTATIONAL ANALYSIS OF SMALL NUCLEAR RNA GENES OF *ORYZA SATIVA* (INDICA AND JAPONICA)

M.SHASHIKANTH, A.SNEHALATHARANI, SK. MUBARAK AND K.ULAGANATHAN
Center for Plant Molecular Biology, Osmania University, Hyderabad, Andhra Pradesh, India.
kulaganathan@yahoo.com

Genome-wide computational analysis for small nuclear RNA (snRNA) genes resulted in identification of 76 and 73 putative snRNA genes from indica and japonica rice genomes, respectively. We used the basic criteria of a minimum of 70 % sequence identity to the plant snRNA gene used for genome search, presence of conserved promoter elements: TATA box, USE motif and monocot promoter specific elements (MSPs) and extensive sequence alignment to rice / plant expressed sequence tags to denote predicted sequence as snRNA genes. Comparative sequence analysis with snRNA genes from other organisms and predicted secondary structures showed that there is overall conservation of snRNA sequence and structure with plant specific features (presence of TATA box in both polymerase II and III transcribed genes, location of USE motif upstream to the TATA box at fixed but different distance in polymerase II and polymerase III transcribed snRNA genes) and the presence of multiple monocot specific MSPs upstream to the USE motif. Detailed analysis results including all multiple sequence alignments, sequence logos, secondary structures, sequences etc are available at http://kulab.org

1. Introduction

Most eukaryotic protein coding genes contain non-translated intronic sequences that are excised from the primary transcripts (pre-mRNA) by the process of splicing.[1] Splicing of nuclear pre-mRNA involves sequential trans-etherification reactions, which take place in a large complex called the spliceosome. The spliceosome is composed of five snRNAs, U1, U2, and U4, U5 and U6 as well as many proteins.[2] Some of these proteins are tightly associated with the snRNAs, forming small nuclear ribonucleoproteins (snRNPs) which assemble in a stepwise manner onto the pre-mRNA to form the spliceosome. Besides the snRNP subunits, a large number of proteins perform various functions during the splicing reaction.[3]

Four of the five snRNAs base pair with the pre-mRNA at various times during the splicing reaction. Interaction between U1-snRNA and the 5' splice site and U2-snRNA and branch site are well established.[4,5,6,7] U5-snRNA interacts with exonic sequences immediately 5' and 3' to the splice junctions, while the U6-snRNA interacts with the 5' splice site.[8,9,7,10] The snRNA genes are present in multiple copies and are synthesized from independent transcription units, which are transcribed either by RNA polymerase II or III. These genes differ from other classes of genes in having unique transcriptional factors.[11] Though transcribed by two polymerases, their promoters are structurally related.[11,12,13] The snRNA promoters are highly conserved within a species but show variation in different organisms except for certain conserved motifs. So analysis of snRNA genes and their expression may help in understanding the eukaryotic transcription.

Majority of the splicing related genes are known in human,[14] Yeast,[15] *Drosophila*[16] and *Arabidopsis thaliana.*[17] Unlike other organisms, in plants, the knowledge on splicing is scanty due to the non-availability of *in vitro* splicing system. In spite of this, the powerful comparative genomic approach can be employed to predict the genes involved in the splicing process which can be analyzed *in silico* to get a comprehensive picture of the splicing process in an organism like rice. Such analysis will help in prioritizing and better planning for wet-lab experiments to understand the process of splicing. Rice is a unique crop plant used as a model plant for genomic research due to the relatively small genome (ca. 450 MB) and the availability of two genomes, indica and japonica which facilitate comparative analysis. We carried out extensive search of indica and japonica rice genomes using other plant snRNA gene sequences and human splicing associated factors as the basis and predicted a total of 149 snRNA genes and more than 800 protein coding genes associated with splicing from indica and japonica genomes together. Here we provide information about the predicted snRNA genes from indica and japonica genomes and their analysis.

2. Methods

A total of 127551 scaffolds from the draft indica rice genome (super hybrid rice) sequence, downloaded from the Beijing Genomics Institute (http://btn.genomics.org.cn/rice/) were used for the analysis of Indica rice genome. Where as, genome sequences accessed from the Rice Genome Project (RGP) was used for the analysis of japonica rice genome. Rice genomic sequences (indica and japonica) in the form of contigs, cDNA and proteins were collected and a local database was created which can be searched by the key word, accession number or by homology search using BLAST[18] algorithm.

Sequences of splicing associated snRNAs of different plants were collected from various online databases and used to query the local database using the local blast search tool with different parameters. The hits with more than 70% identity to the query sequence were collected and used for the analysis. Based on the BLAST alignment, the open reading frame spanning the aligned sequence was extracted from the contig as the putative gene and verified by pairwise alignment with the query sequence using L-Align.[19]

To validate the prediction, the predicted gene sequences were aligned to other known snRNA genes (using ClustalW) from plants, human, mouse and *Drosophila* and the multiple sequence alignment was given as the first validation. Next, the predicted genes were searched against NCBI EST database and if rice plant ESTs were showing significant stretch of alignment with the predicted genes, then the Unigene cluster number was given as the second validation. The third and final validation of the predicted gene was carried out by analysis of the promoter sequence (upstream 500 bp). The extracted upstream sequences were aligned using ClustalW and the conserved TATA box and USE motif sequences, their location and the distance between them were analyzed. Location of monocot specific elements[20] from the upstream 500 bp sequence of putative SnRNA genes was carried out manually using the MSP consensus sequence, RGCCCR and the motif search function of Bioedit sequence analysis tool.

Validated putative snRNA gene sequences were further used for constructing the secondary structure of snRNAs based on published plant RNA secondary structures. Using predicted secondary structure and the multiple sequence alignment as the basis, the nucleotide variations and conserved features were identified. Using the online sequence logo rendering tool, the variations in different stem loop regions of snRNAs were analyzed and displayed.

3. Results and Discussion

3.1. *snRNA genes in the draft rice genome*

Eukaryotic cells contain a large population of snRNAs, which play a major role in splicing of genes that lead to regulation of gene expression, differentiation and development of an organism. snRNAs are broadly classified into two different categories based on the transcription regulation. snRNAs such as U1, U2, U4 and U5 are transcribed by pol II, where as, U3 and U6 are Pol III transcribed. Using comparative genomic approach, we identified a total of 149 putative snRNA coding genes from the rice genome and validated them by identifying upstream conserved motifs and EST matching.

3.2. *Pol II specific snRNA genes*

To predict pol II specific snRNA genes (U1, U2, U4 and U5) in rice genomes, we used the pol II transcribed plant snRNA genes from *Zea mays*, *Triticum aestivum* and *Arabidopsis thaliana*. The U1 (S72336) and U2 snRNA (S72237) gene sequences were from *T.aestivum,* the U4 snRNA gene sequence (X67145) was from *A. thaliana* and the U5 snRNA gene sequence (Z14995) was from *Zea mays*. These sequences were used as queries for BLAST search against rice genomes with 0.01 e-value as the cut off score. In the BLAST results, the alignments with at least 70 % identity to the query sequence were chosen to identify the potential location of snRNA homologues. This resulted in identification of 58 and 52 putative pol II specific snRNA genes from indica and japonica genomes, respectively. There were 10, 23, 6 and 25 paralogs of U1, U2, U4, and U5 snRNA genes, respectively in indica genome. While in japonica genome, there were 16, 23, 6 and 13 paralogs of U1, U2, U4 and U5 snRNA genes, respectively.

The predicted U1-snRNA genes of indica genome ranged in between 160-164 nucleotide long, while, 157-163 nucleotides long in japonica genome. These genes showed 88-93 % identity to the query sequence. They also showed extensive sequence identity with human and *A. thaliana* snRNA genes (Table 1). All the predicted pol-II snRNA genes were finally validated by analyzing their upstream sequences. The genes that lack the characteristic USE motif were discarded (data regarding these genes not included). All the predicted pol-II transcribed U1-snRNA genes, in their upstream sequence contained the characteristic TATA box and one USE motif at a distance of 30-34 nucleotides upstream of the TATA box (Figure 2). The characteristic distance between the USE motif and the TATA box is a plant specific feature conserved in both indica and rice genomes. Unlike plants, the human, mouse and *Drosophila* snRNA genes transcribed by pol II are TATA box less and pol-III transcribed genes has the TATA box in their promoters. There were 1-5 monocot specific promoter elements found upstream to the USE motif of predicted U1-snRNA gene promoters (Table 1).

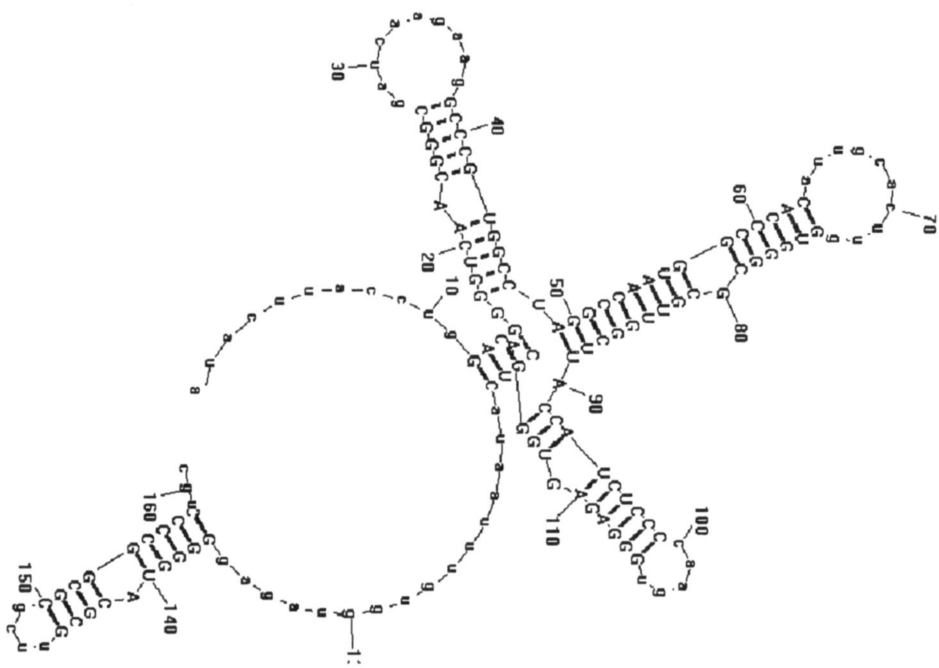

Fig.1. Predicted secondary structure of rice U1-snRNA gene consensus sequence

The secondary structure of *Phaseolus vulgaris* U1-snRNA[21] as the basis rice U1-snRNA secondary structure was drawn with the consensus sequence derived from the nucleotide sequence alignment of indica and japonica U1 snRNA sequence and with the help of RNA structure tool version 4.2. Overall, the structure is conserved in all the rice sequences predicted with variations seen in the stem regions, while the single stranded regions are well conserved (Fig-1). A more detailed explanation regarding the secondary structure is available in our research extension page at http://kulab.org

There were 23 putative U2-snRNA genes in both indica and japonica genomes which were 171-195 nucleotides in length. These genes showed 83.8 to 92.8 per cent sequence identity with the query sequence (*T.aesativum* U2-snRNA gene; S72337). The predicted

U2-snRNA genes in japonica genome showed extensive sequence identity with human (69.1-75.3 %) and *A.thaliana* (83.3-92.8 %) U2-snRNA sequences. EST sequence which belong to the Unigene cluster Os.11638 aligned with the predicted U2-snRNA genes. All the 46 genes contained the characteristic USE motif at 30-34 nucleotides upstream to the TATA box. Upstream to the USE motif there were 1 – 8 Monocot Specific Promoter elements present in the promoters of predicted U2-snRNA genes (Table 1). The secondary structure was drawn using the maize model[22] which showed that the overall structure is conserved. The branch point binding site (GUAGUA) and Sm site (AUUUUUUG) are absolutely conserved in all the 46 genes.

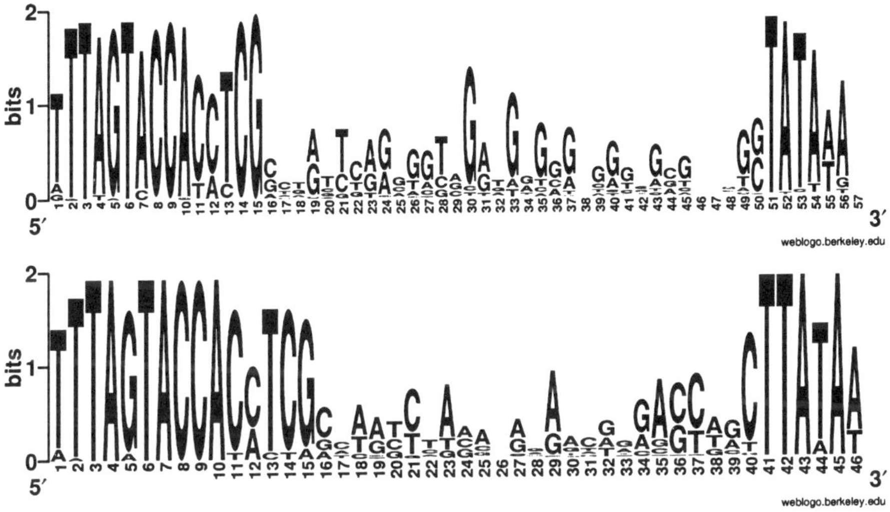

Figure 2. Sequence logos displaying the region between TATA box and the USE motif of pol-II transcribed (U1, U2, U4 and U5)(above) and pol-III transcribed (U3 and U6) (below) snRNA genes of *Oryza sativa* (indica and japonica together).

Each indica and japonica rice genome contained seven U4-snRNA genes (159-160 nucleotide length) showed 80.9 - 83.8 % identity with the query, *A.thaliana* U4-snRNA gene (X67145). All the 14 putative U4-snRNA genes aligned with EST sequences that belong to the Unigene cluster Sbi.6812. The characteristic TATA box and USE motifs were found upstream of all 14 genes. Further upstream to USE motif 1-6 MSPs found (Table 1) (Figure 3). The interacting secondary structure of U4 snRNA consensus and U6 snRNA consensus was made using secondary structure of interecting U4-U6 snRNAs of human.[23] There is overall conservation of the secondary structure of U4-snRNA (Secondary structure image available at http://kulab.org) with three stem loops and three single stranded regions. Two regions were predicted to interact with the U6-snRNA.

A total of 27 U5-snRNA genes were predicted, of which 12 and 15 genes were found in indica (104-107 nucleotides) and japonica (103-107 nucleotides) genomes, respectively. These putative U5-snRNA genes showed 78-95 % sequence identity with maize U5-snRNA gene (Z14995) as the query. These genes aligned with ESTs belonging to the Unigene cluster Os.37121. Upstrem 500 sequences of these putative genes showed the TATA box and the charcteristic USE motif which is located 35 nucleotides upstream to the TATA box. There were 1-9 MSPs located upstream of the USE motif in the promoters of the predicted U5-snRNA genes.

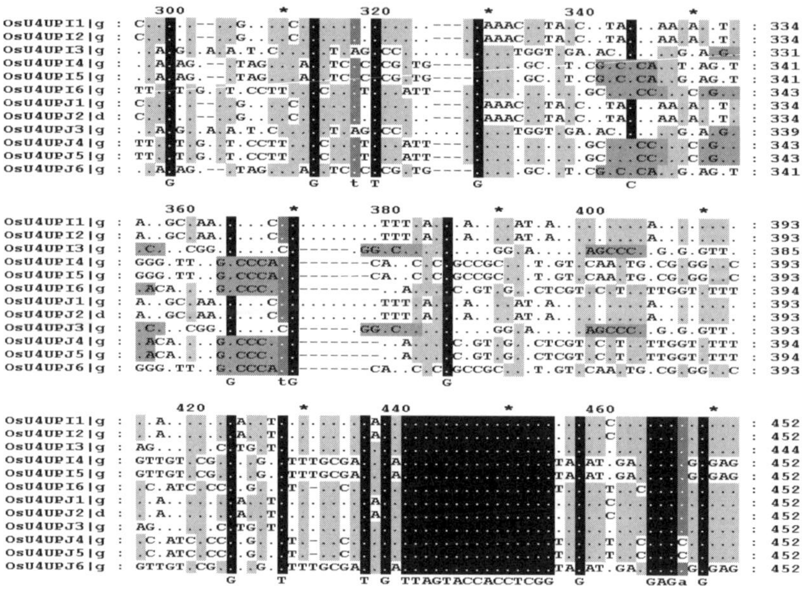

Figure 3. Multiple sequence alignment of upstream-500 sequence of U4 genes showing the conserved USE motif and MSPs.

The secondary structure was drawn using the U5-snRNA consensus sequence which showed that 11 nucleotides of loop-1 is conserved with that of human and yeast U-5 snRNAs. The Sm site and the 3' loop which are essential for Sm protein binding and Cap-tri-methylation are conserved as in human sequences.

Keefe and Newman, 1998, carried out deletion analysis to study the importance of stem loop I of U5 snRNA in yeast. This loop I interacts with 5' exon before first step of m-RNA splicing and with the 5' and 3' exons following the first step. The size of loop I was found to be critical for proper alignment of exons for the second catalytic step of splicing.[24] Rice U5 snRNA showed remarkable conservation of 11 residues (CGCCTTTTACT) among themselves and with other U5 snRNAs indicating that loop size is conserved in rice snRNAs.

3.3. *Pol III specific U3 and U6 snRNA genes*

Homology search for pol III transcribed snRNA genes, U3 and U6 was made using *T.aestivum* U3 (X63065.1) and U6 (X63066) genes as query sequences. This resulted in identification of seven and eight putative U3 snRNA genes in indica and japonica genomes, respectively. These putative U3-snRNA genes showed 80-86% identity and U6 snRNA has 91-99% sequence identity with the query sequence used There were 12 and 15 putative U6 snRNA genes in indica and japonica genomes, respectively (Table 1).

Similar to pol II specific genes, the predicted U3 and U6 snRNA genes of rice also contained three most important upstream promoter elements, the Upstream Sequence Element (USE, TTAGTACCACCTCG), the TATA box and one or more MSPs. The USE and TATA box were 23-26 bp apart (Figure 2) and the latter one lies 21-25 bp upstream of the transcription start site.

The U6 motif had the consensus GTTTAGTACCACCTCG and was present exactly 25 nucleotides upstream of TATA box. In few genes there is a 5 nucleotide deletion in between the TATA and USE motifs. The 14 putative U3 snRNA genes showed the

Table -1. Putative snRNA genes in indica and japonica genomes (Detailed tables for each category of snRNAs are available online at http://kulab.org)

Name of the gene	No. of Copies	Length of the predicted gene	Percent identity to the query	EST matching: Unigene cluster number	Percent identity to Human gene	Percent identity to *A.thaliana* gene	Upstream USE motif	Number of MSPs found
Japonica genome								
U1 snRNA	16	157-163 S72336	(88-93) Ta.28789	J00318 (63-68)	X53175	(72-78)	TTTAGTACCACCTCG	1-5
U2 snRNA	23	171-195	S72337 (87-92)	Os.11638	X01408 (69-75)	X52312 (80-84)	TTTAGTACCACCTCG	1-7
U3 snRNA	6	228-311	X63065 (81-90)	Os.39667	X14945 (52-58)	X52629 (64-73)	TTTAGTACCACCTCG	1-3
U4 snRNA	6	159	X67145 (81-83)	Sbi.6812	X59361 (63-71)	X67145 (81-83)	TTTAGTACCACCTCG	1-5
U5 snRNA	13	107-103	Z14995 (86-91)	Os.37121	X04293 (72-81)	X13012 (86-96)	TTTAGTACCACCTCG	1-9
U6	15	102-105	X52315 (95-99)	At.47201	X07425 (81-86)	X63066 (96-99)	TTTAGTACCACCTCG	1-4
Indica genome								
U1 snRNA	10	160-164	S72336 (88-93)	Ta.28789	J00318 (66-69)	X53175 (70-77)	TTTAGTACCACCTCG	1-5
U2 snRNA	23	192-194	S72337 (94-96)	Os.11638	X01408 (70-84)	X52312 (72-89)	TTTAGTACCACCTCG	1-8
U3 snRNA	6	232-325	X63065 (82-90)	Os.39667	X14945 (52-54)	X52629 (66-75)	TTTAGTACCACCTCG	1-3
U4 snRNA	6	87-92	X67145 (93-96)	Sbi.6812	X59361 (70-73)	X67145 (93-96)	TTTAGTACCACCTCG	1-6
U5 snRNA	25	104-107	Z14995 (89-95)	Os.37121	X04293 (75-83)	X13012 (83-97)	TTTAGTACCACCTCG	1-6
U6 snRNA	12	97-101	X52315 (96-99)	At.47201	X07425 (68-86)	X63066 (76-85)	TTTAGTACCACCTCG	1-4

conserved boxes, A, A^0, B, C and D which were identified in plant U3 snRNA genes[13]. There are no similar conserved boxes seen in U6 snRNA genes but they possessed extensive conserved regions. Both U3 and U6 snRNAs contained many copies of the MSP elements in their promoters, upstream to the USE motif. Unlike pol-II transcribed snRNA genes, in U3 and U6 there were fewer MSPs found (Table 1). Further, the predicted U3 and U6 snRNAs were found to align with EST sequences belonging to the Unigene clusters Os.39667 and At.47201, respectively.

Homology searching of indica and japonica genomes with plant snRNA sequences as queries resulted in the prediction of a number of putative snRNA genes. These putative regions were further analyzed only if they showed at least 70 % sequence identity with the query snRNA sequence used. Further, those predicted genes that lacked the characteristic USE motif were discarded which resulted in identification of 76 and 73 putative snRNA coding genes from indica and japonica rice genomes, respectively. Analysis of these snRNA genes showed that they are conserved as in case of other organisms like human, with respect to the sequence and overall secondary structure. These genes showed the following plant specific features: a) Conform to the overall predicted secondary structures of plant snRNA genes. b) All snRNA genes possessed TATA box in their promoters. c) Plant specific USE motif was found in the promoter upstream to the TATA box. d) Multiple MSP elements were found in the promoter upstream to the USE motif. e) The spacing between the USE and TATA box is the major determinant of pol specificity of plant snRNA genes. In the promoters of Pol II-specific genes the USE motif and the TATA box are centered approximately four DNA helical turns apart (30-34 bp), while in pol III-specific genes these elements are positioned one helical turn closer (20-25 bp). The TATA box is present 23-28 bp upstream in pol II genes and 21-25 bp upstream in pol III genes from the coding region. With respect to rice specific features, there are specific nucleotide variations seen in the snRNAs especially in the single stranded regions and stem of the various stem loops. Wherever variation in nucleotide is observed in the stem of loops, mostly they are compensatory (data not shown). There is variation in number of copies of different snRNA genes, between the two rice genomes with respect to U1, U5 and U6 while the number of copies with respect to U2, U3 and U4 is constant in both genomes. U5 snRNA, especially showed huge variation from 25 copies in indica genome to 13 copies in japonica genome which may be probably due to duplication in indica genome or deletion in japonica genome.

Further more, the predicted secondary structure is similar to the known structure of their counter parts in other plants,[25] indicating the structural conservation. Some of the predicted genes did not have the conserved USE in the promoter region (data not included) but their structure is absolutely conserved. These genes may be pseudogenes and may not be expressed. The Pol II and Pol III snRNA gene promoters of rice, in addition to mammals, frogs, dicot plants, and possibly nematodes,[26] represent one more example of promoters which are remarkably similar despite being recognized by two different RNA Polymerases. Our finding in rice regarding the conserved promoter regions further support that Pol II and Pol III transcription machineries are highly conserved.

Previously, snRNA genes have been predicted mostly from japonica genome of which the Rfam database had the maximum number of snRNAs of japonica rice.[27] Sequence alignment of those snRNAs with the snRNAs predicted in this work showed that many of them aligned perfectly without much variation (data not shown). But our prediction, though started initially with homology search of the genomes, is superior as we did not rely on simple sequence identity alone to predict the gene. We validated the predicted sequences by a) first with the presence of USE motif in the upstream sequences, then b) by the presence of TATA box in the upstream sequences and the distance between the USE motif and TATA box, c) by the presence of multiple MSP elements in the promoters, upstream to the USE motif, d) by alignment to expressed sequence tag sequences, and e) by conservation of structural features (stem loops, protein binding sites and interaction sites between snRNAs) based on the secondary structure.

References

1. C. Burge, T. Tuschl and P.A. Sharp. Splicing of precursors to mRNAs by the spliceosome. In R. F. Gesteland, T. R. Cech and J. F. Atkins (eds), *The RNA World.* 2nd edn. Cold Spring Harbor Laboratory Press, Cold Spring Harbor, NY, 525-560, 1999.
2. M. J. Moore, C. C. Query and P.A. Sharp. Splicing of precursors to mRNAs by the spliceosome. In R. F. Gesteland, T. R. Cech and J. F. Atkins (eds), *The RNA World.* Cold Spring Harbor Laboratory Press, Cold Spring Harbor, NY, 303-357, 1993.
3. C. L. Will. and R. Luhrmann. snRNP structure and function. In Krainer, A.R. (ed.), *Eukaryotic mRNA processing.* IRL press at Oxford University Press, Oxford, UK, 130-173, 1997.
4. A. J. Newman and C. Norman. Mutations in yeast U5 snRNA alter the specificity of 5' splicing – site cleavage. *Cell*, 65: 115-123, 1991.
5. A. J. Newman and C. U5 snRNA interacts with exon sequences at 5' and 3' splicing sites. *Cell*, 68: 743-754, 1992.
6. J. J. Cortes, E. J. Sontheimer, S. D. Seiwert and J. A. Steitz. Mutations in the conserved loop of human U5 snRNA generate use of novel cryptic 5' splice sites in vivio. *EMBO J*, 12: 5181-5189, 1993.
7. E. Sontheimer and J. A. Steitz. The U5 and U6 small nuclear RNAs as active site components of spliceosome. *Science*, 262: 1989-1996, 1993.
8. H. Sawa and J. Abelson. Evidence for base-pairing interaction between U6 small nuclear RNA and 5' splicing site during the splicing reaction in yeast. *PNAS*, 89: 11269-11273, 1992.
9. D. A. Wassarman and J. A. Steitz. Interactions of small nuclear RNA's with precursor messenger RNA during in vitro splicing. *Science*, 257: 1918-1925, 1992.
10. J. S. Sun and J. L. Manley. A novel U2-U6snRNA structure is necessary for mammalian mRNA splicing. *Genes and Dev.*, 9: 843-854, 1995.

11. N.Hernandez. Transcription of vertebrate snRNA genes and related genes. In McKnight,S.L. and Yamamoto,K.R. (eds), *Transcriptional Regulation*. Cold Spring Harbor Laboratory Press, Cold Spring Harbor, NY, Vol. 1, pp. 281-313, 1992.
12. P.C.H.Lo and S.M.Mount. *Drosophila melanogastor* genes for U1 snRNA variants and their expression during development. *Nucleic Acids Res.*, 18:6971-6979, 1990.
13. F.Waibel, W.Filipowicz. RNA-polymerase specificity of transcription of Arabidopsis U snRNA genes determined by promoter element spacing. *Nature*, 346:199–202, 1990.
14. J. Rappsilber,U. Ryder,A. I. Lamond, and M. Mann. Large-Scale Proteomic Analysis of the Human Spliceosome. *Genome Research*, 12:1231–1245, 2002.
15. C. W. Pikielny and M. Rosbash. Specific small nuclear RNAs are associated with yeast spliceosomes. *Cell*, 20: 869-877, 1986.
16. M. M. Stephen and K. S. Helen. Pre-messenger RNA processing factors in the *Drosophila* Genome. *The J. Cell Bio.*, 150: F37-F43, 2000.
17. B. B. Wang and V. Brendel. The ASRG database: identification and survey of *Arabidopsis thaliana* genes involved in pre-mRNA splicing. *Genome Biology*, 5: R102- 102.23, 2004.
18. S. F.Altschul, W.Gish, W.Miller, E.W.Myers and D.J.Lipman. Basic local alignment search tool. *J. Mol. Biol.,* 215: 403-410, 1990.
19. X. Huang. and W. Miller. A time-efficient, linear-space local similarity algorithm. *Adv. Appl. Math.* 12:337-357, 1991.
20. S. Connelly,C.Marshallsay,D.Leader, J.W. S. Brown, and W.Filipowiczi. Small Nuclear RNA Genes Transcribed by Either RNA Polymerase II or RNA Polymerase III in Monocot Plants Share Three Promoter Elements and Use a Strategy To Regulate Gene Expression Different from That Used by Their Dicot Plant Counterparts. *Mol. Cell. Biol.*, 14:5910-5919, 1994.
21. V. L. Van Santen and R. A. Spritz. Nucleotide sequence of a bean (Phaseolus vulgaris) U1 small nuclear RNA gene: Implications for plant pre-mRNA splicing. *PNAS*, 84: 9094-9098, 1987.
22. J.W.S. Brown and R. Waugh. Maize U2 snRNAs: gene sequence and expression. *Nucleic Acid Research,* 17: 8991- 9001, 1989.
23. A. Mougin, A.Gottschalk, P. Fabrizio, R.Luhrmann and C. Branlant. Direct Probing of RNA Structure and RNA-Protein Interactions in Purified HeLa Cell's and Yeast Spliceosomal U4/U6.U5 Tri-snRNP Particles. *J. Mol. Biol.* 317: 631-649, 2002.
24. R. T. O'Keefe and A. J. Newman. Functional analysis of U5 snRNA loop I in the second catalytic step of yeast pre m-RNA splicing. *EMBO J.* 17: 565-574, 1998.
25. A.A.Patel and J.A.Steitz. Splicing double: Insights from second spliceosome. *Nat. Rev. Mol. Cell Biol.,* 4:960-970, 2003.
26. G.J.Goodall and W.Filipowicz. Different effects of intron nucleotide composition and secondary structure on pre-mRNA splicing in monocot and dicot plants. *EMBO J.,*10: 2635–2644, 1991.
27. S. G. Jones, A. Bateman, M. Marshall, A. Khanna and S. R. Eddy. Rfam: an RNA family database. *Nucleic Acids Res.*, 33:439-441, 2003.

RESOVING THE GENE TREE AND SPECIES TREE PROBLEM BY PHYLOGENETIC MINING

XIAOXU HAN

Department of Mathematics and Bioinformatics Program, Eastern Michigan University Ypsilanti, MI 48197, USA

The gene tree and species tree problem remains a central problem in phylogenomics. To overcome this problem, gene concatenation approaches have been used to combine a certain number of genes randomly from a set of widely distributed orthologous genes selected from genome data to conduct phylogenetic analysis. The random concatenation mechanism prevents us from the further investigations of the inner structures of the gene data set employed to infer the phylogenetic trees and locates the most phylogenetically informative genes. In this work, a phylogenomic mining approach is described to gain knowledge from a gene data set by clustering genes in the gene set through a self-organizing map (SOM) to explore the gene dataset inner structures. From this, the most phylogenetically informative gene set is created by picking the maximum entropy gene from each cluster to infer phylogenetic trees by phylogenetic analysis. Using the same data set, the phylogenetic mining approach performs better than the random gene concatenation approach.

1 Introduction

1.1. *Gene tree and species tree problem*

The gene tree/species tree problem is still an important problem in phylogenomics. A gene tree is a phylogenetic hypothesis constructed from one gene; it may not represent the true evolutionary history of the species [1]. On the other hand, a species tree reflects the species evolutionary history correctly, but it is generally unknown to investigators for a group of organisms. Incongruence among species trees and gene trees simply means that gene trees are not isomorphic with species trees. The incongruence occurs due to possible biological or analytical reasons in the phylogenetic reconstruction. The biological reasons include paralogy, lineage sorting and horizontal gene transfer [1,2,3]. The analytic reasons can be the data sampling bias and fit of the mathematical models in the phylogenetic analysis [4,5,6]. Both of them can lead to the artifacts in phylogeny. There are many excellent models and approaches to address the gene tree and species tree problem from different point of view based on these factors [7,8].

There are two most recently proposed approaches to solve the species tree and gene tree problem. The first approach is to use complete genome data in the phylogenetic inference [9,10]. This approach removes the possible gene trees and species tree problem since all information is incorporated in the tree inference by comparing gene contents or gene orders [9,10,11]. Although it shows strong potential, such approach suffers from the primitive mathematical model and temporal data scarcity problem [11]. For example, the main algorithm employed in the gene-order based phylogenomic reconstruction is break-point analysis, a method to minimize the number of breakpoints between genomes, where a NP-hard problem has to be solved in the each iteration. The current genome data needed for genome based approaches are still far from abundant compared with general sequence data although more than 260 genomes already sequenced and thousands of genome sequencing is in progress [9]. Another approach is called gene concatenation [12]. Its main idea is to include more genes involved in gene tree reconstruction by randomly combining a set of widely distributed orthologous genes selected from genome data. Rokas *et.al.* [12] randomly concatenated genes from a set of selected 106 widely distributed orthologous genes for seven *Saccharomyces* species (*S. cerevisiae, S. paradoxus, S. mikatae, S. kudriavavzevii, S. bayanus, S. castellii, S. kluyveri*) and an outgroup *Candida albicans* in their experiment. They showed that the phylogenetic analysis results from maximum parsimony (MP) and maximum likelihood (ML) of the DNA and corresponding protein sequences were the species tree for at least twenty (an experimental number) gene combinations. The random gene combination strategy gives no consideration of gene functionality in the phylogeny. Such a random concatenation strategy may bring noise signals into the phylogenies because the 106 selected genes may not be a sampling bias free data set and different genes may have different evolution history. It is possible that the noise signals play negative roles in the phylogeny. The experimental gene concatenation number in phylogenetic inference has to be reobtained for different data sets each time through large scale simulations. Furthermore, there is little knowledge known about the genes involved in the phylogenetic reconstruction except the basic orthology. The ad-hoc strategy of the random concatenation method inhibits biologists from resolving the species tree and gene tree problem robustly in the phylogenomic era.

1.2. *Gene concatenation under Bayesian analysis*

We define terminology "gene convergence" and "tree posterior probability" in our phylogemomic analysis. A gene is defined as a convenegent gene or a "good" gene if its gene tree is the species tree under a phylogenetic reconstruction model R. Otherwise the gene is called a nonconvergent gene or a "bad gene". In the 106 gene set G used by Rokas *et. al.*(we also use the same gene set), there are 45 convergent genes and 61 nonconvergent genes under a Bayesian analysis with the GTR+Γ model [13, 14]; The tree posterior probability for a evolutionary tree inferred from Bayesian analysis $t_p = \prod_{i=1}^{|B_I|} b_p^i$ is defined as the multiplication of all posterior probabilities of its inferred branches, where the b_p^i is the posterior probability in the ith inferred branch and B_I is the set of all inferred branches (the corresponding branches related to the outgroup taxon are excluded).

We conduct a random gene concatenation under Bayesian analysis as follows. We randomly concatenate the genes according to three cases: good gene concatenation, bad gene combination and random gene combination for total gene set. For each gene combination case, we generate 10 random data sets for Bayesian analysis under the GTR+Γ model and compute the expected tree posterior for the ten trials. We found that simulation results were congruent with the results from those of Rokas *et. al*'s. although different phylogenetic analysis methods were employed. With the increase of the number of genes in the concatenation, we observe that the tree average posterior probability increases for each combination case (Figure 1). The final evolution tree inferred by Bayesian analysis is the species tree with maximum support (Figure 2).

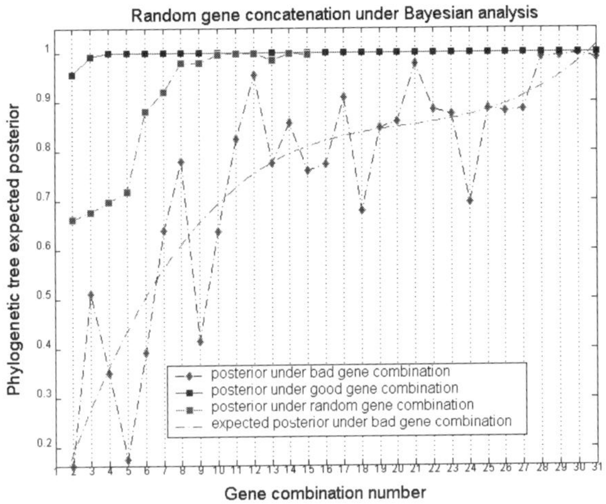

Figure 1. Random gene concatenation under Bayesian analysis.

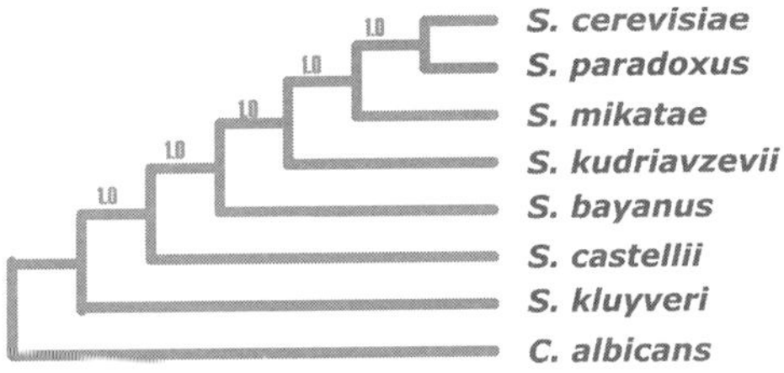

Figure 2. The species tree inferred by Bayesian analysis with the maximum support.

However, the performance from three types of combinations is quite different. To reach the best expected tree posterior probability, it requires at least 4 genes in the good gene combination case, at least 14 genes in the random gene combination case and at least 28 genes in the bad gene combination case, which is "the worst case" in this random gene combination method. In "the worst gene" combination case, there are still generally at least 10% gene combination sets whose gene trees are not the species tree if the gene combination sets have less than 28 genes. After computing the phylogenetic tree posterior probability standard deviations of gene combinations under bootstrap with 1000 bootstrap samples, we found that the bad gene combination had highest level oscillations and the good gene combination had lowest level oscillations in the plots of tree posterior probability standard deviations.

Actually, biologists don not know which genes are good or bad before knowing the species trees for a set of organisms. Thus, the worst gene combination case is unavoidable in the ad-hoc gene combination approach because investigators are not only blind to the inner structure of the gene set but also have little knowledge about which genes are more phylogenetically informative, not to mention the possible situation where noisy data in the phylogenetic inference increase after gene combination.

1.3. *Overview of the methods and results*

In this paper, we develop a robust phylogenomic mining approach to address the gene tree/species tree problem. Self-organizing map (SOM) mining is employed to acquire knowledge from the gene set G using same data set as Rokas *et al.* [12] before the phylogenetic analysis starts. The genes in the gene set G are clustered into different sets according to their gene prototypes obtained from SOM mining. Then the maximum entropy gene from each cluster is selected to create a core gene set G_{core}. A phylogenetic tree will be inferred based on the core gene set. Our experimental results show the phylogenetic analysis (Maximum likelihood, Bayesian analysis) based on G_{core} always infers the desired species tree robustly. Compared with previous approaches, our phylogenomic mining approach can overcome the incongruence between the species tree and gene tree systematically. To our knowledge, this is the first work to integrate SOM mining and entropy theory in phylogenomics.

The paper is organized as follows. In the next section, we describe the basic idea of the phylogenomic mining. The third section describes the detailed phylogenomic mining method and related results. The fourth section presents gene entropy theory and its applications in phylogenetic inference. In addition to exploring the phylogenetic properties of the maximum entropy genes by Shimodaira-Hasegawa test, the Bayesian analysis is conducted to the core gene set created by picking maximum entropy genes from each gene cluster. The Bayesian phylogenetic analysis results are also compared with the random and good gene concatenation cases. Finally, we discuss our results and directions for future work.

2 Phylogenomic Mining: Knowledge Discovery from a Gene Set by Self-Organizing Map (SOM) Learning

The random gene concatenation method treats all genes uniformly in the phylogeny. The basic idea of the phylogenomic mining is to gain some knowledge from a gene set G by some unsupervised data mining algorithms before phylogenetic tree reconstruction. The gene set consists of aligned orthologous genes in our data set (the gene set actually can be rather huge, for example, a set of genomes.) The knowledge learning from gene set G is clustering genes in the gene set to gain insight into the possible structures of the gene set G, to identify some outliers, and to find possible phylogenetically informative genes. These phylogenetically informative genes may contain more evolutionary information based on a gene entropy measure; if so, they will be selected to create a core gene set G_{core}. The final phylogenetic tree will be reconstructed from the gene set carrying the most phylogenetic information. We will demonstrate that the final evolutionary tree from this selected set converges to the species tree with maximum support.

It is highly discouraged to cluster genes directly since different genes have different lengths. To cluster genes directly, Genes have to be encoded into uniform column vectors. That is, zeros have to be filled into the encoded vectors for the short genes. The filled zeros will act as "missing data" in the clustering and it is not advisable to do clustering under such a condition [15]. Thus, we suggest an alternative way to cluster genes. The core idea of the phylogenomic mining is to cluster the prototypes of genes rather than genes themselves directly.

The prototype of a gene is a small dataset containing representative features of genes, compared with a gene (generally a high dimensional dataset). For example, the probability mass function of a gene $x = x_1 x_2 ... x_n$ on the R^2 space can be a gene prototype. Self-organizing map (SOM) is a traditional data mining approach and vector quantization model to map a high dimensional data set D to its representative prototype $W \in R^2 : SOM : D \to W \in R^2$. It takes competitive unsupervised learning (self-organization) to partition original data space into a set of corresponding representative prototypes. With the assumption of no prior knowledge about the data to be mined, the unsupervised learning in SOM is a process of feature extraction and data dimension reduction. SOM mining is widely used in gene expression data processing, vector quantization, visualization and commercial database mining [15,16,17].

3 The Phylogenomic Mining Method in Detail

The phylogenomic mining method to resolve the gene tree and species tree problem mainly consists of the following six steps: 1.Encoding the gene set G into a digital matrix D; 2. Conducting SOM mining for the data set D; 3.Computing the gene distribution on SOM plane P for each gene; 4.Clustering gene hierarchically; 5.Selecting maximum entropy genes from each cluster to build a core gene set; 6.Conducting phylogenetic analysis for the core gene set to infer gene trees.

The first step is to encode an input character matrix to its corresponding digital matrix D before phylogenomic mining. In our analysis, the input character matrix (the transposition of the general character matrix) is a 127026×8 matrix which consists of 106 genes of the eight yeast species: Scer, Spar, Smik, Skud, Sbay, Scas, Sklu, Calb. Each column represents a taxon, each row represents a site and a gene crosses certain number of rows according to its length. The four orthogonal bases are used to encode 'A', 'T',' C', 'G' respectively: A=(1,0,0,0)′, T=(0,1,0,0)′, C=(0,0,1,0)′, G=(0,0,0,1)′. Missing nucleotides and gaps are encoded as a vector with four zeros entries. The input character matrix is encoded as a 508104×8 digit matrix D after such encoding processing.

The digit matrix D is then sent to a self-organization map (SOM) for mining in the second step. The SOM takes an input/output plane with 20×20 neurons and employs sequence learning algorithm. The reference vectors are initialized by the principal component analysis. The neighborhood kernel function used is a Gaussian function. The prototypes of species, which are also called the species patterns, can be obtained immediately after the SOM mining. It is just the final reference vector matrix W. The iteration process is time consuming since the complexity of each training epoch is $O(nmk)$ where the sample number $n = 508104$, species number $m = 8$ and the number of neurons $k = 400$. In our analysis, the SOM mining takes more than 8 hours time to get the species prototype after more than 2000 epochs in a Pentium 4 with a 2.1 GHz CPU.

The third step in our method is to compute the prototypes of genes which are the gene distributions on the SOM plane. The gene prototypes are represented in a matrix $W_1(l \times k)$ in our computing, where l is the number of genes involved in the mining ($l = 106$) and k is the number of neurons on the SOM plane P.

For a multi-species gene $x = x_1 x_2 ... x_n$, its probability mass function on the SOM plane P is its prototype $y = y_1 y_2 ... y_k$ after SOM mining, where k is the number of neurons on the SOM plane. The prototype of a gene is a set of representative features extracted from the dataset x. It will follow the statistical property of the gene x and can be considered as an approximation to the original probability mass function $p(x)$ [18].

The gene distribution on the SOM plane P can be computed as follows. For a gene sample x_i (a site/character), there exists a best match unit j $(j = 1,2,...k)$ whose reference vector is most similar to x_i in the Euclidean distance measure. Then for each neuron j on the SOM plane, there exist a sample set $s_j = \{x_{i_1}, x_{i_2}, ... x_{i_j}\}$, each of its entry acknowledges the neuron j as its best match unit (BMU). That is,

$$s_j = \{x \mid \arg\min_i \| x - w_i \| = j, i = 1,2,...n\} \quad (1)$$

The size of s_j : $|s_j| = h_j$ stands for the number of gene samples of the gene x which are closest to the reference vector in the neuron j in the Euclidean distance; that is, there are h_j gene samples hitting the jth map unit on the SOM plane. So for each gene x in the gene set G, there exists a corresponding hit number sequence $h_x = h_1 h_2 ... h_k$ recording the projection of the gene on the SOM plane P.

Thus, the prototype of the gene x is a sequence of hit rate: $y = y_1 y_2 ... y_k$, where y_i is the hit rate of the gene x on the ith neuron on P: $y_i = h_i / \Sigma_{j=1}^{k} h_j$. The gene distribution on the SOM plane P extracts the representative features from each gene in a uniform format by tracing the projection of each gene on the SOM plane P. Compared with the raw genes, the prototypes of genes are more representative and make gene features and patterns comparable and visualizable. The gene clustering based on the prototypes can help us identify the clusters of genes sharing same features which are biological meaningful. From the visualizations of gene distributions on SOM plane (data not shown), we find that the distributions show interesting patterns: the genes with similar gene distributions are clustered close together on the SOM plane. This is actually the characteristic of the SOM learning, that is, data closer to each other in the high dimensional space are mapped to the neurons in R^2 close to each other topologically [18].

The fourth step in our phylogenomic mining clusters genes hierarchically based on the prototypes of genes. Although SOM itself is a fuzzy clustering algorithm, where similar data samples are mapped in the neighborhoods of BMUs, the data prototype still needs to use hierarchical or partitive clustering to explore the global similarity in the data set [19]. The gene clustering is conducted through the hierarchically clustering the prototypes of genes into natural grouped clusters in our analysis. The natural division of the prototypes is achieved by specifying the inconsistency coefficiency or cutoff number obtained by computing U-matrix for prototype vectors [19]. In our experiment, the 106 orthologous genes from eight species are clustered as 18 naturally grouped clusters. Gene clustering helps us to discriminate between the genes before phylogenetic analysis to overcome the "blindness" in the phylogeny reconstruction where no consideration is given to relationships between genes. It is interesting to see that each cluster shares similar phylogenetic properties in addition to the fact that they have similar gene distributions on the SOM plane. For example, all genes in cluster *4, 5, 12, 15, 18* are nonconvergent ("bad") genes.

4 Exploring Phylogenetically Informative Genes by Gene Entropy

After the gene clustering is completed, a core gene set G_{core} will be built by selecting phylogenetically informative ("important") genes from each cluster before the phylogenetic analysis to infer the final tree. But which genes are phylogenetically informative genes in a cluster and how do we identify them? The initial thought is to employ principal component analysis. But it cannot return to the original gene space since the final data we need are symbolic data in the phylogenomic reconstruction. Thus, another method has to be found to measure the utility of genes in phylogenetic reconstruction. Recalling that a gene can be a set of patterns generated from an alphabet set $\Gamma = \{A, T, C, G\}$, we borrow entropy from information theory to measure the phylogenetic informative potentials of a multiple species gene. The informative genes will selected from each cluster according to gene entropy values to build a core gene set for further phylogenetic analysis.

We give the definition of gene entropy as follows. For a gene $x = x_1 x_2 ... x_n$ (which is a character matrix, a set of aligned sequences; each character x_i is a column in the character matrix), the gene entropy is defined as

$$h(x) = -\sum_{i=1}^{n} p(x_i)\log(p(x_i)) \tag{2}$$

The gene entropy is equivalent to the block entropy definition in DNA entropy if the character matrix is converted to a corresponding one dimensional sequence and the block length is the length of a site. Although the entropy estimate can be conducted by block entropy estimate approaches theoretically, the estimate may not be satisfied since the block length is generally assumed a large number [20]. Since the gene distribution on the SOM plane is the frequency distribution on the SOM based on for each site, such gene prototype is perfectly fitted to estimate the gene entropy, where the gene probability mass function on the SOM plane is the distribution of each site in the gene set.

We predict that a gene with large entropy value will contain more phylogenetic information. Thus, a higher entropy gene is highly likely to be a phylogenetically informative gene, for example, a "good" gene (a gene whose gene tree is nearest to the species tree). To verify such hypothesis, we conduct the Shimodaira-Hasegawa test (SH test) [21] to compute the delta log likelihood and associative p-value for each gene tree with respect to the species tree after ML analysis for each gene. The SH-test is conducted under the GTR+Γ model with reestimated log likelihoods (RELL) approximation for 21 tree topologies for each gene. From the SH-test results, we can see 8/10 genes in the high entropy zone (HEZ) are convergent, (where gene entropy is not less than the sum of mean and standard derivation of all gene entropy in the gene set G). On the other hand, there are 10/12 genes in the low entropy zone (LEZ) that are nonconvergent genes, (LEZ is a set of genes whose gene entropy is not greater than the difference of mean and standard derivation of all gene entropy in the gene set G). There are 31 genes among the 45 convergent genes in the 106 gene set that have entropy greater than the average entropy of the gene set. In sum, higher entropy genes are more likely to be "good" genes in phylogenetic reconstruction.

4.1 Maximum entropy gene concatenation

It is reasonable to combine maximum entropy genes from each cluster into a core gene set and conduct phylogenetic analysis for this core gene set, because a high entropy gene appears to be more likely to be phylogenetically informative. A maximum entropy gene is the "local maximum" gene whose entropy is the maximum for all genes in a particular cluster instead of among the total gene set. Compared with the general random gene concatenation, this phylogenomic mining based approach may be more systematic and robust to resolve the incongruence in the phylogeny because gene combinations are done based on an information measure after phylogenomic mining. There is no experimental gene number computing needed if this new approach is applied to different data set. On the other hand, maximum entropy gene selection from each cluster can remove potential "noise" signals contained in the non-convergent genes, which appear from our analyses, more likely to include low entropy genes. Furthermore, maximum entropy gene concatenation based on clustering prevents from the "data bias" problem if the total gene set obtained has over-representation of one or several types of genes due to data acquisition issues. This property may ameliorate over-support for some branches in the final consensus tree.

We conduct following three experiments to build a core gene set by selecting maximum entropy genes to infer phylogenetic trees. In experiment 1, we build the core gene set G_{core} by selecting a maximum entropy gene from each cluster among 18 clusters. The core gene set consists of following 18 maximum entropy genes selected from available gene clusters: {*YIL125W, YNL220W, YOL049W, YNL082W YDL195W, YPL169C, YDL116W, YJL085W, YDL126C, YMR186W, YOR158W,YPL210C, YJL087C, YNR008W, YLR253W, YIL109C, YFR044C, YGR285C*}. The mean entropy of the core gene set is 5.2325 bits. The gene tree inferred from the Bayesian analysis under GTR+Γ model is the species tree where the posterior probability of 1.0 is on each inferred branch (Figure 2).

In experiment 2, we can further cluster the total gene set G into an arbitrary number of clusters and pick the maximum entropy genes from each cluster. This approach answers the query: suppose the gene set G is clustered as $\#j$ clusters, which genes are the most informative to build a robust phylogenentic tree? If we treat the whole gene set as a cluster (*i.e.* no clustering work), the maximum entropy gene 18 (*YDL126C*) will be the only gene in the core gene set. The corresponding tree probability t_p for gene 18 is 1.0 after Bayesian analysis. Similarly, the core gene set consists of $\#j$ maximum entropy gene selected from $\#j$ clusters if the total gene set G is clustered into $\#j$ sets. We test all cases for $\#j$ from 1 to 14, the corresponding tree posterior probability for each case is 1.0 after Bayesian analysis. Such striking results suggest that our approach can be useful to infer the species tree systematically and robustly. However, such maximum entropy gene selection depends on the gene clustering. Randomly selecting several higher entropy genes may not produce good results because more genes selected from a same cluster will make the sampling bias occurrence which may give "over support" for specific branch patterns.

In experiment 3, we compare phylogentically informative gene (maximum entropy gene selection from each cluster) concatenation under phylogenomic mining with random and good gene concatenations. Just as before, for each gene combination case, we generate 10 random data sets for Bayesian analysis under the GTR+Γ model and compute the expected tree posterior for the ten trials. We found our approach performance is even better than the good gene combination performance from the phylogenetic tree expected posterior probability analysis (Figure 3). Considering the species tree is actually unknown to investigators, we suggest that our method can provide a systematic solution to the gene tree and species tree problem. Moreover, it is well suitable to identify potential phylogenetically informative genes by phylogenomic mining in large genome databases.

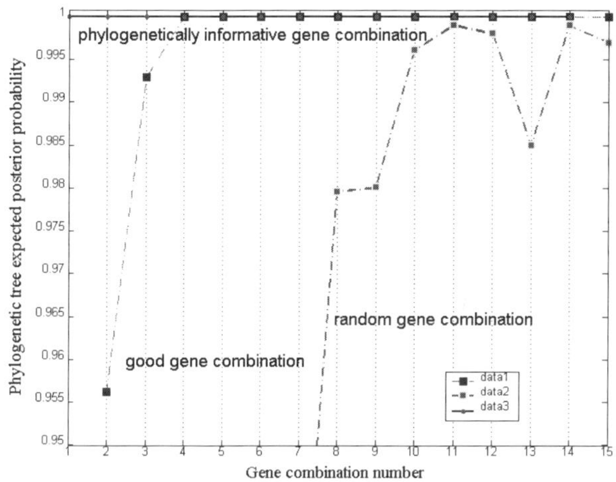

Figure 3. Comparing the performance of random and good gene concatenation cases with phylogenetically informative gene concatenation approaches.

5 Discussion and Future Work

In this paper, we provide a novel analytical solution to resolving the gene tree and species tree problem from phylogenomic mining point of view. Our results show that this approach is more robust than an ad-hoc gene concatenation approach. If generalizable, we also plan to take advantage of the powerful feature extract capability of SOM mining and entropy theory to address paralogy and orthology detection problem. The detection of paralogy and orthology is a key problem in phylogenomics but still in its naïve stage [3,6]. We expect the detection of the paralogous and orthologous genes can be conducted in their corresponding feature spaces and by means of entropy estimations. We are also interested in integrating our phylogenomic mining approach to the Bayesian analysis to explore more powerful tree reconstruction algorithms.

References

1. R. D. Page and E. Holmes. *Molecular evolution, a phylogenetics approach,* Blackwell Science, 1998.
2. W. Maddison. Gene trees in species trees. *Syst. Biol.,* 46:523-536, 1997.
3. J. Cotton. Analytical methods for detecting paralogy in molecular datasets, *Methods in Enzymology,* 395:700-24, 2005.
4. J. Huelsenbeck. Performance of phylogenetic methods in simulation. *Syst. Biol.,* 44:17–48, 1995.
5. Z. Yang, N. Goldman and A. Friday. Comparison of models for nucleotide substitution used in maximum-likelihood phylogenetic estimation. *Mol. Biol. Evol.*11:316–324, 1994.
6. K. Crandall and J. Buhay. Genomic databases and tree of life, *Science*, 306: 1144-1145, 2004.
7. B. Ma, M. Li, and L. Zhang. From Gene Trees to Species Trees. *SIAM Journal on Computing,* 30:729-752, 2000.
8. R. D. Page. Extracting species trees from complex gene trees: reconciled trees and vertebrate phylogeny. *Molecular Phylogenetics and Evolution*, 14:89-106, 2000.
9. F. Delsuc, H. Brinkmann and H. Philippe. Phylogenomics and the reconstruction of the tree of life. *Nat Rev Genet.,* 6:361-375, 2005.
10. B. Snel, M. Huynen and B. Dutilh. Genome Trees and the Nature of Genome Evolution. *Annu. Rev. Microbiol.,* 59:191-209, 2005.
11. B. Moret, J. Tang, and T. Warnow. Reconstructing phylogenies from gene-content and gene-order data. *Mathematics of Evolution and Phylogeny,* Oxford Univ. Press, 321-352, 2005.
12. A. Rokas, B. Williams, N. King and S. Carroll. Genome-scale approaches to resolving incongruence in molecular phylogenies. *Nature,* 425:798-804, 2003.
13. J. Huelsenbeck and F. Ronquist. MRBAYES: Bayesian inference of phylogenetic trees. *Bioinformatics,* 17:754-755, 2001.
14. J. Huelsenbeck and F. Ronquist. Bayesian Analysis of Molecular Evolution using MrBayes, http:// www.csit.fsu.edu/~ronquist/mrbayes/mrbayes_chapter.pdf, 2004.
15. M. Dunham. *Data mining introductory and advanced topics*, Prentice Hall, 2002.
16. S. Haykin. *Neural Networks: A Comprehensive Foundation, 2nd Edition.* Prentice-Hall, 1999.
17. J. Nikkila, P. Toronen, S. Kaski, J. Venna, E. Castren, G. Wong. Analysis and visualization of gene expression data using self-organizing maps. *Neural Networks,* 15:953-966, 2002.
18. T. Kohonen. *Self-Organizing Maps, 3rd edition*, Berlin: Springer-Verlag, 2001.
19. J. Vesanto and E. Alhoniemi. Clustering of the Self-Organizing Map, *IEEE Transactions on Neural networks.* 11:586-600, 2000.
20. J. Lanctot, M. Li, and E. Yang. Estimating DNA sequence entropy, *Proceedings of the eleventh annual ACM-SIAM symposium on Discrete algorithms*. 409-418, 2000.
21. H. Shimodaira and M. Hasegawa. Multiple comparisons of log-likelihoods with applications to phylogenetic inference. *Mol Biol Evol.,* 16:1114-1116, 1999.

CHARACTERIZATION OF THE EXISTENCE OF GALLED-TREE NETWORKS (EXTENDED ABSTRACT)

JÁN MAŇUCH*, XIAOHONG ZHAO, LADISLAV STACHO† AND ARVIND GUPTA‡

School of Computing Science and Department of Mathematics
Simon Fraser University, Canada
Email: arvind,jmanuch,lstacho,xzhao2@sfu.ca

In this paper, we give a complete characterization of the existence of a galled-tree network in the form of simple sufficient and necessary conditions. As a by-product we obtain as simple algorithm for constructing galled-tree networks. We also introduce a new necessary condition for the existence of a galled-tree network similar to bi-convexity.

1. Introduction

With the progress of human genome project[7], large amount of genomic data is available. Analysis of this data requires new methods incorporating events such as recombination, gene conversion, horizontal gene transfer and mobile genetic elements[8,9]. The traditional phylogenetic tree model is not sufficient enymore. In particular, recombination attracts much attention, because of its important role in locating genes influencing complex genetic diseases. A fundamental model which incorporates recombinations, phylogenetic networks, was introduced by Wang et al.[10]. With no restrictions on location of recombinations, they showed that the problem of finding a phylogenetic network with minimum number of recombinations is NP-hard. They also proposed a constrained phylogenetic network model with vertex-disjoint recombination cycles, called a galled-tree network.

Gusfield et al.[6] presented a polytime algorithm for constructing a galled-tree network. The algorithm is based on a number of necessary conditions on the existence of such networks. Some of these conditions are properties of so-called "conflict graph". More necessary conditions were given in the subsequent paper[5]. Surprisingly, unlike in case of phylogenetic trees, no characterization is known for galled-tree networks.

*Research supported in part by PIMS (Pacific Institute for Mathematical Sciences).
†Research supported in part by NSERC (Natural Science and Engineering Research Council of Canada) grant.
‡Research supported in part by NSERC (Natural Science and Engineering Research Council of Canada) grant.

In this paper, we give a complete characterization for the existence of a galled-tree network in the form of simple sufficient and necessary conditions. In particular, we show that two necessary conditions observed by Gusfield et al.[6] are enough to guarantee the existence of a galled-tree network. In our model we assumed that the root of the galled-tree network is labeled by the all-0 sequence. Note that very recently an algorithm for constructing a galled-tree network without any assumption on the label of the root (root-unknown network) was presented[3]. As a by-product, we obtained a simple algorithm for constructing galled-tree networks. Gusfield et al.[6] introduced an interesting necessary condition, called bi-convexity, which they used to design a fast algorithm for the site consistency problem for a matrix A if there exists a galled-tree network explaining A. As another by-product, we present a new necessary condition (bi-inclusiveness) which implies bi-convexity (but not other way around). Gusfield et al.[6] conjectured that the minimum vertex cover of a bi-convex graph can be found in linear time. We show that the cover of a bi-inclusive graph can be found in linear time assuming we know the order of vertices sorted by their degrees. Otherwise we need to add the sorting time to the complexity.

2. Preliminaries

The input to the problem is a haplotype $n \times m$ matrix A with values in $\{0,1\}$ (binary), where each row represents a haplotype sequence of an individual and each column corresponds to a character (an SNP site in the DNA sequence). The set of characters is assumed to be the set $\{1,\ldots,m\}$. For every character c, the sequence in a row contains in column c the state of character c for that individual. We use the terms "column" and "character" interchangeably.

We will assume that the edges of structures used to explain the input matrix (perfect phylogenies, galled-trees) are directed from the root to leaves. An edge (u,v) is a directed edge from u to v, i.e., u is closer the root than v. We will also assume that root is labeled with the all-0 sequence. We can also assume that no column contains only 0-states, as such columns do not affect solution to any of the considered problems. In the following definition we describe two basic operations on the matrices which we will use frequently.

Definition 2.1. Given an $n \times m$ binary matrix A. Let S be a subset of characters of A. The matrix $A[S]$ is the sub-matrix of A restricted to the columns in S. We will assume that the names of columns in $A[S]$ are the same as in the original matrix A. Let x be a binary sequence of length $|S|$. By $A[S]-x$, we denote the sub-matrix of $A[S]$ from which we remove all rows whose strings are identical to x.

2.1. *Perfect phylogeny*

The main combinatorial tool used in evolutionary biology is the concept of perfect phylogeny (phylogenetic tree). In our considerations phylogenetic trees appear in several places (construction of galls, compressed trees for galled-tree networks).

Definition 2.2. (Perfect phylogeny) Given an $n \times m$ binary matrix A. A *phylogenetic tree* on m characters is a rooted tree having each edge labeled with a unique character in the set $\{1, \ldots, m\}$, i.e., no two edges have the same label. Given a phylogenetic tree, we assign to each vertex a binary sequence of length m in top-down fashion as follows: the root is labeled with the all-0 sequence; for every edge (u, v) labeled with a character c, the label of v is obtained from the label of u by changing 0 at position c to 1 (changing state of character c). We say that a phylogenetic tree T *explains* A if each sequence of A (contained in a row) is a label of some vertex in T. If there is such a tree, we sometimes say A has a perfect phylogeny.

Note that the usual definition of phylogenetic tree T requires the sequences of A to be contained in the leaves of T. However, such a definition allows for unlabeled edges along which labels of end vertices do not change. It is easy to convert our phylogenetic tree to a standard phylogenetic tree. We prefer our definition, as our phylogenetic trees are more compact.

The following is the classical characterization of the existence of the perfect phylogenetic tree rediscovered in many papers. Before stating the result we need the following definition.

Definition 2.3. (Conflicting characters) Given an $n \times m$ binary matrix A. Two characters/columns c and c' *conflict* in A if $A[c, c']$ contains three rows with pairs $[0,1], [1,0]$ and $[1,1]$. A character is *unconflicted* if it does not conflict with any other character.

Theorem 2.1. *Given an $n \times m$ binary matrix A. There exists a phylogenetic tree explaining A if and only if no two characters conflict in A.*

Note that if we drop the requirement in the definition of phylogenetic trees to have the root labeled with the all-0 sequence, the above theorem is still true, although we have to redefine conflicts between characters: c and c' conflict in A if $A[c, c']$ contains all 4-possible pairs (so-called *four-gamete test*).

Definition 2.4. Given a tree. If there is a directed path in the tree containing edges e and e', we say that e and e' are *comparable.* Take the shortest such a path. If e is the first edge on the path, we say that e is an *ancestor* of e', and e' is a *descendant* of e, and write $e \preceq e'$. If there is no such path, we say that e and e' are *incomparable.*

Given an $n \times m$ binary matrix M. Let T be a phylogenetic tree explaining M. Define a map $e : \{1, \ldots, m\} \to E(T)$ returning the edge with label c as follows, for every character c, let $e(c) = e$ where e is the edge with the label c. Since we assume that M has no all-0 columns, the map is defined for every character.

2.2. *Definitions of phylogenetic and galled-tree networks*

Definition 2.5. A *phylogenetic network* N on m characters is a directed acyclic graph containing exactly one vertex (the root) with no incoming edges. Each vertex other than the root has either one or two incoming edges. If it has one incoming edge, the edge is called a *mutation edge*, otherwise it is called a *recombination edge.* A vertex x with two incoming edges is called a *recombination vertex.*

Each integer (character) from 1 to m is assigned to exactly one mutation edge in N and each mutation edge is assigned one character. Each vertex in N is labeled by a binary sequence of length m, starting with the root vertex which is labeled with the all-0 sequence. Since N is acyclic, the vertices in N can be topologically sorted into a list, where every vertex occurs in the list only after its parent(s). Using that list, we can define the labels of the non-root vertices, in order of their appearance in the list, as follows:

(1) For a non-recombination vertex v, let e be the mutation edge labeled c coming into v. The label of v is obtained from the label of v's parent by changing the value at position c from 0 to 1.
(2) Each recombination vertex x is associated with an integer $\mathrm{r}_x \in \{2, \ldots, m\}$, called the *recombination point* for x. Label the two recombination edges coming to x by P and S, respectively. Let $P(x)$ ($S(x)$) be the sequence of the parent of x on the edge labeled P (S). Then the label of x consists of the first $\mathrm{r}_x - 1$ characters of $P(x)$, followed by the last $m - \mathrm{r}_x + 1$ characters of $S(x)$. Hence $P(x)$ contributes a prefix and $S(x)$ contributes a suffix to x's sequence.

Recall that, in this paper, the sequence at the root of the phylogenetic network is always the all-0 sequence, and all results are relative to that assumption. More general phylogenetic networks with unknown root were studied in a recent paper by Gusfield[3]. Note also that there are slight differences in the definition of phylogenetic networks from the original definition[6,10]. We assume that each mutation edge has exactly one label. Every phylogenetic network without this assumption can be easily transformed to our model by replacing every mutation edge with multiple labels by a sequence of edges each having one of these labels, and contracting all mutation edges without a label. Our assumption results in more compact phylogenetic networks, however we cannot require that all sequences of an input matrix appear at the leaves of the network.

Definition 2.6. Given an $n \times m$ binary matrix A, we say that a phylogenetic network N with m characters *explains* A if each sequence of A is a label of some vertex in N.

Definition 2.7. (Galled-tree network) In a phylogenetic network N, let v be a vertex that has two paths out of it that meet at a recombination vertex x (v is the lowest common ancestor of the parents of x). The two paths together form a *recombination cycle* Q. The vertex v is called the *coalescent vertex*. We say that Q contains a character c, if c labels one of the mutation edges of Q.

A phylogenetic network is called a *galled-tree network* if no two recombination cycles share an edge. A recombination cycle of a galled-tree network is sometimes referred to as a *gall*.

Note that in the original definition of galled-tree network[6,10] it is required that recombination cycles do not share vertices. It is easy to see that our modification is only a minor difference (one can be transformed to the other easily) introduced for technical reasons.

3. Characterization of the existence of a galled-tree network

In this section we will give a complete characterization of the existence of a galled-tree network explaining a given matrix A. We will show that two conditions (Lemma 4 and Theorem 10) in Gusfield et al.[6]) are also sufficient.

Definition 3.1. Given an $n \times m$ binary matrix A. The *conflict graph* G_A has the vertex set $\{1, \ldots, m\}$ and for every two characters c and c', (c, c') is an (undirected) edge of G_A if they conflict.

Our characterization of galled-tree networks is presented in the following theorem.

Theorem 3.1. *Given an $n \times m$ binary matrix A. There exists a galled-tree network explaining A if and only if every nontrivial component (having at least two vertices) K of the conflict graph G_A satisfies the following conditions:*

(1) K is bipartite with partitions L and R such that all characters in L are smaller than all characters in R; and
(2) there exists a sequence $x \neq 0^{|K|}$ such that $A[K] - x$ has no conflicting characters.

In the rest of this section we will prove several results which will imply the theorem. Throughout the rest of the paper, let A be a given $n \times m$ binary matrix.

The following crucial result shows that if the condition (2) of Theorem 3.1 is satisfied then $A[K] - x$ can be explained by a tree with two edge-disjoint branches.

Lemma 3.1. *If a component K of G_A is bipartite with partitions L and R, and $A[K] - x$ has no conflicting characters for some $x \neq 0^{|K|}$, then any phylogenetic tree T explaining $A[K] - x$ has at most two branches. For $i = 0, 1$, let L_i (R_i) be the set of all $c \in L$ ($c \in R$) such that $x[c] = i$. One possible branch contains all edges labeled with characters in $L_1 \cup R_0$, and the other contains all edges labeled* *with characters in $R_1 \cup L_0$. If T has two branches then they do not share any edge (recall that we assume that a phylogenetic tree has all edges labeled by characters).*[a]

In the following theorem we will show that if a component of the conflict graph G_A satisfies both conditions of Theorem 3.1 then there is a gall explaining $A[K]$.

Theorem 3.2. *If a component K of G_A is bipartite with partitions L and R, $A[K]-x$ has no conflicting characters for some $x \neq 0^{|K|}$ and all vertices in L are smaller than all vertices in R, then $A[K]$ can be explained by a galled tree containing one recombination cycle (gall) rooted in the node with label $0^{|K|}$ and having x as a label of the recombination vertex.*

Proof. By Lemma 3.1, there is a phylogenetic tree T explaining $A[K]-x$ with at most two branches. Let B_P be the branch containing edges labeled with characters in $L_1 \cup R_0$, and B_S the branch containing edges labeled with characters in $R_1 \cup L_0$. If one of these two sets is empty then one of the branches is empty as well. Furthermore, the vertex labeled $0^{|K|}$ is the only vertex shared by B_P and B_S. Now, we will add a recombination vertex z into T. Let y_P (y_S) be the last vertex on the branch B_P (B_S). Add two recombination edges (y_P, z) labeled P and (y_S, z) labeled S, cf. Figure 1. Set the recombination point r_z to any character in $\{p+1, \ldots, q\}$, where p is the maximum character in L and q is the minimum character in R. We will show that the label of recombination vertex z is x, i.e., the gall explains the matrix $A[K]$.

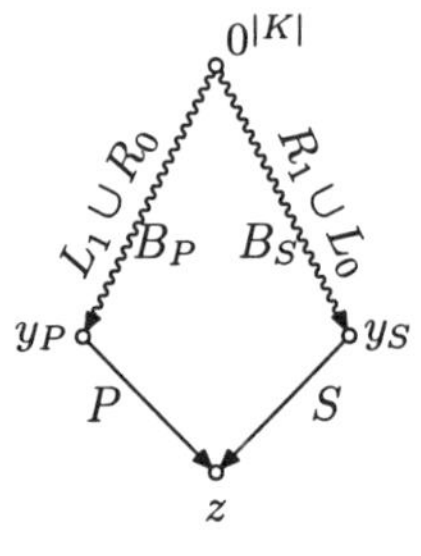

Figure 1. Construction of recombination cycle using two branches B_P and B_S of the phylogenetic tree for $A[K]-x$.

The label of z is formed by concatenating the first $r_z - 1$ characters of $P(z)$ (see Definition 2.5) with the last $|K| - r_z + 1$ characters of S_z. The label $P(z)$ (respectively, $S(z)$) has 0 (respectively, 1) in every position $c \in R_1 \cup L_0$ and 1 (respectively, 0) in every position $c \in L_1 \cup R_0$. The label of z at position $c \in L_0$ comes from $P(z)$, hence it has value 0. Similar arguments show that the label of z agrees with x also on all remaining positions, as required. □

[a]Due to the space limitation the proof will appear in the journal version.

In the following we define a compressed matrix which will be used to build a phylogenetic network. Note that the compressed matrix is similar to the pass-through matrix[4]. However, the pass-through matrix does not contain columns for components of the conflict graph which are singletons.

Definition 3.2. Let $K_1, \ldots, K_k$ be the components of the conflict graph G_A. The *compressed matrix* C_A is the $n \times k$ binary matrix with columns labeled by $K_1, \ldots, K_k$. It has 1 in row $i \in \{1, \ldots, n\}$ and column K_j, $j \in \{1, \ldots, k\}$, if and only if the row i in $A[K_j]$ contains at least one 1.

Lemma 3.2. *The compressed matrix C_A has no conflicting characters.*[b]

It follows that the compressed matrix C_A can be explained by a phylogenetic tree. We will use this tree to construct the galled-tree network explaining A. Recall that a phylogenetic tree with a fixed root is unique up to order of edges labeled with characters having identical columns in the input matrix. From all phylogenetic trees explaining C_A we want to pick one satisfying the following condition:

Definition 3.3. A phylogenetic tree T explaining C_A is called *sorted* if for every two identical columns K_j and $K_{j'}$ such that component K_j is a singleton and component $K_{j'}$ has at least two vertices in the conflict graph, $e(K_j) \prec e(K_{j'})$.

Following lemma shows that sequences in rows of A behave nicely with respect to edges in a sorted phylogenetic tree T explaining the compressed matrix C_A.

Lemma 3.3. *Let T be a sorted phylogenetic tree explaining the compressed matrix C_A. Assume that $e(K_j) \prec e(K_{j'})$ in T for some components K_j and $K_{j'}$ in G_A. Consider all rows containing a 1 in $A[K_{j'}]$, i.e., having 1 in $C_A[K_{j'}]$. Then all sequences in these rows in $A[K_j]$ are identical and different from the all-0 sequence.*[b]

The following algorithm constructs a galled-tree network N_A from a sorted phylogenetic tree for C_A.

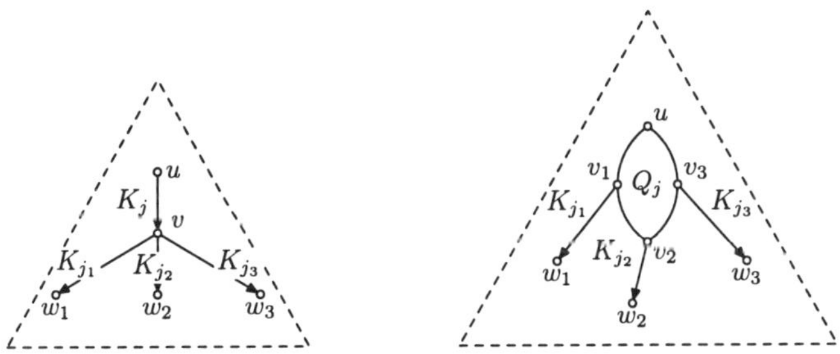

Figure 2. Replacing an edge labeled K_j with a gall Q_j.

[b]Due to the space limitation the proofs will appear in the journal version.

Algorithm 3.1.
Input: An $n \times m$ binary matrix A satisfying assumptions of Theorem 3.2.

(1) Construct a sorted phylogenetic tree T of C_A and for every component K_j, $j \in \{1, \ldots, k\}$, of G_A, construct the gall Q_j explaining $A[K_j]$.
(2) In top-down fashion process every edge (u, v) labeled K_j. If K_j is a singleton, i.e., $K_j = \{c\}$, replace the label of (u, v) by c. Otherwise, replace the edge with a gall Q_j for K_j as follows (cf. Figure 2):
    2.1 Remove edge (u, v).
    2.2 Identify the coalescent node of the gall Q_j with u.
    2.3 For every edge (v, w) labeled $K_{j'}$, consider any row r containing 1 in $C_A[K_{j'}]$. Let s be the sequence in $A[K_j]$ in row r. By Lemma 3.3, $s \neq 0^{|K_j|}$. Since Q_j explains $A[K_j]$, it contains a vertex $v' \neq u$ labeled s. Remove the edge (v, w), add the edge (v', w) and label it $K_{j'}$.
    2.4 Remove vertex v.
(3) To obtain a proper labeling of vertices in N_A, compute new labels of length m using the procedure described in the definition of galled-trees.

The following lemma shows that the algorithm produces essentially unique answer. More precisely,

Lemma 3.4. *After constructing a sorted phylogenetic tree T of C_A and galls Q_j's for every component K_j of G_A in Step 1 of Algorithm 3.1, the remaining construction of the algorithm produces unique result (the resulting galled-tree network depends only on selection of T and Q_j's).*

Proof. The only choice we have in the remaining steps of the algorithm is in Step 2.3 when we can choose any row r containing 1 in $C_A[K_{j'}]$. The selection of vertex v' to which we attach w depends on the sequence s in row r of the matrix $A[K_j]$. However, by Lemma 3.3, for every row r' containing 1 in $C_A[K_{j'}]$, the sequence in row r' of the matrix $A[K_j]$ is also s. □

The question of how many different galls are there for a matrix $A[K_j]$ was studied by Gusfield et al.[6]. It was shown that there are at most three different galls, and if there are enough characters in K_j, there is only one gall explaining $A[K_j]$. Also note that the phylogenetic tree T is unique up to arrangement of characters with identical columns on edges. For our purposes, the fact that Step 2.3 can be performed only in one unique way is sufficient to show that N_A explains A.

Theorem 3.3. *Assume that every non-trivial (with at least two vertices) component K of G_A is bipartite with partitions L and R, $A[K] - x$ has no conflicting characters for some $x \neq 0^{|K|}$ and all vertices in L are smaller than all vertices in R. Then the galled-tree network N_A constructed above explains A.*[c]

---
[c]Due to the space limitation the proof will appear in the journal version.

It is known that the number of galls in any galled-tree network explaining A is at least the number of non-trivial components in the conflict graph G_A[6]. Since the galled-tree network constructed by Algorithm 3.1 has exactly this number of galls, the constructed network is optimal.

Obviously, by Theorem 3.2, Algorithm 3.1 cannot fail to construct a galled-tree network N_A, and by the above theorem, the constructed network explains A. Hence, we have the following corollary.

Corollary 3.1. *If every non-trivial component K of G_A is bipartite with partitions L and R, $A[K] - x$ has no conflicting characters for some $x \neq 0^{|K|}$ and all vertices in L are smaller than all vertices in R, then there exists a galled-tree network explaining A.*

Combining the above corollary with the results of Gusfield et al.[6], Theorem 3.1 follows.

3.1. *Bi-inclusiveness*

Gusfield et al.[6] introduced an interesting necessary condition for the existence of a galled-tree network, called *bi-convexity.*

Definition 3.4. A bipartite graph K with partitions L and R is called *convex for* R if the vertices in R can be ordered so that, for each vertex $i \in L$, $N(i)$ forms a closed interval in R. That is, i is adjacent to j and $j' > j$ in R if and only if i is adjacent to all vertices in the set $\{j, \ldots, j'\}$. A bipartite graph is called *bi-convex* if sets L and R can be ordered so that it is simultaneously convex for L and convex for R.

They used bi-convexity to design a fast algorithm for the *site consistency problem* for a matrix A if there is a galled-tree network explaining A. The site consistency problem for a matrix A is to find a minimum number of columns whose removal from A results in a perfect phylogeny. The problem was introduced and shown to be NP-complete[1]. The problem reduces to finding a minimum vertex cover in the conflict graph G_A. For bipartite graphs, the vertex cover can be found in polynomial time and for bi-convex graphs in $\mathcal{O}(m^2)$ time (recall that m is the number of vertices in the conflict graph)[2]. It was conjectured by Gusfield et al.[6] that to find a minimum vertex cover of a bi-convex graph can be done in linear time. We present a new necessary condition, *bi-inclusiveness*, which is stronger than bi-convexity (it implies bi-convexity but not other way round) and observe that the minimum vertex cover of a bi-inclusive graph can be found in linear time.

Definition 3.5. We say that a collection of sets forms a chain, if there is an order $S_1, \ldots, S_k$ of sets such that $S_1 \subseteq S_2 \subseteq \cdots \subseteq S_k$. A bipartite graph K with partitions L and R is *bi-inclusive* if the sets $N(i_1), \ldots, N(i_k)$ form a chain, where $N(x)$ denotes the neighborhood of x.

Note that it is easy to check that the swapping of partitions does not change the property whether K is bi-inclusive or not.

The next theorem shows that if a matrix A satisfies sufficient and necessary conditions of Theorem 3.1, i.e., A can be explained by a galled-tree network, then every component of the conflict graph G_A is bi-inclusive.

Theorem 3.4. *Given an $n \times m$ binary matrix A. If a component K of G_A is bipartite and $A[K] - x$ has no conflicting characters for some $x \neq 0^{|K|}$, then K is bi-inclusive.*[d]

Since bi-inclusive graphs are chordal bipartite graphs, a minimum vertex cover of a bi-inclusive graph can be found in linear time given some additional information on the graph[2]. Hence we have the following.

Observation 3.1. A minimum vertex cover in a bi-inclusive graph can be found in $\mathcal{O}(m \log m)$ time and in linear time ($\mathcal{O}(m)$) if the chain order of vertices in one partition is given.

References

1. W. H. Day and D. Sankoff. Computational complexity of inferring phylogenies by compatibility. *Syst. Zool.*, 35(2):224–229, 1986.
2. F. F. Dragan. Strongly orderable graphs: A common generalization of strongly chordal and chordal bipartite graphs. *Discrete Appl. Math.*, 99(1-3):427–442, 2000.
3. D. Gusfield. Optimal, efficient reconstruction of root-unknown phylogenetic networks with constrained and structured recombination. *J. Computer and Systems Sciences*, 70:381–398, 2005.
4. D. Gusfield, S. Eddhu, and C. Langley. Powerpoint slides for: Efficient reconstruction of phylogenetic networks (of SNPs) with constrained recombination. http://wwwcsif.cs.ucdavis.edu/~gusfield/talks.html.
5. D. Gusfield, S. Eddhu, and C. Langley. The fine structure of galls in phylogenetic networks. *INFORMS Journal on Computing*, 16(4):459–469, 2004.
6. D. Gusfield, S. Eddhu, and C. Langley. Optimal, efficient reconstruction of phylogenetic networks with constrained recombination. *Journal of Bioinformatics and Computational Biology*, 2(1):173–213, 2004.
7. L. Helmuth. Genome research: Map of the human genome 3.0. *Science*, 293(5530):583–585, 2001.
8. D. Posada and K. A. Crandall. Intraspecific gene genealogies: trees grafting into networks. *Trends in Ecology and Evolution*, 16(1):37–45, 2001.
9. M. Schierup and J. Hein. Consequences of recombination on traditional phylogenetic analysis. *Genetics*, 156:879–891, 2000.
10. L. Wang, K. Zhang, and L. Zhang. Perfect phylogenetic networks with recombination. In *SAC '01: Proceedings of the 2001 ACM symposium on Applied computing*, pages 46–50, New York, NY, USA, 2001. ACM Press.

[d]Due to the space limitation the proof will appear in the journal version.

SEMI-SUPERVISED THRESHOLD QUERIES ON PHARMACOGENOMICS TIME SEQUENCES

J. ASSFALG, H.-P. KRIEGEL, P. KRÖGER, P. KUNATH, A. PRYAKHIN, M. RENZ

Institute for Computer Science, University of Munich

Email: {assfalg,kriegel,kroegerp,kunath,pryakhin,renz}@dbs.ifi.lmu.de

The analysis of time series data is of capital importance for pharmacogenomics since the experimental evaluations are usually based on observations of time dependent reactions or behaviors of organisms. Thus, data mining in time series databases is an important instrument towards understanding the effects of drugs on individuals. However, the complex nature of time series poses a big challenge for effective and efficient data mining. In this paper, we focus on the detection of temporal dependencies between different time series: we introduce the novel analysis concept of *threshold queries* and its semi-supervised extension which supports the parameter setting by applying training datasets. Basically, threshold queries report those time series exceeding an user-defined query threshold at certain time frames. For semi-supervised threshold queries the corresponding threshold is automatically adjusted to the characteristics of the data set, the training dataset, respectively. In order to support threshold queries efficiently, we present a new efficient access method which uses the fact that only partial information of the time series is required at query time. In an extensive experimental evaluation we demonstrate the performance of our solution and show that semi-supervised threshold queries applied to gene expression data are very worthwhile.

1. Introduction

Data mining in time series data is a key step within the study of drugs and their impact on living systems, including the discovery, design, usage, modes of action, and metabolism of chemically defined therapeutics and toxic agents. In particular, the analysis of time series data is of great practical importance for pharmacogenomics.

Classical time series analysis is based on techniques for forecasting or for identifying patterns (e.g. trend analysis or seasonality). The similarity between time series, e.g. similar movements of time series, plays a key role for the analysis. In this paper, we introduce a novel but very important similarity query type which we call *threshold query*. Given a time series database DB, a query time series Q, and a query threshold τ, a threshold query $TSQ_{DB}(Q, \tau)$ returns those time series $X \in DB$ having the most similar sequence of time intervals in which the time series values are above τ. In other words, we assume that each time series $X \in DB \cup \{Q\}$ is transformed into a sequence of disjoint time intervals covering only those values of X that are (strictly) above the threshold τ. Then, a threshold query returns for a given query object Q that object $X \in DB$ having the most similar sequence of time intervals. Let us note that the exact values of the time series are not considered, rather we are only interested in whether the time series is above or below a given threshold τ. In other words, the concept of threshold queries enables us to focus only on the duration of certain events indicated by increased time series amplitudes, while the degree of the corresponding amplitudes are ignored. This advantage is very beneficial, in particular, if we want to compare time

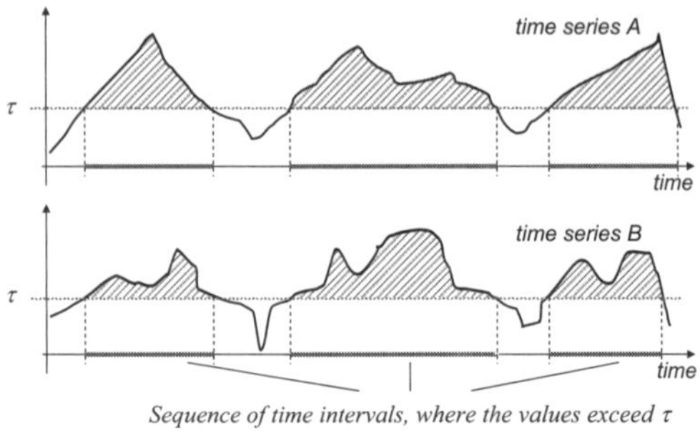

Figure 1. Illustration of transformation of time series into sequences of time intervals.

series reacting on certain stimulations with different sensitivity. The transformation of the time series into interval sequences is visualized in Figure 1. Two time series A and B are each transformed into a sequence of time intervals where the values are above a given threshold τ.

This new query type is very useful for several pharmacogenomics applications. The most straightforward application of threshold queries is the search for similar time series. For example, a common task in pharmacogenomics is the identification of individual drug response or the analysis of the impact of certain environmental influences on gene expression levels or blood values. For this task, the concentration of agents that are suspected to trigger the relevant biochemical reactions is measured over some period of time. Using threshold queries, one is able to efficiently retrieve time series, that are similar to the stimulus time series in terms of threshold crossing events. Note that our technique is able to cope with different thresholds for the stimulus time series and the reacting time series. This is important since usually values of different domains (e.g. chemical concentrations versus gene expression levels) are compared. Another important example where the identification of similar time series is crucial, is the search for similar gene expression patterns. In a time series of gene expression values one can retrieve genes with similar expression levels in order to find genes that are coregulated or showing an interesting response to an external stimulus. In addition, threshold queries can be performed on mixed-type data. Thus, we can correlate data on specific agents such as blood parameter concentrations with gene expression data. Taking an agent time sequence as query object, we can identify genes that are affected by this agent.

In this paper, we propose techniques in order to support these important applications. In particular, we introduce the novel concept of threshold similarity and threshold queries. We propose a suitable data representation method to efficiently support threshold queries. In addition, we present a semi-supervised version of threshold queries which beneficially supports the parameter setting by applying training datasets. Semi-supervised threshold queries automatically detect the best parameter setting for the query process based on a labeled training dataset.

2. Related Work

In general, a time series of length d can be viewed as feature vector in a d-dimensional space, where the similarity between two time series corresponds to their distance in the feature space. Since d is usually large, the analysis of time series data based on the entire time series information is usually very limited. Due to the so-called *curse of dimensionality*, the efficiency and the effectiveness of data analysis methods decrease rapidly with increasing data dimensionality. Thus, it is mandatory to find more suitable representations of time series data for analysis purposes, e.g. by reducing the dimensionality. In the different communities, several solutions have been proposed. Most of them are based on the following indexing approach: extract a few key *features* for each time series and map each time sequence X to a point $f(X)$ in a lower dimensional feature space, such that the (dis)similarity between X and any other time series Y is approximately equal to the Euclidean distance between the two points $f(X)$ and $f(Y)$. For an efficient access any well known spatial access method can be used to index the feature space.

The proposed methods mainly differ in the representation of the time series: for details see the database-oriented survey[11] and the bioinformatics-oriented survey.[3] The database and the bioinformatics communities have successfully applied standard techniques for dimension reduction to similarity search and data mining in time series databases, including Discrete Fourier Transformation[1] and extensions,[13] Discrete Wavelet Transformation,[6] Piecewise Aggregate Approximation,[14] Singular Value Decomposition,[12,2] Adaptive Piecewise Constant Approximation,[11] Chebyshev Polynomials,[5] cubic splines.[4]

However, all techniques which are based on dimension reduction cannot be applied to threshold similarity queries because necessary temporal information is lost. Usually, in a reduced feature space, the original intervals indicating that the time series is above a given threshold cannot be generated. In addition, the approximation generated by dimensionality reduction techniques cannot be used for our purposes directly because they still represent the exact course of the time series rather than intervals of values above a threshold.

The most important issue for any data analysis purposes is the definition of similarity. The most common way to model (dis-)similarity of feature vectors is to measure their Euclidean distance. For many applications, the Euclidean distance may be too sensitive to minor distortions in the time axis. It has been shown that Dynamic Time Warping (DTW), which is conceptually similar to sequence alignment, can fix this problem.[11] Other common distance functions are Pearson's correlation coefficient which measures the global correlation between two time series, or angular separation, also known as cosine distance, which defines the distance in terms of the angle between two feature vectors. All these distance measures are not directly applicable to threshold similarity queries because all of them consider the absolute values of the time series rather than the intervals of values above a given threshold.

3. Threshold Queries

In this section, we introduce the novel concept of threshold queries based on a similarity model which is very promising for the analysis of pharmacogenomics time series data. Furthermore, we present techniques allowing an efficient query processing.

We define a time series X as a sequence of pairs $(x_i, t_i) \in \mathbb{R} \times T : (i = 1..N)$, where T denotes the domain of time and x_i denotes the measurement corresponding to time t_i. Furthermore, we assume that the time series entities are given in such a way that $\forall i \in 1,..,N-1 : t_i < t_{i+1}$. In most cases, when measuring continuously varying attributes at discrete time points, the missing values between two observations are estimated by means of interpolation.[3] In the rest of this paper, if not stated otherwise, $x(t) \in \mathbb{R}$ denotes the (interpolated) time series value of time series X at time $t \in T$.

3.1. *Threshold-Crossing Time-Intervals*

Instead of using time series for the description of the time dependent behavior of pharmacogenomics data, we use sequences of time intervals which are related to a specific user-defined threshold, i.e. for a given threshold τ, the pharmacogenomics data is described by means of disjoint time intervals expressing the points of time when the data values are above τ. We call this description of the time dependent behavior *threshold-crossing time-intervals.*

Definition 3.1. Let $X = \langle (x_i, t_i) \in \mathbb{R} \times T : i = 1..N \rangle$ be a time series with N measurements and $\tau \in \mathbb{R}$ be a threshold. Then the *threshold-crossing time interval sequence* of X with respect to τ is a sequence $TCT_\tau(X) = \langle (l_j, u_j) \in T \times T : j \in \{1,..,M\}, M \leq N \rangle$ of time intervals, such that

$$\forall t \in T : (\exists j \in \{1,..,M\} : l_j < t < u_j) \Leftrightarrow x(t) > \tau.$$

An interval $tct_{\tau,j} = (l_j, u_j)$ of $TCT_\tau(X)$ is called threshold-crossing time interval.

The description of the time dependent behavior of some attribute by means of threshold-crossing time-intervals is a simple method, but generally very promising for the analysis of pharmacological data, in particular for the analysis of pharmacogenomics data. The advantages compared to common methods which consider the entire time series are as follows: time-interval sequences are easy to handle and can be processed and stored very efficiently. Furthermore, threshold-crossing time-interval sequences express the temporal behavior w.r.t. a certain threshold, i.e. the intervals denote the time points when the attribute is above or below a specific threshold value.

The latter advantage is most decisive because threshold-crossing time-interval sequences allow us to infer a causality from the temporal behavior for relevant threshold levels instead of considering the entire time series curve. Let us assume that we have to analyze the temporal behavior of gene expression values by means of fluorescence intensity measurements. Measurable fluorescence intensities are supposed to be proper expression values plus noises. One can imagine that the measurements in the medial florescence spectrum are more precisely compared to the measurements in the extreme spectral areas, because the sensor data from the extreme spectral areas have more noise than the data from the medial spectrum. Traditional analyzing approaches which consider the entire expression level range can lead to incorrect results due to noise.

3.2. *Similarity Model*

Intuitively, two time intervals are defined to be similar if they have "similar" starting and end points, i.e. they are starting at similar times and ending at similar times.

Definition 3.2. Let $t1 = (t1_l, t1_u) \in T \times T$ and $t2 = (t2_l, t2_u) \in T \times T$ be two time intervals. Then the distance function $d_{int} : (T \times T) \times (T \times T) \rightarrow \mathrm{R}$ between two time intervals is defined as:

$$d_{int}(t1, t2) = \sqrt{(t1_l - t2_l)^2 + (t1_u - t2_u)^2}$$

Since for a certain threshold τ a time series object is represented by a sequence or a set of time intervals, we need a distance/similarity measure for sets of intervals. Several distance measures for set based objects have been introduced in the literature.[7] The Sum of Minimum Distances (SMD) most adequately reflects the intuitive notion of similarity between two threshold-crossing time interval sequences. According to the SMD we define the *threshold-distance* d_{TS} as follows:

Definition 3.3. Let X and Y be two time series and $S_X = TCT_\tau(X)$ and $S_Y = TCT_\tau(Y)$ be the corresponding threshold-crossing time interval sequences.

$$d_{TS}(S_X, S_Y) = \frac{1}{|S_X|} \cdot \sum_{s \in S_X} \min_{t \in S_Y} d_{int}(s,t) + \frac{1}{|S_Y|} \cdot \sum_{t \in S_Y} \min_{s \in S_X} d_{int}(t,s),$$

The idea of this distance function is to map every interval from one sequence to the closest (most similar) interval of the other sequence and vice versa. Let us note that the threshold-distance between two time series according to a certain threshold τ is also called τ-similarity.

We are now able to define the *threshold query* which reports the k most τ-similar time series of a database DB for a given query time series Q.

Definition 3.4. Let DB be the domain of time series objects. The threshold query consists of a query time series $Q \in DB$ and a query threshold $\tau \in \mathbb{R}$. The threshold query reports the smallest set $TSQ_k(Q, \tau) \subseteq DB$ of time series objects that contains at least k objects from DB such that

$$\forall X \in TSQ_k(Q,\tau), \forall Y \in DB - TSQ_k(Q,\tau):$$

$$d_{TS}(TCT_\tau(X), TCT_\tau(Q)) < d_{TS}(TCT_\tau(Y), TCT_\tau(Q)).$$

3.3. *Threshold Invariant Data Representation*

In this section, we introduce a novel data representation of the time series objects. Although the time series objects may be very complex, i.e. contain lots of measured values, we need a data representation of the time series objects which allows us to process the threshold queries very efficiently.

The basic idea of our proposed time series representation is that we do not need to access the complete time-series data at query time. Instead, only partial information of the time-series objects which can be efficiently accessed might suffice to report the results. In order to achieve these requirements, the threshold-crossing time-interval-sequences of all possible thresholds must be pre-computed and materialized which is fulfilled by our novel data representation. The key for the novel time series representation is a decomposition of a time series into a set of trapezoids, as depicted in Figure 2. The upper and lower edge of each trapezoid is parallel to the time axis, the left side is bounded by an increasing time series segment and the right side is bounded by a decreasing segment. Each trapezoid represents a set of threshold-crossing time intervals which correspond to a certain range of threshold values. The complete set of required threshold-crossing time intervals w.r.t. an arbitrary threshold τ can be easily computed by all trapezoids crossing τ. We have developed an algorithm which decomposes time series into corresponding trapezoids in linear time w.r.t. the length of the time series, due to the natural ordering of the time series values.

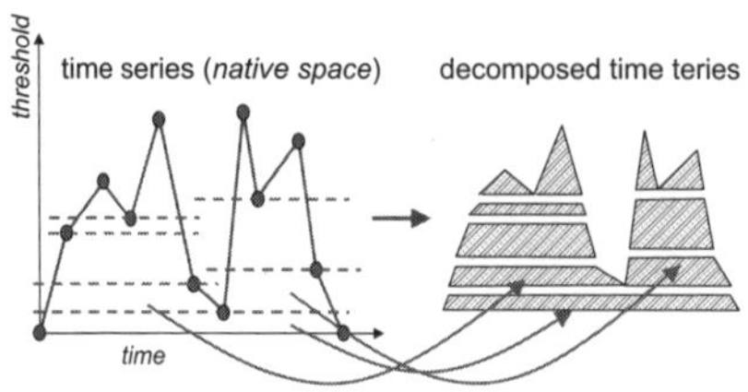

Figure 2. Time Series Decomposition

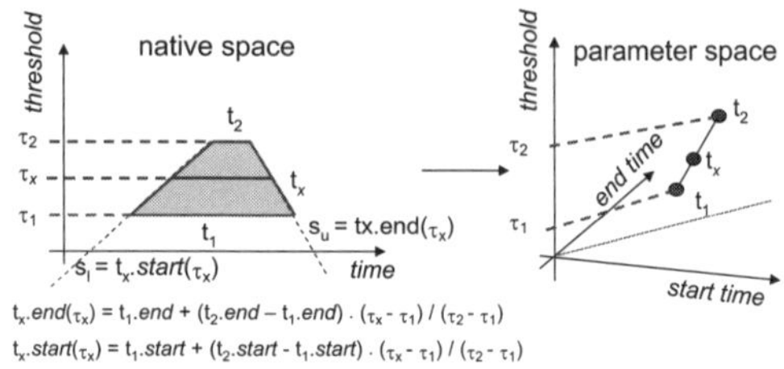

Figure 3. Interval Ranges in Parameter Space

For the management of the trapezoids we transform the trapezoids into a three dimensional space ($start \in T, end \in T, \tau \in \mathbb{R}$), where *start* denotes the start time and *end* denotes the end time for all threshold-crossing time-intervals and τ denotes the corresponding threshold. In the following, we will call this space *parameter space*. A 2-dimensional plane along the threshold axis parallel to the (start,end)-plane at a certain threshold τ in the parameter space is called *time-interval plane* of threshold τ. The time-interval plane has some advantages for the efficient management of intervals. First, the time distances between intervals are preserved. Second, the position of large intervals, which are located within the upper-left region, substantially differs from the position of small intervals. However, the most important advantage is that in this space the Euclidean distance correspond to the (dis-)similarity of intervals according to Definition 3.2.

For the efficient management of the trapezoids we use the following observation: all threshold-crossing time-intervals $tct_{\tau_i,j}(X)$ which start at segment s_l and end on segment s_u lie in the parameter space on the three-dimensional straight line: $g_P : \overrightarrow{x} = \overrightarrow{p_1} + \Delta t \cdot (\overrightarrow{p_2} - \overrightarrow{p_1})$, where $\overrightarrow{p_1} = (tct_{\tau_1,1}(X).start, tct_{\tau_1,1}(X).end, \tau_1)^T$ and $\overrightarrow{p_2} = (tct_{\tau_2,2}(X).start, tct_{\tau_2,2}(X).end, \tau_2)^T$. An example is depicted in Figure 3. At query time, the time-interval plane coordinates of the threshold-crossing time-intervals which correspond to the query threshold τ_q can be easily determined by computing the intersection of all segments of the parameter space with the time-interval plane P at threshold τ_q.

3.4. *Indexing Segments of the Parameter Space*

We apply the R*-tree for the efficient management of the three-dimensional segments representing the time-series objects in the parameter space. As the R*-tree index can only manage rectangles, we represent the 3-dimensional segments by rectangles. The R*-tree efficiently support nearest-neighbor queries[9] which build the basis of the computation of the similarity between two time series objects.

4. Semi-Supervised Threshold Queries

In the previous section (Section 3.3) we proposed a threshold invariant data representation which allows to perform efficient threshold queries where the threshold can be chosen at query time. The question is now, which threshold should be chosen for any analyzing task?

The answer to that question not only depends on the specific characteristics of the data at hand. Quite often the user expects a certain query to yield certain results.

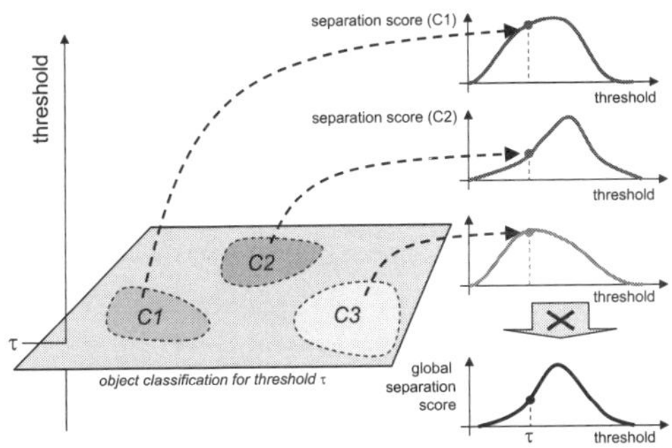

Figure 4. Determination of the threshold dependent class separation score.

Then the first interesting task is to determine a threshold value that yields good results. Furthermore our experimental studies revealed that a shift in the user's expectation frequently makes it necessary to readjust the threshold value the similarity measure is based on.

In the following, we assume that the user's expectation is modeled by means of a small training data set. Results for a given query are considered good, if a lot of results are marked with the same class label as the query time series. The general idea of our technique is to search for threshold values that promise high similarity scores between objects with similar behavior and low scores between objects with different behavior. In the next sections we will explain how we derive quality values for different thresholds and how we use this information to obtain global high quality values.

4.1. *Threshold Value Quality for One Class*

In this section we will outline how we detect suitable threshold values for a fixed class. Let us assume we are given a training data set C_M which consists of k classes, $C_M = C_1, ..., C_k$. For a fixed class C_i, we need to determine these threshold values which yield a good separability of C_i from the remaining classes. To evaluate the score of a query for a fixed class C_i and a fixed threshold τ, we pairwise compute the silhouette width[10] of C_i and all other classes C_j, $j \neq i$, in C_M. The minimal silhouette width of all pairs is used as separation score of a threshold τ. Calculating the separation score for any existing threshold results in a quality measure which estimates the separability score for a training data set C_M, a class $C_i \in C_M$ and a given threshold τ.

4.2. *Derivation of a Global Suitable Threshold Value*

In the last section, we have developed a quality measure which computes the separation score for each class C_i of our training data set C_M. Now, we need a suitable combination of all k separation score functions. For our approach, we chose the sum of all score functions, i.e. compute the silhouette coefficient[10] for C_M. The global separation score function now reflects the overall separability score of our training data set for an arbitrary threshold τ. Based on the idea of semi-supervised learning, the global score function gives the user hints to chose the most promising threshold ranges. The example depicted in Figure 4 points up one step of this procedure for a certain threshold.

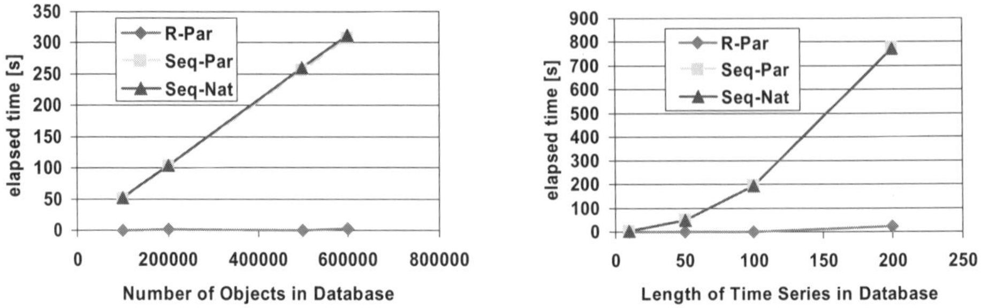

(a) Scalability against database size. (b) Scalability against time series length.

Figure 5. Efficiency Evaluation.

5. Evaluation

In this section we will present the results of our experiments with respect to efficiency and effectiveness.

5.1. *Efficiency*

In this section, we present the results of a large number of experiments performed on a selection of different time series datasets. In particular, we compared the efficiency of our proposed approach (in the following denoted by 'R_{Par}') for answering threshold queries using one of the following techniques. The first competing approach works on native time series. At query time for each database time series the corresponding threshold-crossing time intervals (TCT) are computed for the query threshold and afterwards the distance between the query time series and the corresponding database object are derived. In the following this method will be denoted by 'Seq_{Nat}' as it corresponds to a sequential processing of the entire set of the native data. The second competitor works on the parameter space rather than on the native data. It stores all TCTs without using any index structures. As this storage leads to a sequential scan over the corresponding elements of the parameter space at query time, we will refer to this technique as the 'Seq_{Par}' method.

All experiments were performed on a workstation featuring a 1.8 GHz Opteron CPU and 8GB RAM. We used a disk with a transfer rate of 100 MB/s, a seek time of 3 ms and a latency delay of 2 ms. Performance is presented in terms of the elapsed time including I/O and CPU-time. At first we performed threshold similarity queries against databases of different sizes to measure the influence of the database size. The elements of the databases are time series of fixed length l. Figure 5(a) exhibits the results for each database size averaged over several thresholds and several randomly chosen queries. Second, we explored the impact of the length of the time series to be compared. The results depicted in Figure 5(b) show that our approach significantly outperforms its competitors and is scalable w.r.t. the number and the length of the time series objects.

5.2. *Effectiveness*

At first, we exemplarily show how threshold queries can be beneficially used for pharmacogenomics in practice. An important application of threshold queries is finding

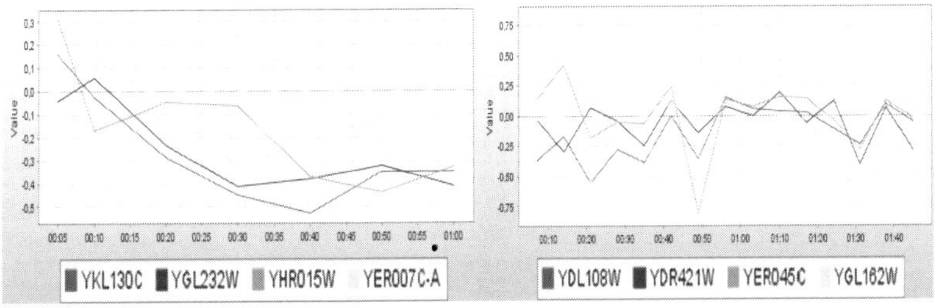

Figure 6. Illustration of sample threshold query results (best seen in color).

genes of similar function or of similar drug response in gene expression data. We used the yeast gene expression time course data sets from Gene Expression Omnibus[a] with accession numbers "GDS30" and "GDS38" and launched for each gene of each data set a threshold query against the rest of the genes. For each gene, we thus computed its three most similar genes based on threshold similarity on each data set. We evaluated the gene functions using Gene Ontology (GO).[b] The results of two sample queries are depicted in Figure 6 (absolute query threshold $\tau = 0$). On the left side of the figure, the 3 most similar genes to ORF "YKL130C" are depicted. All three reported ORFs "YGL232W", "YHR015W", and "YER007C-A" code for proteins with RNA-binding activity. On the right hand side of Figure 6 the 3 most similar genes to ORF "YDL108W" are depicted. All three reported ORFs "YDR421W", "YER045C", and "YGL162W" code for proteins with RNA polymerase II transcription factor activity.

In the example above no information about the dataset characteristics was used to select the threshold value. Different settings of the threshold value can lead to different qualities of the results. In the next experiment, we investigate the effectiveness of semi-supervised threshold queries which are used to find the optimal threshold value by means of a training dataset. At first, we are interested in how the optimal threshold values change when the expected results change, i.e. when the focus of the query changes. The following experiments were performed on the data sets "GDS30" and "GDS38" from Gene Expression Omnibus (GO).[c] For the first experiment we used the GO functional classes on level 4. Afterwards we changed the focus of our queries to the GO level 5. As expected, we obtained different optimal threshold values (cf. Figure 7).

These results gave raise to the question whether the computed optimal threshold values do indeed yield good results on the whole data set. To evaluate this, we clustered the time series for varying threshold values and determined the rand index.[8] For example, the threshold value 0.73 which corresponds to a high separation score on the GDS30 data set for GO level 4 resulted in a rand index equal to 0.94. Contrary, when using a threshold value of 0.2 the rand index decreased to 0.86. Similar results were observed for other levels, for other threshold values, and on other datasets.

6. Conclusions

In this paper, we proposed the novel concept of semi-supervised threshold queries which are suitable to analyze time series data in the area of life science. In addition,

[a] http://www.ncbi.nlm.nih.gov/geo/
[b] http://www.geneontology.org/
[c] http://www.ncbi.nlm.nih.gov/geo/

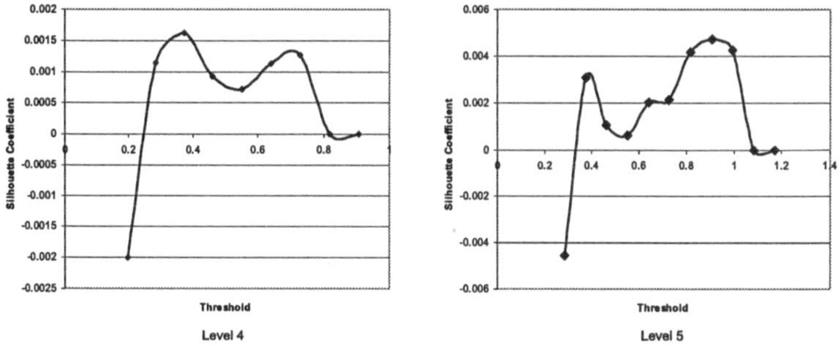

Figure 7. Different Optimal Threshold Values for Different GO Levels.

we presented a data representation method that supports threshold queries efficiently. In our experimental evaluation, we have shown that our proposed data representation accelerates threshold queries drastically. Furthermore, the results of our experimental evaluations have shown that the proposed semi-supervised threshold queries are very effective and are very useful in particular for the analysis of pharmacogenomics time series databases. For future work, we plan to extend our approaches to data mining tasks, such as similarity join and clustering.

References

1. R. Agrawal, C. Faloutsos, and A. Swami. "Efficient Similarity Search in Sequence Databases". In *Proc. 4th Conf. on Foundations of Data Organization and Algorithms*, 1993.
2. O. Alter, P. Brown, and D. Botstein. "Generalized Singular Value Decomposition for Comparative Analysis of Genome-Scale Expression Data Sets of two Different Organisms". *Proc. Natl. Aca. Sci. USA*, 100:3351–3356, 2003.
3. Z. Bar-Joseph. "Analyzing Time Series gene Expression Data". *Bioinformatics*, 20(16):2493–2503, 2004.
4. Z. Bar-Joseph, G. Gerber, T. Jaakkola, D. Gifford, and I. Simon. "Continuous Representations of Time Series Gene Expression Data". *J. Comput. Biol.*, 3-4:341–356, 2003.
5. Y. Cai and R. Ng. "Index Spatio-Temporal Trajectories with Chebyshev Polynomials". In *Proc. ACM SIGMOD*, 2004.
6. K. Chan and W. Fu. "Efficient Time Series Matching by Wavelets". In *Proc. IEEE ICDE*, 1999.
7. T. Eiter and H. Mannila. "Distance Measure for Point Sets and Their Computation". In *Acta Informatica, 34*, pages 103–133, 1997.
8. M. Halkidi, Y. Batistakis, and M. Vazirgiannis. "On Clustering Validation Techniques". In *Intelligent Information Systems Journal*, 2001.
9. G. Hjaltason and H. Samet. "Ranking in Spatial Databases". In *Proc. Int. Symp. on Large Spatial Databases (SSD'95), Portland, OR*, 1995.
10. L. Kaufman and P. Rousseeuw. "Finding Groups in Data: An Introduction to Cluster Analysis". Wiley, New York, 1990.
11. E. Keogh, K. Chakrabati, S. Mehrotra, and M. Pazzani. "Locally Adaptive Dimensionality Reduction for Indexing Large Time Series Databases". In *Proc. ACM SIGMOD*, 2001.
12. F. Korn, H. Jagadish, and C. Faloutsos. "Efficiently Supporting Ad Hoc Queries in Large Datasets of Time Sequences". In *Proc. ACM SIGMOD*, 1997.
13. S. Wichert, K. Fokianos, and K. Strimmer. "Identifying Periodically Expressed Transcripts in Microarray Time Series Data". *Bioinformatics*, 20(1):5–20, 2004.
14. B. K. Yi and C. Faloutsos. "Fast Time Sequence Indexing for Arbitrary Lp Norms". In *Proc. VLDB*, 2000.

STRUCTURE BASED CHEMICAL SHIFT PREDICTION USING RANDOM FORESTS NON-LINEAR REGRESSION

K. ARUN

Department of Biological Sciences,
Carnegie Mellon University,
Pittsburgh, PA 15213
E-mail: arunk@andrew.cmu.edu

CHRISTOPHER JAMES LANGMEAD

Computer Science Department,
School of Computer Science,
Department of Biological Sciences,
Carnegie Mellon University,
Pittsburgh, PA 15213
E-mail: cjl@cs.cmu.edu

Protein nuclear magnetic resonance (NMR) chemical shifts are among the most accurately measurable spectroscopic parameters and are closely correlated to protein structure because of their dependence on the local electronic environment. The precise nature of this correlation remains largely unknown. Accurate prediction of chemical shifts from existing structures' atomic co-ordinates will permit close study of this relationship. This paper presents a novel non-linear regression based approach to chemical shift prediction from protein structure. The regression model employed combines quantum, classical and empirical variables and provides statistically significant improved prediction accuracy over existing chemical shift predictors, across protein backbone atom types. The results presented here were obtained using the Random Forest regression algorithm on a protein entry data set derived from the RefDB re-referenced chemical shift database.

1. Introduction

Any nucleus with spin $I = 1/2$, when placed in an external magnetic field, will exhibit two spin states with an energy differential directly proportional to the strength of the applied magnetic field. Each nucleus, however, is influenced by the electrons in its vicinity and therefore the effective magnetic field at the nucleus is attenuated depending upon this electronic environment. The *chemical shift* (δ) is a measure of the electronic shielding that leads to this magnetic field attenuation, and therefore provides an accurate description of the local electronic environment. Thus, chemical shifts are among the most fundamental of nuclear magnetic resonance (NMR) spectral parameters. Chemical shifts are also among the most accurately measurable quantities in NMR spectroscopy (accuracy up to one part in a billion).

Given these properties of the chemical shift, there has long been an interest in understanding the nature of the relationship between molecular structure and shift, and applying said knowledge to infer additional structural information about the molecule under study. Protein molecules too give rise to NMR spectra in an applied magnetic field in a fashion intricately dependent on their three dimensional structures. The electronic environment around nuclei in the case of proteins is influenced by factors such as neighbor anisotropy, ring current anisotropy, hydrogen bond effects, and through-space electric field effects among others. A graphical representation of the chemical shift measurements from a standard protein NMR experiment (^{1}H/^{15}N HSQC) is depicted in Fig. 1. The center of each of the peaks observed in this two-dimensional plot represents two chemical shifts, the ^{1}H and ^{15}N shifts. The axes of the spectrum are in parts per million (ppm), the standard unit for chemical shifts.

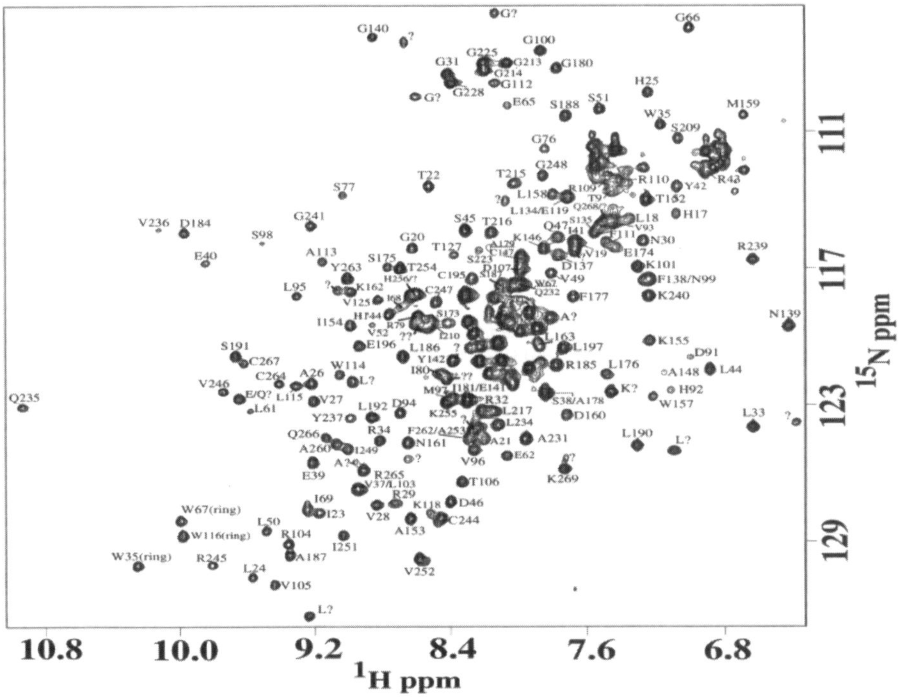

Figure 1: Two-dimensional ^{1}H/^{15}N Heteronuclear Single-Quantum Coherence (HSQC) NMR spectra of an *E. coli* DNA glycosylase, Fpg. (from `http://www.emsl.pnl.gov/homes/msd/bionmr/Buchko_Fpg_poster/Fpg.htm`)

Knowledge of the chemical shift and insight into the structure-shift relationship is useful in many contexts. The most obvious application is resonance assignment in the context of an protein NMR experiment where a model of the target protein's structure is available [10] (either via independent X-ray crystallography experiments or comparative modeling). Predicted shifts may also be used to refine existing structural models. There have also been efforts to infer low-resolution structure models given just the experimental chemical shifts. Examples include techniques for secondary structure prediction [16, 18], backbone torsion angle prediction [5], fold recognition [11, 13, 20], protein-protein docking [7] and modeling ligand interactions [15]. Predicted shifts, subject to their having acceptable accuracy, may be similarly employed.

Existing approaches to chemical shift prediction from protein structure apply quantum mechanical, classical and/or empirical methods to the atomic co-ordinate data. Examples of such algorithms include SHIFTS [19], SHIFTX [14] and PROSHIFT [12]. SHIFTS takes a quantum mechanical approach and employs a pre-calculated database of tri-peptide shifts (via density functional theory), while SHIFTX uses a hybrid empirical/semi-classical approach involving pre-calculated chemical shift hypersurfaces and equations for ring current, electric field, hydrogen bonding and solvent effects. PROSHIFT uses a neural network trained on $\approx$ 69,000 experimentally determined chemical shifts. Each of these shift prediction approaches has unique limitations either in terms of the size and composition of the training and/or test data sets, or due to the general tendency for learning methods such as neural networks to over-fit training data. We hypothesized that a better chemical shift predictor could be built by layering an ensemble machine learning algorithm (Random Forests [4]) capable of non-linear regression on top of these existing predictors in addition to expanding the feature set by taking into account numerous empirical structural features such as solvent accessibility, secondary structure and model quality. This paper presents the results of such an exercise.

In brief, the non-linear regression approach to chemical shift prediction employing the ensemble machine learning Random Forest algorithm outperformed each of the underlying shift prediction programs (*viz.* SHIFTS, SHIFTX, PROSHIFT) across all six backbone atom types. These improvements in prediction accuracies were measured in terms of root mean squared error from experimentally recorded shifts and in the case of the Random Forest algorithm, they ranged between 3% to $\sim$18% when compared to the best performer amongst the aforementioned prediction programs. The decrease in error observed was proven to be statistically significant by comparing the distribution of errors using a standard t-test. Across all atom types, p-values $\ll 0.001$ were observed.

2. Systems and methods

2.1. *Data assembly*

Building a structure-based chemical shift prediction method requires a dataset of protein chains with experimentally recorded chemical shifts matched to structures solved by NMR or X-ray crystallography. The principal community repositories of chemical shift and structural (atomic co-ordinate) data are the BioMagResBank (BMRB) [3] and the Protein Data Bank (PDB) [1] respectively. However, it has been demonstrated that significant chemical shift referencing errors exist for a substantial portion of the BMRB data. Hence, the dataset used in this project is drawn from the RefDB [21] database - a carefully re-referenced set of chemical shifts derived originally from the BMRB. The RefDB also provides a sequence-based mapping to PDB entries for each set of re-referenced shifts. The sub-set of the RefDB entries selected was free of complexes and mapped to 454 PDB entries.

Metadata and structural information from each of the 454 PDB entries were extracted and each entry was split up into its constituent fragments. In this context, a fragment is defined as a single contiguous polypeptide chain present as part of a potentially larger protein structure with multiple such chains. These fragments were then processed through each of the three chemical shift predictors — PROSHIFT, SHIFTS and SHIFTX. STRIDE [8] secondary structural information was obtained for each fragment from the S2C [17] database. Additionally, a per-residue solvent exposure term was calculated using the half-sphere exposure $HSE-\beta$ [9] measure. All structural information and predicted shifts partitioned by protein backbone atom type were stored in a relational database using appropriate schema.

A mapping between the residues in a PDB fragment and those in a RefDB entry with experimental shifts is required to be able to compare the predicted shifts with experimental shifts. Alignments between the corresponding residue sequences were generated using a simple pairwise dynamic programming alignment algorithm provided by Biopython [2].

2.2. *Feature extraction*

Chemical shifts can be predicted from structural models in three ways: using quantum mechanics, classical mechanics, and empirical models. A purely quantum approach is theoretically possible but, in the case of most macromolecules the size of typical protein structures, computationally infeasible. Thus, most protein chemical shift prediction methods employ hybrid techniques, combining quantum, classical and empirical approaches in various ways. Examples of such algorithms include SHIFTS (combines quantum and empirical methods), SHIFTX (combines classical and empirical methods) and PROSHIFT (maps a variety of empirically-determined structural features to chemical shifts using neural networks). Our approach employs each of these individual predictors' final predicted shifts as input to a non-linear regression algorithm. Also, the per-residue quantum mechanical contributions calculated by SHIFTS via density functional analysis of tri-peptides are independently included in the feature array. Additionally, the secondary structural assignments and solvent exposure information obtained in the manner described earlier are incorporated on a per-residue basis. Tables 1 and 2 enumerate the specific features employed in predicting backbone heavy atom and proton shifts respectively. Fig. 2 is a flowchart depicting the assembly of data and feature extraction described herein.

2.3. *Regression using Random Forests*

The proposed regression model has the form :

$$\delta_i = f(\vec{x_i}) \tag{1}$$

where δ_i is the estimated chemical shift for the ith nucleus, $f(\cdot)$ is a non-linear regression function and $\vec{x_i}$ is a vector whose components encode the variables of the regression model. These variables correspond to computable properties in each nucleus' environment and are

Table 1: Feature set employed in regression for protein backbone heavy atoms

Feature	Description
aa	Amino acid residue
sec_str	STRIDE secondary structure
solv_exp	Half-sphere solvent exposure ($HSE - \beta$) terms
$q_{i-1}^{\phi,\psi}$	Contribution from preceding residue's backbone torsion angles
$q_{i}^{\phi,\psi}$	Contribution from target residue's backbone torsion angles
$q_{i+1}^{\phi,\psi}$	Contribution from succeeding residue's backbone torsion angles
q_{i-1}^{χ}	Contribution from preceding residue's type and χ_1 torsion angle
q_{i}^{χ}	Contribution from target residue's type and χ_1 torsion angle
e^{HB}	Hydrogen bond contributions
rand_coil	Random coil reference shift value
pred_shifts	Predicted shifts from SHIFTS, SHIFTX and PROSHIFT

Table 2: Feature set employed in regression for protein backbone protons

Feature	Description
aa	Amino acid residue
sec_str	STRIDE secondary structure
solv_exp	Half-sphere solvent exposure ($HSE - \beta$) terms
e^{RC}	Ring current contributions from neighboring aromatic rings
e^{E}	Electrostatic contributions from nearby point charges
e^{PA}	Peptide group anisotropy
rand_coil	Random coil reference shift value
pred_shifts	Predicted shifts from SHIFTS, SHIFTXand PROSHIFT

essentially the features described in the section above. The algorithm selected for implementing the regression function in this set of experiments is Random Forest regression [4].

A Random Forest is an ensemble of decision trees constructed using bagging (i.e., random instance selection) and random feature selection. Predictions are made by averaging (or voting, in the context of classification) over the predictions made by each tree. The benefits of ensemble methods in machine learning has been studied extensively [6]. Briefly, an ensemble predictor will be more accurate than any of its individual members when the individual predictors are accurate and diverse. Two predictors are diverse if their errors are uncorrelated. Random Forests ensure diversity through random instance and feature selection. The benefits of ensemble predictions can be understood intuitively in terms of the likelihood that two or more trees will make the same incorrect prediction. Let p_i be the probability that the ith tree makes an error. If the trees errors are uncorrelated (i.e., independent), then the probability that k trees make the same error is bounded by $\prod_i^k p_i$.

Breiman has analyzed the properties of Random Forests extensively [4]. Of note is that the generalization error, that is the error on novel instances, converges to a limit as the number of trees grows. In contrast, algorithms such as neural networks have no such guarantee. Additionally, the randomization scheme guards against noise.

In the experiments described, Random Forests were trained for each nucleus type on the given set of features and the accuracy of the final predicted shifts was estimated using 10-fold cross-validation. Chemical shift prediction accuracies are reported for the H^{α}, H^{N}, ^{15}N, $^{13}C^{\alpha}$, $^{13}C^{\beta}$and $^{13}C'$ backbone atom types in terms of root mean squared error (RMSE) from the experimental value. These RMSE values are compared to similar values obtained for the three component chemical shift predictors, PROSHIFT, SHIFTS, and SHIFTX. The p-values of decreases in RMSE are calculated using a standard t-test to assess the significance of improvements in prediction accuracy.

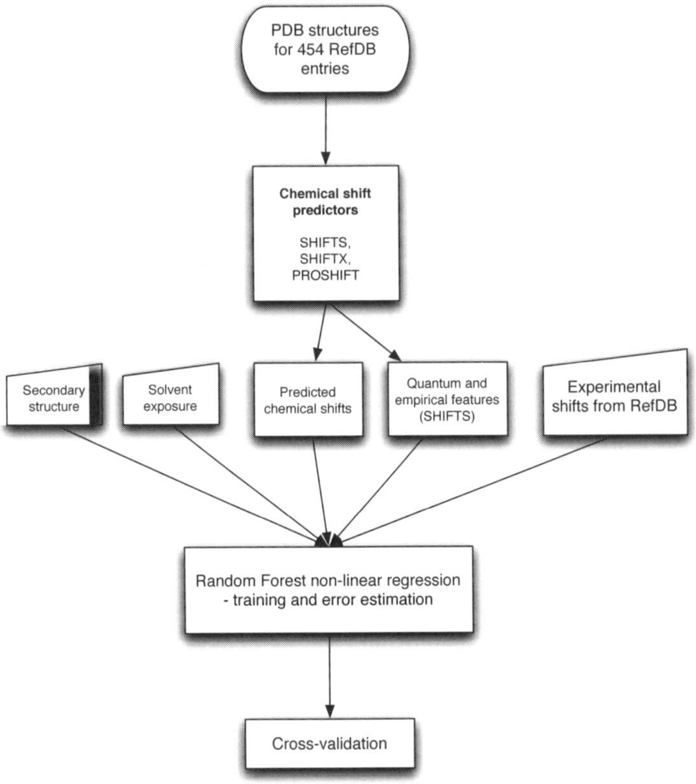

Figure 2: Flowchart depicting the experimental procedure involved in training Random Forest regressors

3. Results and discussion

The database of chemical shifts employed in this exercise consisted of between 24,000 to 47,000 separate chemical shifts depending on the nucleus type. These were mapped to 454 different protein structures from the PDB. The results obtained by training Random Forest regressors for each nucleus type (subject to 10-fold cross-validation) are shown in table 3. Prediction accuracies are reported in terms of root mean squared error (RMSE) from experimental shift values. It is seen that the Random Forest predictions are 15.5%, 17.7%, 14.8%, 7.9%, 3% and 14.9% more accurate than the best of SHIFTS, SHIFTX, and PROSHIFT for H^{α}, H^{N}, ^{15}N, $^{13}C^{\alpha}$, $^{13}C^{\beta}$ and $^{13}C'$ nuclei respectively. The p-values of these decreases in RMSE, based on t-tests on the residuals, are each $\ll 0.001$, thereby indicating that the decreases in error are statistically significant. Note, that although the $^{13}C^{\beta}$ RMSE value shows only modest improvement (3%) when predicted using the Random Forest algorithm, a separate experiment (data not shown) where rotameric configurations served as a feature resulted in an RMSE drop of greater than 7% for the same nucleus. This is to be expected since the configuration of the sidechain and the resultant distribution of the sidechain electrons likely have a significant influence on the $^{13}C^{\beta}$ chemical shift. This also indicates that the same set of regression features may not be optimal for every type of nucleus.

It is clear from these results that the Random Forest-based non-linear regression approach to shift prediction promises significant improvements in prediction accuracy over existing methods. Apart from the resistance of the technique to over-fitting, it is to be noted that the size of the training data set employed in this exercise is significantly larger than any prior comparable effort. This, in turn, will allow this prediction method to better generalize to novel protein structures. Also, given that Random Forests are extremely efficient to train

Table 3: Chemical shift prediction accuracies for individual shift predictors and Random Forest regression in terms of root mean squared error (RMSE) from experimental values. The values in italics identify the least RMSE value amongst the SHIFTS, SHIFTX and PROSHIFT predictors for that atom type. The values in bold type identify the best overall predictor, which is the Random Forest approach for all nuclei. The percentage figures in parentheses in the Random Forest column represent the decrease in RMSE as a percentage of the least RMSE value amongst the underlying predictors.

		SHIFTS	SHIFTX	PROSHIFT	RANDOM FOREST
Nucleus	Instances	RMSE (ppm)	RMSE (ppm)	RMSE (ppm)	RMSE (ppm)
H^{N}	46,991	0.66	0.63	*0.58*	**0.49** (15.5%)
H^{α}	38,767	0.79	0.36	*0.34*	**0.28** (17.7%)
^{15}N	40,166	5.29	3.51	*3.44*	**2.93** (14.8%)
$^{13}C^{\alpha}$	37,006	1.86	*1.64*	2.59	**1.51** (7.9%)
$^{13}C^{\beta}$	29,809	3.13	*3.02*	3.75	**2.93** (3%)
$^{13}C'$	24,253	1.89	*1.40*	2.34	**1.19** (14.9%)

and each tree in the forest can be grown in parallel, additional structural variables may be rapidly tested for their contribution to improvement in shift prediction accuracy. Experiments using B-factors from X-ray crystallographic structures and discrete per-residue rotamer library categories as additional features are currently in progress.

The method reported here is also notable for the fact that it is a hybrid meta-prediction approach, combining quantum, classical and empirical information about protein structures. Purely quantum mechanical approaches to shift prediction work well for small molecules but are computationally infeasible for anything the size of a typical protein structure. Conversely, purely empirical approaches are unlikely to capture all the complexity inherent in the factors affecting the electronic environment which finally dictates the chemical shift. The meta-prediction aspect, wherein predictions from multiple underlying chemical shift predictors (PROSHIFT, SHIFTS and SHIFTX in this case) are incorporated as input to the regression algorithm, allows for a judicious combination of information from both approaches to be incorporated into a single prediction technique. Meta-prediction approaches have been successfully used in secondary and tertiary structure prediction and ligand docking. The results obtained indicate that chemical shift prediction is also a suitable candidate for this approach.

4. Conclusion

We have shown that a non-linear regression approach to chemical shift prediction employing a ensemble machine learning approach has the potential to improve chemical shift prediction accuracy significantly. The ensemble Random Forest algorithm employed is provably resistant to over-fitting the test data and generalizes well to novel test instances. This is demonstrated by the improvement in shift prediction accuracy seen in the 10-fold cross-validation exercise over existing chemical shift predictors across all six protein backbone nuclei. Random Forests allow for rapid training of regressors and are eminently parallelizable, therefore permitting one to explore the protein structural variable space efficiently. They make feasible the potential training of separate regressors for varied partitions of the training data set (all NMR structures versus all X-ray structures, per amino acid type regressors, per secondary structure type regressors etc.). It is possible that a future variant on this method will render predictions by using such different regressors internally to predict on different partitions of the test data. We are in the process of making an implementation of the current method available for public use.

The availability of a rapid, accurate and easily adapted method of chemical shift prediction will make it easier to study the relationship between shift and structure. Any technique that incorporates chemical shift prediction, such as NMR resonance assignment, low resolution structure prediction, fold recognition, protein docking and ligand interaction modeling, will benefit from the increased accuracy provided by this method. Additionally, the speed of training of the Random Forests will permit domain-specific regressors to be trained in these endeavors.

5. Acknowledgment

The authors would like to thank Drs. Robert Murphy and Gordon Rule for enlightening discussions on topics relevant to this work, and Tal Blum for a critical insight into the cross-validation process. K.A. was partially supported by a Merck Computational Biology fellowship for the duration of this work. This work is supported by a Young Pioneer Award to C.J.L. from the Pittsburgh Life Sciences Greenhouse.

References

1. BERMAN, H., WESTBROOK, J., FENG, Z., G. GILLILAND, T. B., WEISSIG, H., SHINDYALOV, I., AND BOURNE, P. The Protein Data Bank. *Nucleic Acids Research 28* (2000), 235–242.
2. The BIOPYTHON project. URL: `http://www.biopython.org`.
3. BioMagResBank (BMRB), a NIH funded bioinformatics resource, Department of Biochemistry, University of Wisconsin-Madison, Madison, WI USA (URL: `http://www.bmrb.wisc.edu`) Grant: LM05799-02.
4. BREIMAN, L. Random forests. *Machine Learning 45*, 1 (2001), 5–32.
5. CORNILESCU, G., DELAGLIO, F., AND BAX, A. Protein backbone angle restraints from searching a database for chemical shift and sequence homology. *J. Biomol. NMR 13*, 3 (1999), 289–302.
6. DIETTERICH, T. G. Ensemble methods in machine learning. *Lecture Notes in Computer Science 1857* (2000), 1–15.
7. DOMINGUEZ, C., BOELENC, R., AND BONVIN, A. Haddock: a protein-protein docking approach based on biochemical or biophysical information. *J. Am. Chem. Soc. 125*, 7 (2003), 1731–1737.
8. FRISHMAN, D., AND ARGOS, P. Knowledge-based protein secondary structure assignment. *Proteins 23*, 4 (1995), 566–579.
9. HAMELRYCK, T. An amino acid has two sides: a new 2D measure provides a different view of solvent exposure. *Proteins 59*, 1 (April 2005), 38–48.
10. HUS, J., PROMPERS, J., AND BRÜSCHWEILER, R. Assignment strategy for proteins of known structure. *J. Mag. Res. 157* (2002), 119–125.
11. LANGMEAD, C., AND DONALD, B. High-throughput 3D homology detection via NMR resonance assignment. In *Proc. IEEE Computer Society Bioinformatics Conference (CSB)* (2004), pp. 278–289.
12. MEILER, J. PROSHIFT: Protein chemical shift prediction using artificial neural networks. *Journal of Biomolecular NMR 26* (2003), 25–37.
13. MIELKE, S., AND KRISHNAN, V. Protein structural class identification directly from NMR spectra using averaged chemical shifts. *Bioinformatics 19*, 16 (2003), 2054–2064.
14. NEAL, S., NIP, A. M., ZHANG, H., AND WISHART, D. S. Rapid and accurate calculation of protein 1H, 13C and 15N chemical shifts. *J. Biomol. NMR 26* (2003), 215–240.
15. PENG, C., UNGER, S., FILIPP, F., SATTLER, M., AND SZALMA, S. Automated evaluation of chemical shift perturbation spectra: new approaches to quantitative analysis of receptor-ligand interaction NMR spectra . *J. Mol. Biol. 29*, 4 (2004), 491–504.
16. SIBLEY, A., COSMAN, M., AND KRISHNAN, V. An empirical correlation between secondary structure content and averaged chemical shifts in proteins. *Biophys J 84*, 2 Pt 1 (Feb 2003), 1223–1227.
17. WANG, G., ARTHUR, J., AND DUNBRACK, R. S2C: A database correlating sequenceand atomic co-ordinate numbering in the Protein Data Bank. URL: `http://dunbrack.fccc.edu/Guoli/s2c/`, 2002.

18. WISHART, D., AND SYKES, B. The 13C chemical-shift index: a simple method for the identification of protein secondary structure using 13C chemical-shift data. *J. Mol. Biol. 4*, 2 (1994), 171–180.
19. XU, X., AND CASE, D. Automated prediction of 15N, 13Calpha, 13Cbeta and 13C' chemical shifts in proteins using a density functional database. *J Biomol NMR 21* (Dec 2001), 321–333.
20. ZHANG, H., LEUNG, A., AND WISHART, D. THRIFTY. URL: `http://redpoll.pharmacy.ualberta.ca/thrifty`, 2005.
21. ZHANG, H., NEAL, S., AND WISHART, D. S. RefDB: A database of uniformly referenced protein chemical shifts. *J. Biomol. NMR 25* (2003), 173–195.

ONBIRES: Ontology-Based Biological Relation Extraction System

Minlie Huang[1], *Xiaoyan Zhu[1], Shilin Ding[1], Hao Yu[1] and Ming Li[1,2]

[1]*State Key Laboratory of Intelligent Technology and Systems (LITS), Department of Computer Science and Technology, Tsinghua University, Beijing, 100084, China,*
[2]*Bioinformatics Laboratory, School of Computer Science, University of Waterloo, N2L 3G1, Ontario, Canada, mli@uwaterloo.ca*

Automated discovery and extraction of biological relations from online documents, particularly MEDLINE texts, has become essential and urgent because such literature data are accumulated in a tremendous growth. In this paper, we present an ontology-based framework of biological relation extraction system. This framework is unified and able to extract several kinds of relations such as gene-disease, gene-gene, and protein-protein interactions etc. The main contributions of this paper are that we propose a two-level pattern learning algorithm, and organize patterns hierarchically.

1. Introduction

Biological data, including both experimental data and textual information, are growing tremendously in these decades. However, most of important biological knowledge, such as protein-protein interaction and gene-disease interaction, is still locked in a large number of literatures, remaining not computer-readable. The heaven burden of accessing, extracting and retrieving biological knowledge of interests is left to the human user. To expedite the process of functional bioinformatics, it is absolutely important to develop information extraction systems to automatically process these online biological documents and extract biological knowledge such as protein-protein interaction (PPI), gene-disease correlation, sub-cellular location of protein and so on. A number of database, for example, DIP for PPI [1], KEGG for biological pathways [2], BIND for molecular interactions [3], accumulate such relations.

The portability is another major problem that impedes the wide use of IE tools in online biological documents. Some systems are aimed to extract PPIs [4,5,6], some are designed to mine gene-disease relation, some are able to discover gene-function correlation [7], but none of them can extract these kinds of relations in a unified framework. In other words, it is not easy or unable to adopt these systems from this kind of relation extraction to another one. Most of the approaches are more focused on a specific application to solve a specific kind of problem.

Ontology is a formal conceptualization of a particular domain that is shared by a group of people [8]. Each concept in an ontology has a canonical and consistent definition, and they are organized in a hierarchical tree, thus knowledge can be easily communicated, shared and reused across applications. In recent decades, a number of biological ontologies have been designed and developed for public usage, including Gene Ontology [9], MeSH [10], and LocusLink [11]. These ontologies

* Corresponding author: zxy-dcs@tsinghua.edu.cn Tel: 86-10-62796831 Fax: 86-10-62771138

provide a controlled vocabulary or conceptualization for biological concepts such as gene, protein, disease and function etc, and thus supply a shared understanding of knowledge among biology communities. When an IE system is structured in an ontology-style way, it is more portable and less dependent on applications.

In this paper, we propose an ontology-based biological relation extraction system to automatically extract biological relations from a huge number of online MEDLINE abstracts. Compared with the previous methods, the main contributions of our method are:

1) External ontology integration. Currently, we have integrated four external ontologies, including GO, MeSH, LocusLink, OMIM [12]. Concepts in these ontologies have been converted into a uniform format, and each concept is described by a set of synonymous terms (i.e. *synset*);

2) Ontology-based semantic annotation of online biological documents. Our method will recognize and identify several categories of biological entities, including GENE, PROTEIN, DISEASE, PROCESS, FUNCTION, CELLULAR COMPONENT (CELLC);

3) Two-level pattern learning, i.e., token pattern learning and syntactic pattern learning. We organize patterns in a hierarchy and then a weighted pattern matching scheme is applied.

The rest of the paper is organized as follows: in Section 2, we present an overview of the architecture of ONBIRES. In Section 3, we state that how external ontologies are integrated into our system and how concepts are organized uniformly. In Section 4, a pattern hierarchy is introduced, followed by the detailed pattern learning algorithm in Section 5. Then, experiments and evaluations are shown in Section 6. At last, we make our conclusion in Section 7.

2. Architecture of ONBIRES

A large number of methods have been proposed and various systems are developed to extract biological knowledge from biological literature such as extracting protein-protein interactions, or integrated systems as [13]. However, most systems do not provide a unified framework and most algorithms are heavily dependent on a specific application. Also there is a lack of a mechanism for automatic learning of pattern for such information extraction tasks.

We proposed a novel framework that can extract several kinds of relations with a mechanism of automatic pattern learning. Our algorithm is much less dependent on a specific problem to be solved. It is able to learn patterns and extract relations in a unified way. The system architecture is shown in Figure 1. Compared with previous methods and systems, our approach has several significant advantages. First, we utilize several external ontologies to try to capture as many synonyms as possible for each type of biological entities, and organize them in a uniform format. Second, a hierarchical pattern structure is introduced, on which a weighted pattern matching scheme is used to balance precision and coverage.

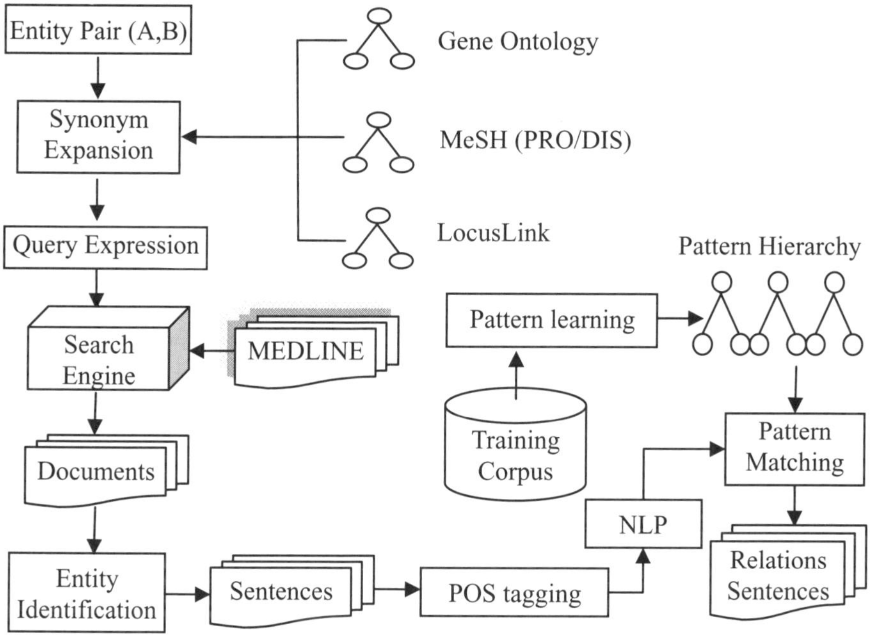

Figure 1. ONBIRES architecture.

There are several steps in our system as follows:

1) For each entity pair (A,B), we will search synonyms in our local ontologies for A and B, respectively. The set of synonyms are later called a synset. The semantic type of each entity is also returned. If no synonym is found, the user has to specify a semantic type.

2) According to the synsets of A and B, a query expression is formed. If the two synset sets are $A=\{a_1,a_2,\ldots a_m\}$ and $B=\{b_1,b_2,\ldots b_n\}$, the query expression is "(a_1 OR a_2 OR ...a_m) AND (b_1 OR b_2 OR ...b_n)". This expression is input into a search engine to retrieve MEDLINE documents (currently only abstracts).

3) For each document, we will do semantic tagging using the synsets of entity A and B. Then, documents are segmented into sentences and we only save those sentences that contain both A and B for the further processing.

4) Sentences are part-of-speech (POS) tagged. At the training stage, patterns are learned from a training corpus, whose sentences have been labeled as positive or negative. At the matching stage, sentences are processed by a natural language processing (NLP) module. We have several shallow parsing techniques in the NLP module as describe in [14].

5) A weighted pattern matching algorithm is applied against the pattern hierarchy. Sentences whose matching scores exceed a threshold are declared to have relations.

6) Both the extracted relations and relevant documents are presented to the user in a user interface. We provide PMID, title and abstract of a relevant document.

3. External Ontology Integration

Biological named entity recognition is a great challenge for IE communities [15]. A number of methods, such as machine learning based ones [16], have been devised to improve the performance, but they are still far away from real applications. In our system, we try to collect external ontologies to enhance the results of entity identification. The first one is Gene Ontology, which has been well-known as a controlled vocabulary for gene annotation of documents. This ontology consists of three subjects, that is, biological process, biological function and cellular component. Accordingly, we extract three kinds of entities, that is, PROCESS, FUNCTION, and CELLC, to form our own synset ontology. The number of the three types of entities amounts to 9852, 7576 and 1679, respectively.

The second ontology we used is MeSH (Medical subject Heading). MeSH models a hierarchical terminology of disease, chemical and drug and so on. In this system we only consider two sub-branches of the hierarchical tree, that is, the disease branch (labeled with C##.###, each '#' is a digit), and the protein branch (labeled with D12.776.###). Totally, we obtain 1610 proteins and 226 diseases from MeSH. 1,303,625 genes are extracted from LocusLink and 9315 from OMIM and another 3125 diseases are obtained from OMIM.

Finally, these data are organized uniformly in the format as shown in Table 1. *UID* is the unique identity of an entity, where this identifier is directly reproduced from the original ontology thus we could search the ontology via this symbol. Synset is a set of terms describing the same entity. Six entity types are defined, that is, PROCESS, FUNCTION, CELLC, PROTEIN, GENE, and DISEASE.

Table 1. Uniform concept format and examples. Synonyms are separated with '#'.

UID	Entity Description	Entity Type	Source	Synset
D12.776.503	Lectins	PROTEIN	MeSH	Animal Lectins# Isolectins#
U_GO:0050285	sinapine esterase activity	FUNCTION	GO	sinapine esterase activity#

4. Pattern Hierarchy

We have defined a pattern hierarchy according to the generalization power of each pattern. An example of the hierarchy is shown in Figure 2.

Syntactic pattern consists of a sequence of part-of-speech tags. This kind of pattern reveals the syntactic constraints that a pattern must conform to. Syntactic patterns are learned by aligning sequences of part-of-speech tags from token patterns.

Token pattern comprises keywords that are commonly used to describe relations. And many token patterns may share the same form of syntactic constraints. They have less generalization power than syntactic patterns, and at the same time, they are more precise. Token patterns are generated by aligning sequences of words from instance patterns.

Instance pattern is a sentence which has been labeled as positive. Token pattern can be learned from positive samples.

We note that the generalization power of a pattern decreases from the top to the bottom along the hierarchy, and the accuracy increases. With a weighted pattern matching scheme at different levels, we could obtain a balance between the accuracy and extensibility. This is the major motivation why we organize patterns hierarchically.

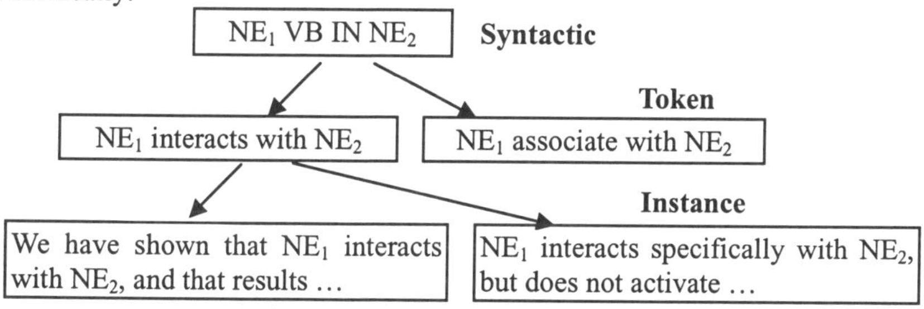

Figure 2. Pattern Hierarchy

5. Automatic pattern learning

The idea of using dynamic programming to automatically learn patterns is used by [5,7,14,17]. The major contribution of our method is that we use a two-level pattern learning algorithm and organize patterns hierarchically, and furthermore we adopt a weighted pattern matching scheme on the structure. We generate patterns at a token level and syntactic level. At each level, sequence algorithm alignment is used to generate patterns. The pattern structure used in our system is Eliza-style [18]. A pattern will be represented in a 5-tuple: <*prefiller*, NE_1, *midfiller*, NE_2, *postfiller*>, where NE_1 and NE_2 are two entities concerned with a specific application. *Prefiller* is a pattern element before the entity NE_1, *midfiller* is a pattern element between NE_1 and NE_2, and *postfiller* is a pattern element after NE_2. They are all lists of words or tags. For instance, given a sentence as "We/NNP found/VBD that/IN NEDD8/PROTEIN modifies/VBZ CUL1/PROTEIN in/IN Drosophila/NN.", the algorithm may learn a token pattern {"", $PROTEIN_1$, modifies, $PROTEIN_2$, ""}, and a syntactic pattern {"", $PROTEIN_1$, VBZ, $PROTEIN_2$,""}. The sentence itself is an instance pattern (may be positive or negative). A sentence is also represented in a similar 5-tuple.

It is well known that local alignment is a dynamic programming algorithm as formula (1a-b).

$$sim(i,0) = sim(0,j) = 0; i = 1,2,\ldots,M, j = 1,2,\ldots,N \tag{1a}$$

$$sim(A_{1,2,\ldots,i}, B_{1,2,\ldots,j}) = \max\begin{cases} 0 \\ sim(A_{1,2,\ldots,i-1}, B_{1,2,\ldots,j-1}) + s(a_i, b_j), \text{ if } a_i = b_j \\ sim(A_{1,2,\ldots,i-1}, B_{1,2,\ldots,j}) + s(a_i, GAP), \\ sim(A_{1,2,\ldots,i}, B_{1,2,\ldots,j-1}) + s(b_j, GAP), \end{cases} \tag{1b}$$

During token pattern learning, we take $s(w,w)=1$ and $s(w_1,w_2)=-1, w_1 \neq w_2$, which means that if two words share the same base form, the score is 1, otherwise the score will be -1. Therefore only those words that have the same base form can be aligned together.

During syntactic pattern learning, the local alignment algorithm is applied again on sequences of part-of-speech tags of token patterns. The scores *s(a,b)* is adopted from [5].

In our pattern learning algorithm, we use a pattern frequency to record how many times each pattern is aligned during the pairwise alignment. Those whose frequencies are less than a user-specified threshold are removed from the pattern set.

When a pattern hierarchy is obtained, a weighted pattern matching scheme is used. The matching score for a sentence is defined in formula (2):

$$Score(S_j) = \arg\max_{P_{tok}}\{w_{tok} * Sim(P_{tok}, S_j) + w_{syn} * Sim(P_{syn}(P_{tok}), S_j)\} \qquad (2)$$

where P_{tok} is a token pattern, $P_{syn}(P_{tok})$ is the syntactic pattern of P_{tok}. w_{tok} is the weight for token pattern, and w_{syn} for syntactic pattern. When this score exceeds a user-specified threshold, we can say definitely that this sentence describes a relation. For those sentences that have more than two entities, all possible combinations of two entities are considered. Since syntactic patterns are much less precise than token patterns, w_{tok} is set to be larger than w_{syn}. We also apply other constraints on syntactic patterns. For example, if two words match in syntactic patterns, but do not match in token patterns, the semantic similarity is computed by using WordNet. The matching score from syntactic pattern is added to the overall score only when the similarity is larger than a threshold (0.7 currently).

The reason why we have to weight between different levels derives from this fact: if we only consider one level of patterns, either the matching precision is quite low, or the coverage is narrow. For example, if we have two patterns, as shown in Figure 2, a token pattern {"", NE_1,"interacts with", NE_2,""} and a syntactic pattern {"", NE_1,VB IN, NE_2,""}, for a sentence "…NE_1 associates with NE_2…", there is no match at the token level, while it can be matched at a syntactic level. Similar cases are also observed for the problem of low-precision.

6. Experiments

Evaluating the precision and recall of ONBIRES is very difficult because a huge collection of online MEDLINE abstracts is involved. For a small number of documents, it is possible to annotate them manually and compute the precision and recall. In the current version of our system, we evaluate our approach on two applications, i.e. gene-disease interactions, and protein-protein interactions.

The first experiment is to extract protein-protein interactions. We collect the training corpus from http://www.biostat.wisc.edu/~craven/ie/ [19]. Each sentence is annotated as either a negative sample or positive sample. Positive samples are labeled with relation tuples which were gathered from the MIPS Comprehensive Yeast Genome Database. We used 1102 positive samples to generate patterns. 1024 sentences from GENIA corpus are used for evaluation [20]. GENIA corpus is available at: http://www-tsujii.is.s.u-tokyo.ac.jp/~genia/topics/Corpus/. These sentences are manually annotated by experts, where there are 238 positive samples.

The second experiment is to extract gene-disease correlations. This corpus is also downloaded from http://www.biostat.wisc.edu/~craven/ie/. The relation tuples were gathered from the Online Mendelian Inheritance in Man (OMIM) database. There are 636 positive samples in this corpus, which are all used for learning patterns. Since the corpus is comparatively small, 100 of the training samples and another 177 negative samples are randomly selected for evaluation.

In each experiment, we compare the performance of token patterns and that of token patterns plus syntactic patterns. These results are shown in Figure 3 and Figure 4. During pattern learning, we provide a vocabulary to restrict which words can be contained by a pattern. Patterns whose frequencies are less than one are removed. The statistics of extracted interactions in these experiments are listed in Table 2.

From these results, we could see that token patterns plus syntactic patterns outperform only token patterns. The two curves converge to the same curve when the threshold becomes larger because sentences with large matching scores have matched token patterns perfectly, and syntactic patterns have tiny contributions to these sentences. With a smaller threshold, the performance is improved remarkably when syntactic patterns are used.

We also investigated into those sentences that can not be extracted correctly. There are three kinds of errors:

1) Incorrect patterns. Although we have limited the vocabulary of patterns, and have removed patterns with low frequencies, a small proportion of incorrect patterns are still left. Unfortunately, they have a fairly high frequency. Therefore, more sophisticated techniques need be developed to assess each pattern.

2) Errors caused by complicated grammatical structures. This method treats a sentence as a linear sequence, thus it is not competent to process complicated grammatical structures. Although we have done long sentence splitting, appositive and coordinative structure recognition as shown in [14], there are more structures that can not be handled. For example, there is a sentence:

The oxygen radical scavenger N-acetyl-L-cysteine, but not an inhibitor of nitric oxide synthase, inhibited LIF -induced HIV replication. where underlined parts are identified as proteins. It matches a pattern "PROTEIN inhibit PROTEIN", which is erroneous.

3) Errors caused by named entity identification. In our system, we used a dictionary-based method to recognize named entities. However, in many cases, this method produces errors. Particularly, it can not discriminate proteins from genes, since most genes and proteins have the same lexical symbols. An example is listed here: *Taken together, our data indicate that MS-2 mediates induction of the CD11b gene as cells of the monocytic lineage mature.* The underlined terms are identified as proteins, but the second should be recognized as a gene.

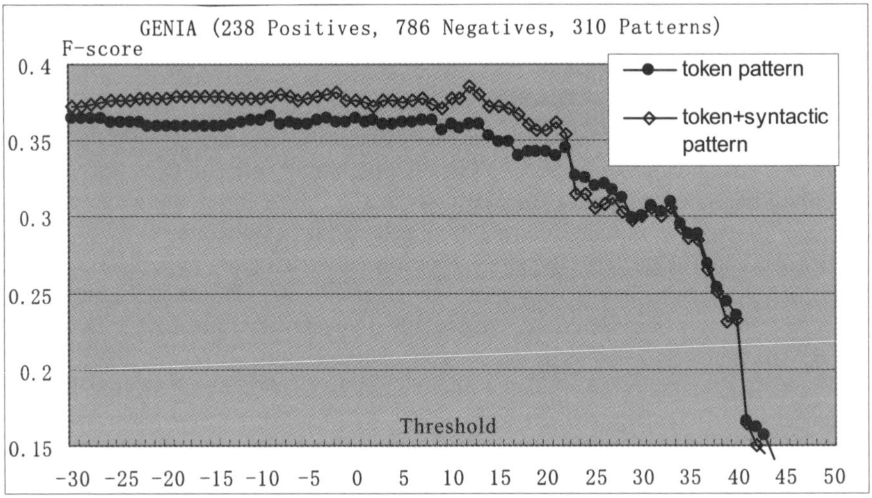

Figure 3. F-Score curve over matching score threshold for GENIA corpus.

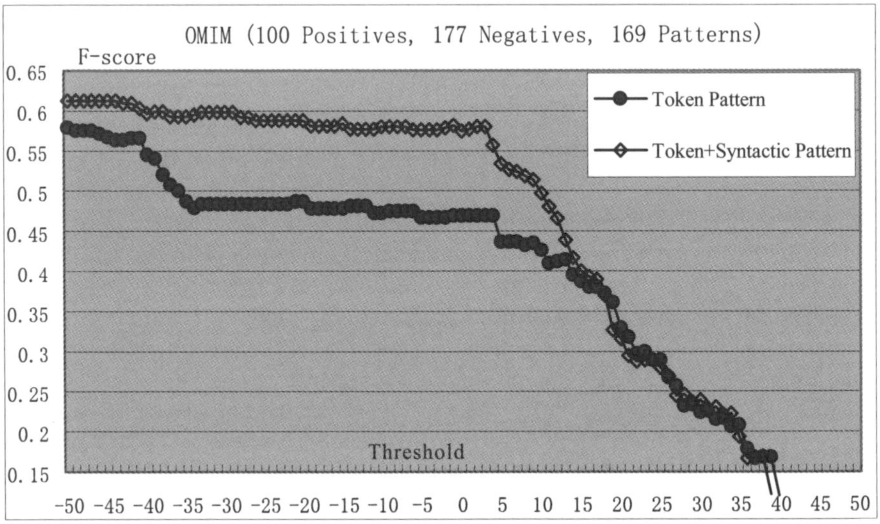

Figure 4. F-Score curve over matching score threshold for OMIM corpus.

Table 2. Statistics of extracted interactions with the best threshold. TP indicates the number of correct samples. ET denotes the number of extracted samples.

Corpus	Experiment	Precision	Recall	F-score	TP	ET
GENIA	Token pattern	0.424	0.235	0.302	56	132
	Token+syntactic	0.411	0.243	0.306	58	141
OMIM	Token pattern	0.532	0.42	0.469	42	79
	Token+syntactic	0.58	0.58	0.58	58	100

7. Conclusion

In this paper, we have proposed an ontology-based information extraction system to search biological relations from online documents. This system, which is ontology-based, has a unified framework and is less dependent on specific applications. Several external ontologies are integrated to improve the structure and organization of concept. A two-level pattern learning algorithm is applied to generate patterns which are then organized in a hierarchy. A weighted matching scheme is devised to balance the accuracy and coverage of the system.

The experimental results show that our system is promising to extract knowledge from a huge number of MEDLINE abstracts. Future work will be focused on how to evaluate patterns more efficiently, process complicated grammatical structures and handle named entity recognition errors.

Acknowledgments

The work was supported by Chinese Natural Science Foundation under grant No. 60272019 and 60321002, the Canadian NSERC grant OGP0046506, CRC Chair fund, and the Killam Fellowship. We also would like to thank Xiaozhe Li and Zhiyuan Liu for coding programs and converting data from several publicly available external resources.

References

1. I. Xenarios, E. Fernandez, L. Salwinski, X.J. Duan, M.J. Thompson, E.M. Marcotte and D. Eisenberg. (2001) DIP: The Database of Interacting Proteins: 2001 update, Nucleic Acids Reg., 29, pp. 239–241.
2. M. Kanehisa and S. Goto. (1997) A systematic analysis of gene functions by the metabolic pathway database. In "Theoretical and Computational Methods in Genome Research" (Suhai, S., ed.), pp. 41–55, Plenum Press
3. G.D. Bader, I. Donaldson, C. Wolting, B.F. Quellette, T. Pawson and C.W. Hogue. (2001) BIND –The Biomolecular Interaction Network Database, Nucleic Acids Research, 29(1), pp. 242–245.
4. T. Ono, H. Hishigaki, A. Tanigami and T. Takagi. (2001) Automated extraction of information on protein–protein interactions from the biological literature, Bioinformatics, 17(2), pp. 155–161.
5. M.L. Huang, X.Y. Zhu, Y. Hao, D.G. Payan, K. Qu and M. Li. Discovering patterns to extract protein-protein interactions from full-texts. Bioinformatics, Dec, 2004; 20(18):3604-3612.
6. E.M. Marcott, I. Xenarios and D. Eisenberg. (2001) Mining literature for protein–protein interactions, *Bioinformatics*, 17(4), pp. 359 363.
7. J.H. Chiang and H.H. Yu. (2003) MeKE: discovering the functions of gene products from biomedical literature via sentence alignment, *Bioinformatics*, 19(11), pp. 1417–1422.

8. T.R. Gruber. "A Translation Approach to Portable Ontology Specifications," Knowledge Acquisition, vol. 5, pp. 199–220, 1993.
9. The Gene Ontology Consortium (2000) Gene Ontology: tool for the unification of biology. Nat. Genet., 25, 25–29. http://www.gene ontology.org/.
10. MeSH: Medical Subject Heading. http://www.nlm.nih.gov/mesh/mesh home.html.
11. K.D. Pruitt and D.R. Maglott. (2001) RefSeq and LocusLink: NCBI gene-centered resources. Nucleic Acids Res., 29, 137–140. http://www.ncbi. nlm.nih.gov/LocusLink/.
12. Online Mendelian Inheritance in Man, OMIM (TM). McKusick-Nathans Institute for Genetic Medicine, Johns Hopkins University (Baltimore, MD) and National Center for Biotechnology Information, National Library of Medicine (Bethesda, MD), 2000. http://www.ncbi. nlm.nih.gov/omim/.
13. X.H. Hu, T.Y. Lin, I.Y. Song, X. Lin, I. Yoo, M. Lechner, M. Song. Ontology-Based Scalable and Portable Information Extraction System to Extract Biological Knowledge from Huge Collection of Biomedical Web Documents. Web Intelligence 2004: 77-83.
14. M.L. Huang, X.Y. Zhu, and M. Li. A hybrid method for relation extraction from biomedical literature. 2005, accepted by International Journal of Medical Informatics.
15. L. Hirschman, J.C. Park, J. Tsujii, L. Wong and C.H. Wu. Accomplishments and challenges in literature data mining for biology. Bioinformatics, 18, 1553-1561, December 2002.
16. G.D. Zhou, J. Zhang, J. Su, D. Shen, and C.L. Tan. "Recognizing names in biomedical texts: a machine learning approach". Bioinformatics Vol. 20 no. 7. 2004, pages 1178-1190.
17. E. Agichtein and L. Gravano. Snowball: extracting relations from large plain-text collections. ACM DL 2000: 85-94.
18. J. Weizenbaum. (1966) ELIZA – A Computer program for the study of natural language communications between men and machine, Communications of the Association for Computing Machinery, 9, pp. 36–45.
19. S. Ray and M. Craven. Representing Sentence Structure in Hidden Markov Models for Information Extraction. IJCAI 2001: 1273-1279. Seattle, USA.
20. J.D. Kim, T. Ohta, Y. Teteisi and J. Tsujii. (2003). GENIA corpus - a semantically annotated corpus for bio-textmining. Bioinformatics. 19(suppl. 1). pp. i180-i182. Oxford University Press.

A MORE ACCURATE AND EFFICIENT WHOLE GENOME PHYLOGENY

P.Y. CHAN T.W. LAM S.M. YIU C.M. LIU
Department of Computer Science
The University of Hong Kong, Hong Kong
E-mail: {pychan, twlam, smyiu, cmliu}@cs.hku.hk

To reconstruct a phylogeny for a given set of species, most of the previous approaches are based on the similarity information derived from a subset of conserved regions (or genes) in the corresponding genomes. In some cases, the regions chosen may not reflect the evolutionary history of the species and may be too restricted to differentiate the species. It is generally believed that the inference could be more accurate if whole genomes are being considered. The best existing solution that makes use of complete genomes was proposed by Henz et al.[13] They can construct a phylogeny for 91 prokaryotic genomes in 170 CPU hours with an accuracy of about 70% (based on the measurement of non-trivial splits) while other approaches that use whole genomes can only deal with no more than 20 species. Note that Henz et al. measure the distance between the species using BLASTN which is not primarily designed for whole genome alignment. Also, their approach is not scalable, for example, it probably takes over 1000 CPU hours to construct a phylogeny for all 230 prokaryotic genomes published by NCBI. In addition, we found that non-trivial splits is only a rough indicator of the accuracy of the phylogeny. In this paper, we propose the followings.
(1) To evaluate the quality of a phylogeny with respect to a model answer, we suggest to use the concept of the maximum agreement subtree as it can capture the structure of the phylogeny.
(2) We propose to use whole genome alignment software (such as MUMmer) to measure the distances between the species and derive an efficient approach to generate these distances.
From the experiments on real data sets, we found that our approach is more accurate and more scalable than Henz et al.'s approach. We can construct a phylogenetic tree for the same set of 91 genomes with an accuracy more than 20% higher (with respect to both evaluation measures) in 2 CPU hours (more than 80 times faster than their approach). Also, our approach is scalable and can construct a phylogeny for 230 prokaryotic genomes with accuracy as high as 85% in only 9.5 CPU hours.

1. Introduction

Reconstructing a phylogeny for a given set of species is a well-known problem in computational biology. The resulting phylogeny can help researchers to understand the evolutionary history and relationship of the species. In the case of viruses, we may be able to identify the origin of the viruses so that precaution can be taken to avoid further spreading of the viruses. Therefore, an accurate and efficient reconstruction method is desirable.

Most of the previous approaches are based on a subset of conserved regions extracted from the corresponding genomes for the inference.[3, 5, 18, 28, 30] The distance between each pair of species is usually derived from the similarity of the selected regions. The accuracy of the produced phylogeny thus depends on the choice of these regions. Not surprisingly, there may be cases that these regions do not truly reflect the whole evolutionary history of the species. Different phylogenies may be obtained by selecting a different set of regions. Or if only a small portion of the genomes is selected, there may be the problem of mutational saturation, that is, the selected regions are not powerful enough to differentiate the phylogenetic relations of some species. It is generally believed that the inference of phylogeny could be more accurate if the whole genomes are being used.[10, 13, 14]

However, there are two concerns for using the complete genomes: the scalability problem and the distance measure. To construct the phylogeny of a given set of species, we need to compute a distance for every pair of species. The amount of computation required grows quadratically with the number of species. Many previous attempts only deal with a small number of species (e.g., only nine and eleven genomes are considered by Hemiou et al.[14] and Fitz-Gibbon and House,[10] respectively). Also, how to derive a good distance measure for every pair of species is not completely resolved as most of the alignment tools are not designed for measuring the similarity (or distance) between two complete genomes.

The best existing solution along this direction was proposed by Henz et al.[13] They are able to construct a phylogeny for 91 prokaryotic genomes in 170 CPU[a] hours with an accuracy of about 70% when compared with phylogeny that is constructed using the taxonomy published by NCBI (we consider this as the *true* phylogeny). The accuracy measure used in their paper is based on the concept of *non-trivial splits*. Each internal edge in the phylogeny is called a non-trivial split. By deleting any of these edges, the species are separated into two groups. If there is a corresponding split in the true phylogeny, the split is considered to be good. The percentage of good splits is used as the accuracy measurement. In fact, using the percentage of good splits as the accuracy measure may not be a good indicator on the quality of the phylogeny. Figure 1 gives an example. The constructed phylogeny given in Figure 1(b) wrongly groups the whole Family B1 with Family A1 in the same subtree, and the Family B2 with Family A2 in another subtree. However, the accuracy based on non-trivial splits is as high as 92.3%. The problem is due to the fact that non-trivial splits do not explicitly capture the topology of the phylogenies.

Moreover, their distance measure is based on the output of BLASTN which is not primarily designed for whole genome alignment. According to their approach, for each pair of genomes, BLASTN is executed to output a set of high-scoring local alignments. The total number of matched nucleotides from these alignments will be used as the similarity measure (and then the value is converted to a distance measure). However, there are examples where closer species may have a low score while two distant species may have a high score. For examples, the species Ralstonia solanacearum (Rs) and Neisseria meningitidis (Nm) should belong to the same group of beta-proteobacteria. On the other hand, the species Chlorobium tepidum (Ct) is from another family Chlorobi. However, based on the score from BLASTN, the distance of Ct from Nm is only 0.206 while the distance of Rs from Nm is about 3.86. That is why Nm and Ct are clustered together instead of Nm and Rs using Henz et al.'s approach (see Figure 2(a), for the mapping of the names of the species with the symbols, please refer to Figure 6). A similar example occurs to Treponema

[a] In their paper, they only report the CPU hours used without mentioning the actual running time which should be longer than the reported CPU hours. For our results, we will report both the CPU hours and the actual running time for comparison.

pallidum (Tp), Borrelia burgdorferi (Bb), and Clostridium perfringens (Cp).

We believe that the problem is due to the design of BLASTN which aims at locating all highly similar local alignments without considering the alignment of the whole genomes globally. Also, their approach is not scalable. It is estimated that to construct a phylogeny for all 230 prokaryotic genomes published by NCBI may take more than 1000 CPU hours which is not practical. To tackle these issues, in this paper, we propose the followings.

- To evaluate the quality of a phylogeny with respect to a true phylogeny, we suggest to use a well-known concept in the computer science community, called *maximum agreement subtree*,[6, 19] which captures the structure of the phylogeny and has been used for comparing the similarity of two given trees. In fact, the same concept has been used to compute a consensus tree given several different phylogenetic trees.[1, 15, 25] Roughly speaking, a maximum agreement subtree is defined as follows. From the constructed phylogeny, we select a maximum subset of species such that the resulting subtree based on these species should have the same topology (structure) as the resulting subtree derived using the same set of species in the true phylogeny. This subtree is called a maximum agreement subtree. The percentage of the selected species is used as the evaluation measurement.

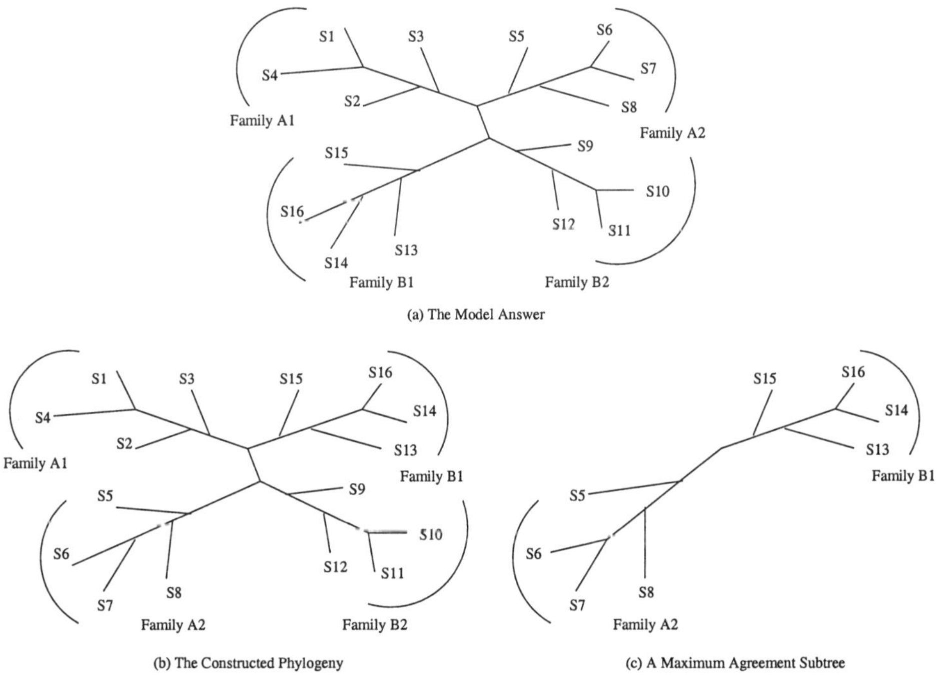

Figure 1. Non-trivial Split may not be a good measure

Referring to the example in Figure 1, Figure (c) shows a maximum agreement subtree and the accuracy of the constructed phylogeny based on this new measure is 50% (contrary to the 92.3% based on non-trivial splits) which reflects the quality of the tree more appropriately. In Section 3, we will highlight the difference of two measures based on the output given in Henz et al.[13]

- For the distance measure, we propose to derive it from the output generated by the whole genome alignment software (such as MUMmer). Basically, we measure the number of matched nucleotides in the conserved regions reported by the software. We believe that the reported regions are more meaningful than the local alignments reported by BLASTN with respect to the comparison of two whole genomes.

  Most whole genome alignment software such as MUMmer are more efficient than BLASTN (note that they report different things). Yet a brute force approach to generate the distances for all pairs of genomes using MUMmer still requires a lot of computation. For example, it takes 9.5 CPU hours (i.e., 11.5 days of actual running time) to execute MUMmer for each pair of the 91 genomes tested by Henz et al. Although it is already much faster than Henz et al.'s approach, it is still not feasible for a larger set of species. So, we derive an efficient approach to speed up the generation of the the pairwise distances, enabling us to have a feasible solution for 230 genomes.

Table 1. Comparison of Two Approaches (Data Set I: 91 Prokaryotic Genomes)

	% of species in Max. Agreement Subtree	% of Good Splits	Running Time in CPU Hours (Actual Time in hours)	
Henz et al.'s Approach	60/91 = 65.9%	72.7% [13]	170 (Unknown)	
Our Approach (Using MUMmer)	81/91 = 89.0%	83/88 = 94.3%	Brute-force	9.5 (276)
			New Approach	2 (7)

Based on the experiments on real data sets, we found that our approach is more accurate and more scalable than Henz et al.'s approach (see Table 1). We can construct a phylogenetic tree on the same set of 91 genomes with an accuracy more than 20% higher (with respect to both evaluation measures) in 2 CPU hours (more than 80 times faster than their approach). The actual running time of our approach is only 7 hours. Our approach is scalable and can construct a phylogeny for 230 prokaryotic genomes with accuracy of 85% and 90% (with respect to our measure and the measure of good splits, respectively) in only 9.5 CPU hours (the actual running time is about 38 hours). In our experiments, we also tried a few different whole genome alignment tools, which all can provide a phylogeny with higher accuracy (details will be given in Section 4). It seems that the output provided by whole genome alignment software should provide a better distance measure than other software (such as BLASTN) that are not designed for whole genome alignment. As a remark, we have also tested two other whole genome alignment software (MSS [22] and Hybrid [4]) the accuracy of the predicted tree is more or less the same.

Organization of the paper: Section 2 discusses our approach, the distance measure we use, and how we speed up the whole procedure. We then describe the details of using maximum agreement subtree as our evaluation measure in Section 3. The experimental results will be presented in Section 4. Section 5 concludes the paper.

2. The Distance Measure and Our Approach

In this section, we describe our approach for generating the phylogenetic tree for a set of given species, in particular, the distance measure we use in the generation process. The following shows the framework of our approach.
Step 1: For each pair of species, perform the whole genome alignment using one of the selected software tools.
Step 2: Based on the output from the whole genome alignment software, we calculate a distance measure for each pair of species.
Step 3: Generate the phylogenetic tree using one of the distance-based phylogeny reconstruction software tools.

The Whole Genome Alignment Tools: The key difference between our approach and Henz et al.'s approach is that our distance measure is derived from the output given by software tools that are specially designed for locating conserved regions in the whole genome alignment. There are a number of software tools for whole genome alignment.[4, 7, 8, 29] They try to report all conserved regions of the given genomes. Most of these tools work as follows. They first identify a set of short substrings that are highly similar and unique in both genomes. These substrings are called *anchors*. These anchors provide a rough guideline on which parts of the genomes we should examine for conserved regions. It is obvious that not all anchors identified in the first step are useful as a lof of them may come from noise. The second step will consider these anchors based on different criteria and techniques (e.g. maximum common subsequence and clustering) so as to eliminate the noise and identify the conserved regions along the whole genomes. The set of anchors reported by the software is believed to be the markers for the conserved regions of the genomes. A common choice for anchors is the maximal substrings that are *exactly* matched and unique in the two genomes (called *MUM*). In this paper we use MUMs as our anchors for all experiments.

The Distance Measure: In order to show that the output from the whole genome alignment software tools is more appropriate for phylogenetic tree generation, we follow the idea of Henz et al.'s approach and use a straightforward distance measure. That is, we derive our measure from the total lengths of all the MUMs reported (that is, the selected anchors) by the software and normalize the value by the length of the shorter genomes. More precisely, we use the following distance measure.

$$\text{Distance Measure} = -\log_2\left(\frac{\text{Total MUM Length}}{\min\{\text{Lengths of Sequences}\}}\right)$$

In Henz et al.'s approach, instead of using the total MUM length, they use the total number of matched base pairs based on the set of high scored non-overlapping local alignments returned from BLASTN.

The Phylogeny Reconstruction Tools: In our research, we focus on distance-based phylogenetic reconstruction tools. Most of these software tools are based on two approaches: UPGMA[24] and Neighbor-Joining.[12, 20, 26] In this paper, our main purpose is not to evaluate the performance of different reconstruction tools. Therefore, based on the experimental results in Henz et al., BIONJ[12] performs the best among all the evaluated tools, so we also perform our experiments using BIONJ. Interested readers can refer to the PHYLIP package developed by Joe Felsenstein[9] for more information on different phylogenetic tree reconstruction software tools.

The Speed Up: In Step 1, for each pair of species, we have to identify a set of MUMs which requires the construction of a suffix tree for one of the species which also dominates the running time of the whole process (when using MUMmer). A brute-force approach would have to construct $O(n^2)$ suffix trees where n is the number of species. From our experiment on 91 prokaryotic genomes, the brute force approach will take about 9 CPU hours and 11.4 days of actual running time. Although it is faster than Henz et al.'s approach, it may not be feasible for a large set of species.

So, instead of constructing a suffix tree for each pair, we speed up the process as follows. We partition the species into groups of x species. We concatenate the genomic sequences of the species in each group and construct one suffix tree for each group, then for each sequence, we search against this suffix tree to locate MUMs for x pairs of species simultaneously. In other words, we avoid constructing the same suffix tree repeatedly as in the brute-force approach and also we speed up the searching process of MUMs by checking x pairs of species for MUMs in one round of searching. We are able to implement this approach in a PC with 4G memory by setting $x = 32$. The running time for the generation process decreases to 2 CPU hours (or 7 hours of actual running time) for 91 genomes.

From the viewpoint of theoretical analysis, we improve the time complexity from $O(mn^2)$ where m is the length of each genome to $O(mn)$ since the number of groups is small and can be considered as a constant in practice. We remark that the number of species (x) in each group should be calculated based on the available amount of memory and the sequence length of the species.

3. The Evaluation Measure

To evaluate the quality of a phylogenetic tree, we compare it to the phylogeny that is constructed using the taxonomy published in NCBI [b] (we call this the *true* phylogeny). One of the common concepts used for the comparison is the *non-trivial splits*.[13, 27] In this section, we formally define the measurement based on non-trivial splits and illustrate by a real example that this measure may not be a good indicator for the quality of the tree. Then, we propose to use the concept of *maximum agreement subtree*, a well-known concept in computer science community used for comparing the similarity of two given trees, to evaluate the quality of a predicted phylogeny.

[b] http://www.ncbi.nlm.nih.gov/entrez/query.fcgi?CMD=search&DB=taxonomy

Non-Trivial Split: Given a phylogenetic tree, each internal edge, that is, the edge without a leaf (a species) attached to it is called a non-trivial split (we simply refer it as a split). For each split, if we delete the split, the tree will be partitioned into two connected components. All species will be divided into two sets according to which component that species belongs to. Intuitively, each split poses a classification on the species. If there is a corresponding split in the true phylogeny so that the species are partitioned exactly the same as that split. It means that the classification is correct and we call that split a *good* split. So, it is natural to define a measurement to evaluate the quality of the predicted tree as the percentage of the number of good splits out of the total number of splits in the predicted tree.

In Section 1 (Introduction), we provide an artificial example to illustrate that counting the percentage of good splits may not be a good indicator of the quality of the tree. In fact, splits do not explicitly capture the topology of the trees which is important in understanding the evolutionary history of the species. Also, some splits should be more important than the others. In particular, the split which separates a big family from another big family should be considered more important than a split which separates a species from the other species inside a subgroup. However, the measurement does not distinguish between these splits. In this section, we try to illustrate this problem using a real example.

Figure 2(a) shows the predicated phylogeny produced by Henz et al.'s approach on 91 prokaryotic genomes and Figure 2(b) is true phylogeny From the figures, one can see that the groups of alpha-proteobacteria, beta-proteobacteria, gamma-proteobacteria, and Spirochaete, are wrongly splitted into two or more subgroups attached to different parts of the phylogenetic tree. However, if we count the percentage of good splits, it is 72.7% which is a rather high score. It seems that this measurement may not be a good indicator.

The Maximum Agreement Subtree: The concept of *maximum agreement subtree* is not new in the computer science community and also, it has been used to reconcile different evolutionary trees and extract the maximum set of species such that the evolutionary relationships among these species are all agreed by these trees.[2,16] Given two trees, T_1 and T_2, with leaves labelled by the same set of species, an *agreement subtree* is defined as follows. Let L_1 be a subset of species (leaves) in T_1. The subtree of T_1 induced by L_1 is an agreement subtree of T_1 and T_2 if this subtree is isomorphic to the subtree of T_2 induced by the same set of species L_1. Intuitively, if there is an agreement subtree induced by the subset L of species, it means that the evolutionary structure of these species are the same in both trees. If the size of L is the largest possible, then the corresponding agreement subtree is called a *maximum agreement subtree*.

Based on this idea, we derive a measure to evaluate the quality of the predicted tree by considering the largest possible size of L such that an agreement subtree exists. The percentage of the species that are selected in L is our proposed measure. Referring to Figure 2(a), if we use the percentage of species in the maximum agreement subtree as our quality measure (the selected species have been bolded in the figure), the evaluation score is 65.9% which we believe is a better score that reflects the quality of the tree.

Remark: In practice, the predicted tree is an unrooted binary tree, however, the true phylogeny is rooted and may not be a binary tree since the exact details of the evolutionary

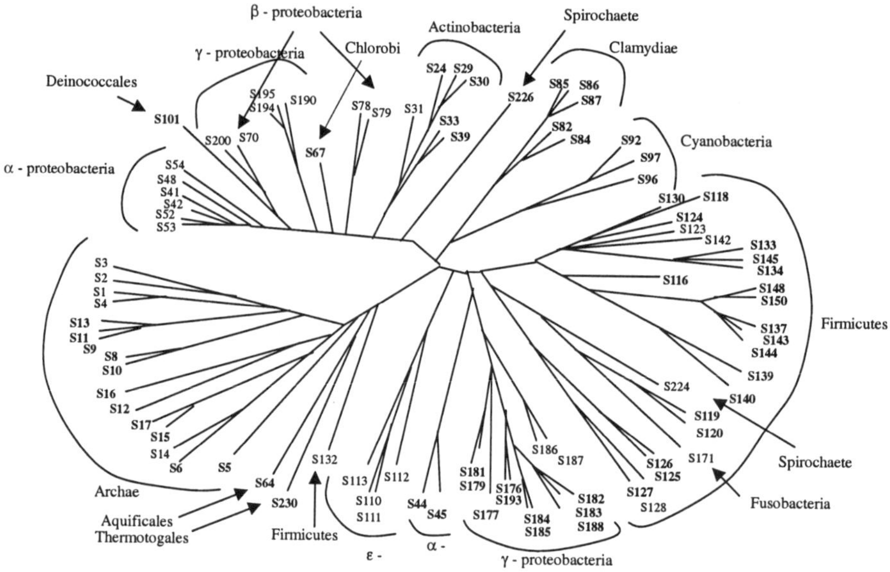

(a) The Best Phylogenetic tree for 91 species produced by Henz et al.'s Approach

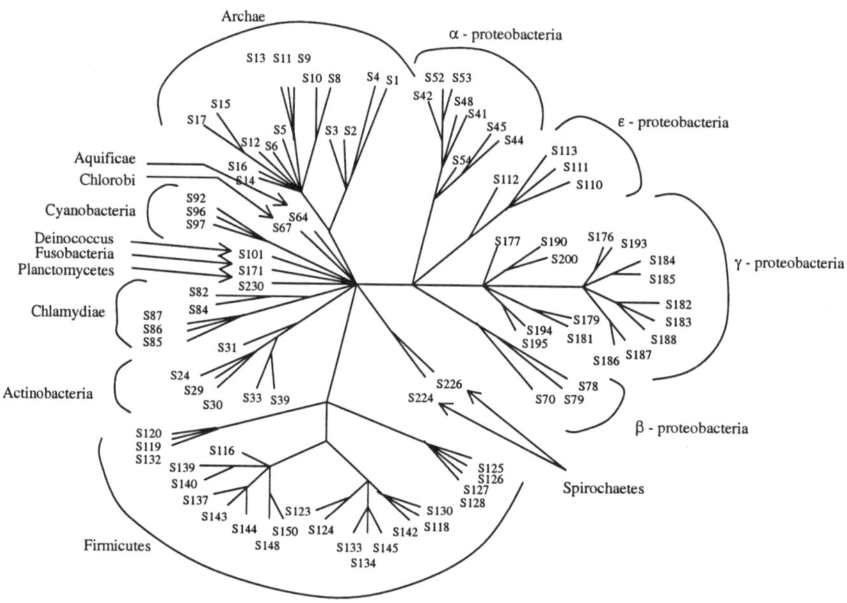

(b) The True Phylogeny based on NCBI taxonomy (June, 2005) for 91 species

Figure 2. Phylogenetic trees produced by Henz et al.'s Approach and NCBI Taxonomy for 91 Species history of the species in a subgroup may not be known. To compute the maximum agreement subtree (it is referred as the *maximum compatible subtree*), if there is a node in the true phylogeny with degree > 3, we allow it to be refined to a binary one by inserting artificial nodes so that deleting all these artificial nodes can get back the original subtree. In other words, we allow the predicted tree to have any evolutionary structure for these set of species. Similarly, the same applies to non-trivial splits. A non-trivial split in the predicted tree will be considered good if it corresponds to an artificial edge added because of the refinement process.

To compute the maximum compatible subtree is not trivial. Ganapathysaravanabavan and Warnow[11] provided a dynamic programming algorithm of $O(n^3 \times 2^{4d})$ time, where n is the number of species, for computing such a subtree for two unrooted trees with bounded degree $d+1$. However, the algorithm takes too long to compute (more than 30 minutes for 91 genomes and 172 hours for 230 genomes). In fact, the algorithm is a straight-forward extension of their algorithm for *rooted* trees. Many entries in the dynamic programming tables are computed more than once. We eliminate this redundancy by deriving a more efficient dynamic programming algorithm and the time complexity can be reduced by an $O(n)$ factor. Also, in our case, one of the trees has degree at most 3, so our algorithm runs in $O(n^2 \times 2^{2d})$ time. It only takes about 20 seconds and 45 minutes for 91 and 230 genomes, respectively.

4. Experimental Results

We have used two data sets for our experiments: Data Set I: 91 prokaryotic genomes that were used in the experiments of Henz et al.[13] Data Set II: all 230 prokaryotic genomes that are published in NCBI [c]. We use MUMmer [23] as the whole genome alignment software and work on the translated protein sequences of the genomes. For the phylogenetic tree reconstruction software, we use BIONJ.[21] The true phylogeny is derived from NCBI taxonomy in both data sets.

For both data sets, we evaluate our predicted phylogenetic tree using both measures. For Data Set I, from Table 1, we can see that our approach achieves an accuracy of more than 20% higher than Henz et al.'s approach in both measurements. Figure 3 shows the our predicted tree. For Data Set II, the accuracy of our predicted tree is 85.2% and 90.3% using the percentage of species in the maximum agreement subtree and good splits respectively. Figure 4 and 5 show the true phylogeny and our predicted tree for Data Set II. To conclude, our approach provides a more accurate method to predict phylogenetic tree.

For running time, our approach only requires 2 CPU hours (or 7 hours of actual running time) for Data Set I and 9.5 CPU hours (or 38 hours of actual running time) for Data Set II. Our approach is much faster (more than 80 times) than Henz et al.'s approach and in fact, their approach is not feasible for Data Set II as the estimated computation required will be more than 1000 CPU hours. So, our approach is more scalable than their approach.

Remark: We have also tried some other whole genome alignment software tools: MSS[22]

[c] http://www.ncbi.nlm.nih.gov/genomes/lproks.cgi

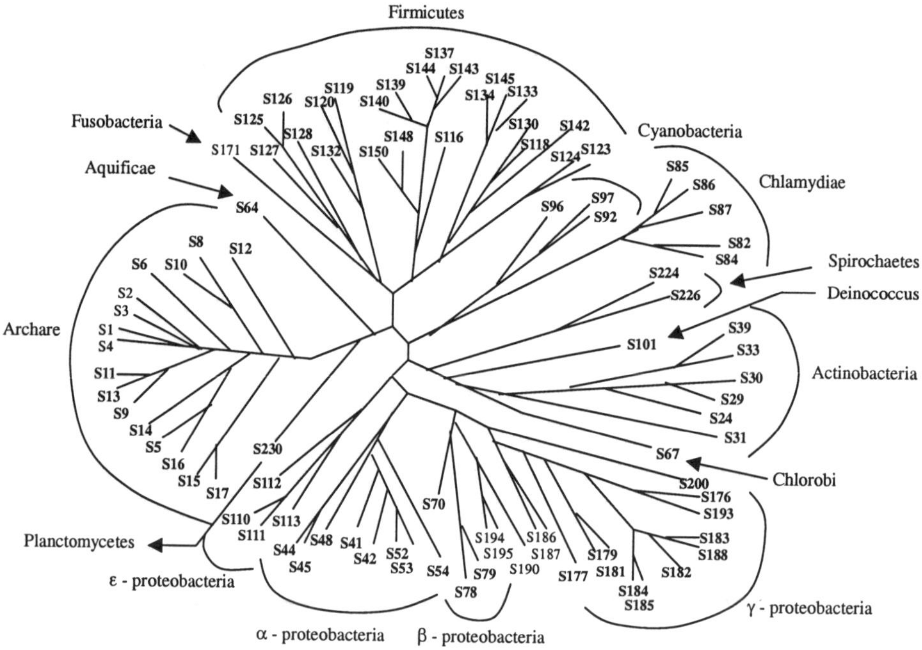

Figure 3. Phylogenetic trees produced by Our approach for 91 Species

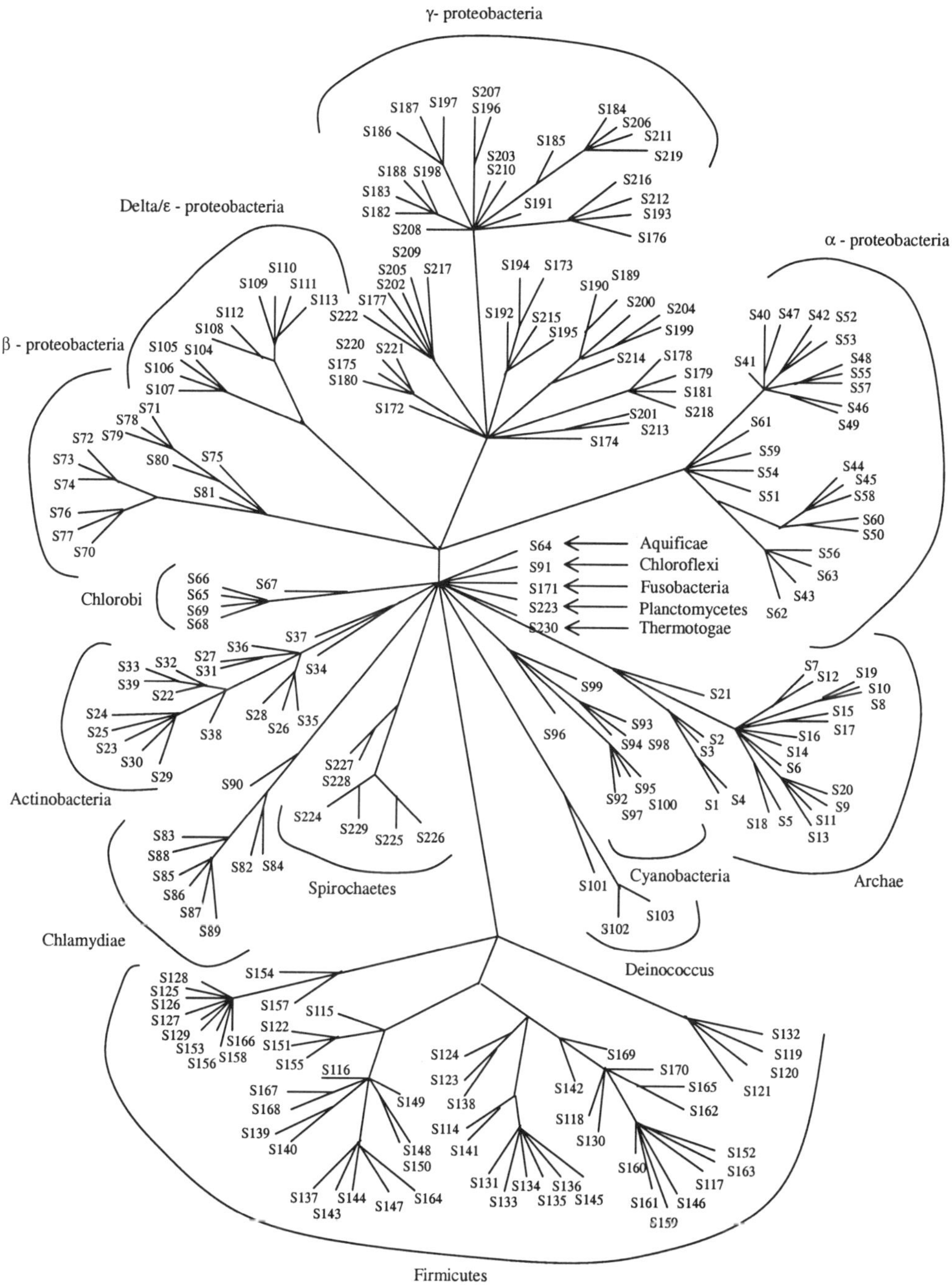

Figure 4. The True Phylogeny based on NCBI Taxonomy (June 2005) for 230 Species

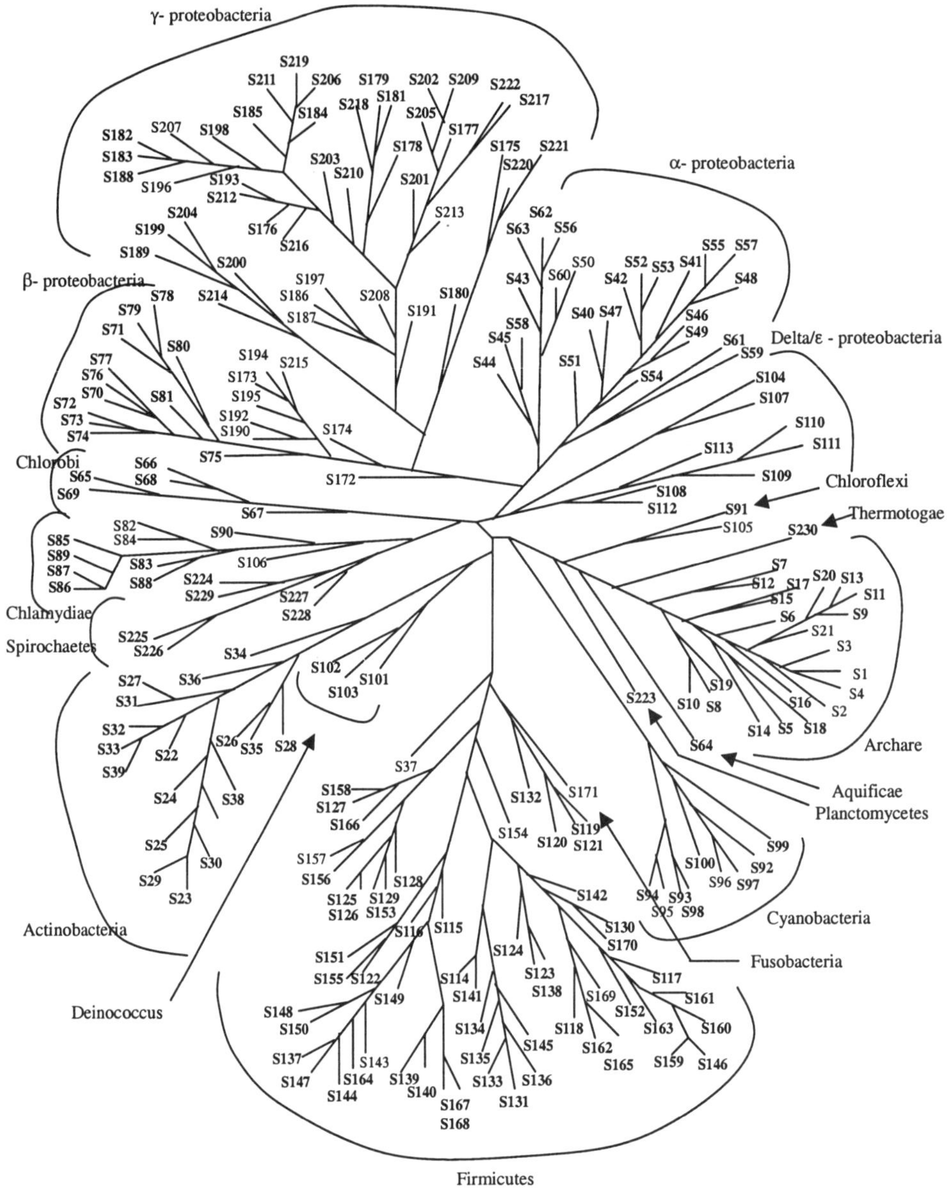

Figure 5. The Phylogenetic tree produced by Our Approach for 230 Species

S1 Sulfolobus solfataricus P2
S2 Pyrobaculum aerophilum str. IM2
S3 Aeropyrum pernix K1
S4 Sulfolobus tokodaii str. 7
S5 Methanocaldococcus jannaschii DSM 2661
S6 Archaeoglobus fulgidus DSM 4304
S7 Haloarcula marismortui ATCC 43049
S8 Thermoplasma acidophilum DSM 1728
S9 Pyrococcus abyssi GE5
S10 Thermoplasma volcanium GSS1
S11 Pyrococcus horikoshii OT3
S12 Halobacterium sp. NRC-1
S13 Pyrococcus furiosus DSM 3638
S14 Methanobacterium_thermoautotrophicum
S15 Methanosarcina acetivorans C2A
S16 Methanopyrus kandleri AV19
S17 Methanosarcina mazei Go1
S18 Methanococcus maripaludis S2
S19 Picrophilus torridus DSM 9790
S20 Thermococcus kodakarensis KOD1
S21 Nanoarchaeum equitans Kin4-M
S22 Corynebacterium diphtheriae NCTC 13129
S23 Mycobacterium bovis AF2122/97
S24 Mycobacterium leprae TN
S25 Mycobacterium avium subsp. paratuberculosis str. k10
S26 Tropheryma whipplei str. Twist
S27 Streptomyces avermitilis MA-4680
S28 Leifsonia xyli subsp. xyli str. CTCB07
S29 Mycobacterium tuberculosis CDC1551
S30 Mycobacterium tuberculosis H37Rv
S31 Streptomyces coelicolor A3(2)
S32 Corynebacterium efficiens YS-314
S33 Corynebacterium glutamicum ATCC 13032
S34 Bifidobacterium longum NCC2705
S35 Tropheryma whipplei TW08/27
S36 Propionibacterium acnes KPA171202
S37 Symbiobacterium thermophilum IAM 14863
S38 Nocardia farcinica IFM 10152
S39 Corynebacterium glutamicum ATCC 13032
S40 Bradyrhizobium japonicum USDA 110
S41 Mesorhizobium loti MAFF303099
S42 Sinorhizobium meliloti 1021
S43 Anaplasma marginale str. St. Maries
S44 Rickettsia conorii str. Malish 7
S45 Rickettsia prowazekii str. Madrid E
S46 Bartonella quintana str. Toulouse
S47 Rhodopseudomonas palustris CGA009
S48 Brucella melitensis 16M
S49 Bartonella henselae str. Houston-1
S50 Wolbachia endosymbiont of Drosophila melanogaster
S51 Silicibacter pomeroyi DSS-3
S52 Agrobacterium tumefaciens str. C58
S53 Agrobacterium tumefaciens str. C58
S54 Caulobacter crescentus CB15
S55 Brucella suis 1330
S56 Ehrlichia ruminantium str. Welgevonden
S57 Brucella abortus biovar 1 str. 9-941
S58 Rickettsia typhi str. Wilmington
S59 Zymomonas mobilis subsp. mobilis ZM4
S60 Wolbachia endosymbiont strain TRS of Brugia malayi
S61 Gluconobacter oxydans 621H
S62 Ehrlichia ruminantium str. Welgevonden
S63 Ehrlichia ruminantium str. Gardel
S64 Aquifex aeolicus VF5
S65 Bacteroides fragilis NCTC 9343
S66 Porphyromonas gingivalis W83
S67 Chlorobium tepidum TLS
S68 Bacteroides thetaiotaomicron VPI-5482
S69 Bacteroides fragilis YCH46
S70 Ralstonia solanacearum GMI1000
S71 Neisseria gonorrhoeae FA 1090
S72 Bordetella bronchiseptica RB50
S73 Bordetella parapertussis 12822
S74 Bordetella pertussis Tohama I
S75 Nitrosomonas europaea ATCC 19718
S76 Burkholderia mallei ATCC 23344
S77 Burkholderia pseudomallei K96243
S78 Neisseria meningitidis MC58
S79 Neisseria meningitidis Z2491
S80 Chromobacterium violaceum ATCC 12472
S81 Azoarcus sp. EbN1
S82 Chlamydia trachomatis D/UW-3/CX
S83 Chlamydophila caviae GPIC
S84 Chlamydia muridarum Nigg
S85 Chlamydophila pneumoniae AR39
S86 Chlamydophila pneumoniae CWL029
S87 Chlamydophila pneumoniae J138
S88 Chlamydophila abortus S26/3
S89 Chlamydophila pneumoniae TW-183
S90 Parachlamydia sp. UWE25
S91 Dehalococcoides ethenogenes 195
S92 Synechocystis sp. PCC 6803
S93 Prochlorococcus marinus subsp. pastoris str. CCMP1986
S94 Prochlorococcus marinus str. MIT 9313
S95 Synechococcus sp. WH 8102
S96 Nostoc sp. PCC 7120
S97 Thermosynechococcus elongatus BP-1
S98 Prochlorococcus marinus subsp. marinus str. CCMP1375
S99 Gloeobacter violaceus PCC 7421
S100 Synechococcus elongatus PCC 6301
S101 Deinococcus radiodurans R1
S102 Thermus thermophilus HB27
S103 Thermus thermophilus HB8
S104 Desulfovibrio vulgaris subsp. vulgaris str. Hildenborough
S105 Geobacter sulfurreducens PCA
S106 Bdellovibrio bacteriovorus HD100
S107 Desulfotalea psychrophila LSv54
S108 Campylobacter jejuni subsp. jejuni NCTC 11168
S109 Helicobacter hepaticus ATCC 51449
S110 Helicobacter pylori 26695
S111 Helicobacter pylori J99
S112 Campylobacter jejuni RM1221
S113 Wolinella succinogenes DSM 1740
S114 Staphylococcus epidermidis RP62A
S115 Enterococcus faecalis V583
S116 Lactococcus lactis subsp. lactis Il1403
S117 Bacillus cereus ATCC 10987
S118 Bacillus subtilis subsp. subtilis str. 168
S119 Clostridium acetobutylicum ATCC 824
S120 Clostridium perfringens str. 13
S121 Clostridium tetani E88
S122 Lactobacillus acidophilus NCFM
S123 Listeria monocytogenes str. 4b F2365
S124 Listeria innocua Clip11262
S125 Mycoplasma genitalium G-37
S126 Mycoplasma pneumoniae M129
S127 Mycoplasma pulmonis UAB CTIP
S128 Ureaplasma parvum serovar 3 str. ATCC 700970
S129 Mycoplasma penetrans HF-2
S130 Bacillus halodurans C-125
S131 Staphylococcus aureus subsp. aureus COL
S132 Thermoanaerobacter tengcongensis MB4
S133 Staphylococcus aureus subsp. aureus Mu50
S134 Staphylococcus aureus subsp. aureus N315
S135 Staphylococcus aureus subsp. aureus MRSA252
S136 Staphylococcus aureus subsp. aureus MSSA476
S137 Streptococcus pyogenes M1 GAS
S138 Listeria monocytogenes EGD-e
S139 Streptococcus pneumoniae TIGR4
S140 Streptococcus pneumoniae R6
S141 Staphylococcus epidermidis ATCC 12228
S142 Oceanobacillus iheyensis HTE831
S143 Streptococcus pyogenes MGAS8232
S144 Streptococcus pyogenes SSI-1
S145 Staphylococcus aureus subsp. aureus MW2
S146 Bacillus anthracis str. Ames
S147 Streptococcus pyogenes MGAS315
S148 Streptococcus agalactiae 2603V/R
S149 Streptococcus mutans UA159
S150 Streptococcus agalactiae NEM316
S151 Lactobacillus plantarum WCFS1
S152 Bacillus cereus ATCC 14579
S153 Mycoplasma gallisepticum R
S154 Onion yellows phytoplasma OY-M
S155 Lactobacillus johnsonii NCC 533
S156 Mycoplasma mycoides subsp. mycoides SC str. PG1
S157 Mesoplasma florum L1
S158 Mycoplasma mobile 163K
S159 Bacillus anthracis str. 'Ames Ancestor'
S160 Bacillus thuringiensis serovar konkukian str. 97-27
S161 Bacillus anthracis str. Sterne
S162 Bacillus licheniformis ATCC 14580
S163 Bacillus cereus E33L
S164 Streptococcus pyogenes MGAS10394
S165 Bacillus licheniformis ATCC 14580
S166 Mycoplasma hyopneumoniae 232
S167 Streptococcus thermophilus LMG 18311
S168 Streptococcus thermophilus CNRZ1066
S169 Geobacillus kaustophilus HTA426
S170 Bacillus clausii KSM-K16
S171 Fusobacterium nucleatum subsp. nucleatum ATCC 25586
S172 Francisella tularensis subsp. tularensis SCHU S4
S173 Xanthomonas campestris pv. campestris str. 8004
S174 Methylococcus capsulatus str. Bath
S175 Legionella pneumophila subsp. pneumophila str. Philadelphia
S176 Yersinia pestis CO92
S177 Vibrio cholerae O1 biovar eltor str. N16961
S178 Haemophilus ducreyi 35000HP
S179 Pasteurella multocida subsp. multocida str. Pm70
S180 Coxiella burnetii RSA 493
S181 Haemophilus influenzae Rd KW20
S182 Escherichia coli K12
S183 Escherichia coli O157:H7
S184 Salmonella enterica subsp. enterica serovar Typhi str. CT18
S185 Salmonella typhimurium LT2
S186 Buchnera aphidicola str. APS (Acyrthosiphon pisum)
S187 Buchnera aphidicola str. Bp (Baizongia pistaciae)
S188 Escherichia coli O157:H7 EDL933
S189 Pseudomonas putida KT2440
S190 Xylella fastidiosa 9a5c
S191 Wigglesworthia glossinidia endosymbiont of Glossina brevipa
S192 Xylella fastidiosa Temecula1
S193 Yersinia pestis KIM
S194 Xanthomonas campestris pv. campestris str. ATCC 33913
S195 Xanthomonas axonopodis pv. citri str. 306
S196 Shigella flexneri 2a str. 301
S197 Buchnera aphidicola str. Sg (Schizaphis graminum)
S198 Escherichia coli CFT073
S199 Pseudomonas syringae pv. syringae B728a
S200 Pseudomonas aeruginosa PAO1
S201 Shewanella oneidensis MR-1
S202 Vibrio vulnificus CMCP6
S203 Erwinia carotovora subsp. atroseptica SCRI1043
S204 Pseudomonas syringae pv. tomato str. DC3000
S205 Vibrio parahaemolyticus RIMD 2210633
S206 Salmonella enterica subsp. enterica serovar Typhi Ty2
S207 Shigella flexneri 2a str. 2457T
S208 Candidatus Blochmannia floridanus
S209 Vibrio vulnificus YJ016
S210 Photorhabdus luminescens subsp. laumondii TTO1
S211 Salmonella enterica subsp. enterica serovar Choleraesuis st
S212 Yersinia pestis biovar Medievalis str. 91001
S213 Idiomarina loihiensis L2TR
S214 Acinetobacter sp. ADP1
S215 Xanthomonas oryzae pv. oryzae KACC10331
S216 Yersinia pseudotuberculosis IP 32953
S217 Vibrio fischeri ES114
S218 Mannheimia succiniciproducens MBEL55E
S219 Salmonella enterica subsp. enterica serovar Paratyphi A str
S220 Legionella pneumophila str. Lens
S221 Legionella pneumophila str. Paris
S222 Photobacterium profundum SS9
S223 Rhodopirellula baltica SH 1
S224 Borrelia burgdorferi B31
S225 Treponema denticola ATCC 35405
S226 Treponema pallidum subsp. pallidum str. Nichols
S227 Leptospira interrogans serovar Lai str. 56601
S228 Leptospira interrogans serovar Copenhageni str. Fiocruz L1-
S229 Borrelia garinii PBi
S230 Thermotoga maritima MSB8

Figure 6. The Mapping between the Species and the Symbols.

and the hybrid approach[4, 22] that combines MaxMinCluster[29] and MSS. We use the same proposed distance measure to construct the distance in all cases. For both measures and data sets, our approach (no matter which software tool is used) is able to produce phylogenies with higher quality (18% higher using MSS and more than 20% for hybrid). It illustrates that the output from the whole genome alignment tools is useful in constructing phylogenetic trees. On the other hand, MSS is quite intensive in computation. So, it takes longer time if we use these two software tools (for Data Set I, about 30 and 95 CPU hours are required, respectively, for MSS and hybrid) although it is already fast than Henz et al.'s approach.

5. Conclusion

In this paper, we study the problem of using whole genomes to reconstruct a phylogeny for a given set of species. We propose to derive the distance from the output reported by software tools that are specially designed for whole genome alignment. Experiments show that our proposed approach outperforms the existing approaches that do not make use of whole genome alignment to derive the distance measure and is able to infer a phylogenetic tree with a much higher accuracy. Moreover, our approach is more scalable and can be used to reconstruct a phylogeny for 230 prokaryotic genomes. Regarding the evaluation of a phylogeny, we point out that the evaluation based on non-trivial splits may not be a good indicator and we propose to use the concept of maximum agreement subtree which can also capture the structure of the tree.

For further work, we will try to apply the same approach to the eurokaryotic genomes and try to derive other distance measures, for example, measures that can capture the number of mutations, in order to further improve the accuracy of the predicted phylogeny. A detailed study on the measures and the related issues, such as the normalization[17] would be carried out.

References

1. A. Amir and D. Kesselman. Maximum agreement subtree in a set of evolutionary trees - metrics and efficient algorithms. In *Proceedings of the 35th IEEE FOCS*, pages 758–769, 1994.
2. Amidhood Amir and Dmitry Keselman. Maximum agreement subtree in a set of evolutionary trees: Metrics and efficient algorithms. *SIAM Journal on Computing*, 26(6):1656–1669, 1997.
3. D.K. Bideshi, Y. Bigot, and B.A. Federici. Molecular characterization and phylogenetic analysis of the *harrisina brillians* granulovirus granulin gene. *Arch. Virol.*, 145:1933–1945, 2000.
4. HL Chan, TW Lam, WK Sung, Prudence WH Wong, and SM Yiu. The mutation subsequence problem and locating conserved genes. *Bioinformatics*, 21(10):2271–2278, 2005. A preliminary version appears in the Proceedings of the IEEE Fourth Symposium on Bioinformatics and Bioengineering (BIBE 2004).
5. X. Chen, W.F.J. Ijkel, C. Dominy, P. Zanotto, Y. Hashimoto, O. Faktor, T. Hayakawa, C.-H. Wang, A. Prekumar, S. Mathavan, P.J. Krell, Z. Hu, and J.M. Vlak. Identification, sequence analysis and phylogeny of the *lef*-2 gene of *helicoverpa armigera* single-nucleocapsid baculovirus. *Virus Research*, 65:21–32, 2001.
6. R. Cole, M. Farach, R. Hariharan, T. Przytycka, and M. Thorup. An $o(n \log n)$ algorithm for the maximum agreement subtree problem for binary trees. *SIAM Journal on Computing*, 30(5):1385–1404, 2000.

7. A.L. Delcher, S. Kasif, R. D. Fleischmann, J. Peterson, O. White, and S.L. Salzberg. Alignment of whole genomes. *Nucleic Acids Research*, 27(11):2369–2376, 1999.
8. A.L. Delcher, A. Phillippy, J. Carlton, and S.L. Salzberg. Fast algorithms for large-scale genome alignment and comparison. *Nucleic Acids Research*, 30(11):2478–2483, 2002.
9. J. Felsenstein. PHYLIP - phylogeny inference package (version 3.2). *Cladistics*, 5:164–166, 1989. http://evolution.genetics.washington.edu/phylip.html.
10. S.T. Fitz-Gibbon and C.H. House. Whole genome-based phylogenetic analysis of free living microorganisms. *Nucleic Acids Research*, 27:4218–4222, 1999.
11. Ganeshkumar Ganapathysaravanabavan and Tandy Warnow. Finding a maximum compatible tree for a bounded number of trees with bounded degree is solved in polynomial time. In *Proceedings of WABI 2001*, pages 156–163, 2001.
12. O. Gascuel. BIONJ: An improved version of the NJ algorithm based on a simple model of sequence data. *Mol. Biol. Evol.*, 14:685–695, 1997.
13. Stafan R. Henz, Daniel H. Huson, Alexander F. Auch, Kay Nieselt-Struwe, and Stephen C. Schuster. Whole-genome prokaryotic phylogeny. *Bioinformatics*, 21(10):2329–2335, 2005.
14. E.A. Herniou, T. Luque, X. Chen, J.M. Vlak, D. Winstanley, J.S. Cory, and D.R. O'Reilly. Use of whole genome sequence data to infer baculovirus phylogeny. *Journal of Virology*, 75(17):8117–8126, 2001.
15. M.-Y. Kao, T.W. Lam, T. Przytycka, W.K. Sung, and H.F. Ting. Efficient algorithms for comparing unrooted evolutionary trees. In *Proceedings of STOC*, pages 54–65, 1997.
16. Chuan-Min Lee, Ling-Ju Hung, Maw-Shang Chang, and Chuan-Yi Tang. An improved algorithm for the maximum agreement subtree problem. In *Proceedings of the 4th IEEE Symposium on Bioinformatics and Bioengineering (BIBE'04)*, page 533, 2004.
17. M. Li, X. Chen, X. Li, B. Ma, and P.M.B. Vitányi. The similarity metric. In *Proceedings of the 14th Annual ACM-SIAM Symposium on Discrete Algorithms*, pages 863–872, 2003.
18. G.J. Olsen, C.R. Woese, and R. Overbeek. The winds of (evolutionary) change: Breathing new life into microbiology. *J. Bact.*, 176:1–6, 1994.
19. T. Przytzcka. Sparse dynamic programming for maximum agreement subtree problem. In *Mathematical Hierarchies and Biology, DIMACS Series in Discrete Mathematics and Theoretical Computer Science*, 1997.
20. N. Saitou and M. Nei. The neighbor-joining method: A new method for reconstructing phylogenetic trees. *Mol. Biol. Evol.*, 4:406–425, 1987.
21. The BIONJ Web Site. http://bioweb.pasteur.fr/seqanal/interfaces/bionj.html.
22. The MSS Web Site, 2004. http://www.cs.hku.hk/m̃ss.
23. The MUMMER Web Site, 2003. http://www.tigr.org/software/mummer/.
24. R.R. Sokal and C.D. Michener. A statistical method for evaluating systematic relationships. *University of Kansas Scientific Bulletin*, 28:1409–1438, 1958.
25. M. Steel and T. Warnow. Kaikoura tree theorems: Computing the maximum agreement subtree. *Information Processing Letters*, 48:77–82, 1993.
26. J.A. Studier and K.J. Keppler. A note on the neighbour-joining algorithm of saitou and nei. *Mol. Biol. Evol.*, 5:729–731, 1988.
27. L.-S. Wang and T. Warnow. Estimating true evolutionary distances between genomes. In *Proceedings of the 33rd Annual ACM Symposium on Theory of Computing (STOC 2001)*, pages 637–646, 2001.
28. C.R. Woese. Bacterial evolution. *Microbiol. Rev.*, 51:221–272, 1987.
29. Prudence W.H. Wong, T. W. Lam, N. Lu, H. F. Ting, and S. M. Yiu. An efficient algorithm for optimizing whole genome alignment with noise. *Bioinformatics*, 20(16):2676–2684, 2004.
30. P.M.D. Zanotto, B.D. Kessing, and J.E. Maruniak. Phylogenetic interrelationships among baculoviruses: Evolutionary rates and host associations. *J. Invertebr. Pathol.*, 62:147–164, 1993.

GENE EXPRESSION DATA CLUSTERING BASED ON LOCAL SIMILARITY COMBINATION*

DE PAN, FEI WANG†

Department of Computer Science and Engineering
220 Handan Road,
Fudan University, Shanghai 200433, P.R.C.
E-mail: {pande,wangfei}@fudan.edu.cn

Clustering is widely used in gene expression analysis, which helps to group genes with similar biological function together. The traditional clustering techniques are not suitable to be directly applied to gene expression time series data, because of the inhered properties of local regulation and time shift. In order to cope with the existing problems, the local similarity and time shift, we have developed a new similarity measurement technique called *Local Similarity Combination* in this paper. And at last, we'll run our method on the real gene expression data and show that it works well.

1. Introduction

The increasingly used microarray techniques generate more and more biological data veryday to the biologists, who reasonably need special computational tools to help. The data obtained from microarray technique records the expression levels of genes in the form of time series points by measuring the concentration of the corresponding mRNAs.[1,2,3] Thus it provides the possibility and opportunity to insight the genes' behaviors and functions indirectly, and to find gene pairs with different kind of regulation relationships and group genes together with similar biological function, which usually demonstrate similar expression profiles against the time series, and at last to construct a biological network.[4,5]

Clustering genes with highly similar expression profiles, locally or globally and time shifted, is one of the most important steps to analyse the microarray data, which usually applies a kind of similarity measurement first, such as the traditional Pearson correlation or Euclidean distance[6] methods, and then a clustering paradigm follows which classifies genes basing on their pairwise similarities into different clusters. Here the time shift means a time lag between the local parts of the two gene profiles, because the first gene——the regulator, usually effects the downstream gene——the regulated, with a time delay. The clustering structure will reveal the transcriptional information of the genes in the biological environment to help the understanding of the biological control mechanism. And in a deeper analysis, the clustering results will serve, usually with many other kinds of biological in-

*This work is supported by grants 60303009, 60496325 of Chinese National Natural Science Foundation.
†Correspondence should be addressed to F. wang email:wangfei@fudan.edu.cn

formation and data, such as transcriptional factors and binding sites, or the protein-protein interaction information, to insight the gene functions in the molecular level.

In order to analyse the time series microarray data, many methods have been developed to measure the similarity between genes. Here we'll give a brief review about the related similarity measurement methods. The aforementioned Euclidean method is widely used in other scientific or engineering fields, which usually directly dismisses the time information and only focuses on the global profile distance calculation, and rarely works well on microarray data. The well-known standard Pearson correlation method also computes the global similarity and ignores the time series characteristic. The modified Pearson correlation method[7] based on the standard Pearson Correlation was developed quickly, which takes the time lag into account, but which also considers the global time lag only and there is no good approach to deciding the size of lag yet. And a particular method introduced by Spellman et al in 1998 applied the Fourier transformation on the time series data,[8] which was proved to be effective for the periodic data. Recently, some more sophisticated methods have been presented. The *Edge Detection Method* tries to find the main changes in expression levels (*edge*) and gives a score by comparing the edges between the two genes,[9] which will lose information when two edges are far apart. Another method is *Dominant Spectral Component Method*,[10] which first decomposes the time series data into frequency components and attains a pair of frequent component with least difference, and then transforms them back to time space and uses the standard Pearson coefficient to calculate the similarity. The *Event method*[11] transforms time series expression level into a string of events—- R(Rising), F(Falling) & C(No changing, and gives each gene pair a score by applying the *Needleman-Wunsch* alignment algorithm, which will aslo lose too much information while only giving a global similarity score.

We propose a new method in our work to measure the similarity between micaoarray gene expression profiles, which takes the local similarity and time shift properties into account, yet simultaneously solved by the previous methods. Our method will discretize the data first and then apply a matrix to find all the local matching information including the time lag. Then a optimal combination of the local matches follows to attain a global similarity.

2. Method

In this section we'll describe the paradigm of our new similarity measurement, in which the original expression data will be discritized by using two equations and, then we'll demonstrate how to use a matrix to discover all the local matching information. And we'll give the definition of the optimal combination of the local match candidates to obtain the optimal alignment between the gene pair, and by defining an equation we'll finally get the similarity.

2.1. *Data Discretization*

Here we'll use a 3-value discretization method to preprocess the original gene microarray expression data matrix $G_{n\times m}$ to get the discretized matrix $E_{n\times(m-1)}$, where n means the gene number and m indicates the time condition points. And before we get the object matrix we'll apply Equ.(1) on G firstly to get a temp matrix $G'_{n\times(m-1)}$. Then the Equ.(2) will be used G' to attain object matrix E.

$$G'_{i,j} = \begin{cases} 1, & \text{if } G_{i,j} = 0 \,\&\, G_{i,j+1} > 0, \\ -1, & \text{if } G_{i,j} = 0 \,\&\, G_{i,j+1} < 0, \\ 0, & \text{if } G_{i,j} = 0 \,\&\, G_{i,j+1} = 0, \\ \frac{G_{i,j+1} - G_{i,j}}{|G_{i,j}|}, & \text{if } G_{i,j} <> 0. \end{cases} \tag{1}$$

$$E_{i,j} = \begin{cases} G'_{i,j}, & \text{if } G'_{i,j} = 0, -1, 1, \\ 1, & \text{if } G'_{i,j} >= t, \\ -1, & \text{if } G'_{i,j} <= -t, \\ 0, & \text{if } -t < G'_{i,j} < t\,. \end{cases} \tag{2}$$

The parameter t is a customized threshold, which we empirically set to be 1.0. This value means that only an apparent change——increasing or decreasing——in expression level can be assigned to the discretized value 1 or -1, and all others should be 0 for the reason to maximally eliminate the potential noise in the original data.

2.2. *Local Matching*

In order to calculate the similarity between two genes, denoted by $S(X, Y)$, we'll use a matching matrix to find all the local matching information. Here X and Y stand for two genes respectively, and represented by sequences $(x_1, x_2, ..., x_{m-1})$ and $(y_1, y_2, ..., y_{m-1})$ derived from the discretized matrix E. And we define $X(i, j)$ to be the subsequence $(x_i, x_{i+1}, .., x_j)$ of X and $X(i)$ is the ith element of X.

In our method, we will calculate similarity by finding all the possible subsequence matching between the X and Y, and then find a optimal combination of the local candidate matches. A local match can be defined to be two subsequences from two genes respectively exactly having the same sequence: $X(i_1, j_1) = Y(i_2, j_2)$ and $X(i_1 + k) = Y(j_1 + k)$ for each $k(0 <= k <= j_1 - i_1)$, where we have $1 <= i_1 < j_1 <= m - 1$, $1 <= i_2 < j_2 <= m - 1$ and $|j_1 - i_1| = |j_2 - i_2|$.

After we have found all the local matches between the genes, we should combine the local matches into an optimal global match with longest length. We should introduce several important parameters. The first is the *minSubLen* which defines the minimum length of a matched subsequence. A local match with too short length means nothing but high random probability in matching. The second parameter, the *maxTimeLag*, is the time shift between the two subsequences which is the difference of i_1 and j_1. A too big time lag is difficult to explain in biology but there always exists time shift when the gene regulation works. So we'll confine the time shift in a limitation. After that, we'll find that some local candidate matches have the problem of overlapping which will not be allowed in the optimal combination.

Here we'll apply a matrix $M_{m \times m}$ to find all the candidate local matches. The first row and first column will be initialed to be 0. And after that we'll fill the matrix as the algorithm in Fig.1 .

```
Begin
 for i = 0 to m − 1
  for j = 0 to m − 1
    if i == 0 or j == 0
      M(i, j) = 0;
    else if X'(i) == Y'(j)
      M(i, j) = M(i − 1, j − 1) + 1;
      M(i − 1, j − 1) = 0;
    else
      M(i, j) = 0
    end if
  end for
 end for
End
```

Figure 1. Algorithm for mining local matches

And now we need to consider the two restrictions: the *minSubLen* and the *maxTimeLag*, which will confine our combination on the cell $M_{i,j}$ satisfying $|i-j| <=$ *maxTimeLag* and $M_{i,j} >=$ *minSubLen*. As a result, some useless candidate subsequence matches will be eliminated, which will greatly reduce the computation complexity to find the optimal combination. Here we give a formal description of the optimal combination problem. Given a set of triples $S = \{s_1, s_2, ..., s_n\}$ with each $s =_{def} (len, rowI, colI)$ in S recoding the information of a cell in M satisfying the aforementioned conditions including the value of the cell and its row and column indexes respectively. Here the value len records the length of matched subsequence, and the match in the two genes ends at positions $rowI$ and index $colI$ respectively. We define the operator *Non-Conflict*(s_i, s_j) to be TRUE if for any $s_i \in S$ and $s_j \in S(i <> j)$ with $s_i.rowI + s_i.len <= s_j.rowI$ and $s_i.colI + s_i.len <= s_j.colI$, or $s_j.rowI + s_j.len <= s_i.rowI$ and $s_j.colI + s_j.len <= s_i.colI$. Now our optimal problem is to find a subset $S' \subseteq S$ that maximizes $\sum_{s' \in S'}(s'.len)$ and any s'_i and s'_j is non-conflict. Then the attained subset S' will serve for the similarity calculation. Here we have proved the general *Optimal Local Combination Problem* to be NPC, which can be reduced from the Weighted Independent Set problem. And we also calculate the size of $|S|$ to obtain 0.6722C as an upper bound, where C is the length of a gene in G when we set*minSubLen* and *maxTimeLag* to be 4 and 3 respectively, which actually makes a brute searching method possible. The solution will list all the legal subset of S with no conflict elements and get the subset with maximized $\sum_{s' \in S'}(s'.len)$ as an optimal combination. We'll sort the elements into a list S in lexicographic order first when implement the brute searching which will reduce the comparison times.

Given gene X and gene Y with the sequences X=(-1,0,1,-1,1,-1,1,-1,1,0,1,-1,1,-1,1,-1,1) and Y=(1,0,0,0,-1,1,-1,1,-1,1,-1,1,-1,1,-1,0,0) respectively, we'll show the matching matrix of gene X and gene Y in Tab.1. It's easy to select the local matches with length greater than the *minSubLen* and we can confine the cells along the diagonal from upper-left to down-right with $|i - j| <=$ *maxTimeLag*. In Tab.1 the candidate cell appears in bold. In the above example, we have

Table 1. Matching matrix

	0	0	0	0	0	0	0	0	0	0	0	0	0	0	0	0	0
0	0	0	1	0	1	0	1	0	0	0	1	0	1	0	1	0	1
0	0	1	0	0	0	0	0	0	0	2	0	0	0	0	0	0	0
0	0	1	0	0	0	0	0	0	0	1	0	0	0	0	0	0	0
0	0	1	0	0	0	0	0	0	0	1	0	0	0	0	0	0	0
0	1	0	0	0	0	0	0	0	0	0	0	0	0	0	0	0	0
0	0	0	0	0	0	0	0	0	2	0	0	0	0	0	0	0	2
0	1	0	0	0	0	0	0	0	0	0	0	0	0	0	0	0	0
0	0	0	0	0	0	0	0	0	**4**	0	0	0	0	0	0	0	4
0	1	0	0	0	0	0	0	0	0	0	0	0	0	0	0	0	0
0	0	0	0	0	0	0	0	0	**6**	0	0	0	0	0	0	0	6
0	1	0	0	0	0	0	0	0	0	0	0	0	0	0	0	0	0
0	0	0	0	0	0	0	0	0	**7**	0	0	0	0	0	0	0	7
0	1	0	0	0	0	0	0	0	0	0	0	0	0	0	0	0	0
0	0	0	0	0	0	0	0	0	7	0	0	0	0	0	0	0	**7**
0	0	0	0	2	0	4	0	6	0	0	0	2	0	**4**	0	**6**	0
0	0	2	0	0	0	0	0	0	0	1	0	0	0	0	0	0	0
0	0	2	0	0	0	0	0	0	0	1	0	0	0	0	0	0	0

S=$\{(4,8,9),(6,10,9),(7,12,9),(4,15,14),(6,15,16),(7,14,17)\}$, where we set *minSubLen* to be 4 and *maxTimeLag* to be 3. Now our job is to find a subset $S' \subseteq S$ satisfying the aforementioned constraints. Here we get $S' = \{(4,8,9),(7,14,17)\}$, which is the optimal local combination. And a triple such as (7,14,17) means a local match with length 7, and ends at index 14 of gene Y and index 17 of gene X.

2.3. *Exact Similarity*

In this section, we'll calculate the similarity of the gene pair after we find the set S'. Assume the size of $|S'|$ to be K, which is the number of elements in the set. Then we'll use the following Equ.(3), where the parameter K is used to adjust the similarity, a higher K meaning more punishment because one long global match obviously has higher similarity than that calculated from the local combination. So a punishment of K with high value will be reflected in the formulation, and 10 in the second multiplication operator under the radical sign is usually set to be half of the gene length. And the constant C usually is set to be the length of gene.

$$S(X,Y) = \sqrt{\frac{\sum_{s' \in S'} s'.len}{C} \left(1 - \frac{K-1}{10}\right)}. \tag{3}$$

Our similarity method focuses on the local similarity, and the regulation between genes often functions locally as well as a time lag exists. The global similarity or distance measurement, the Pearson correlation or the Euclidean distance, have difficulties to solve such problems. Even only one time slice lag will greatly reduce the similarity between two genes, and a local similarity will often not be found when applying a global similarity calculation method. In our method, it is easy to locate the local similarity and even give the time lag between the local matching. The triple in S' fairly records the time lag information, and the value of $|s'.rowI - s'.colI|$ gives the time lag of the local matching. To demonstrate the difference of our method and the *Pearson Correlation* method, Fig.2

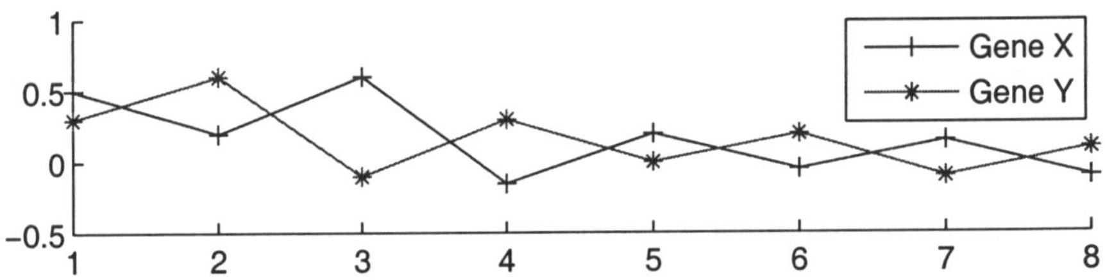

Figure 2. Our similarity method for the profiles is 0.93 whereas the Pearson correlation is -0.38.

shows two profiles with length 8, which have a highly correlated relationship according to our similarity method but almost unrelated when applied the Pearson correlation, and where the -0.38 means a negative regulation——one gene depresses another geneś expression. The reason is that the profiles have a high similar profiles except one slice time offset which the Pearson correlation has difficulty to cope with but our method can easily reflect.

3. Experimental Results

In this section we'll report the experimental results by applying our method on the real gene microarray data.

3.1. *Data & Clustering*

In order to demonstrate the performance and correctness of our new method, we'll run our method on the real time series gene data. We use the earliest gene expression data which is accessible at *Paul T. Spellman's* website [8] and also widely used in academic research. The data is mainly attained by four independent experiments for synchronized reasons: factor arrest, elutriation, arrest of a cdc15 temperature-sensitive mutant and cdc28, which consists of all the 6178 Yeast ORFs. But we'll not directly use this data, and there is too much of them. In the research of Steven Skiena and Vlfilkov(http://www.cs.sunysb .edu/ skiena/gene/jizu/), they searched the Yeast database and got 1007 genes from that in Feb. 2000. And by reviewing the published literatures on these genes, they collected 888 gene regulations, positive or negative. On this basis, we find 288 genes in alpha data set with their known regulation relationship among the 888 gene regulation relationship. Here the alpha data set has 18 time condition points with 7 minutes interval.

We'll run our method on the alpha data set and construct a similarity matrix for the genes in the data set to record the pairwise gene similarity, which is symmetrical and used for later clustering.

Clusters will de attained from the similarity matrix by using the GCLUTO,[12] a clustering tool consisting several clustering analysis methods. Here we use the clustering option based on the graph partition method.[13] GCLUTO will first construct a graph where each gene is represented by a node and edges between nodes are assigned with the corresponding similarities. GCLUTO will cut the graph with an optimal approach recursively until a pre-specified number of clusters been attained.

3.2. *Results Evaluation*

The number of clusters is a critical parameter in the clustering which will greatly effect the robustness of the clustering structure. Fortunately, when the parameter is over 20, the clustering structure is quite robust to the variation of the parameters.

Fig.3 and Fig.4 show cluster 7 in 30-ways clustering and cluster 8 in 25-ways clustering. They have a great high similarity, except in Fig.4 there are a little more genes for there are 5 less clusters. But the genes in Fig.3 can be found in cluster 7 in 25-ways clustering. And Fig.5 and Fig.6 both demonstrate cluster 2 in 30-ways clustering and in 25-ways clustering respectively. They also have a high similarity. And it is the same with the previous cluster pair, that the genes in cluster 2 in 30-ways can all be found in cluster2 in 25-ways. At last, in our experimental results, all the other clusters in the 25-ways clustering except no 2 and 8, they all can find a corresponding cluster in the 30-ways clustering with high similarity except several genes more or less. As a fact, when the gene clusters number is over 20, the structure is rather robust and changes little with the pre-specified cluster number increasing.

In Fig.3 and Fig.4, we obviously find two main profiles of the genes, and there is a time lag between them. This is very difficult for the traditional similarity or distance measurement, Pearson correlation or Euclidean distance, to find such clusters with time lag existing. But our method can give a high similarity between genes with similar profiles even there is a time lag, where the inter similarity in the cluster is 0.787.

In Tab.2, we shows the statistical results of the first 10 clusters in the 30-ways clustering for space reason. The column labeled Size displays the number of objects that belongs to each cluster. The column labeled ISim displays the average similarity between the objects of each cluster. The column labeled ISdev displays the standard deviation of these average internal similarities. The column labeled ESim displays the average similarity of the objects

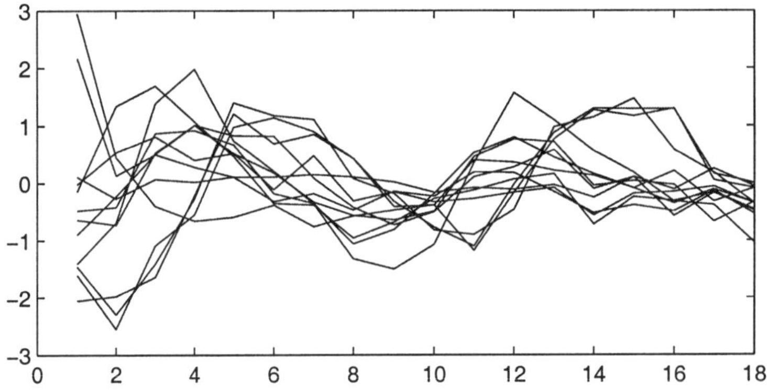

Figure 3. Cluster 8 in 30-ways clustering

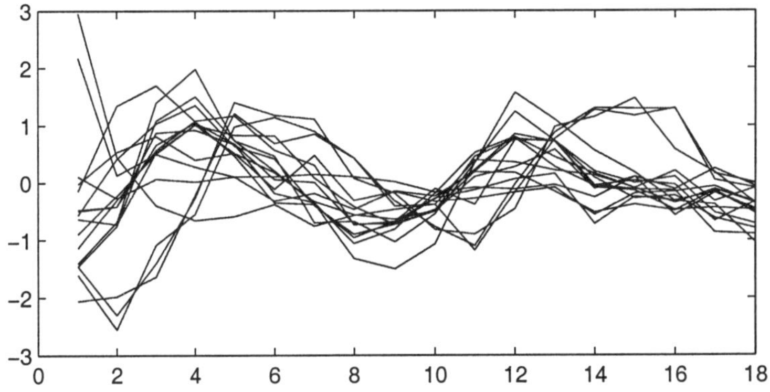

Figure 4. Cluster 7 in 25-ways clustering

of each cluster and the rest of the objects . Finally, the column labeled ESdev display the standard deviation of the external similarities. The ISim is much higher than the ESim and the ISdev and ESdev are much lower compared with the previous values. Now we can conclude that our method has successfully find the clusters with high similarity.

4. Discussion

We have proposed a new gene expression similarity measurement, which has successfully solved the local regulation and time lag problems in microarray data. By discretizing the original data and finding their local match, an optimal combination can be attained from the local matches and get the global match. And we also check our method on the real gene expression data and find that the clustering structure is rather robust. The genes in the same cluster also demonstrate high similarity. In the future work, we can compare our results with the gene database, for instance the MIPS, to check the genes clustered in the same cluster whether have a similar biological function. The data and software is available upon request.

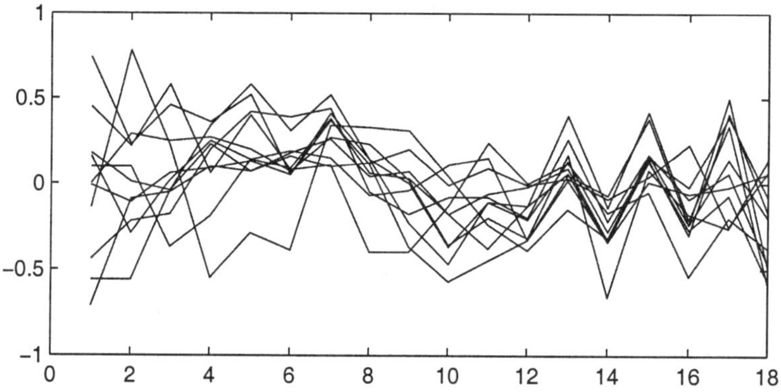

Figure 5. Cluster 2 in 30-ways clustering

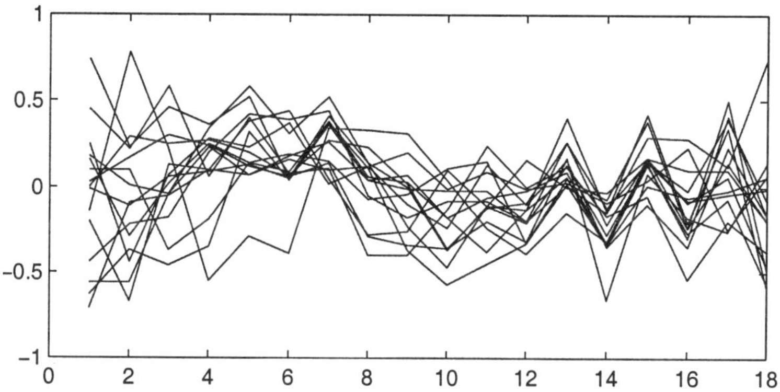

Figure 6. Cluster 2 in 25-ways clustering

Table 2. Statistic Analysis of the 30-ways clustering.

$Cluster$	$Size$	$ISim$	$ISdev$	$ESim$	$ESdev$
0	4	0.640	0.069	0.426	0.060
1	12	0.796	0.009	0.410	0.055
2	11	0.758	0.036	0.421	0.020
3	6	0.758	0.064	0.386	0.054
4	4	0.602	0.029	0.402	0.050
5	6	0.671	0.036	0.423	0.033
6	11	0.717	0.042	0.396	0.027
7	15	0.704	0.039	0.372	0.033
8	16	0.711	0.044	0.328	0.033
9	9	0.606	0.022	0.373	0.043

References

1. Geogre C. Tseng, Min-Kyu Oh, Lars Rohlin, James C. Liao & Wing Hung Wong. Issues in cDNA microarray analysis: qulity filtering, channal normalization, models of variations and assements of gene effects. *Nucleic Acid Research*, 2001, Vol.29, No.12, 2549-2557.
2. Brazma A., Vilo J.. Gene expression data analysis. *Federation of European Biochemical Societies: Letters*, 2000, Vol.480(1),17-24.
3. T. Forster, D. Roy & P. Ghazal. Experiments using microarray technology: limitations and standard operating procedures. *Journal of Endocrinology*, 2003, 178, 195C204.
4. Min Zou & Suzanne D. Conzen.A new Dynamic Bayesian Network (DBN) approach for identifying gene regulatory networks from time course microarray data. *Bioinformatics*, 2005, Vol.21(1) , 71C79.
5. Shoudan Liang. A genereal reverse engineering algorithm for inference of genetic network architectures. *Pacific Symposium on Biocomputing*, 1998, Vol.3, 18-29.
6. Gerstein & R. Jansen. The current excitement in bioinformatics-analysis of whole-genome expression data: how does it relate to protein structure and function? *Curr. Opin. Struct. Biol*, 2000, Vol.10, 574C584.
7. Mamoru Kato, Tatsuhiko Tsunoda, Toshihisa Takagi. Lag Analysis of Genetic Networks in the Cell Cycle of Budding Yeast. *Genome Informatics*, 2001, Vo.12, 266C267.
8. Spellman,P., Sherlock,G., Zhang,M., Iyer,V., Anders,K., Eisen,M., Brown,P., Botstein,D. and Futcher,B. Comprehensive identification of cell cycle-regulated genes of the yeast Saccharomyces cerevisiae by microarray hybridization. *Mol. Biol. Cell*, 1998, Vol.9, 3273C3297.
9. Chen, T., Filkov, V. & Skiena, S. Identifying gene regulatory networks from experimental data.*In Proceedings of the Third Annual International Conference on Research in Computational Molecular Biology* , 1999, 94–103.
10. Yeung, L.K., Szeto, L.K., Liew, A.W., & Yan, H. Dominant spectral component analysis for transcriptional regulations using microarray time-series data. *Bioinformatics*, 2004, Vol.20, 742–749.
11. Andrew T. Kwon, Holger H. Hoos and Raymond Ng, Inference of transcriptional regulation relationships from gene expression data. *Bioinformatics*, 2002, Vol.19(8), 905–912.
12. Karypis, G. GCLUTO - a clustering toolkit, 2002, Available at http://www.cs.umn.edu/gcluto.
13. Hideya Kawaji, Yosuke Yamaguchi, Hideo Matsuda, Akihiro Hashimoto. A Graph-Based Clustering Method for a Large Set of Sequences Using a Graph Partitioning Algorithm. *Genome Informatics*, 2001, Vol.12, 93C102.

AUTHOR INDEX